Concrete, Timber and Metals

Concrete, Timber and Metals

the nature and behaviour of structural materials

J. M. Illston
School of Engineering
The Hatfield Polytechnic

J. M. Dinwoodie
Princes Risborough Laboratory
Building Research Establishment

A. A. Smith
Department of Mechanical Engineering
University of London King's College

VNR VAN NOSTRAND REINHOLD COMPANY
New York · Cincinnati · Toronto · London · Melbourne

**Published by Van Nostrand Reinhold Company Ltd., Molly Millars
Lane, Wokingham, Berkshire, England**

*Published in 1979 by Van Nostrand Reinhold Company, A Division of
Litton Educational Publishing Inc., 135 West 50th Street, New York,
N.Y. 10020, U.S.A.*

*Van Nostrand Reinhold Limited, 1410 Birchmount Road,
Scarborough, Ontario, M1P 2E7, Canada*

*Van Nostrand Reinhold Australia Pty. Limited, 17 Queen Street,
Mitcham, Victoria 3132, Australia*

Library of Congress Cataloging in Publication Data

Illston, J M
 Concrete, Timber, and Metals.

 Bibliography: p.
 Includes index.
 1. Building materials. I. Dinwoodie, J.M., joint author. II. Smith,
A.A., joint author. III. Title. 5—403. 138   624'.18   78—21517
 ISBN 0—442—30144—8
 ISBN 0—442—30145—6 pbk.

Printed in Great Britain by Spottiswoode Ballantyne Ltd.,
Colchester and London

Preface

This book is concerned with the behaviour of certain selected
materials, and its principal aim is to relate the performance of each
material in physical and mechanical terms to the structure of that
material viewed at appropriate orders of magnitude from the molecular to the macroscopic. We believe that our approach has novelty
and see this text as a valuable source of information and stimulation
to both undergraduates and graduates in engineering, materials science
and architecture. The book is addressed primarily to civil engineers,
partly because a large part of the text emerged from experiences in
teaching students of civil engineering, and also because we identified a
particular need in courses on civil engineering materials for a unified
treatment of the kind that we have attempted here.

The range of materials used in civil engineering is wide, and we re-
luctantly concluded that it would be quite impossible to treat them
all adequately in one book. So we decided to concentrate on structural materials; that is, on those materials which are responsible for
the performance of the majority of civil engineering structures. We
regret that in concentrating on concrete, timber and metals we have
omitted asphalts and plastics, and we recognise that soil could usefully
have been included among the structural materials.

Traditionally, materials have been regarded as an adjunct to structural analysis; mechanical testing has been all-important with the
objective of providing the empirical values of the (comparatively few)
materials' constants required in the analyses. Secondly, and often quite
separately, information has been accumulated in the craft skills developed for the processing, handling and placing of materials in practice. Thirdly, and later, have come the deeper studies of the structure
of the materials themselves, in the realm of materials science. The
three themes — empiricism, craft and science — have, in civil engineering, remained largely separate and it has proved difficult to interweave
them. Most courses on materials necessarily have a bias towards one
only. It has been our purpose to try to present a more concerted view
in which the scientific treatment of materials provides a basis for the
understanding of behaviour and may lead to an appreciation of the

logic that underlies much practical empiricism and craft.

Our task was complicated by a second trichotomy. Not only are there three different themes, but the development of the three materials have followed separate paths. This derives from their different origins and structures, and it has led to the accumulated knowledge of the materials differing both in kind and in extent. Here, too, there has been a natural tendency for attention to be confined to one theme only, and in this we are no exception in that we each feel competent to write on one material only. Nevertheless, there are some striking resemblances in performance and we thought it important to bring the three together to present a comparison which we ourselves found to be both fascinating and illuminating. Initially we felt that we could do no better than provide introductory structural material to a three-part book, treating each material on its own. Later, we discovered that the topics, in terms of performance, that we had individually chosen coincided almost completely, and we have therefore dovetailed the chapters on each material within sections covering particular performance topics. We feel that this new arrangement allows the reader to appreciate more accurately the similarities (and differences) that occur, not only in the behaviour of the materials, but also in the underlying reasons for the behaviour. This approach does not preclude the detailed study of any one material since individual chapters relate to a single material only. To this end the text is divided into eight sections. The first contains introductory matters that are common to all three materials, while the remaining seven sections each treat a single performance topic, and consist of separate chapters on each material.

J. M. Illston
J. M. Dinwoodie
A. A. Smith

Acknowledgements

Concrete Section

The views, ideas and subject matter that go into a book are the end
result of a long process of influence on the author by innumerable
writers, colleagues and acquaintances. I am very conscious that I have
been nourished and sustained by my fellow members of staff, and
by my students over a long period of teaching at King's College,
London, and to them I acknowledge a keenly felt debt of gratitude.

I have also received invaluable help from my friends who have
kindly read chapters of the book, and who have made most useful
comments. They have helped me to eliminate some of the worst
deficiencies, and those that remain are entirely my responsibility. In
particular I should like to thank Dr. B. R. Gamble, Dr. S. E. Pihlajavaara, Dr. C. D. Pomeroy, Mr. T. J. Tipler and Dr. G. H. Tattersall.
I am also greatly indebted to Miss K. O'Donnell and Miss A. Rodgers
for cheerfully typing my lengthy and almost illegible manuscript.

J. M. Illston

Timber Section

I wish to express my appreciation to Dr. E. J. Gibson, Head of the
Princes Risborough Laboratory, Building Research Establishment,
for permission not only to reproduce many plates and figures from
the Laboratory's collection, but also to avail myself of the very willing
assistance of the typing, copying, drawing and photographic services
of the Laboratory.

To the many colleagues who so willingly helped me in some form
or other in the production of this part of the text I would like to
record my very grateful thanks. In particular I would like to record my
appreciation to Dr. W. B. Banks, Dr. A. J. Bolton, Mr. J. D. Brazier,
Miss G. M. Lavers, Mr. C. B. Pierce and Professor R. D. Preston FRS,
for reviewing different chapters of the manuscript and offering most
useful comments; to Mrs. A. Miles and Mr. G. Moore for much checking of data, reading of drafts and proofs; to Mrs. R. Alexander for

many long hours of typing; and lastly to my family for their tolerance and patience over many long months.

J. M. Dinwoodie

Metals Section

It is not easy to acknowledge fairly all the direct and indirect debts of gratitude that I have incurred during the gestation of this book. I am grateful to my past mentors, my present colleagues, and my past and present students, all of whom have knowingly and unknowingly thrown up ideas over the years that have helped to shape and clarify my approach to the subject. In particular, I would like to express my gratitude to Professor R. E. Gibson, Mr. W. B. Gosney, and Dr. M. J. Tindal for helpful discussions on various technical matters; to Miss I. E. Smith and Mrs. C. M. Brooks for deciphering and typing a heavy weight of manuscript; to Mr. C. D. Edwards for much metallographic assistance; and to my family for all their help and support over the lengthy period that I have been writing. Finally, for chapters which cannot lay any great claim to originality, I feel that I should record my debt to the many writers of metallurgical and materials texts from which I have profited over the years — writers who have frequently set standards in style and clarity to which I can aspire but which I cannot expect to approach.

A. A. Smith

Contents

CHAPTER 1

Introduction: Materials and the Design of Civil Engineering Structures

1.1 ELEMENTS OF DESIGN

Tredgold's definition (1827) of the profession of civil engineering starts with the oft-quoted statement:

> 'Civil engineering is the art of directing the great sources of power in nature for the use and convenience of man.'

The use of the word art is significant in that it indicates the creative nature of civil engineering design. This is demonstrably true in that the engineer, like an artist, starts with a blank sheet of paper, on which his ideas develop and take conceptual shape. To use a more concrete comparison, the engineering design, in common with the architectural design, starts as an empty site, and finishes as a completed structure. All such structures consist of an appropriate arrangement of materials, put together by means of interlocking geometries, and it can be said that materials are the medium in which the engineer exercises his creative talent.

Although the design process is fundamentally a sequence of creative activities, it is not typical of the group of skills which are taken together under the collective title of Arts. There are three main reasons for this. Firstly, the design process is, to a great extent, based on scientific principles, and is expressed in the language of mathematics. Thus the criteria for safety are stated in terms of factors which are merely numbers relating the performance of the structure under working conditions to that under conditions at failure. For both sets of conditions the performance is assessed in terms of mathematical equations yielding a numerical answer. The civil engineering design process is thus a science-based art.

Secondly, the constraints on the engineer are usually much more onerous than those on the artist, in that the engineering design is produced to meet an explicit need of society. No doubt the average civil engineer would rejoice at the opportunity to express his person-

1

ality and feelings in the form of, say, a flowing suspension bridge; regrettably he is much more likely to have to do his best to meet his creative urges in the realisation of a pipeline which nobody will see or a factory which everybody would rather not see.

Thirdly, the creation of a civil engineering structure requires the efforts of numerous participants, many of whom will be specialists

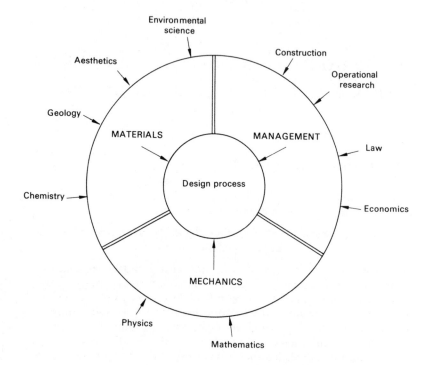

Figure 1.1 The elements of design.

contributing to one aspect only of the total result. Contributions may come from the practitioners of various disciplines, from the sociologist to the mathematician, from the lawyer to the geologist, and the end result of all their disparate labours is a physical grouping of materials, both natural and artificial, which will serve to direct the forces of Nature.

The task of the civil engineer is to contribute his own skills and to coordinate those of other participants in the complex task of bringing the design to fruition. This can be reduced in the broadest terms to the three areas of expertise (shown in Fig. 1.1):

Materials
Mechanics
Management

Expertise in materials is vital in that the fitness for purpose of the structure is directly dependent on the behaviour of the materials of which it is made. Although the structural member may be said to fail, it is, in fact, the material of which the member is composed that fails. Similarly, the finish of a building may be said to aesthetically displeasing, but it is the surface characteristics of the visible materials that are at fault. This sort of argument can be applied to virtually all the design considerations, and it emphasises the need for a thorough and fundamental understanding of materials.

The three areas of expertise interact one with another; thus the laws of mechanics are based on idealisations of material behaviour, and the magnitude of the material response is directly related to the material coefficient (such as elastic modulus) which must be incorporated in the analytical expressions. Similarly, the management of design and construction is concerned with matters of materials processing and quality control, and includes important decisions on the choice of materials.

The above discussion brings to light a number of different aspects of the treatment of materials, and these are examined in more detail in the next section.

1.2   HOW ENGINEERS LOOK AT MATERIALS

1.2.1   Performance and Properties

The selection of the materials of construction is one of the earliest and most important design decisions, and the engineer must consider the fitness of the chosen material for the intended purpose. It is a matter of whether or not the material will perform adequately both during the construction and in service after the structure is built. The engineer makes his choice partly on the basis of his experience, and partly on the scientific evidence that is available to him on the materials at his disposal. Experience is an essential contribution to the engineer's judgement, because it is seldom possible to give anything like a full scientific analysis of all the aspects of an engineering problem. This is hardly surprising but it means that many, if not most, engineering decisions are made on the basis of inadequate data. Even in the age of science, a 'feel' for materials is not to be dismissed lightly, nor should the knowledge of the craftsman, accumulated from long experience, be neglected. Nevertheless, the scientific approach is now the norm. The behaviour of materials has been investigated scientifically for many years, understanding has greatly improved and theoretical treatments have become more realistic and exact.

Many of the results of the scientific approach are available to the engineer in the form of property values. The traditional kind are found in books on strength of materials, in which mechanical tests are described and properties such as the tensile yield stress of steel, elastic moduli, and the cube strength of concrete are quoted. Mechanical testing is not outdated, but much more detailed information

3

is now sought; for example, knowledge is required of the strength of materials under the infinite variety of triaxial states of stress; or more information is needed on the time-dependent deformations of materials under a variety of storage conditions. The rewards can be great in design economy, for knowledge of this kind enables engineers to employ logical factors of safety rather than conservative factors of ignorance.

As well as more detailed information on old property values, new values are required for circumstances previously neglected, or for situations not previously experienced. For instance, properties representing the corrosion resistance of metals, or measuring the effectiveness of timber treatments, have become of greater significance. So too has the behaviour of materials in the radioactive interior of nuclear power plants.

The engineer also seeks to express the behaviour of materials in a general way; he attempts to synthesise the theoretical and experimental evidence into general laws, which can be applied to situations beyond the bounds of immediate experience. The theory of elasticity is a prime example, and it is always the hope that new or modified materials can be shown to obey the laws of elasticity sufficiently well for access to the storehouse of elastic solutions to be justified. However, situations have come to light, many of them recognised long ago, in which other laws are required, and the search for theoretical generality is a continuing process. The theory of plasticity is well developed, laws of viscoelasticity have proliferated and, in another area, the equations of mass transfer enable predictions to be made of moisture distributions in porous materials. For every new law a new set of coefficients is needed for each material to which the law is applied.

Although theoretical ideas on the structure of materials are of great significance, the majority of property values are found experimentally and the importance of materials testing cannot be over-emphasised. It is the foundation of engineering design, providing the basis of the engineer's selection of materials, and determining the accuracy of his analyses. Often a particular property value is related to the chosen test method, or even to an individual test equipment, and thought, expertise and care must go into all stages of making measurements, from the design of the apparatus, through the calibration of the machines and the performance of the experiments, to the interpretation of the results.

1.2.2 Costs and Conservation

The first concern in the selection of materials is fitness for purpose, but it is often the case that more than one material is suitable. Steel, timber and reinforced concrete can all be used for structural members; aesthetically acceptable facing panels can be fabricated from brick, timber or aluminium. In these circumstances other criteria must be invoked, and pre-eminent among these is financial cost. This has the overwhelming advantages of being very important to the client, and

4

of being capable of expression in numerical form. In principle it is simple to decide between one alternative and another as everybody will agree that the scheme with the lowest cost is the most financially attractive, whereas it is much more difficult to reach agreement on the technical, sociological or environmental merits of different designs. In practice, it is not so simple as this in that the total cost is a composite of materials, labour and construction costs, and there are, for example, difficulties in assessing and interpreting the balance between capital costs and running costs. The alternative solutions may, in total, be widely different, and the comparison of financial costs becomes interwoven with the technical assessment. Other criteria may be relevant, among which the question of availability of materials is likely to have increasing significance. This may be a matter of certain imported materials, like timber, being in short supply; or it may be that delivery times are so variable that there is a risk of serious delay to the project as a whole. The financial cost criterion may then be overridden and the decision taken to pay more for greater certainty of supply or for greater certainty of delivery on time.

Financial cost may be rivalled in the future by another quantitative criterion — that of energy cost. No doubt a shortage of energy will be reflected in financial costs, but good design practice may call for a quite separate calculation and comparison. The energy costs of alternative materials in their appropriate structural solutions can be calculated, taking into account both manufacturing energies, for example in the making of cement or steel, and associated energies in transporting and mining the materials. This is not a simple matter, and it is open to argument as to just what should or should not be included. However, assuming that commonsense rules can be drawn up, definite conclusions can be reached which can influence the choice of materials. For instance, there is little doubt about the desirability of using high-quality pulverised fuel ash (a waste product from power stations) as a partial replacement for cement in concrete.

This is just one aspect of conservation, and it is likely that designers will be subjected to increasing nonfinancial constraints in the name of conservation. In general, these tend towards the use of low grade (i.e. low energy cost) materials or to the design of structural forms with a minimum material content. Design for re-use, even though the circumstances of the second or subsequent use are completely unknown, is likely to assume greater importance. There is no very great difficulty in recycling metal scrap but there are problems in relation to re-use of timber, concrete or other composite materials. The demolition of concrete structures can hardly be envisaged as a significant source of aggregates, but it is by no means inconceivable that individual members, beams and columns, could be taken down and re-erected elsewhere.

5

1.3 INFORMATION

1.3.1 Categories of Information

It is readily apparent from the previous section that many of the tasks of the civil engineer require the taking of decisions or the exercise of judgement. It has also been seen that this may well relate to the choice of materials or to the selection of the appropriate numerical expression of their properties. The endeavour must always be to achieve the highest level of competence, and this implies that the decisions and selections must be made on the basis of the best available information. The categories of information in relation to materials can be summarised as follows:

(a) *General Descriptions of Material Structure and Behaviour.* These may range from the quantitative to purely qualitative and they may derive from within the wide extremes of scientific investigation and site experience. An accumulation of such information provides the essential background that will enable the engineer to compare one material with another, or to appreciate how the material will behave in a situation not previously met. It can be termed an aid to understanding.

(b) *Data on Properties.* When the class of material (steel for example) is known and understood so that the laws of mechanics are established, the need is to feed in the correct values of the material coefficients to enable the structural analysis to yield results that truly represent the behaviour of the real structure. The engineer must have access to reliable estimates of the material coefficients and it is preferable for them to be presented in tabular and graphical form, thus allowing rapid and easy abstraction of the data.

(c) *Descriptions of New Materials.* The engineer is always anxious to employ new or radically improved materials, but he must be convinced not only that the new product will meet his immediate specification, but also that it will not be defective in some way that cannot be readily foreseen. Information on a new material must thus give a detailed overall view of its performance, including the manner and results of proving tests done upon it.

(d) *Testing and Quality Control.* Existing techniques for testing and controlling the quality of materials are always open to criticism; they may be too crude, they may take too long to perform or they may be susceptible to human error. The engineer will be glad to adopt new proven methods which will make his task simpler or more rapid.

(e) *Costs and Other Comparative Data.* The previous items covered fitness for purpose, but other criteria must also be met. The most essential information concerns costs, both financial and energetic, but data may also be needed to service other criteria of the kind mentioned earlier.

6

1.3.2 Levels of Information

Freudenthal (see Bibliography) defined the approaches to the study of materials in terms of different levels, depending on the degree of aggregation of the structural elements of the material. For engineering purposes the concept is again useful, taken to apply to levels of information. The three levels discussed here are compared in Table 1.1.

Table 1.1   Levels of Information

1 mm = 10^3 μ (micron) = 10^7 Å (Angstrom unit) = 10^6 nm (nanometres).

Level	Molecular	Materials structural	Engineering
Scale	10^{-7}–10^{-3} mm	10^{-3}–1 mm	Over 1 mm
Elements	Molecules Crystals	Phases Particles Grains	Total material
Examples	Cellulose molecules Calcium silicate hydrates	Wood cells Cement paste and stone	Timber Concrete
Test techniques	X-ray diffraction Electron microscopy	Moisture content Sectioning (small samples)	Mechanical (tensile strength, impact)
Interpretation of results	Structural models Dislocation theory	Multiphase models Mass transfer	Charts Graphs
Use of information	Understanding New materials	Understanding Material coefficients	Material coefficients Costs Testing

1.3.2.1   *Engineering Level*

Here the material is treated on the largest scale. It is normally taken as continuous and homogeneous and average properties are assumed throughout the whole volume of the material body.

The minimum scale to enable the material to be rightly considered in this way is indicated by the size of the representative cell, which may vary considerably from, say, 10^{-3} mm for metals to 100 mm for concrete. The representative cell can only be defined rigorously in statistical terms but it may be taken as the minimum volume of the material that truly represents the entire material system, including its regions of disorder. Properties measured over volumes greater than the unit cell can then be taken to apply to the material at large. If the properties are the same in all directions then the material is isotropic, and the representative cell can be considered as a cube, while if the properties can only be described with reference to orientation, the material is anisotropic, and the representative cell may be regarded as a parallelepiped.

The engineering level derives from the traditional engineering

7

approach to materials implied by the title 'Strength of Materials'. This consists of evaluating the mechanical properties, particularly strength, of large specimens under conditions closely resembling those in the actual engineering structure. It is then possible to make a direct assessment of the structural performance on the basis of the empirical results of the test programme.

The variables that can affect the results, such as the carbon content of the steel or the water/cement ratio of the concrete, are usually of the kind that relate directly to engineering practice, and are thoroughly known and understood by practical men. The data from the tests on the material may well be brought together in the form of empirical equations or charts, and a commentary may be provided which includes the collated experience of those who work with and produce the material on site. It is from these sources of specific and well defined data that most information on testing and costing comes, and it is also the major source of the numerical values of property coefficients for use in structural analysis.

The quality of the information is satisfactory only within the range of the tests that produced it so that it is normally impossible to extrapolate with confidence beyond that range, and hence to develop an acceptable generality for the empirical relationships. Furthermore, the choice of variables often fails to express the physical processes that are going on within the material, and this leads to clumsy equations and a low order of accuracy in the predicted values.

1.3.2.2 *Materials Structural Level*

This refers to the structure of the material and must not be confused with engineering structures. It is a step down in scale from the engineering level, and the material is considered as a composite of different phases.

The properties observed in the material as a whole (i.e. at the engineering level) represent the interactive effects of the different components within the material. This may be a matter of ingrown structure of the material, as in the cells that determine the characteristics of timber, or it may result from the deliberate mixing of disparate parts such as the aggregate and cement paste in concrete. Often the material consists of particles distributed in a matrix of another material; the dimensions of the particles differ enormously, from the wall thickness of the wood cell (10^{-3} mm) to the diameter of a piece of concrete aggregate (10 mm). Size is of no intrinsic importance; what matters is that such a component can be recognised independently, both for its geometry and for its own behaviour. The separate entities making up the material are referred to as phases, and their sum is then a multiphase material. The entities may be different phases in equilibrium, in the strict thermodynamic sense, but this is not necessarily, or even usually, the case, as a mixture of quite unrelated solids, as in concrete or in a fibre-reinforced composite, still comes under the heading of a multiphase material.

The significance of the materials structural level lies in the possibility of developing a more general treatment of multiphase materials than that provided by the empiricism of the engineering level. As implied by the title, the potential exists to express the underlying composition and arrangement of the material, and thus to incorporate the effect of variations in a logical manner. From this results the ability to predict behaviour outside the range of experimental observations with some degree of confidence.

This sort of theoretical treatment is most attractive, but the difficulties are daunting. The behaviour of the multiphase material must be described by means of a multiphase model, with appropriate mathematical expressions to represent the features of the various phases. If the predictions of the model are to be accurate the descriptions of the phases must include all the necessary variables, and the appropriate numerical values must be available. This is by no means easy to achieve, and the following considerations must be remembered.

(a) *Geometry*. The model must recognise that the particulate, or disperse phase, is scattered (or arranged) within the matrix, or continuous phase. It must also take account of the orientation of the dispersed particles, of the shape and size distributions of the particles, and of the overall concentration of the particles.

(b) *State and Properties*. The chemical and physical states of the phases will clearly influence the behaviour of the multiphase material made from them, and these must be incorporated in the model. So, too, must the mechanical properties such as rigidity (elastic modulus) or rheological behaviour, such as flow rate (coefficient of viscosity).

(c) *Interfacial Effects*. The information under (a) and (b) above may not be sufficient because the existence of the interface between the phases may introduce additional modes of behaviour that cannot be predicted from the known behaviour of the individual phases. This is particularly true of strength, the breakdown of the material often propagating from an initial fracture at an interface.

At this level many of the testing techniques are identical to those at the engineering level since the same properties, such as hardness or shrinkage, have to be investigated. However, additional test methods are needed to establish the compositions of multiphase materials, and these include determinations of moisture contents and the examination of thin sections. Further additional tests are concerned with interfacial properties such as bond strength.

At the engineering level the assumption is made of average behaviour throughout the whole volume of the member, but the approach of the materials structural level enables variations throughout the volume to be analysed. Furthermore, the information derived at the materials structural level is valuable for the determination and extension of the range of values of the various material coefficients, and for the increase in understanding of the relationships between the structures and properties of multiphase materials.

9

1.3.2.3 *Molecular Level*

This considers the material at the smallest scale, in terms of discrete particles of molecular size.

The engineering and materials structural levels have it in common that the whole material, or the phases of the material, are treated as homogeneous and continuous, and, as a result, there are strong interconnections between the data produced at each level. The molecular level, on the other hand, with its discrete particles and discontinuities, is generally more difficult to connect with the other two. Where connections can be made they often have all the greater significance, and the enlightenment deriving from understanding at the molecular level is often of the most beneficial kind.

The scale of the discrete particles is very small, ranging from 10^{-7} mm for individual molecules to, say, 10^{-3} mm for aggregations of molecules. The experimental techniques are necessarily sophisticated and specialised; for example, electron microscopy and X-ray diffraction enable knowledge to be gained of the structural forms of aggregations of molecules. Direct observation of individual molecules is virtually impossible, and the traditional atomic models of physics and chemistry are used to describe the molecular behaviour of materials. Thus, the ideas of atoms consisting of basic particles (protons, neutrons and electrons) lead to useful concepts of different types of bonding between atoms, with different relative strengths. The energies required to change the positions of atoms are equally important, and simple principles of the thermodynamics of change are helpful at this level.

Other chemical and physical principles also have their place. Chemical composition is important, for instance, in timber, in which the long-chain cellulose molecules are responsible for many of the features observed at the macroscopic level. Potential weaknesses can be identified in the form of this molecule and it is possible to postulate likely causes of breakdown by animal or mineral agents. Similarly, the hydration reactions of cement and the physical structure of the hydration products are of fundamental importance in concrete technology.

The processing of observations is accompanied by model making, which may be purely descriptive, in the form of sketched geometrical shapes, but which may be amenable to detailed mathematical description. In some circumstances the behaviour of the material as a whole may be structure sensitive, and molecular phenomena may be seen to be the direct cause of an effect at the engineering level. This occurs in the cases of brittle fracture and dislocations in metals.

Although some engineering analyses, like fracture mechanics, come straight from molecular happenings, they are the exception rather than the rule. It is much more likely that information from the molecular level will serve to provide mental pictures of the material which will aid the engineer's understanding. Additionally, and equally important, this sort of information, in the hands of specialists, gives a basis from which new and improved materials can be envisaged and developed.

1.4 OBJECTIVES

In the previous sections an attempt has been made to show the importance to the civil engineer of a wide knowledge of materials, to identify the types of information that the civil engineer needs, and to describe the levels at which this information may be presented. It is the objective of this book to lay the foundation for the study of materials in order to meet these requirements of the civil engineer. The scientific approach to materials is emphasised, and all three levels of information are covered.

In the past there has been a considerable gap between the materials science of the molecular level and the materials technology of the engineering level, and students have been puzzled by their apparent lack of connection. It is one of the main objectives of this book to bridge this gap, and in general to connect the three levels wherever this is possible. The molecular structures of materials are described and the consequent modes of behaviour are pointed out. The phases at the materials structural level are identified and the various geometries are discussed. Theoretical relationships from both the molecular and materials structural levels are developed and related to the properties of the bulk materials. Testing methods and the analysis of experimental observations are given their due and examples are given of investigations at the engineering level. The presentation of information for use by engineers is demonstrated.

The first section of the book deals with the molecular basis that is common to all materials, while in the sections that follow the three main structural materials

Concrete
Timber
Metals

are considered separately. However, the book is divided into sections reflecting particular aspects of materials such as structure, strength and durability. Chapters on each of the three materials are included within each section so that there can be comparison by juxtaposition. In addition appropriate cross-references have been made between the chapters within a section.

It is impossible in one book to cover the field of materials in a fully comprehensive manner and no attempt has been made to include all materials, to provide any kind of manual of good practice or to demonstrate the application of the various design criteria. Nor has it been the intention to provide a compendium of materials data. There is no doubt that for students some introduction to such matters is highly desirable, and the scientific approach adopted here should be supplemented by project work which could profitably be a cooperative venture between the educational institution and engineering practice.

CHAPTER 2

Simple Physical Aspects
of Materials

2.1 THE ATOM

2.1.1 Atomic Structure: A Historical Introduction

In any study of materials aimed at the civil engineer there is a strong
case for commencing with an examination of the structure of the
atom. In a sense this is an arbitrary choice, since it has been known
for many years that there are many more fundamental particles. However, the atom is the smallest unit of matter exhibiting characteristic
physical and chemical properties and may be regarded as the 'brick'
from which all civil engineering edificies are constructed. We have 92
varieties of atom available to us in the earth's crust, but of these only
about a quarter are made use of in engineering to any extent. Scarcity
and adverse properties are immutable reasons why the list of 'useful'
atoms is unlikely to be extended greatly in the foreseeable future.

Of the metallic elements, six are used extensively (in excess of
1×10^6 tonnes per annum): iron, aluminium, copper, zinc, lead and
chromium (with iron an easy first); and another ten are used to an
appreciable extent: nickel, tin, magnesium, manganese, titanium,
molybdenum, cobalt, tungsten, uranium and vanadium. Consumption
figures for nonmetals are not available, but at least four are used
very extensively (albeit sometimes incidentally, as in rocks and clays)
by the engineer: silicon, oxygen, carbon and sulphur. There are of
course many other elements, both metallic and nonmetallic, that are
used in smaller amounts, but which are highly important in specialised aspects of engineering, or that the engineer comes across incidentally (such as calcium, sodium and potassium). We thus have some
23 'building bricks' whose behaviour is of vital importance and which
combine in a multitude of ways to provide the civil engineer with his
bulk materials — and problems.

These atoms can combine with each other in a vast number of
different ways and proportions, but the manner in which they com-

12

bine and resultant properties of that combination are all a logical outcome of their individual structure.

Rutherford's picture of the atom, put forward in 1911, has been
elaborated considerably in the past 60 years, yet it presents a structure
which is still basically correct. In it were fused the independent and
complementary theories of the atom that chemists and physicists had
been slowly assembling. It was the Greeks who made a promising
start to atomic physics by the end of the 4th century BC, although
not all the details were based on a scientific footing. The atoms were
considered subject to astrological influence, and whilst it was postulated (quite correctly) that atoms might on occasion deviate from
rectilinear motion, the reason for this was ascribed to the 'free will'
of the atoms! The theory flourished modestly for a time, and at the
turn of the 2nd century BC Asklepiades was already anticipating
Dalton's work by some 2 000 years when he suggested that atoms
could form into groups — a very early forerunner of the molecule.

Despite the precocious beginning, the Grecian atomic theory fell
into disrepute and oblivion, and it was not until John Dalton in 1807
attempted to systematise the laws of chemical combination of the
then known elements that a fresh impetus was given to the subject.
Attempts at classification of the elements followed, the most notable
being Dobereiner's 'triads' (1829) and Newlands' 'Law of Octaves'
(1858), both of which demonstrated a periodicity in the elements
and, by inference, also in the structure of the various atoms. These
early gropings culminated in the brilliant work of Mendeleeff and
Lothar Meyer in 1869—70; Mendeleeff's Periodic System has remained virtually unchanged since its inception and is a cornerstone of
modern chemistry. This established the 'periodicity' of the elements
beyond doubt and suggested some sort of pattern within the internal
structure of the atom, but the details of that structure were still an
enigma.

It was the discovery that certain elements are radioactive that
stimulated the physicists to atomic research, since it showed that the
atom was not indestructable, as had been thought hitherto. In 1859
the discovery of cathode rays took place, and 20 years later Crookes
showed that they consisted of negatively charged particles. Thomson
in 1897 showed that the particles were similar irrespective of their
source, and hence must be a common constituent of all atoms. They
became known as electrons, the existence of which had already been
postulated by Faraday some 65 years previously. Since the complete
atom is electrically neutral, the part of the atom not consisting of
electrons must have a positive charge, and accordingly Thomson
brought out his theory of the structure of the atom in which the
negative electrons floated in a positively charged matrix. Subsequently,
studies in C. T. R. Wilson's cloud chamber showed that the positively
charged portion of the atom is concentrated into a very small fraction
of the total atomic volume and this led Rutherford in 1911 to suggest
the structure for the atom which in broad principle is that accepted
today. In this theory, the atom consists of a positively charged
nucleus around which circle the appropriate number of electrons in

the manner of a miniature solar system, the total negative charge of the electrons exactly balancing the positive charge on the nucleus. Rutherford's theory was then dovetailed in with Mendeleeff's Periodic System by the work of Moseley, who suggested that the elements could be arranged in order of their 'atomic numbers' (which order agreed with the Periodic System and almost completely with the order of atomic weights), where the atomic number, Z, is equivalent to the number of electrons in the atom. Thus an atom of atomic number Z has a nucleus of charge $+Z \cdot e$ around which Z electrons circulate, where e is the charge on the electron.

2.1.2 Development of the Modern Atomic Theory

The Rutherford atom fitted the general picture built up by the physicists quite well. However, it did not define the way in which the electrons were arranged in their orbits, which alone could explain the periodicity of the chemical behaviour of atoms, and also it was mechanically unsound. If the motion of an electron in orbit was considered in terms of classical mechanics then the electron would be expected to lose energy gradually and eventually fall into the nucleus. When all the electrons associated with the atom had fallen exhausted into the nucleus, the atom itself would be 'run down' and presumably lose its physical and chemical characteristics. All the available evidence suggested conclusively that this was not the case. Such behaviour would require a continuous change in the chemical properties of an atom as one by one the electrons fell into the nucleus, together with simultaneous emission of radiation energy, and presumably some sort of regenerative process for fresh atoms to take the place of exhausted ones. Furthermore, spectroscopic studies had already shown that each element possessed characteristic emission spectra which were associated with the energy levels of the electrons in the atom. These spectra exhibited discreet energy levels (not to be expected on the 'running down' theory, since electrons in all stages of exhaustion would be expected) and the energy levels appeared to be quite stable.

Bohr overcame this impasse in 1913 by applying Planck's Quantum Theory (1900) to the behaviour of the electron. He suggested that the electron could only absorb or emit energy in definite indivisible units or *quanta*, and that an electron in orbit around the nucleus of an atom is in a stable state: that is to say, it neither emits nor absorbs energy and continues to circulate in that orbit indefinitely. If the electron is excited (as, for example, during spectroscopic examination), then by absorbing a quantum of energy it can jump from its stable state (E) to an excited state (E^*), the two states being connected by the relation

$$E^* - E = h\nu$$

where h is Planck's constant, and ν is the frequency of the energy quantum. When the electron falls back to its initial state E, the energy is emitted, and so we obtain discreete emission spectra, since for each electron the values of E and E^* are different. He further suggested

14

that the stable energy states of electrons correspond to orbits for which the angular momentum of the electron is a multiple of $h/2\pi$. Thus the angular momenta available to electrons in any atom are $h/2\pi$, $2h/2\pi$, $3h/2\pi$, $4h/2\pi$, . . ., $nh/2\pi$, where n is an integer known as the *quantum number*. In this way the Bohr atom retained the general structural appearance of the Rutherford atom and at the same time presented a picture consistent with spectroscopic observations.

The next important development took place in 1915 when Sommerfeld showed that elliptical as well as circular orbits satisfy the conditions laid down by Bohr for the electron; the result was the

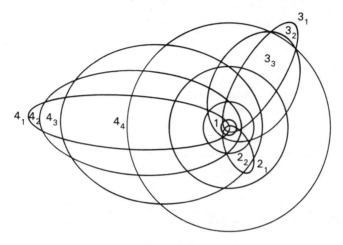

Figure 2.1 Elliptical electron orbits in the hydrogen atom. (From J. R. Partington, *General and Inorganic Chemistry*, by permission of Macmillan, London and Basingstoke.)

Bohr—Sommerfeld atom (Fig. 2.1), in which two quantum numbers are required to define the state of any electron. These are:

n a measure of the total energy associated with the orbit, proportional to the major axis of the ellipse, and

k a measure of the angular momentum associated with the orbit, given by $kh/2\pi$.

This is, of course, similar to the original single quantum number of the Bohr atom. Both k and n are always integers; for any particular value of n, k may take any value from 1 to n. The ratio k/n gives a measure of the eccentricity of the elliptical orbits, the case where $k = n$ indicating a circular orbit. Thus, for $n = 1$, $k = 1$ and there is only one possible orbit, a circle. For $n = 2$, k may take the values 1 or 2, and both a circular and an elliptical orbit become possible, having the same total energy level but differing in their angular momenta. For $n = 3$, three orbits become possible, two of them ellipses, and so on for increasing values of n. Two further quantum states were then added to the above conditions to account for certain behaviours of the atom, and the quantum number l was substituted for k, where $l = k - 1$. Hence, for any one value of n, l may take any value from 0 to $n - 1$.

The third quantum number, m_l, was introduced to explain the splitting of spectral lines when the atom is subjected to a magnetic field, and may be taken to be a measure of the angular momentum of the electron referred to a particular direction. For any one value of l, m_l may take any value from -1 to $+1$, including zero. Finally, a fourth quantum number was added to specify the spin of the electron, denoted by m_s. This spin quantum number can only take the values $\pm\frac{1}{2}$. Thus by 1925 the quantum theory of the atom had accounted for many of the observed effects which were inexplicable according to classical mechanics, and in fact the specification of the state of the electron by means of the four quantum numbers n, l, m_l and m_s has not been modified or elaborated since. However, there were drawbacks to the refined Bohr—Sommerfeld atom, not the least being the absence of any real justification for this treatment of the atom, beyond the fact that it fitted the experimental facts well.

These drawbacks were overcome and the quantum theory began to take its present shape in 1925, when the work of de Broglie and Schrodinger showed that the electron could be considered as a wave front rather than a discreet particle. The validity of this approach was confirmed when it was shown that electrons could undergo diffraction on passing through a crystal lattice in exactly the same way as a beam of light passing through a diffraction grating. A further major development took place in 1927 when Heisenberg showed that one of the implicit assumptions of the Bohr atom was false. The motion of the electron in the Bohr atom was governed by Newtonian mechanics; that is to say, the electron had a definite velocity and angular momentum, which could be determined at any instance. Heisenberg's *Uncertainty Principle* showed this assumption to be unjustified and stated that

$$\Delta p \cdot \Delta x \geqslant \frac{h}{2\pi}$$

where Δp is the uncertainty of the momentum of the electron, Δx is the uncertainty of the position of the electron, and h is Planck's constant.

Thus the more precisely we are able to determine the momentum of the electron, the more indeterminate is its position, and vice versa. The mathematical derivation of the Uncertainty Principle is too complex to be gone into here, but the physical justification for it can easily be seen if it is realized that there is no physical way of measuring the position or momentum of an electron which leaves the electron unaffected. For example, if we attempt to observe the position of an electron by means of a beam of light, the interaction of the energy associated with the light and the electron will completely change the momentum of the electron thereby making it unmeasurable. It must not be thought that the Uncertainty Principle only applies in the realm of atomic physics; it applies to all matter. However, the uncertainty becomes negligible when measuring objects

16

much larger than atoms; the physical analogy of this is that it is perfectly accurate to determine the position of, say, a golf ball by means of a light beam because the interaction between the light and the golf ball has a vanishingly small effect on its position. As a result, when considering the position of an electron, we can only deal in probabilities. It is still convenient to refer to electron orbits when considering the structure of the atom, but the orbits should not be considered to be precisely defined as in Fig. 2.1. Instead, they are best considered as clouds of varying shape round the nucleus, in which

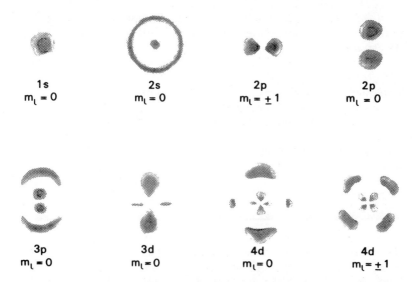

1s	2s	2p	2p
$m_l = 0$	$m_l = 0$	$m_l = \pm 1$	$m_l = 0$

3p	3d	4d	4d
$m_l = 0$	$m_l = 0$	$m_l = 0$	$m_l = \pm 1$

Figure 2.2 The cloud representation of electron orbits, in which the density of the cloud at any point represents the *probability* of finding the electron at that point. The diagrams show sections of the cloud in one plane for various excited states of the hydrogen atom.

the density of the cloud at any point is a measure of the probability of finding the electron at that point (Fig. 2.2).

2.1.3 The Periodic System

We must now consider the way in which the electrons are arranged in the atom. The arrangement is important in that it determines the way in which the atom can bond to other atoms (like or unlike) to form solids, whose physical strength is largely dependent on the nature of the bond. It will be remembered from the previous section that the energy state of the electron is determined by the four quantum numbers n, l, m_l and m_s, and that the first three are inter-related in their possible values. Note also that low values of quantum numbers correspond to low energy states, and as an atom is 'filled up' by the addition of successive electrons the electrons will take up the lowest available energy states. If the state of the electron were to be determined solely by the criterion of minimum energy, it might be thought that all the electrons in an atom would cluster into the lowest possible energy state, i.e. that for which $n = 1$ and therefore $l = 0$. However, from considerations of symmetry in atomic wave functions, Pauli in 1925 put forward his important *Exclusion Principle*, which states that in any atom it is impossible for any two electrons to have all four quantum numbers identical. If we now arrange electrons in states of increasing energy according to their quantum numbers, bearing in mind Pauli's Principle, it will be seen that the general shape of the periodic system follows as a natural result.

At this stage it may be said that for our purposes the important quantum numbers are n and l, since these define the energy level of the electron cloud and its shape. The m_l and m_s numbers define differences in the state of the electron within any particular cloud, and, although they are of significance in the magnetic behaviour of metals, it is usually only necessary to know the number of electrons in the cloud rather than their particular m_l and m_s states. It is customary to use the following notation when referring to the energy states of electrons in the atom.

n is given by the appropriate numerical value (1 to 6), and the l states are referred to by letters as follows:

$$l = 0 \quad s$$
$$l = 1 \quad p$$
$$l = 2 \quad d$$
$$l = 3 \quad f$$

(These apparently random letters are a reference to the relation between the quantum states and the emission spectra of the elements. Thus s refers to the sharp spectrum, p to the principal spectrum, d to the diffuse spectrum, and f to the fundamental spectrum.) Thus the various cloud orbits or energy levels may be written (bearing in mind that l can take any value from 0 to $n - 1$) in ascending order of energy:

$$1s, \ 2s, \ 2p, \ 3s, \ 3p, \ 3d, \ 4s, \ 4p, \ 4d, \ 4f, \ 5s, \text{ etc.}$$

(This energy sequence is strictly true only for an unbound electron; the proximity of other electrons (as in an atom) modifies the order somewhat at the higher levels. See below and Fig. 2.4.) If it is desired to specify the number of electrons in any of these states, then the number is written in superscript outside a bracket, thus: $(3p)^5$, $(1s)^2$, $(2s)^1$, etc.

We commence with the simplest atom, that of hydrogen, which contains only one electron in orbit round the nucleus. Naturally this electron will take up the lowest possible energy level, and therefore it will lie in the orbit given by $n = 1$. For this value l and m_l must both be 0, and thus the only variable quantum number is m_s, which may take the values $\pm\frac{1}{2}$. With hydrogen, having one electron only, it is immaterial whether m_s is positive or negative; the only certainty is that if a second electron is allowed to enter the lowest energy level

18

(giving the helium atom) then the two electrons will have opposing spins. From the previous paragraph, it will be seen that the electronic configuration of hydrogen may be written $(1s)^1$ and that of helium $(1s)^2$. The structures are shown schematically in Fig. 2.3(a) and (b),

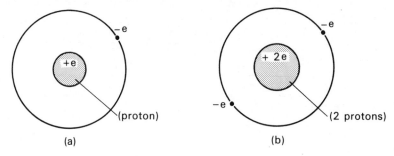

(a)　　　　　　　　　　　　(b)

Figure 2.3　　The schematic structure of (a) the hydrogen atom and (b) the helium atom.

and in tabular form in Table 2.1 (which should be referred to throughout the following discussion of atomic structures).

The next element after helium is lithium, which has three electrons circling the nucleus of the atom. Two of the electrons suffice to fill the $1s$ orbit, and therefore the third must lie in the next higher energy shell for which $n = 2$. At this level there are eight possible quantum states. l may take the values 0 or 1, and for these m_l is 0, and 1, 0 or -1 respectively. For each of the four values of m_l there are two states of m_s ($\pm\frac{1}{2}$), making a total of eight states in all. The level for $l = 0$ is filled first (i.e. the $2s$ state), so that lithium has the configuration $(1s)^2(2s)^1$ and that of beryllium (next in the series) is $(1s)^2(2s)^2$. The next level (the $2p$ state) can accommodate up to six electrons; the order of filling is immaterial, but it is worth noting that the subgroup fills up so that the electron spins are aligned as far as possible. That is to say, as we pass from boron to nitrogen the electrons will have similar m_s quantum numbers and differing m_l quantum numbers; only with oxygen, fluorine and neon do the electrons pair off with opposing spins in the m_l subgroups.

With sodium the extra electron has to occupy the shell for which $n = 3$, and initially this shell is built up in exactly the same way as the 2-quantum shell. Thus two electrons fill the $3s$ level (giving sodium and magnesium) and a further six fill the $3p$ level (giving aluminium, silicon, phosphorus, sulphur, chlorine and argon). At this point the arrangement of electrons around the atoms becomes more complicated. The next level to be filled should be the $3d$ level, for which $l = 2$ and $m_l = 0, \pm1, \pm2$ giving a total capacity of ten electrons. However, the $4s$ level has a lower energy than the $3d$ level and therefore the outermost electrons of the next two elements (potassium and calcium) fill the $4s$ level; only then is the $3d$ level filled with its ten electrons. This is shown clearly in Table 2.1, whilst Fig. 2.4 illustrates the overlapping of the energy levels in atoms of high atomic number

19

Table 2.1 Atomic Structures

Z (n=)	El (l=)	K(1) s	L(2) s	L(2) p	M(3) s	M(3) p	M(3) d	N(4) s	N(4) p	N(4) d	N(4) f	O(5) s	O(5) p	O(5) d	O(5) f	P(6) s	P(6) p	P(6) d	Q(7) s
1	H	1																	
2	He	2																	
3	Li	2	1																
4	Be	2	2																
5	B	2	2	1															
6	C	2	2	2															
7	N	2	2	3															
8	O	2	2	4															
9	F	2	2	5															
10	Ne	2	2	6															
11	Na	2	2	6	1														
12	Mg	2	2	6	2														
13	Al	2	2	6	2	1													
14	Si	2	2	6	2	2													
15	P	2	2	6	2	3													
16	S	2	2	6	2	4													
17	Cl	2	2	6	2	5													
18	A	2	2	6	2	6													
19	K	2	2	6	2	6		1											
20	Ca	2	2	6	2	6		2											
21	Sc	2	2	6	2	6	1	2											
22	Ti	2	2	6	2	6	2	2											
23	V	2	2	6	2	6	3	2											
24	Cr	2	2	6	2	6	5	1											
25	Mn	2	2	6	2	6	5	2											
26	Fe	2	2	6	2	6	6	2											
27	Co	2	2	6	2	6	7	2											
28	Ni	2	2	6	2	6	8	2											
29	Cu	2	2	6	2	6	10	1											
30	Zn	2	2	6	2	6	10	2											
31	Ga	2	2	6	2	6	10	2	1										
32	Ge	2	2	6	2	6	10	2	2										
33	As	2	2	6	2	6	10	2	3										
34	Se	2	2	6	2	6	10	2	4										
35	Br	2	2	6	2	6	10	2	5										
36	Kr	2	2	6	2	6	10	2	6										
37	Rb	2	2	6	2	6	10	2	6			1							
38	Sr	2	2	6	2	6	10	2	6			2							
39	Y	2	2	6	2	6	10	2	6	1		2							
40	Zr	2	2	6	2	6	10	2	6	2		2							
41	Nb	2	2	6	2	6	10	2	6	4		1							
42	Mo	2	2	6	2	6	10	2	6	5		1							
43	Ma	2	2	6	2	6	10	2	6	6		1							
44	Ru	2	2	6	2	6	10	2	6	7		1							
45	Rh	2	2	6	2	6	10	2	6	8		1							
46	Pd	2	2	6	2	6	10	2	6	10									
47	Ag	2	2	6	2	6	10	2	6	10		1							
48	Cd	2	2	6	2	6	10	2	6	10		2							
49	In	2	2	6	2	6	10	2	6	10		2	1						
50	Sn	2	2	6	2	6	10	2	6	10		2	2						
51	Sb	2	2	6	2	6	10	2	6	10		2	3						
52	Te	2	2	6	2	6	10	2	6	10		2	4						
53	I	2	2	6	2	6	10	2	6	10		2	5						
54	Xe	2	2	6	2	6	10	2	6	10		2	6						
55	Cs	2	2	6	2	6	10	2	6	10		2	6			1			
56	Ba	2	2	6	2	6	10	2	6	10		2	6			2			
57	La	2	2	6	2	6	10	2	6	10		2	6	1		2			
58	Ce	2	2	6	2	6	10	2	6	10	1	2	6	1		2			
59	Pr	2	2	6	2	6	10	2	6	10	2	2	6	1		2			
60	Nd	2	2	6	2	6	10	2	6	10	3	2	6	1		2			
61	Il	2	2	6	2	6	10	2	6	10	4	2	6	1		2			
62	Sm	2	2	6	2	6	10	2	6	10	5	2	6	1		2			
63	Eu	2	2	6	2	6	10	2	6	10	6	2	6	1		2			
64	Gd	2	2	6	2	6	10	2	6	10	7	2	6	1		2			
65	Tb	2	2	6	2	6	10	2	6	10	8	2	6	1		2			
66	Dy	2	2	6	2	6	10	2	6	10	9	2	6	1		2			
67	Ho	2	2	6	2	6	10	2	6	10	10	2	6	1		2			
68	Er	2	2	6	2	6	10	2	6	10	11	2	6	1		2			
69	Tm	2	2	6	2	6	10	2	6	10	12	2	6	1		2			
70	Yb	2	2	6	2	6	10	2	6	10	13	2	6	1		2			
71	Lu	2	2	6	2	6	10	2	6	10	14	2	6	1		2			
72	Hf	2	8		18			32				2	6	2		2			
73	Ta	2	8		18			32				2	6	3		2			
74	W	2	8		18			32				2	6	4		2			
75	Re	2	8		18			32				2	6	5		2			
76	Os	2	8		18			32				2	6	6		2			
77	Ir	2	8		18			32				2	6	7		2			
78	Pt	2	8		18			32				2	6	9		1			
79	Au	2	8		18			32				2	6	10		1			
80	Hg	2	8		18			32				2	6	10		2			
81	Tl	2	8		18			32				2	6	10		2	1		
82	Pb	2	8		18			32				2	6	10		2	2		
83	Bi	2	8		18			32				2	6	10		2	3		
84	Po	2	8		18			32				2	6	10		2	4		
85	—	2	8		18			32				2	6	10		2	5		
86	Rn	2	8		18			32				2	6	10		2	6		
87	—	2	8		18			32				2	6	10		2	6		1
88	Ra	2	8		18			32				2	6	10		2	6		2
89	Ac	2	8		18			32				2	6	10		2	6	1	2
90	Th	2	8		18			32				2	6	10		2	6	2	2
91	Pa	2	8		18			32				2	6	10		2	6	3	2
92	U	2	8		18			32				2	6	10		2	6	4	2

(From J. R. Partington, *General and Inorganic Chemistry*, by permission of Macmillan, London and Basingstoke.)

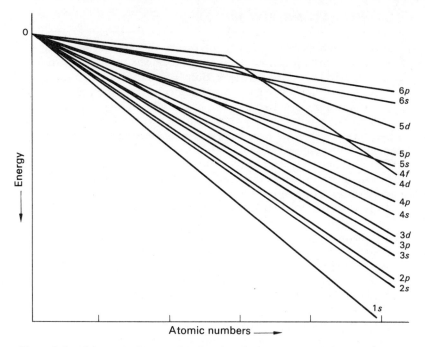

Figure 2.4 Schematic diagram showing the relative energies of electrons in the various energy states as a function of atomic number. This diagram explains the irregularities in the filling of energy states which is apparent in Table 2.1 from element number 19 onwards. (Reproduced with permission, from *The Solid State for Engineers*, by M. J. Sinnott. Copyright © 1958, John Wiley & Sons, Inc.)

and also shows that the overlapping becomes increasingly complex as we approach the highest atomic numbers. The irregular arrangement of electrons in chromium $(3d)^5(4s)^1$ rather than $(3d)^4(4s)^2$ indicates how close are the energies of the $3d$ and $4s$ levels.

From the elementary metallurgical standpoint this effect is not very significant, although it plays a part in determining the magnetic behaviour of some metals. Chemically, however, the result of the build-up of electrons in the $(n-1)$th shell when there are already electrons in the nth shell leads to close similarities between the elements concerned. It is noticeable in elements 21—28 (the 'transition elements') and the analogous groups 39—46 and 71—78, and is particularly prominent in the group 57—71 (the 'rare earths') where it will be seen from Table 2.1 that the 'building' layer of electrons is considerably 'deeper' in the atoms than is the case with the transition elements.

There is not the space here to go over the periodic system atom by atom; the above very brief survey in conjunction with Table 2.1 and Fig. 2.4 explains the general form of the periodic system. We must now consider why the elements divide into two classes, metals and nonmetals, and how the atomic theory discussed in Section 2.1.2 can be extended to explain the diverse physical nature of the elements.

21

2.2   INTERATOMIC BONDING

Atoms in themselves are of little interest to the civil engineer, but they are the units which may link up to form the bulk structures that are the materials of his profession. Sometimes the linkage results in relatively small self-sufficient units (molecules) and the product is then usually a gas or a liquid; if the links build up long chains the result may be a polymer or a glass; but if a regular three-dimensional array develops then we have a crystalline solid, such as a metal or a ceramic. How and why do these varieties of linkage come about?

Bonding between atoms is just one more consequence of the universal tendency of all systems to take up their lowest possible energy state. (For a more detailed discussion on energy see Section 2.7.) In the case of atoms it is found that a symmetrical distribution of electrons in completely filled shells confers particular stability; this is normally achieved by an outer shell of electrons of the form $(ns)^2(np)^6$ where n is the principal quantum number. Atoms achieve a lower energy state, therefore, by the possession of eight electrons in their outermost shell, and it is the urge to make up 'the eight' that is the driving force behind their bonding characteristics. (An exception to this occurs in the first shell, which as we have seen cannot accept any electrons in the p energy levels and is therefore completely and symmetrically filled by two electrons (in the s levels) rather than eight.)

Inspection of Table 2.1 shows that the elements having the above configuration naturally are in fact the inert gases: their chemical inertia (and gaseous state) illustrates that they have no need to combine with either like or unlike atoms to achieve a stable electronic grouping; they have it already.

Because of the 'octet urge', we can make another classification of the elements. Those with 5, 6 or 7 electrons in their outer shells can most readily make up the octet by *accepting* electrons from other atoms; they lie in Groups V, VI and VII of the period system. Those elements with 1, 2 or 3 electrons in the outer shell make up the octet by giving these electrons up, so that they are left with the underlying octet; as such, they readily *donate* electrons, and occur in Groups I, II and III of the period system. The elements in Group IV can behave in either way.

With these general points in mind let us now consider the four principal modes of bonding in more detail.

2.2.1   Ionic Bonding

If an atom (A) with seven electrons in the outermost shell meets an atom (B) with one electron in the outermost shell, then both can attain the octet structure if atom B donates its valency electron to atom A (Fig. 2.5). However, in so doing, the electrical neutrality of the atoms is disturbed: A, with an extra electron, becomes a negatively charged ion and B becomes a positively charged ion. The two ions are then attracted to each other by the electrostatic force between them

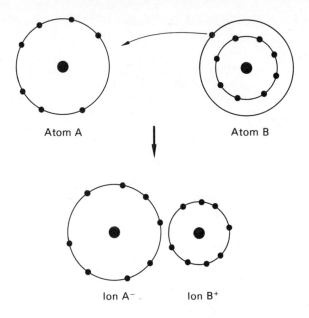

Atom A   Atom B

Ion A⁻   Ion B⁺

Figure 2.5   Schematic representation of ionic bonding.

and an ionic compound results, the strength of the bond being proportional to $e_A \cdot e_B/r$ where e_A and e_B are the charges on the ions
and r is the interatomic distance. The ratio of A ions to B ions need
not be unity as in the above example; thus we can obtain compounds
of the formulae A_2B, AB_2, A_3B_4, A_3B_2, depending on the number of
electrons which must be exchanged to make up the octets.

The bond is strong, as shown by the melting point of ionic compounds, and its strength increases, as might be expected, where two
or more electrons are donated. Thus the melting point of sodium
chloride, NaCl, is 801 °C, that of magnesium oxide, MgO, where two
electrons are involved, is 2 640 °C, and that of zirconium carbide,
ZrC, where four electrons are involved, is 3 500 °C. Since the total
numbers of electrons and protons in the lattice are the same as those
of the individual atoms, the lattice is electrically neutral. The ionic
bond is also nondirectional: that is to say, when a crystal is built up
of large numbers of ions, the electrostatic charges are arranged symmetrically around each ion, with the result that A ions tend to
surround themselves with ions of B and vice versa (see Fig. 2.6). The
crystal pattern adopted then depends on the relative sizes of A and B
ions, i.e. how many B ions can be comfortably accommodated around
A ions whilst preserving the correct ratio of A ions to B ions in the
lattice.

2.2.2   Covalent Bonding

An obvious limitation of the ionic bond is that it can only occur
between *different* atoms, and therefore it cannot be responsible for

23

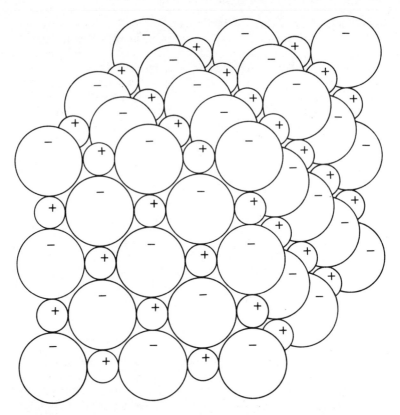

Figure 2.6   Nondirectional nature of the ionic bond: A⁻ ions surrounded by
B⁺ ions, B⁺ ions surrounded by A⁻ ions. This particular arrangement, where each
species of ion taken by itself lies on the sites of a face-centred cubic structure,
is known as the sodium chloride (NaCl) or rock salt structure.

the bonding of any of the solid elements. Where both atoms are of the
electron acceptor type, octet structures can be built up by the
sharing of two or more valency electrons between the atoms. Thus,
to take a simple instance, two chlorine atoms can bond together and
achieve the octet structure by each contributing one electron to share
with the other atom. The interaction is written:

$$:\ddot{C}\!l\cdot \; + \; \cdot\ddot{C}\!l: \longrightarrow :\ddot{C}\!l:\ddot{C}\!l:$$

Where two electrons are required to make up the octet, the interaction
may either be:

$$:\ddot{O}: \; + \; :\ddot{O}: \longrightarrow :\ddot{O}:\ddot{O}:$$

as in the oxygen molecule, or

$$\ldots + \ddot{S}\cdot \; + \; :\ddot{S}: \; + \; :\ddot{S}: \; + \; :\ddot{S}: \; + \ldots \longrightarrow \; \ldots \cdot\ddot{S}:\ddot{S}:\ddot{S}:\ddot{S}\cdot \ldots$$

as in one form of sulphur, where there is an obvious tendency for the
atoms to form up in long chains. Structurally, the elements showing
covalent bonding obey what is known as the *8 − N Rule*. This states

24

that the number of nearest neighbours to each atom is given by $8 - N$ where N is the number of electrons in the outermost shell. Thus chlorine ($N = 7$) has one nearest neighbour and so the atoms pair off as diatomic molecules; with sulphur, selenium and tellurium ($N = 6$) long chains occur; with arsenic, antimony and bismuth ($N = 5$) sheets of atoms occur, and with carbon ($N = 4$) we get a three-dimensional network. Atoms with values of N less than 4 cannot show covalent bonding with their own species, since they would require five nearest neighbours, but have only three electrons available for making up the shared bonds.

The structures are shown schematically in Fig. 2.7, and it will be seen that only carbon gives a *three-dimensional* crystal in which *all*

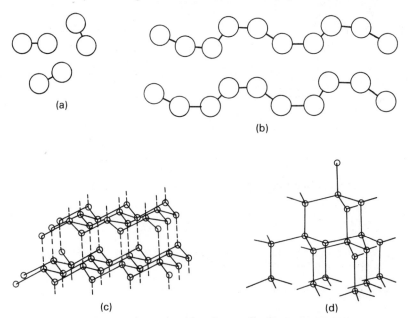

(a)

(b)

(c)

(d)

Figure 2.7   Schematic structure of elements conforming to the $8 - N$ rule. (a) Chlorine ($N = 7$): individual molecules. (b) Tellurium ($N = 6$): spiral chains. (c) Antimony ($N = 5$): corrugated sheets. (d) Diamond ($N = 4$): three-dimensional crystal.

the bonds are covalent. This reveals the major structural difference between ionic and covalent bonding: the covalent bond, unlike the ionic bond, is saturated by the individual atoms participating in it. Thus the chlorine molecule is structually self-sufficient and there is no extension of the covalent bonding between molecules. Similarly, the chains of sulphur and the sheets of bismuth are really large molecules, but there is little bonding between the chains or sheets. The result is that the covalent elements (which effectively means the non-metallic elements) have poor physical strength, not because the covalent bond itself is weak, but because, with the exception of diamond, they do not form a three-dimensional lattice. In fact, the hardness and high melting point of diamond (3 500 °C) show that the covalent bond is extremely strong. Covalent bonding is not limited to elements; many compounds are covalent, some simple examples being HCl, H_2O, CH_4 and NH_3. A large number of compounds show partly ionic and partly covalent bonding. e.g. sulphates such as Na_2SO_4 in which the sulphate radical is covalent whilst the bond to the sodium is ionic. The vast field of industrial polymers is also predominantly concerned with covalent compounds.

Individual bonds may also exhibit hybrid qualities akin to both ionic and covalent bonding; although in the pure form the two bond

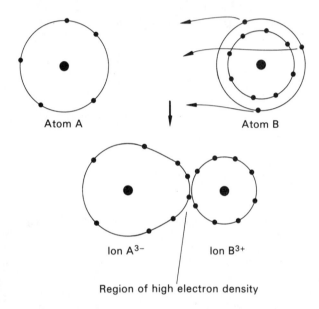

Atom A   Atom B

Ion A^{3-} Ion B^{3+}

Region of high electron density

Figure 2.8   Covalent bias in triple ionic linkage.

types represent different modes of linkage, there is in fact no sharp line of demarcation between them. When elements of high valency combine to form ionic compounds (e.g. the nitrides and carbides of the transition elements), the donor ion may lose three, or even four, electrons to the acceptor ion. The result is a strong polarising pull exerted by the donor ion on the electrons of the acceptor ion, so that the electrons are sucked back towards the donor and spend more time between the two ions than circling each individually (Fig. 2.8). The bond thus acquires some of the characteristics of the covalent linkage. Compounds of this sort are usually extremely hard and have very high melting points, since they combine in some degree the strength of the covalent bond and the nondirectionality of the ionic bond.

2.2.3   Metallic Bonding

The basis of the modern theory of metals was laid as long ago as 1900 when Drude and Lorenz put forward the free-electron theory of metals. This suggested that in a metallic crystal the valency electrons are detached from their atoms and can move freely between the positive metallic ions, as shown in Fig. 2.9. The positive ions are

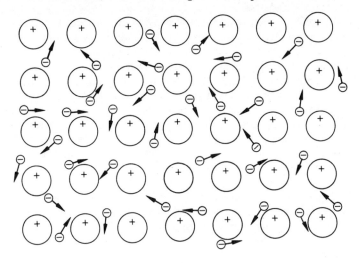

Figure 2.9   Schematic 'free electron' structure for a monovalent metal. In the absence of an electric field the electrons are in ceaseless random motion, but the overall distribution remains uniform over any period of time.

arranged in a crystalline lattice, and the electrostatic attraction between the positive ions and the negative free electrons provides the cohesive strength of the metal. The linkage may thus be regarded as a very special case of covalent bonding, in which the urge to attain the octet grouping is satisfied by a generalised donation of the valency electrons to form a 'cloud' which permeates the crystal lattice, rather than by electron sharing between specific atoms (true covalent bonding) or by donation to another atom (ionic bonding).

The atoms which can take up this structure most readily lie in groups I, II and III of the period system. In these groups the valency electrons are relatively loosely bound to the atom and are therefore readily donated to form an electron 'cloud'. The strength of the bond increases as the shell is filled with electrons so that elements in Groups IV to VIII show a greater tendency to nonmetallic covalent structures. However, the bonding of electrons in the outermost orbits becomes progressively weaker with increasing atomic number and it will be seen on inspection of the periodic system that the members of a group become more metallic in nature with increasing atomic number. Group IV, commencing with carbon, which is a typical non-metal, passing through silicon and germanium, which show increasing metalloid characteristics, and concluding with tin and lead, which are clearly metals at ordinary temperatures, is an excellent example of this tendency.

Since the electrostatic attraction between ions and electrons is nondirectional, i.e. the bonding is not localised between individual

27

pairs or groups of atoms, metallic crystals can grow easily in three dimensions, and the ions can approach all neighbours equally to give maximum density of structure. Crystallographically, these structures are known as 'close-packed' (see Section 2.5.1); they are geometrically simple by comparison with structures of ionic compounds and naturally occurring minerals, and it is this simplicity that accounts in a large part for the ductility of the metallic elements.

This theory explained the high thermal and electrical conductivity of metals. Since the valency electrons are not bound to any particular atom, they can move through the lattice under the application of an electric potential, causing a current to flow, and can also, by a series of collisions with neighbouring electrons, transmit thermal energy rapidly through the lattice. Optical properties are also explained by the theory. If a ray of light falls on a metal, the electrons (being free) can absorb the energy of the light beam, thus preventing it passing through the crystal and rendering the metal opaque. The electrons which have absorbed the energy are excited to high energy levels and subsequently fall back to their original values with the emission of the light energy; in other words the light is reflected back from the surface of the metal, and we can account for the high reflectivity of metals.

The ability of metals to form alloys (of extreme importance to engineers) is also explained by the free electron theory. Since the electrons are not bound, when two metals are alloyed there is no question of electron exchange or sharing between atoms as in ionic or covalent bonding, and hence the ordinary valency laws of combination do not apply. The principal limitation then becomes one of atomic size, and providing there is no great size difference (see Section 5.3), two metals can often form a continuous series of alloys or solid *solutions* from 100%A to 100%B. The Drude theory has since been extensively modified — only about 1% of the valency electrons are in fact sufficiently free to absorb thermal energy — but the modifications are primarily concerned with conductivity and need not be further considered here.

2.2.4 van der Waals Bonds

The three strong *primary* varieties of linkage between atoms all occur because of the urge of atoms to achieve a stable electron configuration. However, even when a stable configuration exists already, as in the case of the inert gases, some sort of bonding force must be present since these elements all liquefy and ultimately solidify, albeit at very low temperatures.

Bonds of this nature are universal to all atoms and molecules, but are normally so weak that their effect is negligible when any primary bonds are present. They are known as van der Waals bonds, and are one reason why real gases deviate from the ideal gas laws. They arise as follows.

In Section 2.1.2 it was shown that the best physical concept of an electron in orbit is a cloud of varying density, the density at any

28

point being related to the probability of finding the electron there. Such a picture implies that the electron charge is 'spread' around the atom, and over a period of time the charge may be thought of as symmetrically distributed within its particular cloud. However, the electronic charge *is* moving, and this means that on a scale of nano-seconds the electrostatic field around the atom is in a state of eternal fluctuation, resulting in the formation of a dynamic dipole. When another atom is brought into proximity, the dipoles of the two atoms

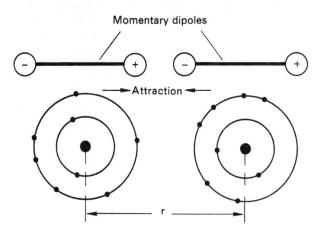

Figure 2.10   Weak van der Waals linkage between atoms arising from fluctuating electronic fields.

may interact cooperatively with one another (Fig. 2.10) and the result is a weak electrostatic bond. This bond is nondirectional.

The attractive force between two atoms is given by the expression

$$F = \frac{\alpha_1 \alpha_2}{r^6} \qquad (2.1)$$

where α_1 and α_2 are the *polarisabilities* of the two atoms, i.e. the ease with which their electronic fields can be distorted by a unit electric field to form a dipole, and r is the distance between them. Polarisability increases with atomic number, since the outermost electrons (which are responsible for the effect) are further removed from the nucleus and therefore more easily pulled towards neighbouring atoms. This is shown by a comparison of the melting points of He (1 K) and Xe (133 K) or fluorine (51 K) and iodine (387 K).

In addition to the fluctuating dipole above, many molecules have permanent dipoles as a result of bonding between different species of atoms. These can play a considerable part in the structure of polymers and organic compounds, where side chains and radical groups of ions can lead to points of predominantly positive or negative charges that will exert an electrostatic attraction to other oppositely charged groups.

The strongest and most important example of dipole interaction occurs in compounds between hydrogen and nitrogen, oxygen or fluorine. It occurs because of the small and simple structure of the hydrogen atom and has in fact become known as the *hydrogen bond.* When hydrogen links covalently with, say, oxygen to form water, the electron contributed by the hydrogen atom spends the greater part of its time between the two atoms, and the bond acquires a definite dipole with hydrogen which becomes virtually a positively charged ion. Since the hydrogen nucleus is not screened by any other electron shells, it can attract to itself other negative ends of dipoles, and the result is the hydrogen bond. It is considerably stronger (roughly x 10) than other van der Waals linkage, but is much weaker (by 10—20 times) than any of the primary bonds. Figure 2.11 shows

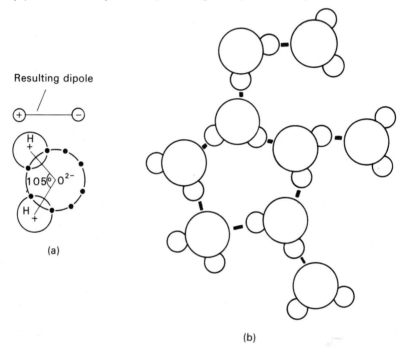

Figure 2.11   The hydrogen bond. (a) Individual water molecule showing dipole resulting from bond angle. (b) Structure of water.

the resultant structure of water where the hydrogen bond forms a secondary link between the water molecules, acting as a bridge between two electronegative oxygen ions. This relatively insignificant bond is then one of the most vital factors for the benefit and survival of mankind; it is responsible for the abnormally high melting and boiling points of water and its high specific heat (which affords an essential global temperature control): in its absence water might well be gaseous at ambient temperatures like ammonia or hydrogen sulphide.

Hydrogen bonds are also important to the engineer in that they play a large part in adsorption and surface reactions. They are thus a

30

major factor in concrete timber and clay technology, and as such will be considered more fully in later chapters.

2.3 STATES OF AGGREGATION

2.3.1 Response to Stress

If asked to classify a variety of substances into the categories solid, liquid or gaseous, most of us would have no difficulty in so doing; although, as we shall see, there are a number of materials that lie somewhere between the solid and liquid states. Such a classification would be carried out for the most part almost without thinking, and must obviously be related to the internal structure of the samples; it is, however, primarily based on response to stress.

When stress is applied to any material, it can respond in one (or sometimes two) of three ways: elastically (by the reversible stretching of interatomic bonds); plastically (by internal shear between planes of atoms or molecules); or by fracture (when the interatomic bonds are ruptured across an internal surface). The exact nature of the response will depend on the degree of stress and the way in which it is applied. For a material to be designated a solid, it must at least be able to withstand indefinitely the stress imposed by its own weight, i.e. it should show a purely *elastic* response. On the other hand, a liquid shears instantaneously under its own weight to take up a position of minimum potential energy, and will therefore never support a shear stress. In other words it shows a purely *plastic* response. The use of the terms 'indefinite' and 'instantaneous' shows that a time factor has to be considered in stress responses, and it is in fact this time factor that provides a means of separating (albeit with a considerable degree of overlap) the liquids from the solids.

Patterns of response are shown in Fig. 2.12. If a shear stress τ is applied at time t_0 to a material, the elastic strain immediately rises to γ, where

$$\tau = G \cdot \gamma \qquad (2.2)$$

and G is the shear modulus of the material. A perfect solid will then support the stress indefinitely (curve 1), but for a liquid such as water relaxation is virtually instantaneous (t_R, the relaxation time, is about 10^{-11} s) and the loading and unloading curves are synonymous. In between these two extremes, elastic stress relaxation may occur in a finite time; depending on the loading conditions there are two possible consequences, both of extreme importance to engineers. If the strain remains constant (as for example in a tension bolt clamping two structural girders together), relaxation may occur without dimensional change by a gradual conversion of elastic or load-carrying strain into plastic strain or *set*. Curve 2 shows the loss of elastic stress in this case. When relaxation is complete, the clamping action of the bolt (due to the elastic tensile stress) ceases to function; the engineer must now beware the onset of fretting and fatigue. If,

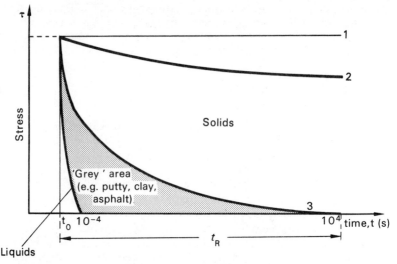

Figure 2.12 Stress relaxation rates in materials.

however, the material is under constant stress (e.g. the structural girders themselves in the above example), then stress relaxation can obviously never occur, and the material will respond by a slowly and steadily increasing degree of strain, which approximates to viscous flow. Under sufficiently severe loading, many normally rigid materials are prone to creep, notably cement and concrete. Polymers and metals are also affected, but in the case of the latter it is normally a high-temperature effect and so of less consequence to the civil engineer, except in the case of high tensile bolts and prestressing rods in concrete (see Section 12.7).

How then may we distinguish between liquids and solids? The answer is that there is no firm dividing line, but a considerable grey area in which materials may be considered equally to be very plastic solids or very viscous liquids. The generally accepted bounds are shown in Fig. 2.12, all materials having $t_R < 10^{-4}$ s being considered as liquids and all having $t_R > 10^4$ s being considered as solids. Many polymers, clays (depending on their moisture content), pitch, putty, asphalt and toffee lie in this grey region and cannot be truly regarded as either solids or liquids since they will show a mixed elastic/plastic response even at low stress levels. General classes of materials that lie in this category include *gels*, *thixotropes* and *dilatants*, and are considered further below.

2.3.2 Gases

Structurally, gases need only concern us very briefly, although they will be considered further in Section 2.7 when introducing some concepts of thermodynamics. Beyond the order associated with individual molecules, a gas exhibits no sort of ordered structure at all; apart from van der Waals interactions (which are too weak to bring about intermolecular bonding except at low temperatures or high

32

pressures) and collisions, the atoms or molecules are in continual random and independent movement at high speeds. Thus a gas can have no fixed shape, and the random molecular velocities will lead it to fill any container into which it is introduced.

2.3.3 Liquids

By contrast, liquids are much closer to solids than gases in their structure; this is born out by the fact that the latent heat of fusion is normally only about 5% of the latent heat of vaporisation. Two structural varieties exist, derived respectively from crystalline and amorphous solids. Those liquids derived from crystals consist of small rafts of atoms still arranged in the crystal structure, i.e. submicroscopic crystallites, but the bonding is not strong enough for the crystallites to link together and form a rigid mass. They are in continuous relative movement, and the atoms on their edges are constantly detaching themselves from one group and attaching themselves to another, although the number of groups and also their individual size remain approximately constant. Structurally, then, a liquid may be said to resemble a well run cocktail party in which the guests are kept circulating from group to group by an assiduous hostess.

Liquids derived from amorphous solids are composed of large molecules — frequently long-chain molecules — which have gained sufficient thermal energy to overcome any linking tendency, and to become flexible and mobile so that they will wriggle past each other under stress. Any reader who has dined on spaghetti will appreciate how easily flexible 'chains' can slither and flow. There is a major difference between the two varieties of liquid described. Those derived from crystalline solids melt at a fixed temperature or over a well defined temperature range; thus on heating the solid there is a fairly abrupt transition to the liquid state. This arises from the regularity of the crystal structure; once the critical temperature is reached the thermal energy is sufficient to rupture a large number of bonds of exactly equal strength, and the result is a complete breakdown of mechanical stability. In an amorphous structure, however, rigidity is maintained by cross-linking and entanglement of rigid long-chain molecules, or is the result of an irregular but well cross-braced array of atoms. The application of heat then leads to a much more gradual rupture of bonds since these will be under widely varying degrees of strain in the solid state. This, coupled with a steady increase in the ability of individual molecules to flex and rotate their inner linkages, results in a gradual softening and almost imperceptible change from the solid to the liquid state. Polymers exhibit this behaviour because they are at best only partly crystalline, and so do glasses. The latter are composed of chains of linked oxide molecules, of which silica, SiO_2, and boron oxide, B_2O_3, are the most frequently used.

2.3.4 Solids

Solids may occur in either crystalline or amorphous form, and if amorphous they may be subdivided into molecular and glassy solids.

Crystalline solids form most readily where the bonding is non-directional, as is the case in metallic and ionic structures. Metals in particular are able to form close-packed structures of high symmetry (see Section 2.4); in the case of ionic solids there is a strong tendency for positive ions to surround themselves with negative ions and vice versa, which should tend towards the development of chequerboard patterns. However, the need to satisfy the requirement of valency in terms of the relative numbers of ionic species, and the widely differing radii of many ions, mean that relatively few ionic crystals have simple structures, and many are exceedingly complex. As explained above (Section 2.2.2) pure covalent bonding does not very readily result in crystalline structures, except in the case of some elements. In Group IV, carbon and silicon can form crystals because the bond angles at each carbon or silicon atom lead to the build-up of three-dimensional structures. Elements in Groups V and VI, however, rely in part on van der Waals linkages for bulk strength, since the covalent bonds are restricted to particular planes or directions, and are thus possessed of very low strength.

Covalently bonded compounds normally have molecules that are electronically complete in themselves; their bulk strength is then dependent on physical entanglement (if they are long-chain compounds such as polymers) and van der Waals bonds. The larger and more complex the molecules, the greater are the opportunities for these forms of linkage; hence the reasonable strength of thermoplastic polymers and the increased melting points as we ascend the aliphatic and aromatic series of organic compounds. They are usually referred to as molecular solids and have the typically indefinite melting point that goes with their structure. Some schematic structures are shown in Fig. 2.13.

By comparison with molecular solids, which are soft and ductile, glassy solids are hard and brittle. True glasses are obtained by cooling liquids which, because of their elaborate random molecular configuration, lack the necessary activation energy at the melting point to rearrange themselves in an ordered crystalline array. The material then retains, even when rigid, a typically glassy structure involving short-range order only and resembling, except for its immobility, the structure of the liquid form. Silica gives rise to the commonest form of glass; if cooled sufficiently slowly from the melt it can be obtained in the crystalline form which is more stable thermodynamically than the vitreous state. However, at ordinary cooling rates crystallisation does not take place, and the amorphous nature of the solid (coupled with the lack of free electrons) gives rise to its most important physical property: its transparency. Most nonmetallic single crystals are transparent, but in the polycrystalline state light is scattered from internal reflections at flaws and crystal boundaries; the material then becomes translucent in thin sections and completely opaque in the mass, as with naturally occurring rocks. Totally amorphous polymers may also be transparent, but most owe their opacity to the presence of crystalline regions (spherulites) within their structure that act as light scatterers.

34

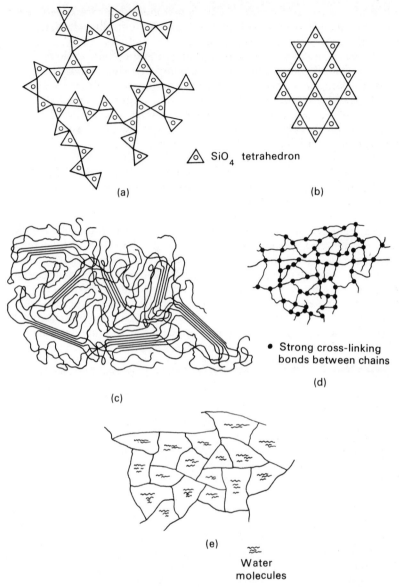

Figure 2.13 Schematic structures of some covalent materials: (a) vitreous silica (glass); (b) crystalline silica; (c) thermoplastic polymer showing crystalline regions (spherulites); (d) thermosetting polymer; (e) gel — by comparison with (d), the links between the chains are weak and easily broken.

Thermosetting polymers, although not classified as glasses (which are oxides), exhibit a glassy structure: a rigid three-dimensional random latticework of bonds, so that the whole structure is really a giant molecule. Figure 2.13(d) gives a two-dimensional representation of such a structure, and it will be seen that, as in glasses, there is a degree of short-range order, but as a whole the structure is random.

35

It is the presence of strong interatomic bonds coupled with the lack of any possible shear planes that imparts the typical hardness and brittleness to structures of this nature.

We have already emphasised the lack of any definite dividing line between solid and liquid, the reason being that the response of materials to shear stresses can vary continuously over a wide range of rate. There are, however, more complex structures involving both solid and liquid phases which can give the same effect. One of the most familiar of these structures is the *gel*, known to most of us from childhood in the form of jellies and lozenges.

Gels are formed when a suspension of very fine particles (usually of colloidal dimensions) in a liquid bonds in a loose 'spillikins' or 'card-house' structure, trapping liquid in its interstices. Depending on the number of links formed, gels can vary from very nearly fluid structures to rigid solids. If the links are few or weak, then individual particles have considerable freedom of movement around their points of contact, and the gel deforms easily (Fig. 2.13(e)). A high degree of linkage gives (as with thermosetting polymers) a structure that is hard and rigid in spite of all its internal pores. The most important engineering gel is undoubtedly *cement*, which develops a highly rigid structure. When water is added to the cement powder, the individual particles take up water of hydration, swell and link up with each other to give rise to a high-strength but permeable gel of complex calcium silicates. A feature of many gels is their very high specific surface area; if the gel is permeable as well as porous, the surface is available for adsorbing large amounts of water vapour, and such a gel is an effective drying agent. Adsorption is a reversible process (see Section 2.7.2); when the gel is saturated, it may be heated to drive off the water and its drying powers regained. *Silica gel* is a familiar example of this.

If a gel sets by the formation of rather weak links, the linkages may be broken by vigorous stirring and the gel liquefies again. When the stirring ceases, the bonds will gradually link up and the gel will thicken and return to its original set. Behaviour of this sort, in which an increase in the applied stress causes the material to act in a more fluid manner, is known as *thixotropy*. Not all gels behave in this manner, or, if they do, it is only at a certain stage of their setting procedure: concrete if cracked will not, alas, heal itself spontaneously, although it exhibits a marked degree of thixotropy at an early stage of setting.

The most familiar application of thixotropy concerns the manufacture of nondrip paints, which liquefy when stirred and spread easily when being brushed on, but which set to a gel as soon as brushing is completed so that dripping or streaks on vertical surfaces are avoided. Clays can also exhibit thixotropy. This is turned to advantage in the mixing of drilling muds for oil rigs: the thixotropic mud serves to line the shaft with an impermeable layer, whilst in the centre it is kept fluid by the movement of the drill and acts as a medium for removing the rock drillings. On the other hand a thixotropic clay underlying major civil engineering works could be highly hazardous.

36

The inverse effect of thixotropy occurs when an increase in the applied stress causes a viscous material to behave more in the manner of a solid, and is known as *dilatancy*. It is a less familiar but rather more spectacular phenomenon. Cornflour-water mixtures demonstrate the effect over a rather narrow range of composition, when the viscous liquid will fracture if stirred vigorously; it is of short duration, however, since fracture relieves the stress, and the fracture surfaces immediately liquify and run together again. Silicone putty (marketed as 'Potty Putty') is also dilatant; it flows very slowly if left to itself, but fractures if pulled suddenly, and will bounce like a rubber ball if thrown against a hard surface. So far, no engineering applications of dilatancy have been developed.

2.4   THE CRYSTALLINE STRUCTURE

It will be shown in the following section that the equilibrium spacing r_0 between two ions bonded together is determined by a balance between the attraction due to the bond and the repulsion which develops when their outer electron shells begin to overlap each other.

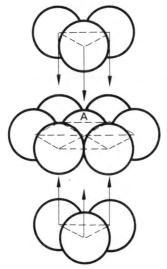

Figure 2.14   Twelvefold coordination in close-packed structures. The atom A is touched by six atoms in its plane, three in the plane above, and three in the plane below.

We can thus say that pairs of atoms 'touch' when in spatial equilibrium, and there is therefore a strong natural tendency for three-dimensional structures to be *close packed*, i.e. for each ion to touch as many other ions as possible. The ideal degree of close packing is realized in many metallic structures; in these the ions may have 12 nearest neighbours, or a *coordination number* of 12, arranged as in Fig. 2.14. The extension of this pattern results in a *regular* three-dimensional array of ions, in which each has six closest neighbours

37

in its plane, three in the plane above, and three in the plane below. Such an assembly is an example of a *crystal*, which term implies a three-dimensional assembly of atoms or ions exhibiting some sort of regularly repeating pattern. Crystals of this nature exhibit a high degree of symmetry, but it is only metals and alloys that are able to exhibit maximum close packing, and then only if the ions do not vary in size too greatly; hence the high density that is a general feature of the metallic state. Other materials, held together by ionic or covalent or van der Waals bonds, are subject to restrictions in the degree to which they can coordinate, such as widely differing ionic sizes, or directed rather than general bonds, or the need to satisfy valency requirements; maximum close packing then becomes an impossibility and a large proportion of the bonds will be extended beyond their equilibrium lengths. In these materials the stable configuration of ions is the one which has the lowest overall strain energy in the bonds, and this criterion again favours the development of a regular crystalline structure, albeit one with a much lower degree of symmetry than is usually the case with metals.

Despite the tendency to form crystals, the occurrence of single large crystals in solids is rare and largely confined to geological systems that have cooled very slowly from the liquid state. Metals and ceramics are normally polycrystalline, the size of individual crystals or *grains* varying from about 1μ in diameter to the very large (but thin) zinc crystals that can frequently be seen on the surface of galvanized mild steel sheet. The size of crystals plays a considerable part in determining material strengths, particularly in the case of ductile metallic systems, and will be further considered in Section 12.4. It should also be noted that the shape of crystals in most materials bears no relation whatever to their internal structure; each crystal grows from a nucleus until it collides with other growing crystals, and it is the random juxtaposition of these growing crystals that determines their shape. Only when a crystal grows freely until it has used up all the available atoms or molecules to its maximum size are the facets that most of us have seen on quartz or copper sulphate crystals likely to be developed.

2.4.1 Elementary Classification of Crystal Structures

Classification of crystal structures is based on the concept of *space lattices*. These are regular three-dimensional arrays of points in space extending infinitely in all directions; in a real crystal the points are replaced by one or more species of atom or ion, and in complex structures each point may be represented by a group of atoms. Depending on the arrangement of the points, the degree of symmetry of the space lattice can vary considerably, but the only basic essential common to all space lattices is shown in Fig. 2.15. If we take any point P selected at random, vectors **a**, **b** and **c** can be drawn connecting P with its three nearest neighbours; it should be noted that **a**, **b** and **c** do not have to be equal, nor have they to be mutually at right angles. Continuing along any of these vectors, we shall then find the

38

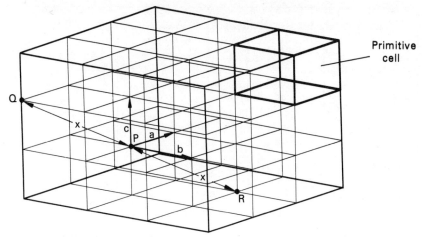

Figure 2.15   A space lattice.

points evenly spaced out, separated by distances a, b and c respectively in the three directions. The heavily outlined parallelepiped in Fig. 2.15 is known as the *primitive cell* of the space lattice, and is also repeated infinitely in three directions to make up the lattice. Two features of all space lattices follow from this construction: firstly, the

Table 2.2   Bravais Space Lattices

Crystal system	Lattice type	Essential features
Cubic	Simple Body-centred Face-centred	Three axes mutually at right angles $x = y = z$
Tetragonal	Simple Body-centred	Three axes mutually at right angles $x = y \neq z$
Orthorhombic	Simple Body-centred Base-centred Face-centred	Three axes mutually at right angles $x \neq y \neq z$
Hexagonal	Simple	Three co-planar axes x, y, q at $120°$ One axis (z) at right angles to these $x = y = q \neq z$
Rhombohedral	Simple	Three axes equally inclined, $\neq 90°$ $x = y = z$
Monoclinic	Simple Base-centred	Two axes *only* at right angles $a \neq b \neq c$
Triclinic	Simple	Three axes inclined at any angles $a \neq b \neq c$

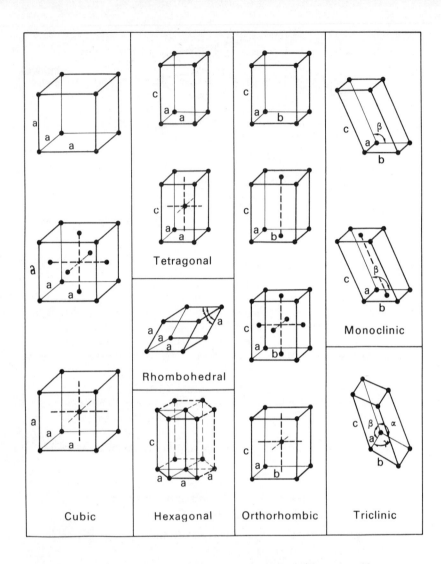

Figure 2.16 Unit cells of the 14 Bravais space lattices. (Reproduced by permission, from *The Structure and Properties of Materials*, by W. G. Moffatt, G. W. Pearsall and J. Wulff (Vol. I). Copyright © 1964 by John Wiley and Sons, Inc.)

view of the lattice *in the same direction* is identical at all lattice points; there is therefore no way of establishing the position of any individual point in an infinite lattice. Secondly, for every point Q at a certain distance *x* and in a certain direction from a point P, there will be another point R at an exactly similar distance from P in the opposite direction, so that P is the midpoint of the line joining Q and R.

40

At first sight it might appear that there must be a multitude of space lattices that can be arranged to satisfy the rather modest essentials outlined above. However, it was established by Bravais in 1848 that there are in fact only fourteen possible varieties of space lattice and these can be classified into a mere seven crystal systems; they are defined in Table 2.2 and their *unit cells* illustrated in Fig. 2.16. The *unit cell* of a lattice is the smallest cell which conveniently demonstrates the lattice characteristics; it is not necessarily the smallest cell which can be multiplied to build up the lattice. The latter is known as a *primitive cell*, contains only one lattice point ($\frac{1}{8}$ of a point at each corner), and often shows little indication of the symmetry elements present in the lattice. For example, Fig. 2.17 illustrates the

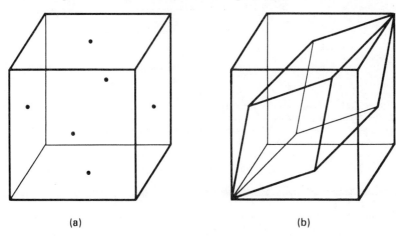

| (a) | (b) |

Figure 2.17   The face-centred cubic structure: (a) the conventional unit cell; (b) the primitive unit cell.

unit cell and the primitive cell of the face-centred cubic structure; the difference between the two is very apparent.

The features in a crystal that result in symmetry are principally *planes of symmetry* and *axes of symmetry* and *inversion* through a lattice point. The latter feature is simply a restatement of the second basic essential for a space lattice mentioned above. Since it applies to all points in all space lattices, it is a common feature of symmetry — and is in fact the *only* symmetrical feature of the triclinic system. Reflection of the lattice points from one side of a plane of symmetry to the other results in exact congruence with the points already on the other side. In the case of an axis of symmetry, rotation about the axis leads to exact superimposition of lattice points; both these operations are depicted in Fig. 2.18. Axes of symmetry may be *twofold (diads)*, *threefold (triads)*, *fourfold (tetrads)*, or *sixfold (hexads)* only; a twofold axis requires a rotation of 180° to achieve congruence, a fourfold axis a rotation of 90°, and a sixfold axis a rotation of 60°.

To be valid, rotation about an axis of symmetry must achieve coincidence for *all* lattice points. If we consider two points P and Q, which have the closest spacing in the lattice, and apply rotation about

41

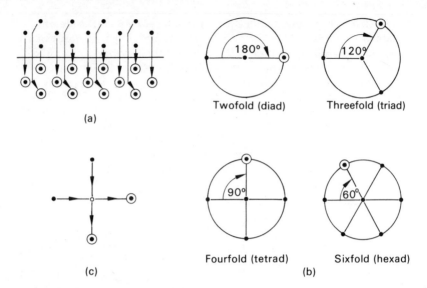

Twofold (diad) Threefold (triad)

(a)

Fourfold (tetrad) Sixfold (hexad)

(c) (b)

Figure 2.18   Features of symmetry in crystal lattices. (a) Reflection. The points reflected across the plane (depicted as circles) are coincident with the points below the planes. (b) Rotation: axes of symmetry. For clarity, only one point is shown undergoing rotation. (c) Inversion.

an axis passing through P and perpendicular to the paper, then it becomes apparent that a greater than sixfold symmetry axis is impossible. For if Q becomes congruent with R (Fig. 2.19) and the

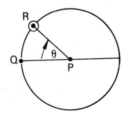

Figure 2.19   If the symmetry axis through P is greater than 6, then $\theta < 60°$. But if $\theta < 60°$ then QR $<$ QP, which contradicts the initial premise that P and Q are the nearest neighbouring points in the lattice.

necessary rotation is less than 60°, then Q and R must be closer than Q and P, which is impossible from our initial selection of P and Q. Fivefold axes are also ruled out.

Structures exhibiting a high degree of symmetry, such as the body-centred or face-centred cubic systems, show many elements of symmetry. For example, the unit cell of the face-centred cubic system has 6 diads, 4 triads, 3 tetrads, 3 face-centred planes of symmetry and 6 diagonal planes of symmetry (the reader may care to work this out for himself using a die or a sugar cube as a model). By contrast, the triclinic has no axes or planes of symmetry, but merely one centre of symmetry.

One of the most important features of any crystal structure concerns the various planes that can be drawn through it. If, for the sake of simplicity, we consider a simple two-dimensional array of points, or *net*, as in Fig. 2.20, it is clear that an infinite number of different

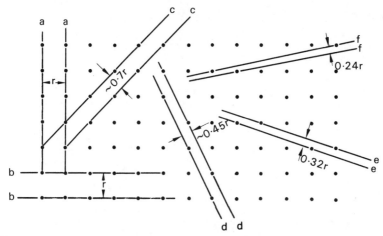

Figure 2.20 Spacing of lines in a net. The spacing between points in lines is inversely proportional to the spacing between neighbouring parallel lines.

(i.e. nonparallel) lines can be drawn through it. However certain lines appear much more prominent than others; these are the lines containing the greatest number of lattice points for a given length, such as the lines a—a and b—b. It will also be apparent that these are the lines that are most widely spaced from similar parallel lines, and that as less and less obvious lines are drawn the density of points in each line decreases whilst similar parallel lines become closer and closer spaced. (Compare d—d and e—e with a—a and c—c in Fig. 2.20.) If we extend these concepts to three-dimensional reality exactly the same considerations apply, except that we are now dealing with *planes* instead of *lines*.

2.4.2 The Miller System of Indices

Planes are important in crystals. Shear may occur between neighbouring planes, permitting the crystal to shear and conferring *ductility*, as in metals. If bonding is weak across a set of parallel planes, brittle fracture may occur readily by the parting of the planes, and they are then referred to as *cleavage planes*. Many naturally occurring minerals exhibit this behaviour very readily (e.g. calcite, mica and slate) and most ionic and covalent crystals will tend to fracture more readily along specific planes. It then becomes necessary to be able to specify individual crystal planes and, in the case of shear, to specify directions within a particular plane. Such identification is carried out by means of *Miller indices*. These are assigned as follows.

43

Three axes X, Y and Z are selected following the edges of the unit cell from an origin at one of the lattice points (since all points have exactly similar surroundings it does not matter which point is taken). Units along the axes are usually taken as the respective dimensions of the unit cell. The normal convention is that the Z axis should be vertical, or near vertical in the case of the low-symmetry systems, and of the two horizontal axes the X axis should point towards the viewer and the Y axis be aligned parallel to the viewer, as shown in Fig. 2.21.

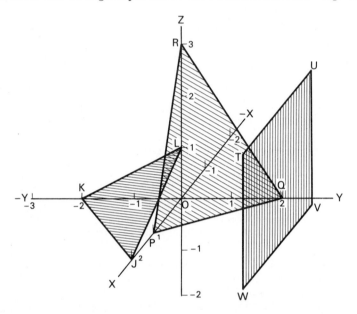

Figure 2.21 Identification of planes by means of axial intercepts. The Miller indices are the reciprocals of the direct intercepts, adjusted to give the lowest possible set of integers. The plane TUVW lies parallel to both X and Y axes.

Any plane may then be uniquely defined in terms of its intercepts on these three axes. Consider the plane PQR in Fig. 2.21. This has intercepts 1, 2 and 3 on the X, Y and Z axes respectively. The Miller indices are calculated by taking the *reciprocals* of these intercepts and putting the resultant fractions over their lowest common denominator. The *numerators* of the fractions then give the required indices. In the above example, the direct intercepts 1, 2, 3 have reciprocal values of 1, $\frac{1}{2}$ and $\frac{1}{3}$ respectively. Putting these over their lowest common denominator, we obtain $\frac{6}{6}$, $\frac{3}{6}$, $\frac{2}{6}$, and the Miller indices of the plane are 6, 3, 2, normally expressed in brackets, thus: (632) and referred to as 'a six-three-two plane'. When considering them in general terms, the indices are conventionally given the letters (hkl). When one or more of the intercepts are negative (plane *JKL* in Fig. 2.21), then the negative sign is placed *over* the appropriate indice and referred to as a 'bar'. Thus *JKL* has direct intercepts, 2, −2, 1, giving Miller indices for the plane of 1, −1, 2, but these are written (1 1 2) and referred to as 'one-bar-one-two'. Where a plane is parallel to one or more axes, then

44

its intercepts on those axes are ∞, and the Miller indices 0. Thus plane TUVW in Fig. 2.21 is an (010) plane.

Two other important points to note are firstly that all parallel planes are normally regarded as the same plane crystallographically speaking, so that (to give simple example) the (111), (222), (333), ..., (nnn) planes are all considered as (111) planes. Secondly, various nonparallel planes in crystal systems have obvious similarities, and are referred to as *families of planes*. The planes containing the faces of the unit cell of the cube system (Fig. 2.22) which have the indices

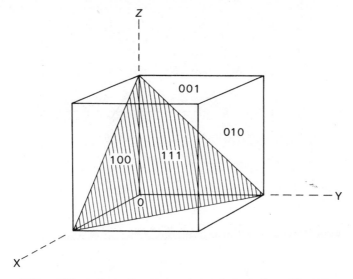

Figure 2.22   Miller indices of some prominent planes in the cubic system.

(100), (010) and (001) are called the 100 family of planes and, when it is required to refer to the family, the indices are expressed in wavy brackets, thus: {100}. Inspection of the cubic system will show that there are, in addition to the above example, 6 members of the {110} family and four members of the {111} family of planes. It is instructive to attempt to list all the members with their appropriate individual indices. These planes are important in respect of the mechanical properties of metals and alloys since they are the planes on which shear may occur preferentially.

The above system of indexing applies to all crystal systems, but it has been found convenient to modify it somewhat by the addition of a fourth axis in the case of the hexagonal system. For crystals of this habit, the axes are arranged as in Fig. 2.23, the X, Y and U axes being coplanar and inclined at $120°$ to each other. As a result, every plane has *four* indices, $(hkil)$ and it can be found by inspection that h, k and i are always related by the expression $h + k + i = 0$. Thus the plane QRST has the indices (1120).

Directions in crystals are also referred to by indices, but in this case the indices of a direction line passing through the origin are given by the simplest integral coordinates of a point on the line. Thus the line

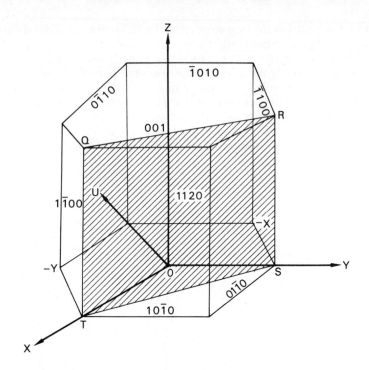

Figure 2.23   Some prominent planes in the hexagonal system.

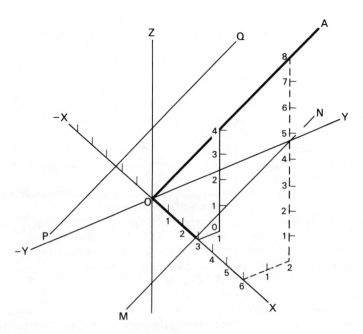

Figure 2.24   Direction indices for a line.

OA in Fig. 2.24 has direction indices 314, although the indices 628 would define an exactly similar line. The indices of a direction line that does not pass through the origin are taken to be identical with those of a parallel line that does pass through the origin; hence PQ and MN in Fig. 2.24 will also have indices 314. Direction indices are expressed in square brackets [110] and if a family of crystallographically similar directions is referred to, the indices are written ⟨313⟩. Negative indices are expressed exactly as with planes. It should also be noted that the line or direction having the same indices as a plane is normal to that plane, but only in the case of cubic crystals.

2.4.3 Water of Crystallisation

It is well known that many ionic crystals contain water molecules locked up in their structure as *water of crystallisation*; such crystals are known in general as hydrates, and their formation can be very important in the development of bulk strength. The ratio of water molecules to ionic molecules is usually 2 : 1, 4 : 1 or 6 : 1, but 5 : 1, 7 : 1, 8 : 1 and 12 : 1 ratios are also encountered. Both cement and plaster of paris owe their commercial importance to their ability to take up water and form a rigid mass of interlocking crystals, although in the case of cement the crystals are so small that it becomes almost a philosphical problem to decide whether to classify the structure as a crystalline hydrate or a hydrated gel.

Water and ammonia are the only two molecules that can be taken up up as a structural part of crystals, and both for the same reasons: they are small molecules which are strongly polar in nature. The small size permits them to penetrate into the interstices of crystal structures where close packing of ions is not possible; this is particularly the case where the anion is a radical, such as $-SO_4$ (sulphate), SiO_4 (silicate) or $-B_4O_7$ (borate). It must be emphasised that this process is not to be thought of as a capillary action, analogous to the take-up of large amounts of water by clays; the water molecules are bonded into definite sites within the crystal structure, and the crystal will only form a stable hydrate if ions of the appropriate sign are available and correctly placed to form bonds with the positively and negatively charged regions of the water molecule. We have already mentioned (Section 2.2.4) the abnormal properties of water as arising from the ability of the molecules to link up by means of hydrogen bonds; the formation of crystalline hydrates is an extension of the same behaviour. Normally the water molecules cluster round the cations in the crystal forming hydrated ions; this has the effect of making the small cation behave as if it were a good deal larger. As a result, the size difference between anion and cation is effectively reduced, thus making for simpler and more closely packed crystal structures. Water bonded in this manner is very firmly held in many instances, so that hydration becomes virtually irreversible. Copper sulphate in the anhyorbus form readily absorbs water to form crystals of the pentahydrate, but the hydrate has to be heated to 260 °C before the water of crystallisation is expelled, and traces of water are retained up to about 650 °C. Cement also retains its water of crystallisation up to about 300 °C.

2.4.4 Allotropy

A number of materials may exist in more than one crystal structure; such behaviour is known as *allotropy* or *polymorphism*, and the different crystalline species are known as *allotropes* or *polymorphs*. Carbon, which occurs in such widely different forms as graphite, lampblack and diamond, is the most flamboyant example; tin, which transforms from its normal metallic (body-centred tetragonal) structure to a *nonmetallic* covalent structure with a diamond-type lattice below 13 °C, is perhaps the most interesting. Other elements exhibiting allotropy include phosphorous, sulphur, and — a very important example for the engineer — iron. Alloy systems may also undergo changes in crystal structure as a result of their composition, which can not be strictly termed allotropy. These changes are of great importance in manipulating the mechanical strength of alloys, and arise because of the freedom from valency restrictions that alloy systems enjoy; the composition of ionic and covalent compounds are constant and their basic strength is therefore difficult to adjust.

2.5 BONDING FORCES AND THE STRENGTH OF SOLIDS

2.5.1 Strength of Ionic Crystals

We must now consider the relation between the interatomic forces that result in a crystalline structure and the strength of such a structure. All solids rely for their strength and hardness on the degree of cohesion between individual atoms (and perhaps molecules) within the structure, and, whatever the nature of the bonding, the final cohesion results from a state of equilibrium between two opposing forces. We will examine these forces briefly in terms of ionic crystals, which are perhaps the simplest in their ionic interactions, but the general argument is the same for other bonds and runs as follows.

If two ions, carrying charges $z_1 e$ and $z_2 e$ respectively, are separated by a distance r, the potential energy of the interaction between them is given by the expression

$$\phi_E = \frac{z_1 z_2 e^2}{r} \tag{2.3}$$

where e is the electronic charge and z_1 and z_2 are the number of charges associated with each ion. Conventionally, ϕ_E is taken to be negative when the interaction of the ions is attractive; thus the potential energy of the bond is lowered as the ions approach each other, and attraction will only occur when z_1 and z_2 are of opposite sign. Clearly this is the case we must consider, since a mutual repulsion will prevent any bond forming at all. For simplicity, let us take

48

the case of monovalent ions of opposite sign, in which case Equation 2.3 reduces to

$$\phi_E = -\frac{e^2}{r} \tag{2.4}$$

However, if the ions are brought so close together that the electron shells begin to overlap (i.e. if $r < R_1 + R_2$, where R_1 and R_2 are the respective ionic radii), a strong repulsion is developed, rising steeply as r decreases further, and given by the expression

$$\phi_R = \frac{B \cdot e^2}{r^n} \tag{2.5}$$

where B is a constant and the exponent n is usually about 9, indicating the very short-range nature of the repulsion. The total bond energy associated with a single pair of ions is therefore given by

$$\phi = \phi_E + \phi_R = -\frac{e^2}{r} + \frac{Be^2}{r^n} \tag{2.6}$$

For ions arranged in a crystal lattice, however, bonds will be developed between all the coordinating ions, and in fact ϕ_E is of a sufficiently long-range nature that we cannot only consider nearest neighbours; more distantly spaced ions must also be considered. In an ionic crystal this involves ions of like as well as unlike charges, and therefore it is necessary to take into account both attractive and repulsive energies when summing up the overall value of ϕ_E acting on an individual ion. Thus Equation 2.4 can be better expressed as

$$\Sigma\phi_E = -\frac{Ae^2}{r} \tag{2.7}$$

where A is a constant known as the *Madelung number* which has a characteristic value for each ionic solid. (For example, $A = 1.75$ for sodium chloride.) Thus the total cohesive energy associated with an ion in a crystal is given by

$$\phi = -\frac{Ae^2}{r} + \frac{Be^2}{r^n} \tag{2.8}$$

Curves showing the relationship of ϕ, ϕ_E and ϕ_R with r are depicted in Fig. 2.25. They are known as *Condon–Morse* curves. The equilibrium spacing is shown by r_0 and corresponds to a minimum potential energy value, ϕ_0, for the bond.

Differentiation of the bond energy ϕ with respect to r gives the force acting on the ion, i.e.

$$F = \frac{d\phi}{dr}$$

This force is similarly made up of attractive and repulsive components, and is shown plotted in Fig. 2.26. r_0 is now given by the point for which $F = 0$, and the gradient of the curve at this point (which is very nearly constant for small variations in r) is a measure of *Young's Modulus* for the crystal, since it gives the restoring force acting on the

49

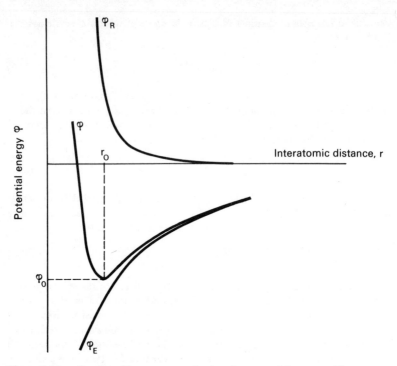

Figure 2.25 Condon–Morse curves plotting the potential energy of interaction ϕ against the interatomic spacing r.

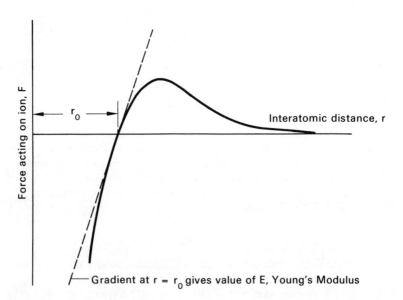

Gradient at $r = r_0$ gives value of E, Young's Modulus

Figure 2.26 The force F acting on an ion as a function of the interatomic spacing r.

50

ion for small elastic displacements from the equilibrium position. Thus

$$E \propto \left(\frac{d^2\phi}{dr^2}\right)_{r=r_0} \qquad (2.9)$$

The shape of the ϕ curve in Fig. 2.25 also shows a relationship with physical properties of the crystal. The depth of the potential well is a measure of the energy required to detach an ion from the crystal and is therefore related to the boiling point or sublimation temperature of the crystal. At normal temperatures the ions will have an energy $\phi_1 > \phi_0$ (ϕ_0 is strictly the value of ϕ at 0 K) and will therefore oscillate between two bond lengths given by r_1 and r_2 in Fig. 2.27

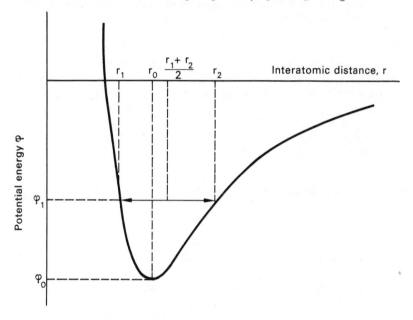

Figure 2.27   A deep trough with a sharp minimum on the ϕ, r curve (indicating strong attractive energy ϕ_E) gives a more symmetrical Curve about r_0 and a lower coefficient of thermal expansion.

when their energy is raised to ϕ_1, and have a mean spacing given by $(r_1 + r_2)/2$. However, the curve for ϕ rises more steeply when $r < r_0$ than when $r > r_0$, so that $(r_1 + r_2)/2 > r_0$, and increases in value as the temperature rises. Thus do materials expand when heated; the more symmetrical the trough in the ϕ curve, the lower will be the thermal expansion of a material.

The above treatment of cohesion has been given in relation to ionic crystals, but the same general arguments hold for all solids, although the attractive energy may differ somewhat in origin and considerably in intensity. Thus, in covalent bonding, attraction is mainly due to the lowered energy associated with a resonant sharing of the bonding electrons; however a small amount (roughly 5%) of the

51

energy is a result of momentary ionic attraction as when the configuration shown in Fig. 2.28 occurs, and a further 10–15% is due to spin

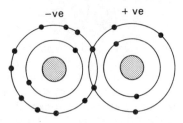

Figure 2.28   Covalent bonding: owing to the majority of the shared valency electrons circling the left hand atom at a particular point in time, there is a momentary ionic (electrostatic) attraction between the atoms.

interactions of the shared electrons. In the case of metals, cohesion results from the attraction between the negatively charged free electron cloud and the positively charged lattice of ions. Either component by itself would disintegrate by the mutual repulsion of its component parts; together they are stable. Molecular solids, bonded in bulk by van der Waals linkage and physical entanglement, have shallow ϕ curves and we may therefore expect low melting points and high thermal expansions.

2.5.2   Theoretical Fracture Strength of Solids

If the stress applied to a solid is greater than its cohesive strength then fracture will occur. Fortunately for engineers, metallic materials and some polymers show an extensive *plastic* response to stress (by internal shear) in addition to the purely elastic response that is implicit in the above treatment of cohesive strength. It is this additional response that confers upon metals and alloys their invaluable quality of *toughness*. However, there are some highly important civil engineering materials that, because of their structure, are unable to make a plastic response to stress; examples are cement, concrete and glass. In such cases fracture under tension intervenes directly and without warning on the elastic response by the rapid propagation of a crack or cracks, and is said to be *brittle*.

We may make an estimate of the theoretical brittle fracture strength of a solid as follows.

Figure 2.29 shows the relationship between applied stress, σ, and interatomic spacing, a, in our solid (it is, of course, the same curve as that depicted in Fig. 2.26). The fracture stress is given by σ_f, since σ_f is the maximum stress needed to increase the separation of the atoms; for spacings greater than a_c the stress to cause further separation drops rapidly. Now the work to cause fracture is given by the area under the curve, and the curve may be approximated to half a sine wave (see dotted line) in which case we may say:

$$\sigma = \sigma_f \cdot \sin \frac{2\pi x}{\lambda}$$

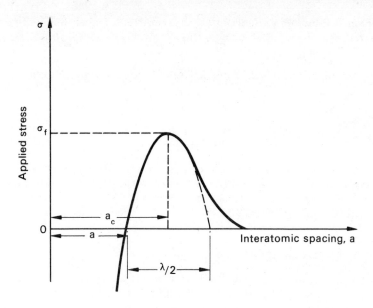

Figure 2.29 Diagram for calculating the theoretical fracture strength of a crystal. (The curve is the same as that in Fig. 2.27, but is approximated to a sinusoidal form to simplify the calculation.)

and the work to fracture, W, is therefore given by

$$W = \int_0^{\lambda/2} \sigma_f \cdot \sin \frac{2\pi x}{\lambda} \, dx$$

$$= \frac{\lambda \sigma_f}{\pi} \tag{2.10}$$

If we assume that the fracture is totally brittle, then all the work is used in creating fracture surfaces; hence

$$W = 2\gamma \tag{2.11}$$

where γ is the specific surface energy of the solid. Combining Equations 2.10 and 2.11 we then have

$$\sigma_f = \frac{2\pi\gamma}{\lambda} \tag{2.12}$$

For *elastic* deformation

$$\sigma = \frac{E \cdot x}{a}$$

where E is Young's Modulus and therefore

$$\frac{Ex}{a} = \sigma_f \cdot \sin \frac{2\pi x}{\lambda}$$

$$\simeq \sigma_f \cdot 2\pi x$$

for small values of x. Hence

$$\sigma_f \simeq \frac{Ex}{a} \cdot \frac{\lambda}{2\pi x} \qquad (2.13)$$

Combining Equations 2.12 and 2.13 by multiplication, we obtain

$$\sigma_f^2 \simeq \frac{2\pi\gamma}{\lambda} \cdot \frac{E\lambda}{2\pi a}$$

$$\sigma_f \simeq \left(\frac{E\gamma}{a}\right)^{\frac{1}{2}} \qquad (2.14)$$

This estimate gives values for σ_f that are too high by a factor varying from 10 to 1 000 when compared with experimental results obtained from brittle materials. However, the discrepancy was resolved when Griffith in 1920 suggested that the low strength was due to the presence of small cracks in the material.

2.5.3 Behaviour of Cracks in Brittle Solids

The presence of a small crack acts to concentrate the overall stress at its tips. In Fig. 2.30(a) the crack of width $2c$ and tip radius ρ, or a

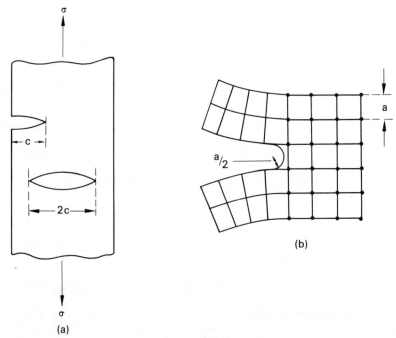

Figure 2.30 Cracks in solids. (a) Griffith cracks in an ideal elastic solid. (b) Idealised crack tip in an eleastic solid of interatomic distance a.

surface notch of depth c and similar tip radius, both result in an effective stress σ_{max} at their tips given by

$$\sigma_{max} = 2\sigma \cdot \left(\frac{c}{\rho}\right)^{\frac{1}{2}} \qquad (2.15)$$

54

Failure will then occur when σ_{max} rises to the magnitude of σ_f, i.e. when

$$\left(\frac{E\gamma}{a}\right)^{\frac{1}{2}} = 2\sigma\left(\frac{c}{\rho}\right)^{\frac{1}{2}}$$

Solving for σ, we may then say that the failure stress in the presence of a crack is given by

$$\sigma_f = \left(\frac{E\gamma\rho}{4ac}\right)^{\frac{1}{2}} \tag{2.16}$$

If the crack in a perfectly brittle material is to grow, the energy to produce the additional surface area of the fracture surfaces must be provided by the relaxation of the elastic strain in the material due to the growth of the crack. For a crack of length $2c$, the surface energy is $4\gamma c$ and the elastic strain energy released by the presence of the crack under an applied stress σ is $\pi\sigma^2 c^2/E$. The critical condition for crack growth is therefore obtained by differentiating these expressions with respect to c and equating them, i.e.

$$\frac{d}{dc}(4\gamma c) = \frac{d}{dc}\left(\frac{\pi\sigma^2 c^2}{E}\right)$$

From which, simplifying and solving for σ, we obtain the expression

$$\sigma_f = \left(\frac{2E\gamma}{\pi c}\right)^{\frac{1}{2}} \tag{2.17}$$

$$\simeq \left(\frac{E\gamma}{c}\right)^{\frac{1}{2}} \tag{2.18}$$

It will be seen that Equation 2.17 takes no account of the sharpness of the crack; it sets different criteria for failure to Equation 2.16 and both expressions must be satisfied for failure to occur. If we put $\rho = 8a/\pi$ in Equation 2.16 the two expressions become identical and this value for ρ thus becomes critical. For cracks where $\rho < 8a/\pi$, crack growth and failure will commence when expression 2.17 is satisfied; when $\rho > 8a/\pi$, the stress must rise until Equation 2.16 is satisfied before the crack starts to run. It should be noted that in a brittle material even a blunt crack or flaw will sharpen once it starts to propagate, since the mode of failure cannot by definition include any plastic blunting of the crack tip. Thus ρ should ideally decrease to the value $\rho \approx a/2$ (see Fig. 2.30(b)) and in practice drops sufficiently to satisfy Equation 2.17.

Certain valuable general conclusions result from inspection of the relationships given in Equations 2.15–2.17. In the first place it will be seen that σ_{max} in Equation 2.15 must increase as a crack propagates in brittle material since c increases (by definition), and ρ will not increase and may very well decrease if the initial flaw has blunt ends. On the other hand, σ_f decreases as c increases (Equations 2.16 and 2.17) and thus if the stress on a brittle material once rises to the critical level to start a crack propagating, failure is almost certain to result. It also follows that the only crack taking part in the failure mechanism will be that one with values of c and ρ giving the lowest magnitude of σ_f in the material; all other cracks will remain inert.

The strength of a *brittle* solid therefore depends upon the size of the largest suitably orientated flaw, which size may be largely a matter of chance. Strength assessments of brittle materials must therefore be based in part on the *probability* of occurrence of a flaw of particular size and orientation, and testing procedures should take account of this. Metals, however, are normally immune to the presence of small flaws, and a single test carefully performed will suffice to give perfectly adequate information. This is because the stress concentration developed at the tip of a crack or notch will

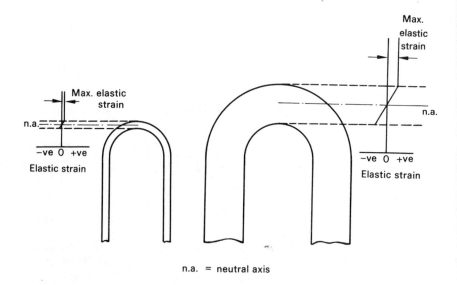

n.a. = neutral axis

Figure 2.31   Variation in elastic strain at a bend in thin and thick sections.

bring about intense plastic deformation in the vicinity of the crack-tip, blunting it and causing ρ in Equation 2.15 to increase more rapidly than c. Thus σ_{max} is reduced rather than raised to the danger level. Only when a metal fails by fatigue (see Section 15.3) does a crack spread in an apparently brittle fashion. In these circumstances the same criteria apply as in the fracture of brittle materials; the crack is found to originate from a surface flaw, and the fatigue life becomes in part a statistical parameter, depending on the probability of formation of a surface flaw of a particular depth and sharpness.

One further point should be noted, and that is that a size factor enters into the strength of brittle materials because of their sensitivity to flaws. This operates in two ways. Firstly, fine filaments of brittle materials are more flexible than thick sections because the variation in strain from surface to surface is much smaller (Fig. 2.31). Glass fibres may be flexed elastically through 180°, but the same could

56

hardly be expected of window glass. Secondly, if the filament is very small in cross section the largest flaw (i.e. the strength-determining one) cannot but be small also, and will therefore require a high stress to make it run. Such filaments, even when perfect, cannot support a high *load* but they have a high ultimate tensile stress and, if a large number of them are bonded together with a suitable adhesive to make a thick section, the resultant structure can support a very considerable load and at the same time show considerable elastic resilience. This is the principle behind the development of fibreglass as a structural material; Nature, however, has forestalled human engineers in the use of this concept by several millions of years, in developing the structures of timber and bamboo.

2.6 IMPERFECTIONS IN CRYSTALS

2.6.1 Theoretical Shear Strength of Solids

Hitherto, we have considered only the ideal structures of materials, but, in many ways fortunately for engineers, the materials we use in practice fall very far short of this ideal. In the previous section it was pointed out that plastic deformation occurs by shear along slip planes within metals. Let us consider the forces acting on the ion A in Fig. 2.32. At the position $x = 0$ the displacement it is in equilibrium and therefore the force is zero; similarly so when it has slipped one place along to the next equilibrium position, when $x = b$, the inter-atomic spacing within the slip plane. The ion is also in a state of unstable equilibrium when it is exactly midway between the two equilibrium positions, i.e. when $x = \frac{1}{2}b$. Between these three points the force curve may be reasonably approximated to a sine curve, the positive part indicating that we have to do work on the ion to displace it, and the negative part indicating that the ion will continue to move without further application of force and would indeed need a restraining force to hold it stationary. In other words, we have an energy barrier to slip, as shown in Fig. 2.32(c). It will also be seen that the critical shear stress τ_c is given by the maximum of the curve at $x = \frac{1}{4}b$; in other words, if we apply a stress only sufficient to move the ion a quarter of the distance towards its next position, slip will occur.

Let τ be the applied shear stress

then
$$\tau = \tau_c \sin \frac{2\pi x}{b}$$

For small displacements we may say sin $x \simeq x$ and write

$$\tau \simeq \frac{2\pi x \tau_c}{b}$$

but $\tau = Gx/a$, where a is the spacing between the slip planes, hence

$$\tau_c \simeq \frac{Gx}{a} \cdot \frac{b}{2\pi x}$$

57

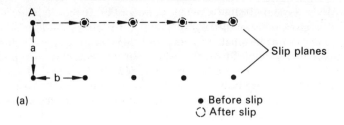

(a)

● Before slip
○ After slip

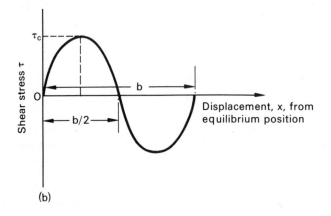

(b)

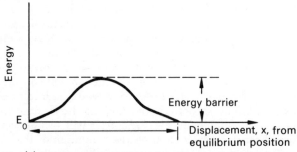

(c) Relation between energy of ion and displacement

Figure 2.32 Estimation of the theoretical shear strength of a metal. (a) Schematic movement of ions during slip. (b) Relation between shear stress and the displacement of an ion during slip. Note that b is not on the same scale as in (a). (c) Relation between the energy of an ion and its displacement.

and if the metal is cubic $(a = b)$ then

$$\tau_c \simeq \frac{G}{2\pi} \tag{2.19}$$

A more precise calculation puts $\tau_c = G/15$, but if this is compared with experimental values of τ_c for metals it will be seen that it is too great by a factor of up to 10^4. This is similar in magnitude to the discrepancy observed in Section 2.5.2 between the theoretical and

58

practical cohesive strengths of solids, and both discrepancies arise from the presence of imperfections in all normally prepared materials. Where particular care has been taken to grow perfect crystals (as with 'whiskers' of metals) the agreement between theory and practice becomes good — but materials of this quality are expensive and have so far been prepared only in very small sizes.

Defects may reasonably be classified in dimensional terms:

(a) Point defects (zero dimension) : impurity atoms, vacancies.
(b) Line defects (one dimension) : dislocations.
(c) Surface defects (two dimensions) : stacking faults, twins, grain boundaries, cracks.

It should be noted that all these defects (with the exception of cracks) apply only to regular (i.e. crystalline) materials.

2.6.2 Point Defects

These occur principally as impurity atoms (interstitial or substitutional) or as vacancies in 'pure' elements. The simple vacancy is referred to as a *Schottky defect*, but if the vacancy has an associated interstitial atom, then it is known as a *Frenkel defect*. An interstitial atom of the same species as the bulk crystal is known as a *self-interstitial*.

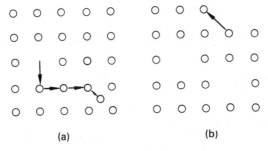

(a)   (b)

Figure 2.33   Point defects in a crystal lattice. (a) Frenkel defects. An atom moves to an interstitial position. Usually this involves a series of cooperative movements of several atoms, as indicated. (b) Schottky defect. A vacancy. (From A. H. Cottrell (1965) *Theoretical Structural Metallurgy, 2nd edition*, by permission of Edward Arnold (Publishers) Ltd.)

When considering ionic compounds, the presence of a vacancy upsets the local valency balance, and an anion vacancy is usually balanced by a cation vacancy nearby. These various species are depicted in Figs. 2.33 and 2.34. All these defects have an equilibrium concentration which is given by the relation

$$n_e = N \cdot e^{-E_f/kT} \tag{2.20}$$

where n_e is the equilibrium number of vacancies, N is the number of sites in the crystal, E_f is the energy of formation of the defect in question, and T is the absolute temperature. The number of vacancies thus increases exponentially with increasing temperature, helping to weaken the lattice until it collapses as the solid melts. In fact the concentration of vacancies at the melting point is surprisingly low: about 1 in 10^4 sites. Interstitial atoms have about five times the energy of formation of a vacancy, and are therefore rare in occurrence.

Point defects are of little importance in concrete and timber since small irregularities in a structure already highly irregular will have

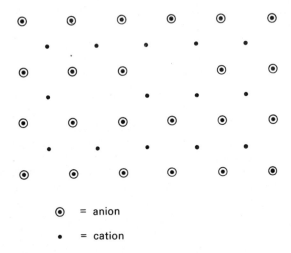

<table>
<tr><td>◉</td><td>=</td><td>anion</td></tr>
<tr><td>•</td><td>=</td><td>cation</td></tr>
</table>

Figure 2.34   'Balanced' vacancies in an ionic lattice.

little further effect and the mechanical effects in any case are negligible by comparison with larger imperfections such as cracks and pores. In metallic structures they aid crystal transformations and chemical reactions by accelerating diffusion rates. Mechanically they only become of direct importance when, in the form of substitutional or interstitial 'impurity' atoms, they are added in large proportions to form *alloys*, frequently being sufficient in content to bring about a crystallographic change in the metal.

2.6.3   Line Defects

The only line defect to be considered is the *dislocation*; as with point defects, it only has real engineering importance in metallic structures and so will be considered in detail in Chapter 12. Dislocations are the reason for the discrepancy between the ideal and the experimentally determined yield strengths of metals, and they achieve this remarkable result by permitting slip to occur progressively between two slip planes rather than simultaneously over the entire slip surface. Thus the dislocation loop in Fig. 2.35 is in fact the boundary between the slipped region of the crystal and the unslipped region. Each time a loop nucleates within the crystal and ripples through it to vanish at the surface a quantum of slip results in the appropriate slip direction. The magnitude of slip and its direction is

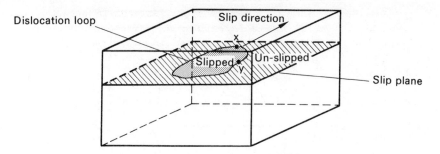

Figure 2.35   Dislocation loop in a slip plane in a crystal.

a characteristic of the loop and constant at all points on the loop: it is known as the Burgers vector of the dislocation, symbol b. The magnitude of the slip is normally the distance from equilibrium position of an ion to the next, i.e. the distance b in Fig. 2.32. Since the bonding in metals and alloys is nondirectional and general in nature, there is no difficulty in the ions moving one step along into the next equilibrium position (as there would be in a covalent solid, where specific bonds would have to be broken for slip to occur). However, each ion has to surmount the energy barrier shown in Fig. 2.32 before it can settle in its new position; there is thus some resistance to the movement of a loop, even though the crystal lattice is to all intents and purposes unchanged after its passage. The stress required to overcome this resistance is known as the Peierls—Nabarro stress, and is given by

$$\tau_{PN} = \frac{2G}{1-\nu} \cdot e^{\frac{-2\pi b}{a(1-\nu)}} \tag{2.21}$$

where a and b are lattice constants (see Fig. 2.32), G is the shear modulus and ν is Poisson's ratio. Substitution of an appropriate value for ν in Equation 2.21 and putting $a = b$ gives $\tau_c \simeq 2 \times 10^{-4}\ G$, which is a much more respectable agreement than that given by Equation 2.19.

If we examine the ionic configuration at the dislocation loop at points x and y in Fig. 2.35 we shall find the structures shown in Fig. 2.36. These are known respectively as *edge* and *screw* dislocations, and the presence of both is necessary in all dislocation loops. It should be noted that the relationships between the dislocation line, the dislocation movement and the slip generated are complementary for the two types. Thus edge dislocations generate a slip movement *at right angles* to the dislocation line but *parallel to* the movement of the line, whereas screw dislocations generate a slip movement *parallel to* the dislocation line but *at right angles to* the movement of the line. Inspection of any dislocation loop, given the direction of slip, will therefore quickly show which lengths are made up of edge components, which of screw components and which of a mixture of the two.

The transition at the dislocation line from the unslipped to the slipped state is not necessarily abrupt, but spread over a number of

61

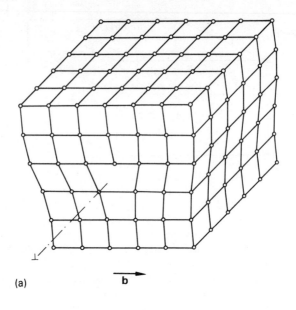

(a)

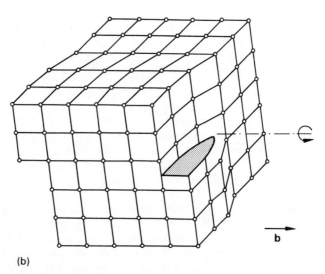

(b)

Figure 2.36 (a) Edge and (b) screw dislocations in a simple cubic crystal. b is the Burgers vector. (From K. J. Pascoe, *An Introduction to the Properties of Engineering Materials, 3rd edition*, by permission of Van Nostrand Reinhold.)

ionic spacings; this is illustrated in Fig. 2.36 and in greater detail in Fig. 2.37. In this way the energy barrier to slip has a more gentle slope, and movement of the loop is accompanied by small cooperative movements of ions for some distance on either side of the centre of the dislocation. Figure 2.37 illustrates the ionic movements involved when the centre of the line moves forward by one ionic spacing. Ions 1, 2 and 3 for example are moved a little further from their nearest

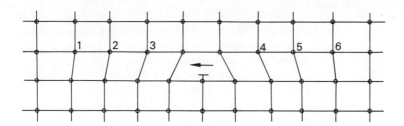

Figure 2.37   Ionic displacements associated with dislocation movement.

equilibrium position as the dislocation moves in the direction of the arrow; therefore they will oppose this movement with a force that could be roughly estimated from the Condon—Morse curves, Fig. 2.25. However, ions 4, 5 and 6 will be moved *towards* their nearest equilibrium position by the self-same movement, and will therefore tend to pull the dislocation forward with a force approximately equal to the drag exerted by ions 1, 2 and 3. Thus the position of the ions on either side of the centre result in a near-equilibrium of forces, so that very little additional applied stress is sufficient for it to move rapidly through the crystal. It is these two features, the 'spreading' of the dislocation to minimise the gradient of the energy barrier and the balance of forces acting on the dislocation from the surrounding crystal lattice, that contribute most to the outstanding ductility of many metals. The wider the dislocation, the more easily it will move, and therefore the more plastic the structure.

By their very nature, dislocations can occur only in an ordered (i.e. crystalline) structure, and in theory the sort of imperfections shown in Fig. 2.36 should be able to occur in any variety of crystal. Dislocations have in fact been studied extensively in some ionic crystals, notably lithium fluoride, and even materials such as marble have been made to deform plastically at sufficiently high pressures, showing that dislocations must be present in the structure. Yet under normal conditions it is only in metals that dislocations play a commanding role in determining mechanical behaviour. Why should this be so?

In covalent crystals the directional bonding means that valencies have to be unsatisfied at a dislocation; this is not so in ionic or metallic structures. This, coupled with the high energy required to strain the nearby covalent bonds, means that covalent dislocations are unlikely to form because of their very high strain energy; when they do form they are very narrow and require a high stress to move them. Hence covalent crystals are brittle.

Ionic crystals have a generalised bonding, so that the question of broken bonds does not arise, but the creation and movement of dislocations is difficult for a number of other reasons. Where the structure is complex (e.g. where each space lattice point is represented by a group of ions rather than a single ion) the distance from one equilibrium position to the next in a possible slip direction may be large. In other words b, the Burgers vector for a dislocation, is large. The energy of dislocations is proportional to b^2, so a large Burgers vector renders dislocations improbable of occurrence. In simpler structures, many possible slip planes cannot have dislocations on them because they would involve the juxtaposition of similarly charged ions (Fig. 2.38); if on the other hand a dislocation forms on the plane $(y - y)$ in Fig. 2.38, the ratio b/a is such as to give a high Peierls–Nabarro force, and the dislocation is extremely difficult to move.

Plasticity in polymers, it should be noted, is due to entirely different causes, although dislocations can occur in the crystalline

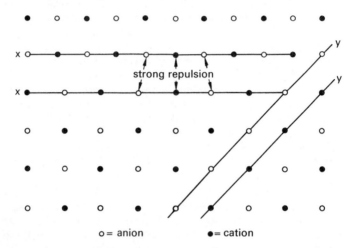

o = anion $\bullet$ = cation

Figure 2.38   Dislocations on crystallographically possible slip planes in an ionic crystal. The strong repulsive forces between like ions in the centre of the dislocations between planes $x-x$ makes dislocations of this habit impossible. Dislocations on $y-y$ are possible, but are relatively immobile because of the high Peierls–Nabarro force necessary to move them.

regions. It arises from the viscous movement of entire long chain molecules past one another, and imperfection of the sort under discussion play no part in deformation procedures.

2.6.4   Surfaces

All materials are bounded by interfaces of varying nature. For the engineer the most important are the liquid-vapour, solid-vapour, solid-liquid and solid-solid interfaces, the last-mentioned existing as the boundary either between two differing solid phases in a material (e.g. cement gel and aggregate in concrete) or between two similar crystals which differ only in orientation (e.g. the *grain boundaries* in a pure metal). Surfaces owe their interest and importance to two simple features: they are areas of abnormality in relation to the structure that they bound (in which sense they may be regarded as a form of imperfection), and they are the only part of a material *accessible* to chemical change; that is to say all chemical change and, for that matter, most temperature changes take place through the media of surfaces. The importance of surfaces in determining the bulk behaviour of materials naturally depends on the ratio of surface area to total mass, and this in turn will depend partly on the size and partly on the shape of the individual particles making up the bulk material. Surface influence probably reaches its zenith with *clays*, which are composed of platelets that may be as small as 0.1 μ across by 0.01 μ thick. Platelets have a high surface area/volume ratio by comparison with spheres of equal volume: 1 g of montmorillonite clay (rather smaller than a lump of cube sugar) may contain a total surface area of over 800 square yards! Porous structures such as cement and wood also contain enormous internal surface areas which exercise a considerable effect on their engineering properties, and the risk of explosions with dusts (composed of materials which may be perfectly innocuous in bulk form, e.g. aluminium or titanium) is well known.

All surfaces have this in common: the atoms or ions in the surface will be subjected to asymmetric or unsaturated bonding forces (this is particularly the case with solid-vapour and liquid-vapour interfaces), and since bonding is taken to lower their energy (Fig. 2.25) they will be in a state of higher energy than interior ions. This excess energy is known as the *surface energy* of the material.

2.6.4.1 *Surface Energy*

A rough estimation of the energy of a solid-vapour interface E_{sv} may be made by relating it to the heat of vaporisation of the material, ΔH_v. Consider an ion situated in a close-packed plane forming the surface of a close-packed solid. The ion is bound to six contiguous ions in the surface plane and three immediately below, but three out of the maximum of twelve bonds remain unsatisfied. If the ion is then excited into the vapour phase, the nine existing bonds will be ruptured, creating 18 unsatisfied half bonds (nine on the ion and one on each of its former neighbours). Thus vaporisation should require six times the energy necessary to bring the ion from the interior to the surface, and we may expect the relation: $\Delta H_v \simeq 6E_{sv}$, which is roughly born out by experiment. Similar relationships may be derived for solid-liquid and liquid-gas interfaces if we make suitable estimates of the degree of coordination in the liquid phase.

In solids the presence of a surface energy is not immediately apparent, since the atoms in the surface are firmly held in position; with liquids, however, the mobile structure permits the individual atoms to respond, and the result is the well known *surface tension* effect. Because surfaces are high-energy regions they will always act to minimise their area (and thus lower their energy) when possible; if a soap film is stretched across a frame with a movable wire as in Fig. 2.39, the wire exerts a force

$$F = 2\gamma l$$

where l is the length of the wire, γ is the surface tension of the soap film/air interface, and the factor 2 is introduced because the film has two surfaces. It is important to note that surface tension differs from an elastic force acting between the surface atoms in that it remains constant whether the film in Fig. 2.39 is forced to expand or allowed

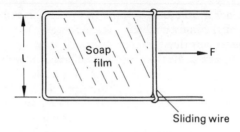

Figure 2.39 Equilibrium between soap film and applied force F.

to contract. This is because the work done in expanding the film is used to bring additional atoms to the surface rather than to increase the interatomic spacing in the surface. Only when the film has become so thin that the two surfaces interact with each other will the force show partial elastic behaviour, by which time the film is on the point of rupture.

2.6.4.2 *Wetting*

There are other ways besides reduction in area in which a system of interfaces may reduce its total surface energy; one of the most important technological aspects concerns the behaviour of liquids on solids. If a droplet of liquid is placed on a solid, the immediate behaviour of the liquid depends on the relative magnitudes of three surface energies: liquid-solid γ_{ls}, liquid-vapour γ_{lv}, and solid-vapour γ_{sv}. At the periphery of the droplet the three energies operate as depicted in Fig. 2.40 and the final contour of the droplet will result

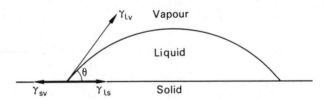

Figure 2.40 Surface forces acting at the periphery of a droplet.

when they are in equilibrium resolved parallel to the solid surface, i.e. when

$$\gamma_{sv} = \gamma_{ls} + \gamma_{lv} \cos \theta \tag{2.22}$$

If $\gamma_{sv} > \gamma_{ls} + \gamma_{lv}$, then $\theta = 0°$, and complete *wetting* of the solid surface occurs; the energy of such a system is obviously lowered when the solid-vapour interface is replaced by a solid-liquid and a liquid-vapour interface.

When $\gamma_{ls} + \gamma_{lv} > \gamma_{sv} > \gamma_{ls}$, then $\theta < 90°$ and partial wetting occurs; the resultant force tending to spread the droplet (from Equation 2.22) is given by $\gamma_{sv} - \gamma_{ls} = \gamma_{lv} \cos \theta$.

When $\gamma_{sv} < \gamma_{ls}$ (this is comparatively rare, provided that the surfaces

66

are clean) then $\theta > 90°$, and $\gamma_{lv} \cos \theta$ becomes negative, in which case there is little or no tendency to wetting. The rise of water in a glass capillary tube is thus a consequence of the ability of water to *wet* glass. If, in Fig. 2.41, θ is the angle of contact between water and

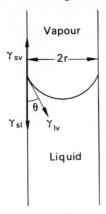

Figure 2.41 Capillary rise of liquid up a tube.

glass, the water is drawn up the tube by a circumferential force equal to $2\pi r \gamma_{lv} \cos \theta$. At equilibrium this is balanced by the weight of water in the column; hence

$$2\pi r \gamma_{lv} \cos \theta = \pi r^2 h \rho$$

For water the density $\rho = 1$, and therefore (neglecting the mass of water contained in the curve of the meniscus)

$$h = \frac{2\gamma_{lv} \cos \theta}{r} \tag{2.23}$$

2.6.4.3   *Adhesives*

The ability of adhesives to spread and thoroughly wet surfaces is highly important, particularly where (say) molten solder or brazing alloy has to penetrate into a thin joint. Furthermore, the adhesion of a liquid to a solid surface is relevant to the performance of adhesives. The work to break away the adhesive (which may be considered as a viscous liquid) from the solid is the work required to create a liquid-vapour and a solid-vapour interface from an equivalent area of liquid-solid interface, i.e. it is the work to totally 'de-wet' the solid surface. Hence the work to cause breakage at the interface is given by

$$W = \gamma_{lv} + \gamma_{sv} - \gamma_{ls}$$

But $\gamma_{sv} - \gamma_{ls} = \gamma_{lv} \cos \theta$ (Equation 2.22); therefore

$$W = \gamma_{lv}(1 + \cos \theta) \tag{2.24}$$

Thus the liquid-solid adhesion increases with the ability of the adhesive to wet the solid, reaching a maximum given by

$$W = 2\gamma_{lv} \tag{2.25}$$

when $\theta = 0°$ and wetting is complete. Under these conditions fracture will then occur within the adhesive, since the energy necessary to form two liquid-vapour interfaces is less than that to form a liquid-vapour and solid-vapour interface.

Surface tension is also the cause of the adhesion between two flat surface separated by a thin film of fluid. Where the surface of a liquid is curved (as for example in Fig. 2.41), there will be a pressure difference p across it; if the curvature is spherical of radius r, then

$$p = \frac{2\gamma}{r}$$

In the case of two circular discs, however, the surface of the film has two radii of curvature, as shown in Fig. 2.42; r_1 is approximately

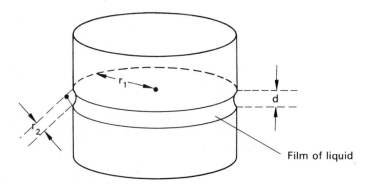

Figure 2.42 Adhesive effect of a thin film of liquid between two flat plates. If the liquid wets the surface of the discs, then $r_2 \simeq d/2$.

equal to the radii of the discs and presents a convex surface to the atmosphere whilst $r_2 \simeq d/2$, where d is the width of the film between the plates, and presents a concave surface to the atmosphere. The pressure difference between the liquid film and its surroundings is now given by

$$p = \gamma\left(\frac{1}{r_1} - \frac{1}{r_2}\right) \tag{2.26}$$

If r_2 is small compared with r_1, we can say

$$p \simeq -\frac{\gamma}{r_2} + \frac{2\gamma}{d}$$

the negative sign indicating that the pressure is *lower* within the liquid than outside it. Since the pressure acts over the whole surface of the discs, the force to overcome it and separate the plates is given by

$$F = \frac{2\pi r_1^2 \gamma}{d} \tag{2.27}$$

The magnitude of F thus depends on the factor r_1^2/d, and it is therefore important that joint surfaces should be as flat and closely spaced as possible. Any reader who has tried to pull apart two wet glass plates — even objects as small as microscope slides — will confirm how tenaciously they cling to each other; by contrast, however, they can easily be slid apart since liquid films have little resistance to shear. If we take $d = 0.01$ mm and γ for water as 72 dynes, then $F \simeq 2.8$ Kgf; for two panes of window glass size 30 cm x 20 cm, $F \simeq 89$ Kgf. The magnitude of these values of F for a *liquid* film gives some idea of the potential of adhesives which gain additional strength and rigidity by setting to highly viscous materials on polymerisation or solvent evaporation.

2.6.4.4 Adsorption

The ability of liquids to wet solids depends very much on the *cleanliness* of the solid, as anyone with any experience of soldering will appreciate. The presence of any dirt, such as oxide or grease films, will totally alter the balance of surface tensions discussed above and usually results in complete prevention of wetting. Clean surfaces, in fact, are so rare as to be virtually nonexistent, since the broken bonds will readily attract to themselves any foreign atoms or molecules that have a slight affinity for the surface material. This effect is known as *adsorption*, and by satisfying or partially satisfying the unsaturated surface bonds serves to lower surface energy.

Adsorption is a dynamic process; that is to say molecules are constantly alighting on and taking off from the surface and at equilibrium the rate of adsorbed molecules at any instant has been shown to vary with atmospheric pressure according to the Langmuir isotherm:

$$f = \frac{aP}{1 + aP} \tag{2.28}$$

where f is the fraction of surface covered, P is the pressure, and a is a constant. Thus at high pressures $f \to 1$ and the surface should be completely covered. In actual fact the position is more complex because a second layer will usually commence before the first layer is complete, but at a changed rate, since the adsorbing surface is now different and the value of a in the isotherm is therefore also different. In this way layers several molecules thick may be built up.

The quantitative value of the Langmuir isotherm at any pressure depends on the species of adsorbing molecule and the nature of the surface onto which it is adsorbed. Different molecules adsorb with varying degrees of intensity, depending on the nature of the bond that is able to form at the interface, and the strength of the bond may be expressed in terms of ϕ_a, the energy of adsorption. As in the case of interatomic bonds, a *negative* value of ϕ_a is taken to indicate a *positive* adsorption; that is to say the molecules are attracted to the interface, and the surface tension is lowered thereby. A positive value of ϕ_A indicates a repulsive interaction and the molecules avoid the surface.

Typical plots of ϕ_a against the spacing of the adsorbed layer away from the surface are given in Fig. 2.43; they closely resemble

Condon—Morse curves (Fig. 2.25) and their shape is due to the same circumstance of equilibrium between attractive and repulsive forces, although the attraction is far weaker than that of the principal inter-atomic bonds. If the adsorbing molecule is nonpolar and does not react chemically with the surface, adsorption (if it occurs) will be by van der Waals bonds, and the minimum value of ϕ_a is small (curve 2 in Fig. 2.43). If on the other hand the molecule is strongly polar (as

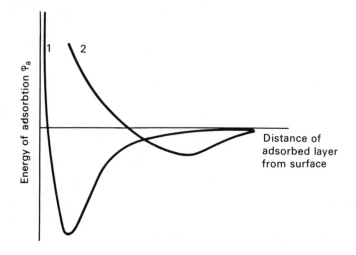

Figure 2.43   Energies of adsorption for different adsorption mechanisms: Curve 1, chemisorption; Curve 2, physical adsorption.

is the case with water or ammonia), the electrostatic forces between the surface and the charged portion of the molecule give rise to stronger bonding; if a chemical reaction occurs as part of the bonding mechanism (e.g. when fatty acid in a lubricant forms an adsorbed layer of metallic soap on a metal surface), the bonding is still stronger (curve 1) and the effect is referred to as *chemisorption*.

The behaviour of water is of particular importance in this context. Because of its ability to form hydrogen bonds with neighbouring molecules, water adsorbs rapidly and strongly on most solid surfaces. Despite the tenacity with which such a layer is held (clay does not lose all its adsorbed water until heated to 300 °C), the interaction cannot be thought of as chemisorption; rather it is in a sense a half-way stage to solution or alternatively to the taking up of crystalline water of hydration. Bonding is strong enough to maintain a surface layer, perhaps several molecules thick, but the affinity is not sufficient for the molecules to penetrate into the interstices of the structure. The physical nature of such a film is difficult to visualise; it cannot be thought of as a fluid in the accepted sense of the term even when more than one molecule thick (as in the case of clays and cements). Yet the molecules are mobile in this situation. They will not desorb readily, but they can diffuse *along* the surface under the impetus of

70

pressure gradients. Such movements, occurring over the vast internal surface area of cement gels, are primarily responsible for the slow *creep* of concrete under stress. The ability of water molecules to penetrate solid-solid interfaces in clays and build up thick adsorbed layers results in the *swelling* of clays and has caused (and may well continue to cause) considerable structural damage to buildings erected on clays that are liable to behave in this manner. The readiness with which water will adsorb on surfaces is turned to account in the use of porous silica gel and molecular sieves as drying agents.

Adsorption also plays a major part in the functioning of *catalysts*, which are substances that accelerate chemical reactions without themselves being used up in them. No chemical reaction will proceed unless the reactant molecules can come into contact, and the chance of contact where the reactants are gaseous is greatly increased if both are positively adsorbed onto a surface; for a satisfactory outcome it is necessary that the products do not adsorb but are released from the surface so that the sites are available for repetition of the process.

2.7 CHANGE AND ENERGY

2.7.1 Introduction

It is a truism that the world as we know it is totally dependent on change; physical, chemical and spacial changes are the basis of all life and the nature of the environment. It is equally true that no change of any sort will take place without an associated absorption or emission of energy. Whatever the actual outward nature of the change — whether it be a pebble rolling down the hillside, or the setting of concrete or a dam, or the ripening of a strawberry — it is invariably a change of energy that supplies the driving force for the physical change. On a personal level, all of us owe our existence to continuing energy changes in our muscles and brains, and in the processes of digestion and respiration. No-one is more involved with energy changes than the civil engineer, who is concerned with harnessing natural energy resources for his purposes.

The universal principles underlying all energy changes are worked out in the science of thermodynamics. It is the purpose of this section to examine briefly, at a qualitative and elementary level, the part of thermodynamics which is relevant to a study of materials. If the principles involved are well understood, the logic of the great variety of material treatments immediately becomes apparent. Thermodynamics, however, tells us little about the *rate* of change taking place. This may be almost as important to the engineer as the nature of the change, if a process is to be economically viable. As a simple example, there is a considerable difference between the combustion rate of a cigarette and a flask of nitroglycerine, which is of great importance to an engineer handling these materials. In fact, as we shall see, there are a number of examples of changes in materials that are thermodynamically correct, being totally suppressed, i.e.

71

having a zero rate of change, and some of these are of extreme technological importance.

2.7.2 Mixtures: The Statistical Approach

Everyday observation suggests that changes may take place in a variety of ways. There are changes which take place readily and permanently, such as the mixing of milk and water if these are placed in the same container. There are changes that depend on an input of energy, such as the transformation of water to steam. These changes are impermanent or reversible in the sense that the steam will revert back to water by condensation if the thermal energy is dissipated. There are also changes that will proceed spontaneously and irreversibly, as in the first example, if a sufficient *activation* energy is supplied. Most familiar examples of combustion fall into this category since they depend on raising the fuel to a suitable ignition temperature, after which combustion will proceed until complete without further input of energy.

Let us commence by considering the first instance of change mentioned above, i.e. that which will occur spontaneously and irreversibly without requiring any activation energy. The mixing of two suitable fluids placed in contact affords a good example. Suppose we have an apparatus consisting of a gas-tight container divided exactly in half by a gas-tight removable shutter. With the shutter in place we fill one half with gas A and the other half with gas B, ensuring that both gases are at identical pressures and temperatures. The shutter is then withdrawn. Observation and analysis demonstrate that the two gases will rapidly intermingle, and once mixed will stay mixed. Furthermore, whilst mixing the two is easy, it is usually difficult to separate them, and if after separation they are again brought into contact, they will mix again. We may therefore say that in this experiment an irreversible change will always take place in the system which may be denoted:

$$A + B \rightarrow AB_{mixt}$$

Countless observations have shown that the reverse of this process never occurs. Never? Well, to quote W. S. Gilbert, 'hardly ever'. Let us examine in more detail what is happening.

The conditions of the experiment are that two different species of atoms, moving, colliding and impinging on the container walls entirely at random, are allowed to intermingle by random movements between the two chambers. A simple small-scale analogy (since the number of atoms involved in the above experiment is astronomically large) would be to take a box as shown in Fig. 2.44 containing eight white and eight black balls, the two colours being separated by a central shutter. Such an arrangement is of course totally static; let us simulate the continual movement of real atoms by assuming that we can shake the box so that a complete random rearrangement of the contents takes place after each shaking.

As long as the shutter remains in place, no amount of shaking can alter the macroscopic distribution of black and white balls; within

72

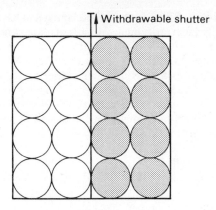

Figure 2.44   A box with eight black and eight white balls, before mixing.

each compartment the balls will interchange positions, but the overall state of the system remains the same. Although it may at this stage appear to be labouring a very obvious point, it is worth pointing out that if we assume all balls of similar colour are indistinguishable from one another, then with the shutter in position there is only *one* possible way of arranging all 16 balls in the box; no matter how much shaking the box suffers, the distribution remains the same. (This is paralleled in the experiment using gases; in their separate compartment the atoms are perpetually changing position at high speed, but their overall distribution remains the same and macroscopically the gases are considered to be in equilibrium.)

If the shutter is now removed and the box shaken up, the initial arrangement with all the black balls on one side and all the white balls on the other is only one of a large number of equally probable distributions. Let us calculate the number of possible distributions now, when every ball is free to pick any position in the box, that is to say in the absence of any attraction or repulsion between the different species of balls.

Assume that the balls come to rest one by one after the shaking. Then the first ball to settle has 16 possibilities for its position (we are neglecting the effects of gravity here). Once it has settled, the second ball has only 15 possibilities, since one space is already occupied. Thus the total number of ways in which these two balls alone could be arranged in the box is 16 x 15. Extending the argument to the third ball, we see that this has 14 possibilities, and the total number of ways of arranging the three balls is given by 16 x 15 x 14. If we carry on until all 16 balls are settled, the total number of possible arrangements is given by 16 x 15 x 14 x . . . x 3 x 2 x 1 or 16! This is a very large number: approximately 2.1×10^{13}. We may modify this result if we assume (as above) that the balls of similar colour are indistinguishable. In this case we may say that the 8! ways in which the white balls can fill their particular positions can be considered as just one way, and of course a similar argument applies to the black balls. Our expression for the total number of possibilities then becomes

16!/8! x 8! = 12 870. There is thus one chance in 12 870 that we shall find the original arrangement of balls after shaking; if they are shaken and surveyed once a minute continuously, the initial arrangement should reappear about once every nine days. Note that the initial arrangement is just as likely as any other particular arrangement, but there is a large number of 'mixed' arrangements and only one separated arrangement. Even with this very small number of balls, then, there is an almost overwhelming likelihood of achieving a mixed distribution of black and white, whether we start with the colours separated or already mixed.

If the number of balls is increased, the probability of mixing becomes a virtual certainty. With 25 balls of each colour, the probability of the initial separate distribution re-appearing again (or any other *particular* distribution for that matter) becomes 1 in $50!/25! \times 25! = 1$ in 1.264×10^{14}. If the balls are now shaken and surveyed *once a second* the original separation will occur on average once in 4 008 454 years! We are therefore safe to say that under conditions of continual agitation the balls will mix and will stay mixed; even though the exact pattern of mixing is constantly changing, the chances of unmixing are vanishingly small, whatever the duration of the experiment.

Finally we must return from a consideration of coloured balls to the original experiment with gases. If each chamber contains 1 mol of gas, then the probability of the gases remaining unmixed, even for a short space of time after the removal of the shutter, becomes one in $(12.06 \times 10^{23})!/((6.03 \times 10^{23})!)^2$, i.e. one in $10^{4.6} \times 10^{23}$ in the absence of any affinity or repulsion between the atoms. The calculation shows conclusively that when we are considering atoms in normal quantities mixing becomes an absolute certainty and is permanent. It should be noted that the possibility of segregation dwindles to vanishing point so rapidly with increasing numbers because the total number of distributions increases on a factorial scale, whilst there can only be one way of achieving the initial distribution however many atoms we are dealing with.

2.7.3 Entropy

We can express the conclusions reached above in a slightly different way, which has more relevance to materials. The segregated distribution of balls or atoms is only one of a very large number of possible distributions, but it is also a rather special example in that it has a definite pattern. It is in fact the distribution that we might expect if the two species of atom were mutually repellent. We could consider other possible 'special' distributions; for example, if unlike atoms exerted a definite attraction between each other, a chequerboard distribution (Fig. 2.45) might well result. Another possible arrangement would be the separation of the atom species into alternate vertical or horizontal layers.

All these arrangements (and numerous others that could be devised) are special in that the atoms are arranged in a definite pattern; they

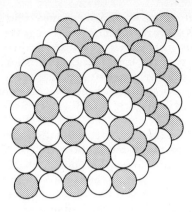

Figure 2.45 A chequerboard pattern; one of the relatively very few *ordered* mixtures possible.

are 'ordered' distributions rather than random mixtures. However, even when all possible ordered distributions have been taken into account, their numbers are utterly negligible by comparison with the numbers of disordered random distributions. It is thus apparent that, in the absence of any attractive or repulsive forces between the atoms, there will be a spontaneous change from order to disorder when two gases are allowed to mix, simply because there are so many more ways of achieving disorder than order that disorder is overwhelmingly more possible. Numerically this may be expressed by saying that if W is the number of ways in which a given macrostate may be achieved, then W *will tend to a maximum*, since the maximum number of ways of achieving a given state (in this case disorder) must correspond to the most probable state (in terms of atomic configurations, the most overwhelmingly probable state).

Most engineers will have at the least a nodding acquaintance with the concept of *entropy* in classical thermodynamics. This is normally introduced as a capacity property of a system, which increases in proportion to the heat absorbed at a given temperature. It is therefore usually defined by the relation

$$dS \geqslant \frac{dQ}{T}$$

where S is the entropy, Q is the heat, and T the absolute temperature. (The inequality refers to irreversible changes in the system.) This relation is a cornerstone of thermodynamics, but it does not give a clear picture of the nature of entropy, and thus often proves a stumbling block to full understanding of the concept. However, entropy has a physical as well as a thermodynamic image, which was realized by Boltzmann in the 19th century. He showed that the disorder of a system, which has a close relation to its heat content, is directly related to its entropy by the expression

$$S = k \cdot \log_e W$$

where $k = R/N_0$ and is known as Boltzmann's constant, R being the gas constant and N_0 Avogadro's number. This is the statistical definition of entropy and shows that it is a measure of the disorder of a system. The use of the logarithm of W satisfies the requirement that S should be a capacity property, i.e. that the total entropy of two systems is given by the sum of their separate entropies, since the total number of distributions W_1 and W_2 of two systems is *not* $W_1 + W_2$ but $W_1 \times W_2$.

It therefore follows directly from the statistical background that the entropy of any closed system will tend to a maximum, since this merely means that it will take up the most probable configuration. This is in fact one way of stating the second law of thermodynamics.

Finally we should note the differing approach of the classical and statistical methods. The classical method deals with the overall nature of systems and formulates its conclusions as absolute laws; the statistical method, on the other hand, recognises that a system may be continuously passing through an immense series of microstates which are for all practical purposes indistinguishable, but which nevertheless gives rise to the possibility, however remote, of a totally abnormal distribution occurring at any instant. It may be discounted by the practical engineer, but it is a fascinating thought. For example, there is a finite possibility that all the air molecules in a room might cluster in one corner into 1% of the available space, leaving the remainder of the room in vacuo.

2.7.4 Free Energy

It must be clear that the fundamental tendency for entropy to increase can only be a part of the science of energy, and is only applicable to systems consisting of particles which are totally indifferent to each other. If this were not so, the entire universe would break down into total disorder and chaos. Complete disintegration of atoms — the final stage of disorder — is only realized at ultrahigh temperatures, probably only in some stars; the nearest approaches in our technology occur during nuclear fission or plasma-arc welding. The reason for the existence of solids and liquids lies in the fact that no atoms are totally indifferent to other atoms. In Section 2.2 we outlined the various ways in which atoms may associate to give stable structures; all these are dependent on definite attractive forces between individual atoms.

This tendency to association is obviously proportional to the strength of the individual bonds, and therefore to the total bond energies within a volume of material. To take a simple example, the formation of a simple ionic compound between monovalent elements A and B will lead to the likelihood of a chequerboard structure. This is a highly ordered configuration; if for any reason the ordering is disturbed so that either the interatomic distances cease to be ideal, or so that some atoms of one species are not surrounded by six nearest neighbours of the other species, the internal energy of the structure will rise above its minimum equilibrium value. This amounts to some degree of disordering (it might well be achieved by heating the material) and will energetically favour the entropy tendency at the same time. What then determines the way in which the system moves? Clearly some balance must be involved, which for example must tilt in favour of entropy increase above the melting point, and in favour of minimum internal energy below the melting point. To establish this balance, we must have recourse briefly to classical thermodynamics.

If we take a system of internal energy E and introduce a small amount of heat dQ into it so that it performs a small amount of work dW on its surrounding, then the change in internal energy is given by

$$dE = dQ - dW \qquad (2.29)$$

Also, for irreversible changes,

$$dS > \frac{dQ}{T}$$

and therefore, substituting for dQ in Equation 2.29, we have

$$dE - TdS < - dW$$

If we put $dW = PdV$, then we can say that

$$dE + PdV - TdS < 0$$

for irreversible changes, and that at equilibrium

$$dE + PdV - TdS = 0$$

For the majority of materials changes, the work done on the surroundings may be neglected and we can therefore write

$$dE - TdS < 0 \qquad (2.30)$$

For changes at constant temperature,

$$SdT = 0$$

hence

$$dE - (TdS + SdT) < 0$$
$$d(E - TS) < 0 \qquad (2.31)$$

The term $(E - TS)$ is known as the *Helmholtz Free Energy*, denoted by the symbol F.

From the above expressions, it can be seen that dF is always negative for irreversible changes and only becomes zero at equilibrium. Hence we can now use the criterion of *minimum free energy* as a guide to the way in which energy changes will take place and note that F will be at a minimum when the system is at equilibrium.

It is now apparent why materials become more ordered as the temperature drops; at sufficiently low temperatures even the inert gases first liquefy and ultimately solidify. At low temperatures the term TS is small compared with E and so order prevails; as the temperature rises, so TS increases until eventually the transformation

77

to a more disordered state occurs (e.g. at the melting point) because the increase in E occasioned by the change is more than offset by the negative increase in TS.

Finally, it is important to realize that in many materials systems the equilibrium state as predicted by free energy considerations is never reached, the commonest reason being that the atoms or molecules that make up the structure are physically unable to rearrange themselves in a new equilibrium array. In the case of window glass, for example, the chains of silica tetrahedra are too complex and unwieldy to take up readily a crystalline habit at the freezing point (see Fig. 2.13). The result is a supercooled metastable 'liquid' structure, which will persist indefinitely at room temperature since the molecules are then completely immobile. Steels cooled suddenly from 900 °C (quenched) are also in a metastable condition since the carbon atoms do not have time to rearrange themselves to the equilibrium structure during cooling. The quenched structure, which is of great technological importance, can be retained for any desired period of time at ordinary temperatures, but the application of heat ('tempering') permits the structure gradually to approach true equilibrium.

Other anomalies appear in chemical reactions, some of which are important in matters of corrosion, but these are more apparent than real since they result from the formation of protective films which separate the reactants and bring the reaction to a halt. The well known resistance of iron to concentrated nitric acid, due to the formation of an iron oxide film on the surface of the metal, is a good example. Lead owes its immunity to attack by hydrochloric acid to the fact that lead chloride is sparingly soluble and forms a protective barrier on the lead. The excellent corrosion resistance of pure aluminium and stainless steel arises paradoxically from the chemical reactivity of these materials: in the presence of oxygen an inert oxide barrier is rapidly formed on their surfaces and prevents further attack.

CHAPTER 3

Structure of Concrete

3.1 CONSTITUENTS OF CONCRETE

Concrete is made by mixing together a number of constituents. They include:

cement
water
sand
stones   } which together comprise the aggregate

Sometimes there are additives in the cement or admixtures included when the ingredients are mixed together, and air is present, usually unintentionally.

The cement (and additives) react with water to form *hardened cement paste* — which will be referred to frequently in the following pages and will be abbreviated to h.c.p. The h.c.p. is the fundamental material because it provides the strength that allows concrete to be used structurally. The aggregate is also essential in that it gives rigidity and dimensional stability to concrete, and its cheapness relative to cement assures its use for reasons of economy. It normally occupies 60–80% of the volume of the concrete.

Concrete consisting of h.c.p. and aggregate is thus a two-phase material in which the aggregate is dispersed in a matrix of h.c.p. The h.c.p. is a complex material in itself and as well as providing the essential strength it exhibits many other idiosyncratic characteristics which have been, and still are, a matter of continuing fascination and employment to many research workers. The first and longer part of this chapter is devoted to the structure of pastes made of Portland cement. Aggregates and additives are covered in the second part.

3.2 HARDENED CEMENT PASTE

In the course of the transformation from raw materials to h.c.p. the cement goes through two separate stages of chemical change, as indicated in Fig. 3.1. In the first stage, *manufacture*, the raw

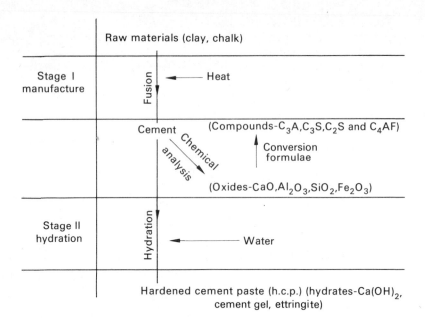

Figure 3.1 Stages in the formation of hardened cement paste (h.c.p.).

materials are fused to form the ingredient cement, and in the second stage, *hydration*, the cement as part of the concrete mix is hydrated to form h.c.p. The chemical reactions are described in the next two sections.

3.2.1 Manufacture

The essential ingredients of Portland cement are lime, silica and alumina. Fortunately, all three occur abundantly in nature as chalk (or limestone) and clay (or shale). Occasionally they even occur as a single raw material such as marl. Chalk and clay also contain smaller proportions of other minerals, the most important of which is iron oxide, but these are not generally deleterious to the cement.

The manufacturing process is basically simple, its main drawback being the need to raise the temperature of the raw materials to 1 500 °C. Initially the chalk and clay are intimately mixed in a slurry (wet process) or blended and transported in an air stream (dry process) which is fed to the top of a long, sloping and rotating kiln. The material moves steadily down the kiln undergoing successive changes as it becomes hotter. The water is evaporated, then the calcium carbonate of the chalk is decomposed to give quicklime, carbon dioxide being driven off up the kiln. Finally, fusion occurs (at 1 500 °C) and the raw materials combine to form calcium silicates, calcium aluminates and smaller proportions of other compounds. This is the cement clinker, and it is cooled and ground to a fine powder. Gypsum (calcium sulphate) is added and the final product, Portland cement, is packaged for despatch.

80

The chemical analysis of the cement is complicated by the presence of impurities, and to make matters worse it is not possible to determine the main compounds by conventional chemical methods. Instead the cement is subject to an oxide analysis, and the proportions of the compounds are estimated by means of the Bogue formulae (Bogue, 1955) quoted below. The four main constituents are:

Tricalcium silicate	$3CaO . SiO_2$	in short	C_3S
Dicalcium silicate	$2CaO . SiO_2$		C_2S
Tricalcium aluminate	$3CaO . Al_2O_3$		C_3A
Tetracalcium aluminoferrite	$4CaO . Al_2O_3 . Fe_2O_3$		C_4AF

Strictly, C_4AF is not a compound, but it typifies the composition of a solid solution.

In addition there are small quantities of magnesium oxide, free lime, the added gypsum, and a total of some 2% of other minor compounds.

The oxide analysis of a typical Portland cement is given in Table 3.1(a) and the importance of the lime (CaO) and the silica (SiO_2)

Table 3.1   The Chemical Composition of a Typical Portland Cement

(a) Oxide analysis	%
SiO_2	21.8
Al_2O_3	4.9
Fe_2O_3	2.4
CaO	64.7
CaO free	2.0
Na_2O	0.2
K_2O	0.4
MgO	1.1
Ignition loss	1.4
SO_3	2.6
(b) Calculated compound composition	
C_3S	45.7
C_2S	28.0
C_3A	8.9
C_4AF	7.3

stands out clearly. Similarly, the alumina (Al_2O_3) and iron oxide (Fe_2O_3) are seen to be of lesser importance but still of significance. The loss on ignition is found by heating to 1 000 °C and is caused by the driving off of moisture and carbon dioxide.

The corresponding composition is given in Table 3.1(b); it is found from the following (Bogue) conversion formulae:

$$(C_3S) = 4.07\,CaO - (7.60\,SiO_2 + 6.72\,Al_2O_3 + 1.43\,Fe_2O_3$$
$$+ 2.85\,SO_3)$$
$$(C_2S) = 2.87\,SiO_2 - 0.754\,C_3S$$
$$(C_3A) = 2.65\,Al_2O_3 - 1.69\,Fe_2O_3$$
$$(C_4AF) = 3.04\,Fe_2O_3$$

The symbols SiO_2, etc., are here taken to represent the percentages of silica, etc., in the oxide analysis. The Bogue formulae provide a guide to the way in which the proportions of the compounds can be adjusted by changes in the mixing of the raw materials, but the calculated values cannot be regarded as more than an approximate estimate of the constitution of the actual cement.

3.2.2 Hydration

The second stage of chemical change takes place when water is added to the dry cement, initiating the processes of hydration. The early reactions are controlled by the gypsum present and the later rates of reaction of the four main compounds differ greatly. The products of hydration are both crystalline, with calcium hydroxide ($Ca(OH)_2$) as a main constituent, and microcrystalline, consisting of hydrated calcium silicates and referred to as cement gel.

Study of the hydration of cement is difficult; not only are there a number of different compounds undergoing hydration simultaneously but the hydration products also interact one with another. Inevitably, the reactions given below must be considered as representative of average behaviour, with other possibilities and compounds either unknown or omitted. The hydrations of the four main compounds (C_3S, C_2S, C_3A and C_4AF) have been investigated individually, and it is apparent that the results are relevant to the hydration of cement in which all four occur together.

The C_3A reacts very rapidly with water and, after passing through intermediate compositions, eventually reaches a stable, cubic crystalline form in which six molecules of water are taken up.

$$3CaO.Al_2O_3 + 6H_2O \longrightarrow 3CaO.Al_2O_3.6H_2O$$

The reaction is so rapid that, without the addition of gypsum, a flash set would occur in which the cement paste would stiffen immediately the water was added. The gypsum ($CaSO_4$) reacts with the C_3A to form calcium sulphoaluminate, which is often referred to as ettringite.

$$3CaO.Al_2O_3 + 3CaSO_4 + 32H_2O \longrightarrow 3CaO.Al_2O_3.3CaSO_4.32H_2O$$

Although the sulphoaluminate is insoluble and crystallises out it does not cause flash setting of the cement. Usually 3–4% of the weight of cement gypsum is added, and this is all used up within 24 hours of mixing. After that, the straightforward hydration reaction takes over.

The gypsum also reacts with the C_4AF to form calcium sulpho-ferrite and calcium sulphoaluminate. The straightforward hydration reaction for C_4AF is in doubt but it has been suggested that lime may take part as well as water

$$4CaO . Al_2O_3 Fe_2O_3 + 2Ca(OH)_2 + 10H_2O \longrightarrow 3CaO . Al_2O_3 . 6H_2O$$
$$+ 3CaO . Fe_2O_3 . 6H_2O$$

The C_4AF is a small part of the total cement and it contributes little to any significant behaviour. It may be said to provide its modicum to the crystalline hydration products.

The two calcium silicates form the bulk of the hydrated material once the early reactions are completed, and they are also responsible for many of the most unusual facets of behaviour of h.c.p. Both have the same end product after hydration, tricalcium disilicate hydrate. For C_2S

$$2(2CaO . SiO_2) + 4H_2O \longrightarrow 3CaO . 2SiO_2 . 3H_2O + Ca(OH)_2$$

and C_3S

$$2(3CaO . SiO_2) + 6H_2O \longrightarrow 3CaO . 2SiO_2 . 3H_2O + 3Ca(OH)_2$$

The calcium silicate hydrate (C–S–H) is often given the name tober-morite after a naturally occurring mineral. The difference between the reactions of the two silicates lies firstly in the greater quantity of lime released by the C_3S, and secondly in that the rate of reaction for C_3S is much greater than that for C_2S. The lime contributes considerably to the crystalline products while the silicates are microcrystalline, of colloidal dimensions, and form the cement gel.

Table 3.2   Rates of Reaction of the
Components of Portland Cement
(80% Hydration)

(From H. Goetz, 'The Mode of Action of Concrete Admixtures', *Symposium on the Science of Admixtures*, 1969, by permission of The Concrete Society.)

Type of reaction	Time (days)
$C_3S + H_2O$	10
$C_2S + H_2O$	100
$C_3A + gypsum + H_2O$	—
$C_3A + H_2O + Ca(OH)_2$	6
$C_4AF + H_2O + Ca(OH)_2$	50

The hydration reactions of the main cement compounds proceed at different rates, from the very rapid of the C_3A to the distinctly slow of the C_2S as summarised in Table 3.2. This is also demonstrated in Fig. 3.2, in which the development of strength of the individual compounds is taken as a fair representation of the progress of the hydration reactions.

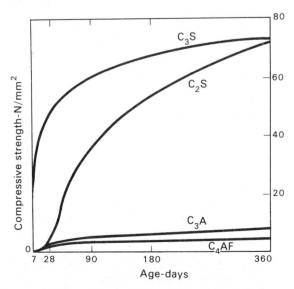

Figure 3.2 Development of strength of pure compounds. (From R. H. Bogue (1955) *Chemistry of Portland Cement*, by permission of Van Nostrand Reinhold Co.).

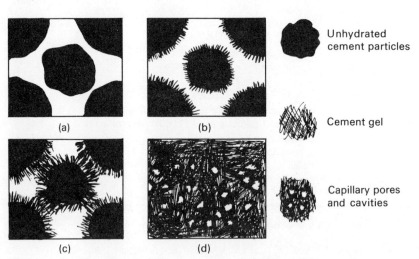

Figure 3.3 Diagramatic representation of the hydration process and the formation of cement gel. (From K. Newman (1966) in L. Holliday (Ed.) *Composite Materials*, Elsevier, based on T. C. Powers (1960) *Proceedings of the 4th International Symposium on the Chemistry of Cement*, by permission on the National Bureau of Standards, Washington D.C.)

84

Many attempts have been made to monitor the hydration reactions under the electron or stereoscan microscope and in spite of misgivings over the interpretation of what is observed a credible picture has emerged.

The development of hydration is shown diagrammatically in Fig. 3.3, in which four critical stages are treated:

(a) Immediately after mixing, the cement paste is in its most fluid state. The cement grains are dispersed in the mixing water, the spacing being determined by the water/cement ratio. Cement is a fine powder with a large specific surface area of the order of $300 \ m^2/kg$. This is sufficiently great for bleeding not to be a significant problem with the water/cement ratios used for most structural concrete. That is, the cement grains remain reasonably in suspension until setting occurs, rather than settling and allowing the mixing water to rise to the surface.

(b) After two hours the cement paste is much less fluid but can still be worked. There is clear evidence of hydration on the surface of the cement grains, both rods of ettringite and acicular or crumpled foils of calcium silicate hydrate being observed. At the same time the water becomes saturated with lime and needle-like formations of calcium silicate hydrate also appear in the intergranular water.

(c) After a day, the cement paste has set, thus turning into h.c.p., but it has no real strength. The hydration on the surface of the grains has penetrated further outwards and inwards. The hydrates in the intergranular spaces have grown and interconnected, thus forming a continuous gel establishing a solid skeletal structure, which is reinforced by the other crystalline products, such as ettringite, C_3A and C_4AF hydrates and, above all, large crystals of calcium hydroxide.

(d) After 7 days, the h.c.p. has achieved considerable strength, though the hydration reactions are far from complete. The skeletal structure had been further developed by infilling between the original hydration links to produce a denser gel structure. Nevertheless some larger capillary pores remain unfilled, and the original cement grains are not completely used up. Nor are they ever likely to be.

The hydration process is thus a continuous chemical reaction which presents the engineer with three different materials — fresh concrete that is workable, stiffened concrete that is soft, and hard concrete that has structural strength. To suit the various applications in practice he may want to alter the times taken to reach a given stage of hydration. This can be done by adjustments to the proportions of the compounds taking part in the chemical reactions. For example, quick setting could be encouraged by reducing the gypsum content, and this is quite different from rapid-hardening cement which may be more finely ground than normal, or may have a chemical admixture included.

The hydration reactions of all four components are exothermic, and the total quantity of heat given out is about 120 cal/g of cement. Most of this is liberated in the first few days, and, concrete being a poor conductor of heat, the temperature in massive structures can rise considerably. The four cement compounds have differing heats of hydration, and the following oft-quoted figures give a good idea of the relative quantities of heat liberated.

$$C_3S \quad 500 \ kJ/kg$$
$$C_2S \quad 260$$
$$C_3A \quad 865$$
$$C_4AF \quad 420$$

As will be demonstrated later in Section 3.3, some contribution to the solution of temperature problems is offered by control of the proportions of the four compounds.

The rate of liberation of heat is a direct function of the rate of hydration and it can immediately be deduced that the C_3A with its high rate of hydration and its high heat of hydration is the main contributor to the early temperature problems.

There are several methods for determining heat of hydration, including the use of an adiabatic calorimeter in which all the evolved heat is retained within the calorimeter, and the temperature rise is measured. Typical curves are given in Fig. 3.4, and if the vertical

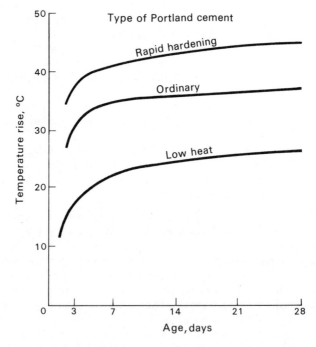

Figure 3.4 Temperature rise in 1:9 (by weight) concrete under adiabatic conditions. (From F. M. Lea (1970) *The Chemistry of Cement*, by permission of Edward Arnold (Publishers) Ltd.)

scale for each curve is normalised to cover a range between zero and unity, the curves may be taken as a measure of degree of hydration. Adiabatic temperature rises can occur in the middle of large concrete masses, but usually heat escapes from the concrete and the temperature rise is much less. Higher temperature accelerates the hydration reaction and it follows that the degree of hydration under isothermal conditions will be less, at any given age, than under the adiabatic conditions shown in Fig. 3.4. Hydration continues even when the temperature drops below the freezing point of water, and $-10°$ C is the accepted temperature below which hydration is completely inhibited. The effect of temperature is discussed further in relation to strength in Section 13.6.2.

The degree of hydration can also be measured by, for instance, determining the proportion of the cement grains that remain unhydrated, or by finding the quantity of water that combines with the cement. There are difficulties in all the methods but the concept of degree of hydration is vital to the proper consideration of the structure of the h.c.p. The variations between cements and between the storage conditions are numerous and the attempt must be made to measure degree of hydration if an accurate answer is required. As a guide, the development of hydration is given empirically by the factor, g (Rastrup, 1956) which corresponds to the degree of hydration for the given conditions. Then

$$g = \exp(-d/t^{\frac{1}{2}}) \qquad (3.1)$$

where t is the age of the h.c.p. in days, and d is a parameter depending on the type of cement, water/cement ratio and moisture conditions. It has been suggested (Pihlajavaara, 1963) that for sealed h.c.p. of water/cement ratio 0.5 stored at 20 °C, the value of d is 0.85.

3.2.3   Physical Structure of Hardened Cement Paste

The confused nature of h.c.p. has prevented the successful outcome of crystallographic investigations, so that there is virtually no direct evidence on its molecular structure. Special techniques have enabled observations to be made of the structures of aggregations of molecules and it has been concluded that cement gel exists in the form of fibres or sheets that are distorted, crumpled or rolled up. If the compounds found in cement are taken individually and in pure form, and hydrated, the structures are much more ordered than that of h.c.p. and examination by such methods as X-ray diffraction has met with some success. The calcium silicate hydrates have been shown to have layered or fibrous form, which connects with the observations on the gel, and points to the formation of the same or similar hydrates in h.c.p. Correspondence between other properties reinforces this deduction, and it seems not unreasonable to associate the sheet-like appearance of the bodies of gel with a layered or laminar molecular formation.

The molecules within any layer are held together strongly by chemical bonds, covalent and ionic, and the interlocking of bodies of gel indicates that chemical bonds also form where the gel aggregates meet; that is, the lattice of one crumpled sheet orients with and connects to the lattice of another closely adjacent crumpled sheet. This interlocking throughout the skeletal structure of the gel prevents the unlimited take-up of water that characterises a true gel, and hence cement gel is referred to as a limited swelling gel.

The specific surface of h.c.p. has been calculated from measurements of adsorbed water to be of the order of 250 m^2/g. This is a very large figure, especially when compared with the specific surface of unhydrated cement — which is, after all, a fine powder — at about 1 000 times less, 0.3 m^2/g. The importance of the high specific surface lies in the possible occurrence of surface forces, often called van der Waals forces. These physical forces of attraction between surfaces are much weaker than the chemical bonds, but, in h.c.p., the surfaces are very close to each other, and the areas of the surfaces are very large, so that the total attractive force could be sufficient to give very significant strength to the material. In addition, as suggested by T. C. Powers, the large surface area creates a correspondingly great capacity for attracting water molecules, that is for holding water in an adsorbed state.

Powers was also responsible for the simple, admittedly approximate, calculation of gel dimensions that follows.

Cement gel is considered to be plate-like with a typical area b by L and a typical thickness t. Then the specific surface, s_s, is given by

$$s_s = 2 \frac{(bt + tL + bL)}{btL \cdot \rho} \tag{3.2}$$

where ρ is the density of gel = 2.5 x 10^6 g/m^3.

From the micrographs,

$$b \doteqdot 10t \qquad L \doteqdot 300t$$

and hence

$$s_s = 250 = 2 \frac{(10 + 300 + 3\,000)t^2}{3\,000t^3 \times 2.5 \times 10^6}$$

and t = 35 Å.

The basal spacing, or c-spacing between adjacent solid layers has been found for C–S–H to be between 9 and 14 Å, depending on the moisture state, so the thickness of the solid gel represents some two to three layers of C–S–H.

Similarly, the hydraulic radius of gel pores is estimated at 7 Å. Assuming that the space between the gel solids is in the form of a rectangular slit, of dimension b broad by t^1 deep, the hydraulic radius

$$r_h = 7 = \frac{t^1 b' L}{(2t^1 + 2b') L} \tag{3.3}$$

and for $b' = 10t^1$

$$t^1 = 15 \text{ Å}$$

which is estimated to be enough to accommodate some five molecules thickness of water.

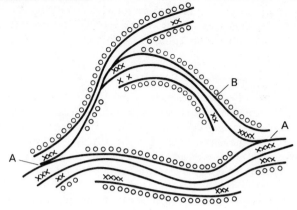

A – Interparticle bonds

× – Interlayer hydrate water

B – Tobermorite sheets

o – Physically adsorbed water

Figure 3.5 Simplified model for hydrated cement. (From R. F. Feldman and P. J. Sereda (1969) *Proceedings of the 5th International Symposium on the Chemistry of Concrete, Vol. 3*, by permission of the Cement Association of Japan.)

In fact, the thickness of gel solid and pores vary very greatly from place to place, as shown in the Feldman—Sereda model drawn in Fig. 3.5. This represents a section through the gel, cutting through the solids and showing a typical large gel pore as an enclosed area. Note the layers in the gel solid and their connection at meeting points by interparticle bonds, which may be chemical or physical. The gel pore enlarges from narrow trumpet-shaped spaces at the ends to enclose a volume with dimensions of some tens of Å. The h.c.p. as a whole should be imagined as consisting of aggregations of gel

Table 3.3   Typical Dimensions of Features of Hardened Cement Paste

Entity	Size (Å)
Adsorbed water molecule	2.6
c-spacing for CS hydrates	9–14
Gel pore thickness	15
Gel solid thickness	30
Capillary pore diameter	500
Gel body diameter	5 000
Cement grain diameter	30 μm = 300 000 Å
Air entrainment pore	50 μm = 500 000 Å

(including pores) growing out from unhydrated remains of cement grains and interspersed with larger capillary cavities, the volume of which depends on the initial water/cement ratio and the degree of hydration. As a summary on the solid structure, the dimensions of the various entities are given in Table 3.3.

3.2.4 The State of the Water in Hardened Cement Paste

As will be seen in later chapters, the water in the h.c.p. has a big influence on its properties. It is one of the essential constituents, taking part in the strength-giving hydration reactions, and, once the skeletal structure has been formed, some of it is attracted to the solid surfaces within the gel. Water is thus held in a number of different states in the h.c.p., and these can be listed conveniently in order of increasing bond energies, that is in order of difficulty in removing the water by drying and heating. The sites of the water in the gel are shown in Fig. 3.5.

(a) *Water vapour.* Hardened cement paste very seldom contains no voids at all, and the voids will often contain air from the outside environment or from the water in the h.c.p. In addition, the voids also contain water vapour exerting the vapour pressure appropriate to the relative humidity and temperature.

(b) *Free water.* The free water is mostly located in the capillaries, but also in the larger gel pores, and it is, by definition, sufficiently far from the solid surfaces to be free from the forces of attraction to those surfaces.

(c) *Adsorbed water.* In contrast to free water, adsorbed water is under the influence of the surface forces; that is, it is adsorbed onto the surface. The first layer is often said to be chemisorbed in that it is rigidly held and can be considered as part of the solid. Up to four or more molecular layers are physically attracted, their bonding energies diminishing with their distance from the surface.

(d) *Interlayer water.* As with some clays, water can penetrate the lattice layers of the gel solids, or into the intercrystalline spaces. This is interlayer water, sometimes referred to as zeolitic water, and its removal results in a closing of the interlayer spaces. The reduction of the c-spacing of CS hydrate from 14 Å to 9 Å, observed when the material is dried, has been attributed to the loss of interlayer water.

(e) *Chemically combined water.* An alternative name is water of hydration. This is the water combining with the unhydrated cement in the hydration reactions, and which forms an integral part of the solid products.

It is impossible to isolate the water of any one state from that of any other. If h.c.p. is dried at normal temperature, the waters in categories (a) to (d) are removed by drying, the capillary water in general disappearing at higher humidities and the interlayer water at the lower. However, there is a considerable overlap between the removal of waters of different states, and this leads to the continuous loss of water as the relative humidity is reduced (see Chapter 8).

3.2.5 The Powers Model

Many of the features of the physical structure discussed earlier are incorporated in the overall view formed by Powers to fit his own impressive experimental results (see Bibliography). His model has the great merit of providing a simple, consistent and acceptably accurate representation; its utility lies in the ability to predict porosity, which is a most important link between structure and properties. The model is founded on a number of definitions and observations:

(1) Hardened cement paste has three volume-forming constituents — unhydrated cement, cement gel and capillary space. The gel incorporates all the hydration products, both crystalline and microcrystalline. Both the gel and the capillary space may contain water.

(2) The water in the h.c.p. is divided into two categories only — nonevaporable, which is the water combined in the gel solids, and evaporable, which is the water that is removed by drying at a temperature of $105\,°C$. The evaporable water is held in the gel and the capillaries.

(3) The gel is always the same regardless of the stage of hydration at which it is formed, the type of cement, the water/cement ratio, and the water content. It follows that the combined water is a constant proportion by weight of the cement with which it combines, or

$$w_n/c = \text{constant} = 0.23$$

where w_n is the weight of combined water, c is the weight of cement with which it combines.

Similarly, the gel porosity is constant, so that the specific volumes of gel including pores and of gel solids are constants. Then

the specific volume of unhydrated cement	$v_c = 0.315$
the specific volume of gel solids	$v_{hc} = 0.411$
the specific volume of gel including pores	$v_g = 0.567$

v_g is very much greater than v_c and this implies that the gel fills a much greater space than the cement grain from which it grows.

(4) The volume of the solid products of hydration is less than the total volume of cement plus water before combination. Since there is no significant change in the overall volume of cement paste during hydration, this means that water must be drawn into the gel from the capillaries as combination occurs. If the paste is sealed voids must form, but if it is kept wet water will be drawn in from outside. To keep the paste in a saturated state, 0.25 g of water are drawn in for every gram of water reacting with the cement.

(5) Hydration stops for one of three possible reasons:

   (a) All the unhydrated cement is used up; that is, the degree of hydration, m, reaches unity.
   (b) The cement gel, growing out from the cement grains, fills the space available between the grains. This is the condition of insufficient space, and it may occur in h.c.p. kept in a saturated condition.
   (c) The hydration process, drawing in water, may empty the capillary pores. Gel water is not able to take part in further hydration, so the limit of hydration is reached. This is the condition of insufficient water, and it may occur in sealed h.c.p.

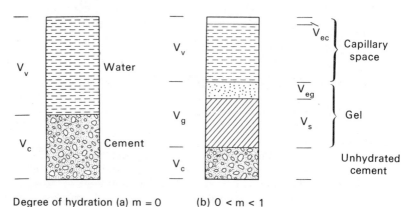

Degree of hydration (a) m = 0   (b) 0 < m < 1

Figure 3.6   Cement paste structure (a) on mixing, (b) sealed and partially hydrated.

The development of hydration is demonstrated in Fig. 3.6

(a)   immediately after mixing, with $m = 0$;
(b)   at a later stage with $0 < m < 1$.

If the water/cement ratio on mixing is w_0/c, and assuming that the h.c.p. is sealed, then for a weight of unhydrated cement of 1 g, the volume of the h.c.p. at any stage

$$V = v_c + v_w w_0/c$$

where v_w is the specific volume of the capillary water, usually taken as unity.

After hydration has started, the volume of gel solids

$$V_s = v_{hc} m (1 + w_n/c)$$

The volume of gel including pores

$$V_g = v_g m (1 + w_n/c)$$

so that the volume of gel pores

$$V_{eg} = V_g - V_s = m(1 + w_n/c)(v_g - v_{hc})$$

92

The volume of uncombined cement

$$V_c = v_c(1 - m)$$

Hence the volume of capillaries

$$V_v = V - V_g - V_c$$

and the volume of voids in the capillary pores

$$V_{ec} = \text{volume of water drawn into the gel}$$

$$= 0.25 \, m v_w w_n / c$$

3.2.5.1 Gel Properties

Expansion Factor. The expansion factor relates the volume of gel including pores to the volume of unhydrated cement from which it was formed. From the previous formulae,

$$\text{expansion factor} = \frac{V_s}{(V_c)_{m=0}} = \frac{v_g(1 + w_n/c)}{v_c}$$

$$= \frac{0.567 \, (1 + 0.23)}{0.315}$$

$$= 2.2$$

Porosity. The gel porosity is given by

$$\frac{\text{volume of gel pores}}{\text{volume of gel including pores}} = \frac{V_g - V_s}{V_g} = \frac{v_g - v_{hc}}{v_g}$$

$$= \frac{0.567 - 0.411}{0.567}$$

$$= 0.27$$

According to Powers, as stated in condition 3 earlier in Section 3.2.5, these values are characteristic of gel whenever it is formed, and they can be used as property values instead of those for v_{hc} and v_g listed earlier.

3.2.5.2 Limits of Hydration

The Condition of Insufficient Space. The h.c.p. is kept saturated so that there is no question of hydration ceasing because of lack of water. The space available for gel growth is determined by the initial water/cement ratio. If it is high the hydration will reach completion and there will be capillary space unreached by the gel. Conversely, at low water/cement ratios the gel fills all the available space before hydration is complete so that some cement remains unhydrated. The necessary condition for the end of hydration is that the capillary space is reduced to zero.
i.e.

$$V = V_c + V_g$$

or

$$v_c + v_w w_0/c = (1 - m_{Ls})v_c + m_{Ls}v_g(1 + w_n/c)$$

where m_{Ls} is the limiting degree of hydration.

i.e.
$$m_{Ls} = \frac{v_w w_0/c}{v_g(1 + w_n/c) - v_c} \qquad (3.4)$$

The highest water/cement ratio for which the condition of insufficient space applies is given by $m_{Ls} = 1$.
Substitution gives

$$(w_0/c)_{max} = 1[0.567(1 + 0.23) - 0.315] = 0.38$$

This is a unique case in that, at full hydration, an h.c.p. with an initial water/cement ratio of 0.38 would have no unhydrated cement and no capillary pores; it would consist entirely of cement gel.

The proportions of constituents when mixes of differing water/cement ratios reach their limits of hydration are shown in Fig. 3.7(a).

The Condition of Insufficient Water. The h.c.p. is sealed so that there is no loss or gain of water. In these circumstances hydration will always cease because of lack of water before it might be affected by lack of space for the gel to grow into.

Here the governing condition is that the capillary pores are empty while the gel pores are full. That is, the volume of capillary pores is equal to the volume of water drawn in by the progress of hydration. i.e.

$$V - V_g - V_c = 0.25 m_{Lw} v_w w_n/c$$

where m_{Lw} is the limiting degree of hydration for this case, i.e.

$$(v_w w_0/c + v_c) - m_{Lw}v_g(1 + w_n/c) - (1 - m_{Lw})v_c = 0.25 m_{Lw} v_w w_n/c$$

$$(3.5)$$

or

$$m_{Lw} = \frac{v_w w_0/c}{v_g - v_c + (0.25 + v_g)w_n/c}$$

Again, the higher water/cement ratio for which the condition (of insufficient water) applies is given by $m_{Lw} = 1$.
i.e.

$$(w_0/c)_{max} = 1[0.567 - 0.315 + 0.23(0.25 + 0.567)] = 0.44$$

At the limit of hydration there is always capillary space remaining, as shown by the constituents diagram in Fig. 3.7(b).

3.2.5.3 Porosity

Two porosity calculations are possible; firstly the capillary porosity in terms of the capillary space only, and secondly the total porosity,

94

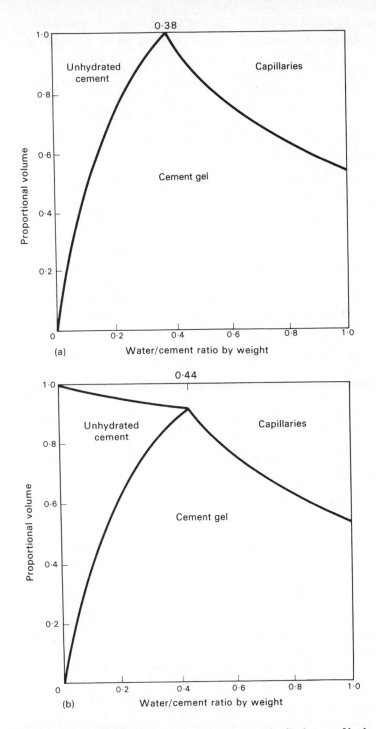

Figure 3.7 Composition of hydrated cement paste at the final state of hydration after prolonged storage (a) in water, (b) sealed. (From T. C. Hansen (1970) *Proceedings of the American Concrete Institute*, 67, 404, by permission of the American Concrete Institute.)

which includes the gel pores as well as the capillaries. When considering the properties of h.c.p. there is no agreed preference for one rather than the other.

Capillary porosity p_c

$$p_c = \frac{V_v}{V} = \frac{V - V_g - V_c}{V} = \frac{(v_w w_0/c + v_c) - m v_g(1 + w_n/c) - (1 - m)v_c}{(v_w w_0/c + v_c)}$$

(3.6)

Total porosity, p_t

$$p_t = \frac{V_v + V_{eg}}{V} = \frac{(V - V_g - V_c) + (V_g - V_s)}{V} = \frac{(V - V_s - V_c)}{V}$$

$$= \frac{(v_w w_0/c + v_c) - v_{hc}(1 + w_n/c)m - (1 - m)v_c}{(v_w w_0/c + v_c)}$$

(3.7)

The subdivision of the model space into gel pores and capillary pores gives the impression of two size ranges which might well be associated with a double-peaked distribution of pore sizes. In reality, measurements of pore sizes have often shown a continuous distribution, thus demonstrating that this model, like all others, has its limitations.

The importance of porosity will be demonstrated in later sections, when it will be shown to be the variable that best represents the structure of the h.c.p. and to which most of the important properties (such as strength and deformation) are closely related. Furthermore it is the characteristic of the structure that is responsible for the similarity between concrete and timber (see Section 11.4). In both materials, water is held in a variety of states and is removed in a continuous manner as the drying humidity is decreased. In both materials there are consequent dependences of properties on moisture content. Conversely, metals (see Chapter 5) have a well ordered structure, unadulterated by voids and uncomplicated by imbibed water. This relative simplicity may explain why the structures of metals are better understood and are more clearly described.

3.3 TYPES OF PORTLAND CEMENT

The four main compounds in Portland cement (C_2S, C_3S, C_3A and C_4AF) hydrate at different rates and their hydration reactions are affected differently by added chemicals. Different quantities of heat are evolved during hydration and their strengths are different. The deliberate variation of the proportions of the compounds enable cements to be produced which have different properties to suit different circumstances of construction.

Four oxide and compound compositions are given in Table 3.4, each typical of one type of Portland cement.

Table 3.4　Typical Oxide Analyses and
Calculated Compounds' Proportions for
Different Types of Portland Cement

| | Type of cement | | | |
	A	B	C	D
Oxide analysis (%)				
CaO	66	65	62	61
SiO_2	23	21	25	21
Al_2O_3	6	5	5	4
Fe_2O_3	2	3	2	7
Compounds				
C_2S	34	21	59	31
C_3S	41	50	17	40
C_3A	13	9	9	0
C_4AF	5	9	6	20

Cement A is the happy medium, ordinary Portland cement, and it
serves here as a standard of comparison for the other three cements.

Cement B has the highest proportion of C_3S and can be expected to
develop early strength more rapidly than the others. It may be described as a rapid-hardening cement, though in practice the attainment of rapid hardening is more often achieved by finer grinding of
the clinker, allowing the access of more hydration water to the surface of the cement grains, and hence a more rapid rate of early hydration. The surface area of rapid-hardening cement is at least 325 m²/kg,
in comparison with a minimum of 225 m²/kg for ordinary Portland
cement. Very rapid hardening can be stimulated by grinding even
finer, and cement with a surface area of around 800 m²/kg has been
produced by elutriation for that purpose.

Rapid-hardening cement is popular where the early removal of
formwork is important for economy of operation. It would be contrary to human expectation if this advantage were not accompanied
by a corresponding disadvantage, and here the penalty for high early
strength is a higher rate of evolution of heat.

Cement C has the lowest proportion of C_3S, which, by reference to
the arguments applied to cement B, indicates a low early strength. It
is difficult to imagine the circumstances in which this is advantageous,
and the merit here lies in the low rate of heat evolution attributable
to small proportions of C_3S, and the below average proportion of
C_3A. Cement C is thus *low-heat* cement and it was originally
developed for use in dams. The problem arises in any large mass of
concrete in which the rate of evolution of heat greatly outweighs the
rate of its dissipation. The internal temperature rises, and temperature
gradients occur through the structure. This induces strain and stress
gradients, and there are real dangers of excessive tensile stresses developing; that is, cracks may occur in the concrete.

Cement D has the lowest (possible) proportion of C_3A, the compound that is attacked by sulphates. So cement D is *sulphate-resisting* cement. The sulphates react with both the calcium hydroxide and the hydrated calcium aluminates to form calcium sulphoaluminate; as previously mentioned, gypsum is added to increase the setting time by encouraging this very reaction, but the quantities of gypsum are carefully controlled so that there is no danger to the strength of the concrete. The reaction gives a product of greater volume than the original solids, so that a prolonged attack by sulphates penetrating from outside causes internal expansion and disintegration of the material. Ground water in gypsum-bearing soils poses the most widespread threat, but the magnesium sulphate of seawater is also a danger, and ammonium sulphate is the most deleterious of all.

The oxide analyses of the four cements do not differ all that much — the proportion of lime varies only between 61 and 66%. This high sensitivity of the compound proportions to the relative quantities of the raw materials indicates that the production of cement with precisely specified properties demands a very high level of control.

3.4 ADMIXTURES

An admixture is a material other than water, sand, coarse aggregate and cement that is used as an ingredient of concrete and is added to the mix immediately before or during mixing. Admixtures are chemicals, that, in spite of being only a small proportion of the cement content, modify the hydration reactions so as to introduce a significant change in behaviour. Their popularity has greatly increased so that the number of admixtures is measured in hundreds, and their functions have proliferated to include, for instance, the improvement of permeability as well as the more obvious acceleration of strength development, protection against fungicidal attack as well as the improvement of workability, the colouring of the concrete as well as the retardation of setting. Admixtures can justifiably be called the fifth ingredient, even though, despite intense research effort, the chemical and physical basis of their action is often imperfectly understood.

Four well established categories of admixtures are dealt with below, namely accelerators, retarders, water-reducers and air entraining agents. Mention will be made of the most common chemicals, some of which can be categorised under more than one of these four titles.

3.4.1 Accelerators

The rate of *hardening* of the cement paste is increased by the addition of an accelerator, thus enhancing the early strength and allowing the earlier removal of formwork, or reducing the time required for elaborate curing. The most popular chemical has been calcium chloride ($CaCl_2$) which is added in solution during mixing, but other chloride-free accelerators are available.

98

The $CaCl_2$ reacts with the C_3A, but its presence also results in the more rapid hydration of the C_3S. It may be said that the $CaCl_2$ acts as a catalyst with respect to the hydration of the C_3S. The effect on hydration and strength is shown in Fig. 3.8, in which the rate of

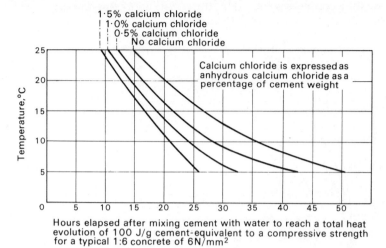

Figure 3.8   The effect of calcium chloride on the hardening of a typical 1:6 concrete at various temperatures. Derived from conduction calorimetry. (From K. E. Fletcher (1973) *Civil Engineering and Public Works Review*, **68**, 793, by permission of Morgan-Grampian (Professional Press) Ltd.)

hydration is represented by the time taken for the evolution of a standard quantity of heat. As can be seen, the rate of hydration depends on both the calcium chloride content (measured as the weight percentage of anhydrous $CaCl_2$ to cement) and the temperature.

The presence of chlorides reduces the protection of any included steel against corrosion (see Section 19.6) and the use of $CaCl_2$ in prestressed concrete has long been prohibited. More recently, its use in reinforced concrete has also been strongly discouraged.

Although accelerators are intended to improve the early strength, they often accelerate the setting time also. Calcium chloride is no exception, and there is even the possibility of a flash set if an excess of $CaCl_2$ is used. Most admixtures have secondary effects which may or may not be helpful. Calcium chloride improves abrasion resistance, but the resistance to sulphate attack is reduced.

3.4.2   Retarders

The title of this category usually refers to retardation of the *setting* of the cement paste, and the uses include counteracting the accelerating effect of high temperature, or avoiding the embarrassment of a lorry full of hard concrete when transporting over a long distance. A number of chemicals are effective, the best known and most popular of which are sugar, lignosulphonates and hydroxycarboxylic acids. Their performance is demonstrated for a lignosulphonate in Fig. 3.9, in which the three left-hand lines represent different stages of setting. The fourth line shows a stage of early hardening and it is clear that this, too, is retarded. The apparent reversal of behaviour at low solid contents is caused by the versatility of this admixture which also acts as a water-reducer (see below).

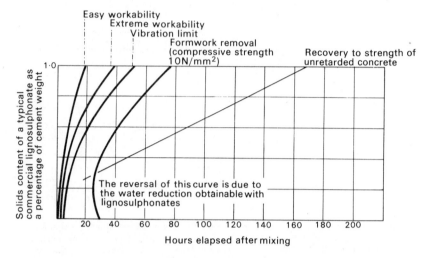

Figure 3.9 The retarding effect of a lignosulphonate admixture in a typical 1:6 concrete at 18 °C. (From K. E. Fletcher (1973) *Civil Engineering and Public Works Review*, **68**, 793, by permission of Morgan-Grampian (Professional Press) Ltd.)

The action of retarders is explained by the adsorption of the admixture onto the surface of the C_3A, with a consequent delay in the precipitation of the early hydration products. The hydration of the C_3S is also delayed, and this may be attributed to the inhibition by the admixture of the nucleation of calcium hydroxide.

It is impossible to be precise about the period of retardation, especially as temperature, the fineness and composition of the cement, and time of adding the admixture can all influence its performance.

3.4.3 Water-Reducers

The alternative title of plasticisers gives an indication of the function of water-reducers; it is to improve the workability of the mix. This allows economies to be made by increasing the aggregate content and hence saving cement. Alternatively, a stronger and more durable concrete can be produced by reducing the water content. The most common chemicals are the lignosulphonates and hydroxycarboxylic acids, which have already appeared as retarders. As their name implies, super-plasticisers (Working Party, 1976) are especially effective; they are chemically distinct from normal plasticisers, one being, for example, sulphonated melamine formaldehyde condensate.

The action of the plasticisers derives from their adsorption onto the surface of the cement grains. This leads to a build-up of the forces of repulsion between the grains, which is equivalent to saying that the cohesive forces between the grains are reduced. The particles become more dispersed, that is, the admixture causes deflocculation and the viscosity is reduced. In engineering terms, the workability is improved.

3.4.4 Air-Entraining Agents

Entrained air is intentional and consists of small bubbles (less than 0.1 mm in diameter), whereas entrapped air is unintentional and forms in larger more irregular volumes. Both increase the porosity of the h.c.p. or concrete, with a consequent drop in strength. Entrained air must be dispersed throughout the whole volume of the material, and the size of individual bubbles must be kept small for effective frost resistance and if an unacceptable loss of quality is to be avoided. It has two functions. Firstly, and most important, it provides space for water to expand into when it freezes, thus preventing the occurrence of large internal disintegrating forces and improving durability (see Chapter 17). Secondly, of less importance, entrained air is an aid to workability.

The air-entraining agents are soap-like substances or foaming agents such as vinsol resin, which reduce the surface tension of the mixing water and allow stable air bubbles to exist in the mix. The greatest protection is provided by about 5% by volume of entrained air. Compared with the same mix without entrained air, the reduction of strength would be about 25%, but in practice the improvement in workability allows the use of a smaller water content, and the loss of strength can be kept below 10%.

3.5 AGGREGATES

Hardened cement paste has cohesiveness and strength, and could conceivably perform on its own as a structural material, but for the matter of cost. It is both energy and money expensive, and cheaper materials, sand and coarse aggregate, are added for reasons of economy. The objective must be to use as much of the cheaper materials as possible, binding them together with the more expensive h.c.p. in sufficient quantity to ensure technical efficiency. In practice this means that the largest possible stones should be included, and the interstices between the stones should be filled for the most part by similar stony particles and sand. The h.c.p. then fills the interstices of the sand. This is a process of maximising the volume concentration of the aggregates (within the constraints imposed by technical requirements) by employing a material with a continuous grading, or, at least, two or more sizes of aggregate.

The aggregates are usually inert in comparison with the h.c.p., in that they are not gel-like, they do not hydrate, and they do not exhibit shrinkage and swelling. They are distributed as particles within the h.c.p. and it is convenient and proper to regard concrete as a two-phase material, either as coarse aggregate dispersed in a matrix of mortar, or as aggregate dispersed in h.c.p. Most of the properties of concrete derive from the properties of the h.c.p., modified in accordance with the simple behaviour of the aggregate, in accordance with a suitable law of mixtures. Appropriate models are derived in the relevant sections of later chapters, but mention is made here of the more obvious characteristics of the aggregates.

3.5.1 Properties of Aggregates

There are two main types of aggregates — normal and lightweight — and, to complete the symmetry, one subsidiary type, heavyweight aggregate. Normal aggregates are those that have been traditionally used, consisting of natural materials, including gravels, igneous rocks such as granite and basalt, and hard sedimentary rocks such as limestones and sandstones. Of importance to engineers are their specific gravities, which influence the selfweight of engineering structures; their rigidity, which can be expressed in terms of their elastic moduli, and which influence both the deformations and the forces in engineering structures; their strength, which should be at least as great as the h.c.p. with which they are mixed; and their absorptivity which influences their thermal conductivity.

The range of values for the different types of aggregate are given in Table 3.5 and it is noticeable that, in spite of the many different rock types, the normal aggregates have a sufficiently narrow range of property values for the concretes of which they are made to be generally similar.

Table 3.5   Properties of Aggregate

Type	Aggregate		Concrete		
	Type	Loose bulk density (kg/m^3)	Density (kg/m^3)	Typical compressive strength (N/mm^2)	Thermal conductivity $(Jm/m^2 \, s \, ^\circ C)$
Lightweight	pumice	700	1 000	18	0.5
	expanded clay	400	1 600	25	0.6
Normal	granite limestone	1 400	2 400	35	1.5 3.5
Heavyweight	barytes	2 500	3 400	45	1.4

Lightweight concretes have been developed for several reasons. The much lower specific gravity of the lightweight aggregates reduces the selfweight of structures made of lightweight concrete, thus enabling smaller cross-sections to be designed, and savings of materials

102

to be made. Similarly, lightweight blocks are advantageous in building, not only because they are easy to handle, but also because of their good insulation against heat losses.

Although pumice and other volcanic materials are used, most lightweight aggregates are artificial materials; they include expanded clay pellets produced by heating suitably clays in a rotary furnace, clinker, and sintered pulverised fuel ash from power station boiler plants. The conservationist approach seems to be to suggest that any waste product arising in large quantity should be disposed of either as a road fill or as a concrete aggregate. Surprisingly, that is almost always possible — even sintered domestic waste has been included in concrete — and the economic attractiveness of deploying waste in this way is likely to continue to increase.

Heavyweight aggregates include barytes (barium sulphate) and steel shot. They are used in concrete for radiation shielding.

3.5.2 Interfacial Effects

The laws of mixtures are appropriate and effective when there is no over-riding happening at the interface between the aggregate and the h.c.p. For example, at low stresses, the bond between aggregates and h.c.p. remains intact and the creep and elastic strain are not enhanced by any interfacial slip; the laws of mixtures turn out to be valuable in predicting these strains.

In the mixing and casting of the fresh concrete there are several changes, actual or potential, introduced in the mix because of the characteristics of the interface. Thus the shape of the aggregate affects the workability of the mix. There is less interference and friction between the particles of rounded aggregate than there is for irregular or angular aggregates, which also have a higher void ratio. This means that the mix proportions must be adjusted to counteract the lower workability of the less rounded aggregate.

Another effect is water gain, in which, as shown in Fig. 3.10, sand and larger cement particles settle out from the larger stones, to be replaced by finer particles and water moving up to form lenses under the stones. These become voids after the concrete has hardened; that is, the proportion of entrapped air is increased with a consequent weakening of the concrete. Thirdly, if the aggregate is dry it will absorb water from the surrounding h.c.p., thus changing the local value of that very important quantity, the water/cement ratio. The water content of the aggregate is a vexed question, and, in the attempt

Figure 3.10   Water gain in fresh concrete. (From B. P. Hughes and J. E. Ash (1969) *Concrete*, 3, 494, by permission of the Cement and Concrete Association.)

103

to rationalise it, four possible states have been identified:

(a) completely (oven) dry,
(b) air-dry,
(c) saturated, and surface dry,
(d) wet.

In the first and second states the aggregate is likely to take in water from the h.c.p., and in the fourth state it will give up water to the h.c.p. While it is relatively simple to attain the desirable third state in the laboratory, it is usually impracticable under the rigours of the site. Nevertheless, it is often used as the reference state and the water added to aggregate in this third state is called the free water (it is not absorbed by the aggregate). The corresponding mix ratio is then the free water/cement ratio.

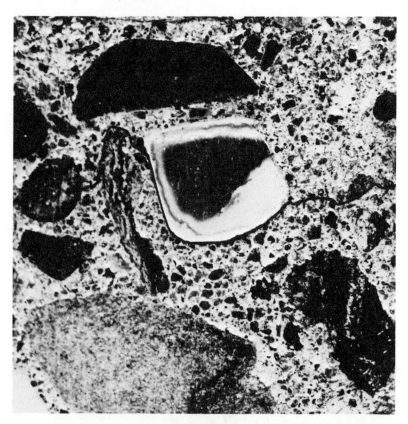

Figure 3.11   The path of a crack in concrete.

The most important interfacial effect occurs in the hardened concrete, and concerns cracking and failure (see Chapter 13). The aggregate and h.c.p. may both be strong, but the concrete will usually be considerably weaker because it fails through a breakdown of the bond between the two. As shown in Fig. 3.11, the cracks in concrete usually form and propagate along the boundaries of the aggregate particles. In these circumstances the laws of mixtures are of little relevance, and a new approach, such as that based on the concepts of fracture mechanics, must be sought.

3.6 REFERENCES

Bogue, R. M. (1955) *Chemistry of Portland Cement*, Reinhold, New York.

Feldman, R. F. and Sereda, P. J. (1969) discussion of paper III—1, *Proceedings of the 5th International Symposium on the Chemistry of Concrete, Vol. 3.*

Fletcher, K. E. (1973) Admixtures for concrete, *Civil Engineering and Public Works Review*, 68, 793.

Goetz, M. (1969) The mode of action of concrete admixtures, *Symposium on the Science of Admixtures*, The Concrete Society, London.

Hansen, T. C. (1970) Physical composition of hardened Portland cement paste, *Proceedings of the American Concrete Institute*, 67, 404.

Hughes, B. P. and Ash, J. E. (1969) Water gain and its effects on concrete, *Concrete*, 3, 494.

Lea, F. M. (1970) *The Chemistry of Cement and Concrete*, Edward Arnold, London.

Newman, K. (1966) *Concrete Systems*, Chapter VIII, Composite Materials, Elsevier.

Pihlajavaara, S. E. (1963) *Notes on the Drying of Concrete*, The State Institute of Technical Research, Finland.

Powers, T. C. (1960) Proceedings of the 4th International Symposium on Chemistry of Cement, Washington, p. 577.

Powers, T. C. (1964) The physical structure of Portland cement pastes, in *The Chemistry of Cements*, H. F. W. Taylor (Ed.), Academic Press, Vol. 1, Ch. 10, p. 392.

Rastrup, E. (1956) The temperature function for heat of hydration in concrete, *RILEM Symposium on Winter Concreting*, Copenhagen.

Working Party (1976) Superplasticising admixtures in concrete, *Report of a Working Party of the Cement Admixtures Association and the Cement and Concrete Association*, London.

CHAPTER 4

Structure of Timber

4.1 INTRODUCTION: TIMBER AS A MATERIAL

From earliest recorded times timber has been a ubiquitous material; the ancient Egyptians produced furniture, sculptures, coffins and death masks from it as early as 2500 BC: elaborate wooden couches and beds were produced in the days of the Greek empire (700 BC). The Ancient Briton, somewhat less sophisticated in his requirements, used wood for the handles of his weapons and tools and for the construction of his huts and rough canoes. Considerably more diversity in utilisation appeared in Mediaeval times when, in addition to the use of timber for the longbow, and later the butt of the crossbow and the chassis of the cannon, timber found widespread use in timber-frame housing and boats; musical instrument manufacture based on wood advanced significantly during this period.

In the industrial era of the 19th century timber was used widely for the construction not only of roofs but also of furniture, water-wheels, gearwheels, rails of early pit railways, sleepers, signal poles, bobbins and boats. The present century has seen an extension of its use in certain areas and a decline in others due to its replacement by newer materials. Despite competition from the lightweight metals and plastics, whether foamed or reinforced, timber continues to be used on a massive scale. The present world production is about 10^9 tonnes, an amount surprisingly similar to the annual production of iron and steel. With an average cost of about £150 per tonne (1974 prices) the annual consumption is therefore valued at £150 000 million.

From the early part of the industrial era the UK has been a large importer of timber. In 1974 the value of timber and timber products (excluding pulp and paper) imported into this country was £850 million and, as such, constituted about 3.7% of our total imports. Home production of timber and timber products is small in comparison to the volume of imports; only 5% of our softwood requirements and 35% of our hardwood demands are met by home production, which is valued at the present time at about £50 million.

In the UK, timber and timber products are consumed by a large range of industries, but the bulk of the material continues to be used in construction, either structurally such as roof trusses or floor joists, or nonstructurally, e.g. doors, window frames, skirting boards and external cladding. Of a total value in 1974 of about £900 million, about 60% of this was used in the construction industry. On a volume basis, annual consumption continues to increase slightly and there is no reason to doubt that this trend will be maintained in the future, especially with the demand for more houses, the increasing price of plastics and the favourable strength-weight and strength-cost factors for timber.

Timber is cut and machined from trees, themselves the product of nature and time. The structure of the timber of trees has evolved through millions of years to provide a most efficient system which will support the crown, conduct mineral solutions and store food material. Since there are approximately 30 000 different species of tree it is not surprising to find that timber is an extremely variable material. A quick mental comparison of the colour, texture and density of a piece of balsa and a piece of lignum vitae, used to make playing bowls, will illustrate the wide range that occurs. Nevertheless, man has found timber to be a cheap and effective material and, as we have seen, continues to use it in vast quantities. However, he must never forget that the methods by which he utilises this product are quite different from the purpose that nature intended and many of the criticisms levelled at timber as a material are a consequence of man's use or misuse of nature's product. Unlike so many other materials, especially those used in the construction industry, timber cannot be manufactured to a particular specification. Instead the best use has to be made of the material already produced, though it is possible from the wide range available to select timbers with the most desirable range of properties. Timber as a material can be defined as a low-density, cellular, polymeric composite, and as such does not conveniently fall into any one class of material, rather tending to overlap a number of classes. In terms of its high strength performance and low cost timber remains the world's most successful fibre composite.

Four orders of structural variation can be recognised — macroscopic, microscopic, ultrastructural and molecular — and in subsequent chapters the various physical and mechanical properties of timber will be related to these four levels of structure. In seeking correlations between performance and structure it is tempting to describe the latter in terms of smaller and smaller structural units. Whilst this desire for refinement is to be encouraged, a cautionary note must be recorded for it is all too easy to overlook the significance of the gross features. This is particularly so where large sections of timber are being used under practical conditions; in these situations gross features such as knots and grain angle are highly significant factors.

107

4.2 STRUCTURE OF TIMBER

4.2.1 Structure at the Macroscopic Level

The trunk of a tree has three physical functions to perform; firstly, it must support the crown, a region responsible for the production not only of food but also of seed; secondly, it must conduct the mineral solutions absorbed by the roots upwards to the crown; and thirdly it must store manufactured food (carbohydrates) until required. As will be described in detail later, these tasks are performed by different types of cell.

Whereas the entire cross-section of the trunk fulfils the function of support, and increasing crown diameter is matched with increasing diameter of the trunk, conduction and storage are restricted to the outer region of the trunk. This zone is known as *sapwood*, while the region in which the cells no longer fulfil these tasks is termed the *heartwood*. The width of sapwood varies with species and with age of the tree, but it is seldom greater, and is usually much less than, one third of the total radius (Figs. 4.1 and 4.2). The advancement of the

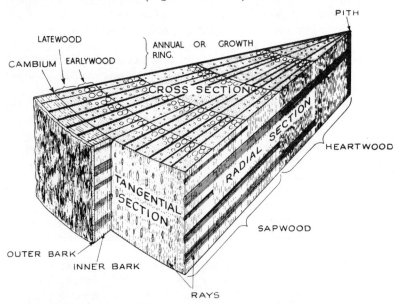

Figure 4.1 Diagramatic illustration of a wedge-shaped segment cut from a five-year-old hardwood tree, showing the principal structural features. (From the Princes Risborough Laboratory, Building Research Establishment. © Crown copyright.)

heartwood to include former sapwood cells results in a number of cell changes, primarily chemical. The acidity of the cells increases slightly, though certain timbers have heartwoods of very high acidity. Substances, collectively called *extractives*, are formed in small quantities and these impart not only colouration to the heartwood but also resistance to both fungal and insect attack. Different substances

108

are found in different species of wood and some timbers are devoid of them altogether: this explains the very wide range in the natural durability of wood about which more will be said later. Many timbers develop gums and resins in the heartwood while the moisture content of the heartwood of most timbers is appreciably lower than that of the sapwood in the freshly felled state.

Figure 4.2 Cross-section through the trunk of a Douglas fir tree. The annual growth rings, the darker heartwood and the lighter sapwood can all be clearly seen. (From the Princes Risborough Laboratory, Building Research Establishment. © Crown copyright.)

With increasing radial growth of the trunk commensurate increases in crown size occur, resulting in the enlargement of existing branches and the production of new ones; crown development is not only outwards but upwards. Radial growth of the trunk must accommodate the existing branches and this is achieved by the structure that we know as the *knot*. If the cambium of the branch is still alive at the point where it fuses with the cambium of the trunk continuity in growth will arise even though there will be a change in orientation of

the cells. The structure so formed is termed a *green* or *live* knot
(Fig. 4.3). If, however, the cambium of the branch is dead, and this
frequently happens to the lower branches, there will be an absence
of continuity, the trunk growing round the dead branch often
complete with its bark. Such a knot is termed a *black* or *dead* knot
(Fig. 4.4), frequently dropping out of planks on sawing. The grain
direction in the vicinity of knots is frequently distorted and in a
later section the loss of strength due to different types of knots will
be discussed.

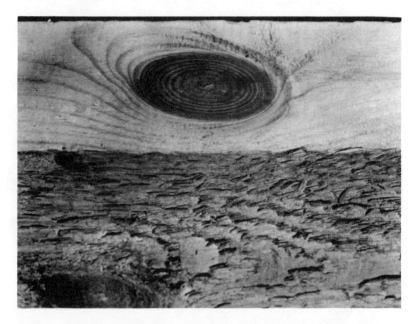

Figure 4.3   *Green* or *live* knot showing continuity in structure between the
branch and tree trunk. (From the Princes Risborough Laboratory, Building
Research Establishment. © Crown copyright.)

4.2.2   Structure at the Microscopic Level

The cellular structure of wood is illustrated in Figs. 4.5 and 4.6.
These three-dimensional blocks are produced from micrographs of
samples of wood 0.8 x 0.5 x 0.5 mm in size removed from a coniferous tree (Fig. 4.5) — known technically as a *softwood* — and a broad-
leaved tree (Fig. 4.6) — a *hardwood*. In both samples it will be
observed that 90—95% of cells are aligned in the vertical axis, while
the remaining percentage is present in bands (rays) aligned in one
of the two horizontal planes known as the radial plane or quarter-
sawn plane (Fig. 4.1). This means that there is a different distribution
of cells on the three principal axes and this is one of the two principal
reasons for the high degree of anisotropy present in timber.

It is popularly believed that the cells of wood are living cells: this

110

Figure 4.4 *Black* or *dead* knot surrounded by the bark of the branch and hence showing discontinuity between branch and tree trunk. (From the Princes Risborough Laboratory, Building Research Establishment. © Crown copyright.)

is certainly not the case. Wood cells are produced by division of the *cambium*, a zone of living cells which lies between the bark and the woody part of the trunk and branches (Fig. 4.1). In the winter the cambial cells are dormant and generally consist of a single circumferential layer. With the onset of growth in the spring, the cells in this single layer subdivide radially to form a cambial zone some ten cells in width. This is achieved by the formation within each dividing cell of a new vertical wall called the primary wall. During the growing season these cells undergo further radial subdivision to produce what are known as daughter cells and some of these will remain as cambial cells while others, to the outside of the zone, will develop into bark or, if on the inside, will change into wood. There is thus a constant state of flux within the cambial zone with the production of new cells and the relegation of existing cambial cells to bark or wood. Towards the end of the growing season the emphasis is on the relegation and a single cambial layer is left for the winter period.

To accommodate the increasing diameter of the tree the cambial zone must increase circumferentially and this is achieved by the periodic tangential division of the cambial cells. In this case the new wall is sloping and subsequent elongation of each half of the cell results in cell overlap, often frequently at shallow angles to the vertical axis, giving rise to spiral grain formation in the timber. The rate at which the cambium divides tangentially has a significant effect on the average cell length of the timber produced (Bannan, 1954).

111

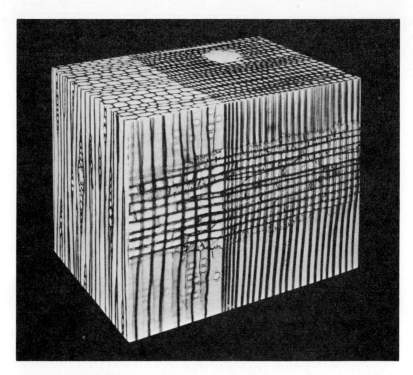

Figure 4.5 Cellular arrangement in a softwood (*Pinus sylvestris* — Scots pine, redwood). (From the Princes Risborough Laboratory, Building Research Establishment. © Crown copyright.)

Table 4.1   The Functions and Wall Thicknesses of the Various
Types of Cell Found in Softwoods and Hardwoods

Cells	Softwood	Hardwood	Function	Wall thickness
Parenchyma	+	+	Storage	
Tracheids	+	+	Support Conduction	
Fibres		+	Support	
Vessels (pores)		+	Conduction	

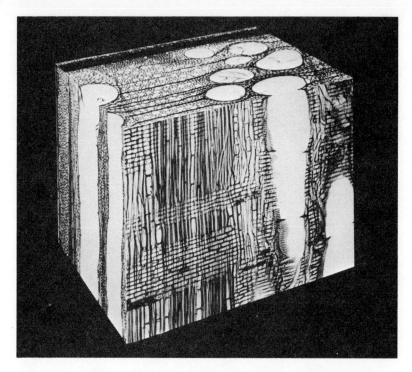

Figure 4.6 Cellular arrangement in a ring-porous hardwood (*Quercus robur* – European oak). (From the Princes Risborough Laboratory, Building Research Establishment. © Crown copyright.)

The daughter cells produced radially from the cambium undergo a series of changes extended over a period of about three weeks; this process is known as differentiation. Changes in cell shape are paralleled with the formation of the secondary wall, the final stages of which are associated with the death of the cell; the degenerated cell contents are frequently to be found lining the cell cavity. It is during the process of differentiation that the standard daughter cell is transformed into one of four basic cell types (Table 4.1). Chemical dissolution of the lignin-pectin complex cementing together the cells will result in their separation. In the softwood (Fig. 4.7) two types of cell can be observed. Those present in greater number are known as *tracheids*, some 2–4 mm in length with an aspect ratio (L/D) of about 100 : 1. These cells, which lie vertically in the trunk, are responsible for both the supporting and conducting roles. The small block-like cells some 200 × 30 μm in size, known as *parenchyma*, are mostly located in the *rays* and are responsible for the storage of food material.

In contrast, in the hardwoods (Fig. 4.8), four types of cell are present albeit that one, the tracheid, is present in small amounts. The role of storage is again primarily taken by the parenchyma, which can be present horizontally in the form of a ray, or vertically, either scattered or in distinct zones. Support is effected by long thin cells

113

with very tapered ends, known as *fibres*; these are usually about 1–2 mm in length with an aspect ratio of about 100 : 1. Conduction is carried out in cells whose end walls have been dissolved away either completely or in part. These cells, known as *vessels* or *pores*, are usually short (0.2–1.2 mm) and relatively wide (up to 0.5 mm) and when situated above one another form an efficient conducting tube. It can be seen, therefore, that while in the softwoods the three functions are performed by two types of cell, in the hardwoods each function is performed by a single cell type (Table 4.1).

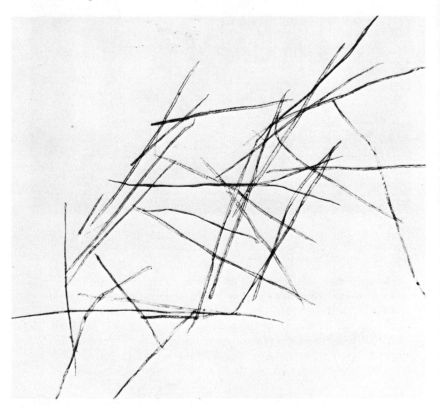

Figure 4.7   Individual softwood cells (x 20). (From the Princes Risborough Laboratory, Building Research Establishment. © Crown copyright.)

Although all cell types develop a secondary wall this varies in thickness, being related to the function that the cell will perform. Thus the wall thickness of fibres is several times that of the vessel (Table 4.1). Consequently, the density of the wood, and hence many of the strength properties as will be discussed later, will be related to the relative proportions of the various types of cell. Density, of course, will also be related to the absolute wall thickness of any one type of cell, for it is possible to obtain fibres of one species of wood several times thicker than those of another. The range in

114

density of timber is from 120 to 1 200 kg/m³ corresponding to pore volumes of from 92% to 18%.

Growth may be continuous throughout the year in certain parts of the world and the wood formed tends to be uniform in structure. In the temperate and subarctic regions and in parts of the tropics growth is seasonal, resulting in the formation of *growth rings*; in this country where there is a single growth period each year these rings are referred to as *annual rings* (Fig. 4.1).

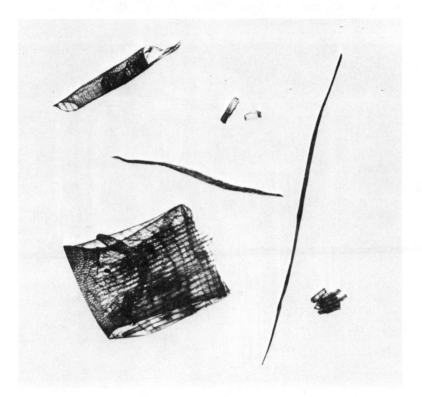

Figure 4.8 Individual cells from a ring-porous hardwood (x 50). (From the Princes Risborough Laboratory, Building Research Establishment. © Crown copyright.)

When seasonal growth commences, the dominant function appears to be conduction, while in the latter part of the year the dominant factor is support. This change in emphasis manifests itself in the softwoods with the presence of thin-walled tracheids (about 2 μm) in the early part of the season (the wood being known as *earlywood*) and thick walled (up to 10 μm) and slightly longer (10%) in the latter part of the season (the *latewood*) (Fig. 4.1).

In some of the hardwoods, but certainly not all of them, the earlywood is characterised by the presence of large-diameter vessels surrounded primarily by parenchyma and tracheids; only a few fibres are present. In the latewood, the vessel diameter is considerably

115

smaller (about $\frac{1}{5}$) and the bulk of the tissue comprises fibres. It is not surprising to find, therefore, that the technical properties of the earlywood and latewood are quite different from one another. Timbers with this characteristic two-phase system are referred to as having a *ring-porous* structure (Fig. 4.6).

The majority of hardwoods, whether of temperate or tropical origin, show little differentiation between earlywood and latewood. Uniformity across the growth ring occurs not only in cell size, but also in the distribution of the different types of cells (Fig. 4.9): these timbers are said to be *diffuse-porous*.

Figure 4.9   Cellular arrangement in a diffuse-porous hardwood (*Fagus sylvatica* – beech). (From the Princes Risborough Laboratory, Building Research Establishment. © Crown copyright.)

In addition to determining many of the technical properties of wood, the distribution of cell types and their sizes is used as a means of timber identification.

Interconnection by means of pits occurs between cells to permit the passage of mineral solutions and food in both longitudinal and horizontal planes. Three basic types of pit occur. *Simple pits*, generally small in diameter and taking the form of straight-sided holes with a transverse membrane, occur between parenchyma and parenchyma, and also between fibre and fibre. Between tracheids a complex structure known as the *bordered pit* occurs (Fig. 4.10; see also Fig. 9.3(a) for sectional view). The entrance to the pit is domed and the internal chamber is characterised by the presence of a dia-

116

phragm (the *torus*) which is suspended by thin strands (the *margo*). Differential pressure between adjacent tracheids will cause the torus to move against the pit aperture, effectively stopping flow. As will be discussed later, these pits have a profound influence on the degree of artificial preservation of the timber. Similar structures are to be found interconnecting vessels in a horizontal plane. Between parenchyma cells and tracheids or vessels there occur *semi-bordered* pits, often referred to as ray pits. These are characterised by the presence of a dome on the tracheid or vessel wall and the absence of such on

Figure 4.10 Electron micrograph of the softwood bordered pit showing the margo strands supporting the diaphragm (torus), which overlaps the aperture (x 3600). (From the Princes Risborough Laboratory, Building Research Establishment. © Crown copyright.)

the parenchyma wall: a pit membrane is present, but the torus is absent. Differences in the shape and size of these pits is an important diagnostic feature in the softwoods.

4.2.3   Molecular Structure and Ultrastructure

4.2.3.1   *Chemical Constituents*

One of the principal tools used to determine the chemical structure of materials is X-ray diffraction analysis. In this technique a beam of X-rays is pinpointed onto the substance and the diffracted beam is recorded on a photosensitive emulsion. The nature, position and

117

intensity of the image of this beam are indications of the degree and type of crystallinity present in the material. An analysis of this type on a small block of timber produces an image similar to that illustrated in Fig. 4.11. The diffuse zone in the centre of the image is

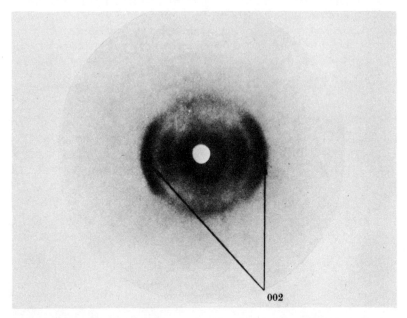

Figure 4.11 X-ray diagram of wood showing diffuse central halo due to the amorphous regions, and the lateral arcs resulting from the crystalline regions. A densitometric trace across the 002 ring provides information from which microfibrillar angle can be determined. (From D. R. Cowdrey and R. D. Preston (1966) *Proceedings of the Royal Society*, B 166, 245–272, by permission of the Royal Society.)

characteristic of an amorphous or noncrystalline material, while the band of higher intensity indicates the presence of some crystalline material. The fact that the band is segmented rather than continuous shows that this crystalline material is orientated in a particular plane, the angle of orientation being determined by measurement of the intensity distribution on either the paratropic planes (002), $(10\bar{1})$ and (101) or the diatropic plane (040); these techniques are reviewed by El-osta *et al.* (1973).

Chemical analysis reveals the existence of four constituents and provides data on their relative proportions. The information revealed by X-ray diffraction and chemical analyses may be summarised as in Table 4.2: proportions are for timber in general and slight variations in these can occur between timber of different species.

Cellulose. Cellulose $(C_6H_{10}O_5)_n$ occurs in the form of long slender filaments or chains, these having been built up within the cell wall from the glucose monomer $(C_6H_{12}O_6)$. Whilst the number of units

118

per cellulose molecule (the degree of polymerisation) can vary considerably even within one cell wall it is thought that a value of 8 000–10 000 is a realistic average (Goring and Timell, 1962). The anhydroglucose unit $C_6H_{10}O_5$, which is not quite flat, is in the form

Table 4.2   Chemical Composition of Timber

	Percent weight	Polymeric state	Molecular derivatives	Function
Cellulose	40–50	Crystalline highly oriented large molecule	Glucose	'fibre'
Hemicelluloses	20–25	Semi-crystalline smaller molecule	Galactose Mannose Xylose	'matrix'
Lignin	25–30	Amorphous large 3-D molecule	Phenyl propane	
Extractives	0–10	Some polymeric; others nonpolymeric	e.g. Terpenes Polyphenols	extraneous

of a six-sided ring consisting of five carbon atoms and one oxygen atom (Fig. 4.12); the side groups play an important part in intermolecular bonding as will be noted later. Successive glucose units are covalently linked in the 1,4 positions giving rise to a potentially straight and extended chain; i.e. moving in a clockwise direction around the ring it is the first and fourth carbon atoms after the oxygen atom that combine with adjacent glucose units to form the long-chain molecule. Glucose, however, can be present in one of two forms dependent on the position of the —OH group attached to carbon 1. When this lies above the ring the unit is called α-glucose and when this combines with an adjacent unit with the removal of H—O—H (known as a condensation reaction) the resulting molecule is called starch, a product which is manufactured in the crown and stored in the parenchyma cells.

When the —OH group lies below the ring, the unit is known as β-glucose and on combining with adjacent units again by a condensation reaction a molecule of cellulose is produced: it is this product which is the principal wall-building constituent. Usually the linkage of β-units is stronger than that of α-units but, in order to achieve the former, alternate anhydroglucose units must be rotated through 180°.

Cellulose chains may crystallise in many ways, but one form, namely cellulose I, is characteristic of natural cellulosic materials. In studies on cellulosic materials including wood, X-ray diffraction analyses have been interpreted as indicating that the cellulose crystal is characterised by a repeat distance of 1.03 nm, equivalent to two anhydroglucose units, in the chain direction (b-axis), with the other edges of the unit cell having lengths of 0.835 nm in the [100] crystallographic direction (a-axis) and 0.790 nm in the [001] direction (c-axis): the a and c axes are inclined at 84° to each other and

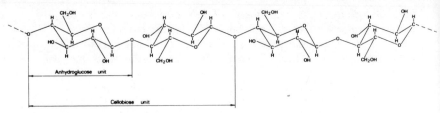

Figure 4.12 Structural formula for the cellulose molecule. (From the Princes Risborough Laboratory, Building Research Establishment. © Crown copyright.)

both are perpendicular to the b-axis (i.e. the crystal is monoclinic). These spacings and angle have been fitted more or less satisfactorily into a number of unit cells over the years; the one which has found

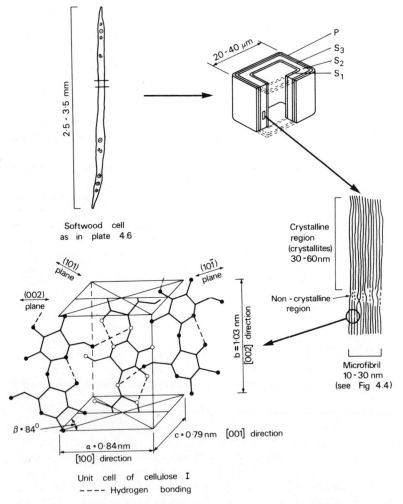

Figure 4.13 Relationship between the structure of timber at different levels of magnitude. (Adapted from J. F. Siau (1971) *Flow in Wood,* reproduced by permission of Syracuse University Press.)

120

most acceptance is that proposed by Meyer and Misch (1937) incorporating four glucose units, but later modified to depict the chains in a slightly bent configuration (Hermans, 1943) (Fig. 4.13). Although there is no direct evidence from studies on timber to indicate that adjacent chains lie in opposite directions, there is considerable indirect evidence to suggest this arrangement. The presence of an antiparallel arrangement has been widely adopted in the past, though during the last decade some workers have come to dispute this hypothesis and to indicate that the chains could all lie in the same direction.

Recent investigations on the structure of cellulose have been carried out on the alga *Valonia* and these have indicated that in this simple plant there is a very high probability that the cellulose chains are all facing the same way (Gardner and Blackwell, 1974; Sarko and Muggli, 1974). Whereas the former workers interpret the unit cell still as monoclinic, though with dimensions twice that of the Meyer and Misch cell, thereby incorporating eight chains, the latter workers propose a triclinic cell with new dimensions, but comprising only two chains.

So far these results have not been confirmed for the structure of crystalline cellulose in timber, but it is difficult to see why the structure of cellulose in timber should vary from that in *Valonia*. The adoption of the crystalline structure of *Valonia* for cellulose in timber will necessitate a revision of the established hydrogen bonding pattern and the recalculation of the crystal unit cell elastic constants. It is to be hoped that the passage of time will clarify the situation.

Whether the chains are parallel or antiparallel, steric considerations require that the chain at the centre of the unit cell is displaced longitudinally by 0.29 nm, a distance equivalent to $b/4$; this facilitates interchain bonding. An antiparallel displacement does not infer that the cellulose molecule is folded, as is characteristic of most of the man-made crystalline polymers; although a folded-chain model was proposed of cellulose some time ago the results of more recent work have refuted this hypothesis.

Cellulose which has regenerated from a solution displays a different crystalline structure and is known as cellulose II: in this case there is complete agreement that the unit cell possesses an antiparallel arrangement of the cellulose molecule. Within the structure of cellulose I both primary and secondary bonding is represented and many of the technical properties of wood can be related to the variety of bonding present. Covalent bonding both within the glucose rings and linking together the rings to form the molecular chain contributes to the high axial tensile strength of timber. There is no evidence of primary bonding laterally between the chains: rather this seems to be a complex mixture of the fairly strong hydrogen bonds and the weak van der Waals forces. Whereas Meyer and Misch (1937) had placed the hydrogen bonds within the (002) plane, it was later thought that these bonds united the cellulose chains of the (101) and (10$\bar{1}$) planes (Jaswon *et al.*, 1968) (Fig. 4.13), thereby providing the main mechanism for stabilising the crystal against

121

relative displacement of the chain and consequently contributing considerably to the axial stiffness of wood. The same OH groups that give rise to this hydrogen bonding are highly attractive to water molecules and explain the affinity of cellulose for water. Gardner and Blackwell (1974) on cellulose from *Valonia* identify the existence of both intermolecular and intramolecular hydrogen bonds all of which, however, are interpreted as lying on the (020) plane; they consider the structure of cellulose as an array of hydrogen-bonded sheets.

The degree of crystallinity is usually assessed by X-ray and electron diffraction techniques and has been shown to be at least 67%, though some workers have assessed wood as up to 90% crystalline. This range is due in a large extent to the fact that wood is comprised not just of the crystalline and noncrystalline constituents but rather a series of substances of varying crystallinity. Regions of complete crystallinity and regions with a total absence of crystalline structure (amorphous zones) can be recognised, but the transition from one state to the other is gradual.

The length of the cellulose molecule is about 5 000 nm (0.005 mm) whereas the average size of each crystalline region determined by X-ray analysis is only 60 nm in length, 5 nm in width and 3 nm in thickness. This means that any cellulose molecule will pass through several regions of high crystallinity — known as crystallites or micelles — with intermediate noncrystalline or low-crystalline zones in which the cellulose chains are in only loose association with each other (Fig. 4.13). Thus the majority of chains emerging from one crystallite will pass to the next creating a high degree of longitudinal coordination (Fig. 4.13); this collective unit is termed a *microfibril* and has infinite length; it is clothed with chains of cellulose mixed with chains of sugar units other than glucose (see below) which lie parallel, but are not regularly spaced. This brings the microfibril in timber to about 10 nm in breadth, and in some algae such as *Valonia* to 30 nm. The degree of crystallinity will therefore vary along its length and it has been proposed that this could be periodic.

Hemicelluloses and lignin. In Table 4.2 reference was made to the other constituents of wood additional to cellulose. Two of these, the hemicelluloses and lignin, are regarded as cementing material contributing to the structural integrity of wood and also to its high stiffness. The hemicelluloses, like cellulose itself, are carbohydrates and differ in composition depending on whether the wood is from a conifer or a broad-leaved tree. Both the degree of crystallisation and the degree of polymerisation are low, the molecule containing less than 150 units; in these respects and also in their lack of resistance to alkali solutions the hemicelluloses are quite different from true cellulose.

Lignin, present in about equal proportions to the hemicelluloses, is chemically dissimilar to these and to cellulose. Lignin is a complex aromatic compound composed of phenyl groups, but the detailed structure has still not been established. It is noncrystalline and the structure varies between wood from a conifer and from a broad-leaved tree. About 25% of the total lignin in timber is to be found in the middle lamella, an intercellular layer composed of lignin and pectin. Since the middle lamella is very thin, the concentration of lignin is correspondingly high.

The bulk of the lignin (about 75%) is present within the cell wall, having been deposited following completion of the cellulosic frame-work; the termination of the lignification process towards the end of the period of differentiation coincides with the death of the cell. Most cellulosic plants do not contain lignin and it is the inclusion of this substance within the framework of wood that is largely respon-sible for the stiffness of wood, especially in the dried condition.

4.2.3.2 *The Cell Wall as a Fibre Composite*

In the introductory remarks, wood was defined as a natural com-posite and the most successful model used to interpret the ultra-structure of wood from the various chemical and X-ray analyses ascribes the role of 'fibre' to the cellulosic microfibrils while the lignin and hemicelluloses are considered as separate components of the 'matrix'. The cellulosic microfibril is interpreted therefore as conferring high tensile strength to the composite owing to the pres-ence of covalent bonding both within and between the anhydro-glucose units. Experimentally it has been shown that reduction in chain length following gamma irradiation markedly reduces the tensile strength of timber (Ifju, 1964); the significance of chain length in determining strength has been confirmed in studies of wood with inherently low degrees of polymerisation. While Ifju considered slippage between the cellulose chains to be an important contributor to the development of ultimate tensile strength, this is thought to be unlikely due to the forces involved in fracturing large numbers of hydrogen bonds.

Preston (1964) has shown that the hemicelluloses are usually intimately associated with the cellulose, effectively binding the microfibrils together. Bundles of cellulose chains are therefore seen as having a polycrystalline sheath of hemicellulose material and consequently the resulting high degree of hydrogen bonding would make chain slippage unlikely: rather it would appear that stressing results in fracture of the C—O—C linkage.

The deposition of lignin is variable in different parts of the cell wall, but it is obvious that its prime function is to protect the hydro-philic (water-seeking) cellulose and hemicelluloses which are mechanically weak when wet. Experimentally, it has been demon-strated that removal of the lignin markedly reduces the strength of wood in the wet state, though its reduction results in an increase in its strength in the dry state calculated on a net cell wall area basis. Consequently, the lignin is regarded as lying to the outside of the fibril forming a protective sheath.

Since the lignin is located only on the exterior it must be responsible for cementing together the fibrils and in imparting shear resistance in the transference of stress throughout the composite. The role of lignin in contributing towards stiffness of timber has already been mentioned.

Two schools of thought exist on the possible location within the microfibril of these two components of the matrix. These are illustrated in Fig. 4.14; the more widely held view is that depicted on the

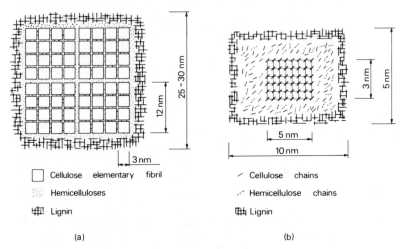

(a)

(b)

Figure 4.14 Models of the cross-section of a microfibril. In (a) the crystalline core has been subdivided into elementary fibrils, while in (b) the core is regarded as being homogeneous. ((a) adapted from D. Fengel (1970) *The Physics and Chemistry of Wood Pulp Fibres*, reproduced by permission of the Technical Association of the Pulp and Paper Industry; (b) adapted from R. D. Preston (1974) *The Physical Biology of Plant Cell Walls*, reproduced by permission of Chapman and Hall.)

left where cellulosic subunits some 3 nm in diameter are thought to exist. These units, comprising some 40 cellulose chains, are known as elementary fibrils or protofibrils. Gaps (1 nm) between these units are filled with hemicellulose while more hemicellulose and lignin form the sheath. This subdivision of the microfibril is in dispute (Fig. 4.14(b)) and it has been suggested that the evidence to support such a subdivision has been produced by artefacts in sample preparation for electron microscopy. In this second model the crystalline core is considered to be about 5 nm x 3 nm containing about 48 chains in either 4- or 8-chain unit cells; the latter configuration is now receiving wider acceptance. However, both models are in agreement in that passing outwards from the core of the microfibril the highly crystalline cellulose gives way first to the partly crystalline layer containing mainly hemicellulose but also some cellulose, and then to the amorphous lignin: this gradual transition of crystallinity from fibre to matrix results in high interlaminar shear strength which contributes considerably to the high tensile strength and toughness of wood.

124

When a cambial cell divides to form two daughter cells a new wall is formed comprising the middle lamella and two primary cell walls, one to each daughter cell. These new cells undergo changes within about three days of their formation and one of these developments will be the formation of a secondary wall. The thickness of this wall will depend on the function the cell will perform, as described earlier, but its basic construction will be similar in all cells.

Early studies on the anatomy of the cell wall used polarisation microscopy, which revealed the direction of orientation of the crystalline regions (Preston, 1934). These studies indicated that the secondary wall could be subdivided into three layers and measurements of the extinction position was indicative of the angle at which the microfibrils were orientated. Subsequent studies with transmission electron microscopy confirmed these findings and provided some additional information with particular reference to wall texture and variability of angle. The relative thickness and mean microfibrillar angle of these layers in a sample of spruce timber are illustrated in Table 4.3.

Table 4.3 Microfibrillar Orientation and Percentage Thickness of the Cell Wall Layers in Spruce Timber (*Picea abies*)

Wall layer	% thickness	Angle to longitudinal axis
P	5	Random
S_1	9	$50°-70°$
S_2	85	$10°-30°$
S_3	1	$60°-90°$

The middle lamella, a lignin-pectin complex, is devoid of cellulosic microfibrils while in the primary wall (P) the microfibrils are loosely packed and interweave at random (Fig. 4.15); no lamellation is present. In the secondary wall layers the microfibrils are closely packed and parallel to each other. The outer layer of this wall, the S_1, is again thin and is characterised by having from four to six lamellae, the microfibrils of each alternating between a left- and right-hand spiral both with a pitch to the longitudinal axis of from $50°$ to $70°$ depending on the species of timber.

The middle layer of the secondary wall (S_2) is thick and is composed of $30-150$ lamellae, the microfibrils of which all exhibit a similar orientation in a right-hand spiral with a pitch of $10°-30°$ to the longitudinal axis as illustrated in Fig. 4.16. Since over three quarters of the cell wall is composed of the S_2 layer it follows that the ultrastructure of this layer will have a very marked influence on the behaviour of the timber. In later sections, anisotropic behaviour, shrinkage, tensile strength and failure morphology will all be related to the microfibrillar angle in the S_2 layer.

The S_3 layer, which may be absent in certain timbers, is very thin with only a few lamellae; it is characterised, as in the S_1 layer, by

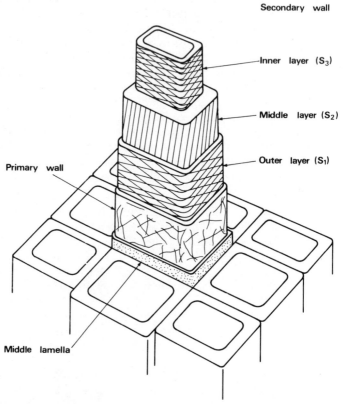

Figure 4.15 Simplified structure of the cell wall showing orientation of micro-fibrils in each of the major wall layers. (From the Princes Risborough Laboratory, Building Research Establishment. © Crown copyright.)

alternate lamellae possessing microfibrils orientated in opposite spirals with a pitch of 60–90°. Generally this wall layer has a looser texture than the S_1 and S_2 layers and is frequently encrusted with extraneous material. The layer is also characterised by a relatively high proportion of lignin.

Further investigations have indicated that the values of micro-fibrillar angle quoted above are only average for the layers and that systematic variation in angle occurs within each layer. The inner lamellae of the S_1 tend to have a smaller angle, and the outer lamellae a larger angle than the average for each layer: a similar but opposite situation occurs in the S_3 layers. Electron microscopy has also revealed the presence of a thin warty layer overlaying the S_3 layer in certain timbers.

126

Microfibrillar angle appears to vary systematically along the length of the cell as well as across the wall thickness. Thus the angle of the S_2 layer has been shown to decrease towards the ends of the cells, while the average S_2 angle appears to be related to the length of the cell, itself a function of rate of growth of the tree. Systematical differences in microfibrillar angle have been found between radial and

Figure 4.16 Electron micrograph of the cell wall in Norway spruce timber (*Picea abies*) showing the parallel and almost vertical microfibrils of an exposed portion of the S_2 layer. (From the Princes Risborough Laboratory, Building Research Establishment. © Crown copyright.)

tangential walls and this has been related to differences in degree of lignification between these walls. Openings occur in the walls of cells and many of these pit openings are characterised by localised deformations of the microfibrillar structure.

127

Before leaving the chemical composition of wood, mention must be made of the presence of extractives (Table 4.2). This is a collective name for a series of highly complex organic compounds which are present in certain timbers in relatively small amounts. Some, like waxes, fats and sugars, have little economic significance, but others, for example rubber and resin, from which turpentine is distilled, are of considerable importance. The heartwood of timber, as described previously, generally contains extractives which, in addition to imparting colouration to the wood, bestow on it its natural durability, since most of these compounds are toxic to both fungi and insects.

4.2.4 Variability in Structure

Variability in performance of wood is one of its characteristic deficiencies as a material. It will be discussed later how differences in mechanical properties occur between timbers of different species and how these are manifestations of differences in wall thickness and distribution of cell types. However, superimposed on this genetical source of variation is both a systematic and an environmental one.

There are distinct patterns of variation in many features within a single tree. Length of the cells, thickness of the cell wall, angle at which the cells are lying with respect to the vertical axis (spiral grain), angle at which the microfibrils of the S_2 layer of the cell wall are located with respect to the vertical axis, all show systematic trends outwards from the centre of the tree to the bark and upwards from the base to the top of the tree. This pattern results in the formation of a core of wood in the tree with many undesirable properties including low strength and high shrinkage. This zone, usually some ten growth rings in width, is known as the *core* wood or *juvenile* wood as opposed to the *mature* wood occurring outside this area.

Environmental factors have considerable influence on the structure of the wood and any environmental influence, including silviculture, which changes the tree's rate of growth will affect the technical properties of the wood. However, the relationship is a complex one; in softwoods, increasing growth rate generally results in a decrease in density and mechanical properties. In diffuse-porous hardwoods increasing growth rate, provided it is not excessive, has little effect on density, while in ring-porous hardwoods, increasing rate of growth, again provided it is not excessive, results in an increase in density and strength.

There is a whole series of factors which may cause defects in the structure of wood and consequent lowering of its strength. Perhaps the most important defect with regard to its utilisation is the formation of *reaction wood*. When trees are inclined to the vertical axis, usually as a result of wind action or growing on sloping ground the distribution of growth-promoting hormones is disturbed, resulting in the formation of an abnormal type of tissue. In the softwoods, this reaction tissue grows on the compression side of the trunk and is characterised by having a higher than normal lignin content, a higher

128

microfibrillar angle in the S_2 layer resulting in increased longitudinal shrinkage, and a generally darker appearance (Fig. 4.17): this abnormal timber, known as *compression wood*, is also considerably more brittle than normal wood. In the hardwoods, reaction wood forms on the tension side of trunks and large branches and is therefore called *tension wood*. It is characterised by a higher than normal cellulose content which imparts a rubbery characteristic to the fibres resulting in difficulties in sawing and machining (Fig. 4.18).

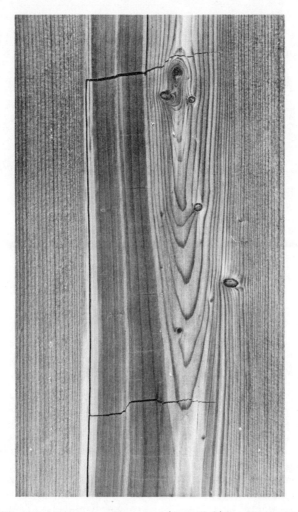

Figure 4.17   A band of compression wood (centre left) in a Norway spruce plank, illustrating the darker appearance and higher longitudinal shrinkage of the reaction wood compared with the adjacent normal wood. (From the Princes Risborough Laboratory, Building Research Establishment. © Crown copyright.)

One other defect of considerable technical significance is *brittle-heart*, which is found in many low-density tropical hardwoods. Due to the slight shrinkage of cells after their formation, the outside layers of the tree are in a state of longitudinal tension resulting in the cumulative increase of compression stress in the core. A time is reached in the growth of the tree when the compression stresses due to growth are greater than the natural compression strength of the wood. Yield occurs with the formation of shear lines through the cell wall and throughout the wood. Compression failure will be discussed in greater detail in a later section.

Figure 4.18 Board of African mahogany showing rough surface and concentric zones of well developed tension wood (the lighter zones) on the end grain. (From the Princes Risborough Laboratory, Building Research Establishment. © Crown copyright.)

4.3 APPEARANCE OF TIMBER IN RELATION TO ITS STRUCTURE

Most readers will agree that many timbers are aesthetically pleasing and the various and continuing attempts to simulate the appearance of timber in the surface of synthetic materials bear testament to the very attractive appearance of most timbers. Although a very large proportion of the timber consumed in the UK is used within the construction industry, where the natural appearance of timber is of little consequence, excepting the use of hardwoods for flush doors, internal panelling and wood-block floors, a considerable quantity of timber is still utilised purely on account of its attractive appearance particularly for furniture and various sports goods. The decorative appearance of many timbers is due to the *texture* or to the *figure* or to the *colour* of the material and, in many instances, to combinations of these.

130

4.3.1 Texture

The texture of timber depends on the size of the cells and on their arrangement. A timber such as boxwood in which the cells have a very small diameter is said to be *fine-textured*, while a *coarse-textured* timber such as keruing has a considerable percentage of large-diameter cells. Where the distribution of the cell-types or sizes across the growth ring is uniform, as in beech, or where the thickness of the cell wall remains fairly constant across the ring, as in some of the softwoods, e.g. yellow pine, the timber is described as being *even-textured*: conversely, where variation occurs across the growth ring, either in distribution of cells as in teak or in thickness of the cell walls as in larch or Douglas fir, the timber is said to have an *uneven texture*.

4.3.2 Figure

Figure is defined as the 'ornamental markings seen on the cut surface of timber, formed by the structural features of the wood' (BS 565. 1972 — Glossary of Terms Relating to Timber and Woodwork) but the term is also frequently applied to the effect of marked variations in colour. The four most important structural features inducing figure are grain, growth rings, rays and knots.

4.3.2.1 *Grain*

Grain refers to the general arrangement of the vertically aligned cells. It is convenient when examining timber at a general level to regard these cells as lying truly vertical; however, in practice these cells may deviate from the vertical axis in a number of different patterns.

Where the direction of the deviation is consistent, the direction of the cells assumes a distinct spiral mode which may be either left- or right-handed. In young trees the helix angle is frequently of the order of 4°, though considerable variability occurs both within a species and also between different species of timber. As the trees grow, so the helix angle in the outer rings decreases to zero and quite frequently in very large trees the angle in the outer rings subsequently increases but the spiral has changed hand. Although spiral grain does not produce any figure effect, it has very significant technical implications: strength is lowered, while the degree of twisting on drying and amount of pick-up on machining increase as the degree of spirality of the grain increases.

In certain hardwood timbers, and the mahoganies are perhaps the best example, the direction of the helix alternates from left to right hand at very frequent intervals along the radial direction; grain of this type is said to be *interlocked*. Tangential faces of machined timber will be normal, but the radial face will be characterised by the presence of alternating light and dark longitudinal bands produced by the reflection of light from the tapered cuts of fibres inclined in different directions (Fig. 4.19). This type of figure is referred to as *ribbon* or *stripe* and is desirous in timber for furniture manufacture.

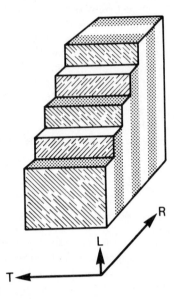

Figure 4.19 Diagramatic illustration of the development of interlocked grain in certain hardwoods. The fibres in successive radial zones are inclined in opposite directions thereby imparting a striped appearance on the longitudinal-radial plane. (Adapted from F. W. Jane (1962) *The Structure of Wood*, by permission of A & C Black Ltd.)

If instead of alternating in a longitudinal radial plane the helix alternates in a longitudinal tangential plane, a *wavy* type of grain is produced. This is very conspicuous in machined tangential faces where it shows up clearly as alternating light and dark horizontal bands (Fig. 4.20); this type of figure is described as *fiddleback*, since timber with this distinctive type of figure has been used traditionally for the manufacture of the backs of violins: it is to be found also on the panels and sides of wardrobes and bookcases.

4.3.2.2 *Growth Rings*

Where variability occurs across the growth ring, either in the distribution of the various cell types or in the thickness of the cell walls, distinct patterns will appear on the machined faces of the timber. Such patterns, however, will not be regular like many of the man-made imitations, but will vary according to changes in width of the growth ring and in the relative proportions of early and latewood.

On the radial face the growth rings will be vertical and parallel to one another, but on the tangential face a most pleasing series of concentric arcs is produced as successive growth layers are intersected. In the centre part of the plank of elm timber illustrated in

Fig. 4.21 the growth rings are cut tangentially forming these attractive arcs, while the edge of the board with parallel and vertical growth rings reflects timber cut radially: in the case of elm it is presence of the large earlywood vessels which makes the growth ring so conspicuous, while in timbers like Douglas fir or pitch pine the striking effect of the growth ring can be ascribed to the very thick walls of the latewood cells.

Fig. 4.20   Fig. 4.21

Figure 4.20 'Fiddleback' figure due to wavy grain. (From the Princes Risborough Laboratory, Building Research Establishment. © Crown copyright.)
Figure 4.21 The effect of growth rings on figure in elm (*Ulmus* sp.). (From the Princes Risborough Laboratory, Building Research Establishment. © Crown copyright.)

4.3.2.3 *Rays*

Another structural feature which may add to the attractive appearance of timber is the ray, especially where, as in the case of oak, the rays are both deep and wide. When the surface of the plank coincides with the longitudinal radial plane these rays can be seen as sinuous light-coloured ribbons running across the grain.

4.3.2.4 *Knots*

Knots, though troublesome from the mechanical aspects of timber utilisation, can be regarded as a decorative feature; the fashion of knotty-pine furniture and wall panelling in the early seventies is a very good example of the decorative feature of knots. However, as a decorative feature knots do not possess the subtlety of variation in grain and colour that arises from the other structural features described above.

Exceptionally, trees produce a cluster of small shoots at some point on the trunk and the timber subsequently formed in this region contains a multitude of small knots. Timber from these *burrs* is highly prized for decorative work, especially if walnut or yew.

4.3.3 Colour

In the absence of extractives, timber tends to be a rather pale straw colour which is characteristic of the sapwood of almost all timbers. The onset of heartwood formation in many timbers is associated with the deposition of extractives, most of which are coloured, thereby imparting colouration to the heartwood zone. In passing, it should be recalled that although a physiological heartwood is always formed in older trees extractives are not always produced; thus the heartwood of timbers such as ash and spruce is colourless.

Where colouration of the heartwood occurs, a whole spectrum of colour exists among the different species. The heartwood may be yellow, e.g. boxwood; orange, e.g. opepe; red, e.g. mahogany; purple, e.g. purpleheart; brown, e.g. African walnut; green, e.g. greenheart; or black, e.g. ebony. In some timbers the colour is fairly evenly distributed throughout the heartwood, while in other species considerable variation in the intensity of the colour occurs. In zebrano distinct dark brown and white stripes occur, while in olive wood patches of yellow merge into zones of brown. Dark gum-veins, as present in African walnut, contribute to the pleasing alternations in colour. Variations in colour such as these are regarded as contributing to the 'figure' of the timber.

It is interesting to note in passing that the noncoloured sapwood is frequently coloured artificially to match the heartwood, thereby adding to the amount of timber converted from the log. In a few rare cases, the presence of certain fungi in timber in the growing tree can result in the formation of very dark coloured heartwood: the activity of the fungus is terminated when the timber is dried. Both *brown oak* and *green oak*, produced by different fungi, have always been prized for decorative work.

4.4 MASS-VOLUME RELATIONSHIPS

4.4.1 Density of Timber

The density of a piece of timber is determined not only by the amount of wood substance present, but also by the presence of both extractives and moisture. In a few timbers extractives are completely absent, while in many they are present, but only in small amounts and usually less than 3% of the dry weight of the timber. In some exceptional cases the extractive content may be as high as 10% and in these cases it is necessary to remove the extractives prior to the determination of density.

The presence of moisture in timber not only increases the weight of the timber, but also results in swelling of the timber, and hence both weight and volume are affected. Thus in the determination of density where

$$\rho = \frac{m}{v} \qquad (4.1)$$

both the mass (m) and volume (v) must be determined at the same moisture content. Generally, these two parameters are determined at zero moisture content but, as density is frequently quoted at moisture contents of 12% since this level is frequently experienced in timber in use, the value of density at zero moisture content is corrected for 12% if volumetric expansion figures are known, or else the density determination is carried out on timber at 12% moisture content.

Thus if

$$m_x = m_0(1 + 0.01\,\mu) \qquad (4.2)$$

where m_x is the mass of timber at moisture content x, m_0 is the mass of timber at zero moisture content, and μ is the moisture content %; and

$$v_x = v_0(1 + 0.01\,s_v) \qquad (4.3)$$

where v_x is the volume of timber at moisture content x, v_0 is the volume of timber at zero moisture content, and s_v is the volumetric shrinkage/expansion %, it is possible to obtain the density of timber at any moisture content in terms of the density at zero moisture content thus:

$$\rho_x = \frac{m_x}{v_x} = \frac{m_0(1 + 0.01\,\mu)}{v_0(1 + 0.01\,s_v)} = \rho_0\left(\frac{1 + 0.01\,\mu}{1 + 0.01\,s_v}\right) \qquad (4.4)$$

As a very approximate rule of thumb the density of timber increases by approximately 0.5% for each 1.0% increase in moisture content up to 30%. Density therefore will increase, slightly and curvilinearly, up to moisture contents of about 30% as both total mass and volume increase; however, at moisture contents above 30%, density will increase rapidly and curvilinearly with increasing moisture content, since, as will be explained in Chapter 7, the volume remains constant above this value, whilst the mass increases.

135

In Section 4.2.2 timber was shown to possess different types of cell which could be characterised by different values of the ratio of cell-wall thickness to total cell diameter. Since this ratio can be regarded as an index of density, it follows that density of the timber will be related to the relative proportions of the various types of cells. Density, however, will also reflect the absolute wall thickness of any one type of cell, since it is possible to obtain fibres of one species of timber the cell wall thickness of which can be several times greater than that of fibres of another species.

Density, like many other properties of timber, is extremely variable: within timber it can vary by a factor of ten ranging from an average value at a moisture content of 12% of 176 kg/m^3 for balsa to about 1 230 kg/m^3 for lignum vitae. Balsa then has a density similar to that of cork, while lignum vitae has a density slightly less than half that of concrete or aluminium. The values of density quoted for different timbers, however, are merely average values: each timber will have a range of densities reflecting differences between early and latewood, between the pith and outer rings, and between trees on the same site. Thus, the density of balsa can vary from 40 to 320 kg/m^3.

In certain publications, reference is made to the *weight* of timber, a term widely used in commerce; it should be appreciated that the quoted values are really densities.

4.4.2 Specific Gravity

The traditional definition of specific gravity (G) can be expressed as:

$$G = \frac{\rho_t}{\rho_w} \tag{4.5}$$

where ρ_t is the density of timber, and ρ_w is the density of water at $4\,°C = 1.0000$ g/cc. G will therefore vary with moisture content and consequently the specific gravity of timber is usually based on the oven-dry weight and volume at some specified moisture content. This is frequently taken as zero though, for convenience, green conditions are sometimes used when the term *basic specific gravity* is applied. Hence:

$$G_\mu = \frac{\text{oven-dry weight of timber}}{\text{weight of displaced volume of water}} \tag{4.6}$$

$$= \frac{m_0}{V_\mu \rho_w}$$

where m_0 is the oven-dry weight of timber, V_μ is the volume of timber at moisture content μ; ρ_w is the density of water, and G_μ is the specific gravity at moisture content μ.

At low moisture contents specific gravity decreases slightly with increasing moisture content up to 30%, thereafter remaining constant. In research activities specific gravity is defined usually in terms of oven-dry weight and volume. However, for engineering applications specific gravity is frequently presented as the ratio of oven-dry weight to volume of timber at 12% moisture content; this can be derived from the oven-dry specific gravity, thus:

$$G_{12} = \frac{G_0}{1 + 0.01\mu G_0/G_{s12}} \tag{4.7}$$

where G_{12} is the specific gravity of timber at 12% moisture content, G_0 is the specific gravity of timber at zero moisture content, μ is the moisture content %, and G_{s12} is the specific gravity of bound water at 12% moisture content. The relationship between density and specific gravity can be expressed as:

$$\rho = G(1 + 0.01\,\mu)\rho_w \tag{4.8}$$

where ρ is the density at moisture content μ, G is the specific gravity at moisture content μ, and ρ_w is the density of water. Equation 4.8 is valid for all moisture contents. When $M = 0$ the equation reduces to

$$\rho = G_0$$

i.e. density and specific gravity are numerically equal.

4.4.3 Density of the Dry Cell Wall

Although the density of timber may vary considerably, as already discussed, the density of the actual cell wall material remains constant for all timbers with a value of approximately $1\,500$ kg/m³ (1.5 g/cc) when measured by volume-displacement methods.

The exact value for cell-wall density depends on the liquid used for measuring the volume: when a polar or swelling liquid such as water is used the apparent specific volume of the dry cell wall is lower than when a nonpolar (nonswelling) liquid such as toluene is used. Densities of 1.525 and 1.451 g/cc have been recorded for the same material using water and toluene respectively. This disparity in density of the dry cell wall has been explained in terms of, firstly, the greater penetration of a polar liquid into microvoid spaces in the cell wall which are inaccessible to nonpolar liquids; and, secondly, the compaction or apparent reduced volume of the sorbed water compared with the free-liquid water (Stamm, 1964). The voids are considered to occupy about 5% of the volume of the cell wall and consequently it has been shown that about 85% of the difference in specific volume of the dry cell wall (and hence density) by polar and nonpolar liquid displacement is caused by the lower accessibility of the latter to the microvoids in the cell wall and only 15% of the difference is credited to the apparent compression of the sorbed water (Weatherwax and Tarkow, 1968).

Cell wall density can be measured by optical techniques as well as by volume displacement. Generally, density is calculated from cell wall measurements made on microtomed cross-sections, but the values obtained are usually lower than those by the volume-displacement method. The lack of agreement is usually explained in

137

terms of either very fine damage to the cell wall during section cutting, or to inaccuracies in measuring the true cell wall thickness due to 'shadow effects' resulting from the use of a point source of light (Petty, 1971).

4.4.4 Porosity

In Section 4.2.2 the cellular nature of timber was described in terms of a parallel arrangement of hollow tubes. The *porosity* (p) of timber is defined as the fractional void volume and is expressed mathematically as

$$p = 1 - V_f \qquad (4.9)$$

where V_f is the volume fraction of cell wall substance. Provided both the density of the cell wall substance and the moisture content of the timber are known, the volume fraction of the cell wall substance can be determined as follows:

$$V_f = G\left(\frac{1}{\rho_c} + \frac{0.01}{G_s}\mu\right) \qquad (4.10)$$

where G is the specific gravity at moisture content μ, G_s is the specific gravity of bound water at moisture content μ, μ is the moisture content less than 25%, and ρ_c is the density of cell wall material (1.46 for helium displacement). Knowing V_f, the porosity (p) can be calculated from Equation 4.9.

A good approximation of p can be obtained if ρ_c is taken as 1.5 and G_s as 1.0: then

$$p = 1 - G(0.667 + 0.01\,\mu) \qquad (4.11)$$

As an example of the use of Equation 4.11 let us calculate the porosity of both balsa and lignum vitae, the densities of which were quoted in Section 4.4.1.

Let $\mu = 0$, then $G_0 = \rho$ numerically. Volumetric shrinkages from 12% to 0% for balsa and lignum vitae are approximately 1.8% and 10% respectively. From Equation 4.4,

$$\rho_0 = \rho_{12}\frac{(1 + 0.01\,S_v)}{(1 + 0.01\,\mu)}$$

Therefore for (a) balsa

$$\rho_0 = 176\frac{(1 + 0.018)}{(1 + 0.12)}$$

$$= 160 \text{ kg/m}^3$$

(b) lignum vitae

$$\rho_0 = 1230\frac{(1 + 0.10)}{(1 + 0.12)}$$

$$= 1208 \text{ kg/m}^3$$

$\therefore G_0$ balsa $= 0.160$ and lignum vitae $= 1.208$.

138

Therefore from Equation 4.11 porosity (p) of balsa $= 1 - 0.160$ $(0.667) = 0.89$ or 89% at zero moisture content and of lignum vitae $= 1 - 1.208\ (0.667) = 0.19$ or 19% at zero moisture content.

4.5 REFERENCES

Bannan, M. W. (1954) Ring width, tracheid size and ray volume in stem wood of *Thuja occidentalis*, *Canadian Journal of Botany*, 32, 466–479.

Cowdrey, D. R. and Preston, R. D. (1966) Elasticity and microfibrillar angle in the wood of Sitka spruce, *Proceedings of the Royal Society*, B166, 245–272.

El-Osta, M. L., Kellogg, R. M., Foschi, R. O. and Butters, R. G. (1973) A direct X-ray method for measuring microfibril angle, *Wood and Fiber*, 5(2), 118–128.

Gardner, K. H. and Blackwell, J. (1974) The structure of native cellulose, *Biopolymers*, 13, 1975–2001.

Goring, D. A. I. and Timmell, T. E. (1962) Molecular weight of native cellulose, *TAPPI*, 45, 454–459.

Hermans, P. H., De Booys, J. and Maan, C. H. (1943) Form and mobility of cellulose molecules, *Kolloid Zeitschrift*, 102, 169–180.

Ifju, G. (1964) Tensile strength behaviour as a function of cellulose in wood, *Forest Products Journal*, 14, 366–372.

Jaswon, M. A., Gillis, P. P. and Marks, R. E. (1968) The elastic constants of crystalline native cellulose, *Proceedings of the Royal Society*, A306, 389–412.

Meyer, K. H. and Misch, L. (1937) Position des atomes dans le nouveau modele spatial de la cellulose, *Helvetica Chimica Acta*, 20, 232–244.

Petty, J. A. (1971) The determination of fractional void volume in conifer wood by microphotometry, *Holzforschung*, 25(1), 24–29.

Preston, R. D. (1934) The organisation of the cell wall of the conifer tracheid, *Philosphical Transactions*, B224, 131–174.

Preston, R. D. (1964) Structural and mechanical aspects of plant cell walls, in *The Formation of Wood in Forest Trees*, H. M. Zimmermann (Ed.), Academic Press, New York, p. 169.

Sarko, A. and Muggli, R. (1974) Packing analyses of carbohydrates and polysaccharides, III: *Valonia* cellulose and cellulose II, *Macromolecules*, 7, 4, 486–494.

Stamm, A. J. (1964) *Wood and Cellulosic Science*, Ronald Press, New York.

Weatherwax, R. C. and Tarkow, H. (1968) Density of wood substance: Importance of penetration and adsorption compression of the displacement fluid, *Forest Products Journal*, 18(7), 44–46.

CHAPTER 5

Crystal Structure of Metals

It will be recalled from Chapter 2 that the generalised nature of the metallic linkage results in the development of crystalline structures in which each ion is bound to all its immediate neighbours. Metal ions tend, therefore, to pack together as closely as possible, which is convienient for engineers since it limits the common crystal structures of pure metals (in which all the ions are of equal size) to three as compared with the enormous number of structures met with in ionic and covalent solids. These metal structures can be realised by considering the ions as incompressible spheres and finding how to pack them together to a maximum density. Let us consider two ways in which this may be achieved.

If we commence with one sphere, the maximum number of similar spheres which can be packed around it in *one* plane is six (Fig. 5.1).

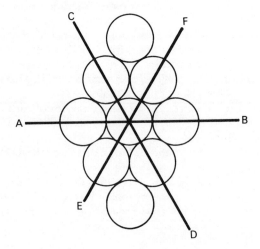

Figure 5.1 Close-packed directions in a close-packed plane of spheres.

By continuing this pattern indefinitely, a close-packed plane can be built up, since the plane always contains the maximum number of spheres for a given area. Furthermore, within this plane we can see that there are three close-packed directions, AB, CD and EF, i.e. directions in which the spheres touch. To extend this structure to three dimensions, it is only necessary to stack these close-packed planes regularly on top of each other and as closely as possible. There are two ways of doing this. The maximum degree of close packing between adjacent planes is achieved if any sphere in a plane touches *three* spheres in the plane above and *three* spheres in the plane below; this is shown in Fig. 2.14, and can easily be verified by experiment. Let us denote the positions of the first layer of spheres as 'A' positions. If a second plane of spheres is placed on top of the first, these spheres will readily take up (say) the close-packed positions labelled B in Fig. 5.2. It is when the third layer is placed on top of the second

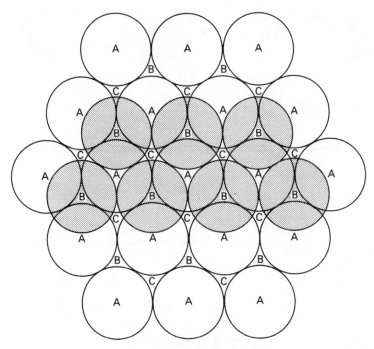

Figure 5.2 The B spheres (shaded) rest on the A spheres. A third close-packed layer may then rest on the B spheres either directly above the A spheres or in C positions.

that a choice presents itself. The third layer may be placed in either the A positions (i.e. vertically above the *first* layer) or in a *third* set of positions, labelled C, which is vertically above neither the first nor the second layer. This uses up the number of possible positions in space which the spheres may take up, and so, to achieve crystalline regularity, we may stack close-packed layers either in the sequence ABABABAB . . . or the sequence ABCABCABC . . . Faults in the

sequence may occur so that we get (say) ABABABCBCBCACA . . .,
but they will be considered in Chapter 10 together with other crystalline imperfections.

The difference between the two possible sequences may appear to
be rather trifling, but in fact it has far-reaching effects on the mechanical properties. Metals deform by a process of slip between the close-
packed planes and between the close-packed rows in any one close-
packed plane, and so as a very general proposition those structures
with the greatest number of close packed planes tend to result in the
most ductile metals. The ABABAB . . . sequence gives rise to the
hexagonal structure (commonly abbreviated to h.c.p. for 'hexagonal
close-packed'); the unit cell is shown (expanded for clarity) in
Fig. 5.3(a), and it will be found that there is only *one* angle in which

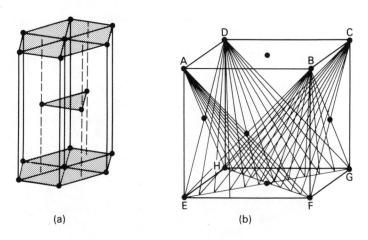

(a)   (b)

Figure 5.3   (a) Expanded unit cell of hexagonal close-packed structure. There
are three parallel slip planes in the cell, shown lightly shaded. (b) Expanded unit
cell of face-centred cubic structure. The cell has four different (i.e. nonparallel)
slip planes, shown shaded. These are the planes AHF, CHF, BEG and DEG, sometimes known as the octahedral planes.

this structure can be sectioned to show a close-packed plane, In the
figure, this plane is horizontal, whilst there are, of course, a very
large number of exactly similar planes in any one crystal, we say that
the structure has only one set of slip planes. On the other hand, when
the planes are stacked ABCABCABC . . ., in addition to the set of
planes used for stacking, *three* other sets of parallel close packed
planes are developed, making a total of *four* sets within the crystal.
An expanded diagram of the structure (known as the face-centred
cubic structure, f.c.c.) is shown in Fig. 5.3(b) with the four slip
planes outlined to show their angles relative to the unit cell. Sphere
models of the unit cells of both the above structures are shown in
Fig. 5.4, from which it can easily be seen how they got their names.

142

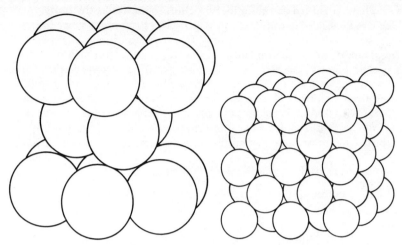

Figure 5.4 (a) The close-packed hexagonal structure. (b) The face-centred cubic structure. (From Hume-Rothery and Raynor (1956) *The Structure of Metals and Alloys*, by permission of the Institute of Metals.)

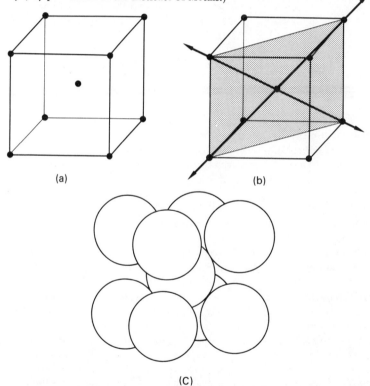

(a)

(b)

(C)

Figure 5.5 The body-centred cubic structure. (a) Expanded unit cell. (b) Unit cell showing one of the six 110 slip planes and the two close-packed 111 slip directions within the plane. (c) Sphere model of unit cell (from Hume-Rothery and Raynor (1956) *The Structure of Metals and Alloys*, by permission of the Institute of Metals).

143

There is a third common metal structure known as the body-centred cubic structure (b.c.c.), which does not in fact show the maximum degree of close packing. Figure 5.5(a) shows an expanded diagram of the structure and outlines the nearest approach to a close packed plane in the structure. From Fig. 5.5(b) we can see that there are only *two* close packed directions within this plane, rather than the three in a truly close-packed plane. When these layers are stacked in an ABABAB . . . fashion, the structure gains five other similar sets of planes at different orientations, making six in all. Figure 5.5(c) shows the unit cell and again the derivation of the name is clearly seen.

Metals crystallising in the above forms are:

f.c.c.   aluminium, copper, nickel, iron (above 910 °C), lead, silver, gold

h.c.p.   magnesium, zinc, cadmium, cobalt, titanium, zirconium, beryllium

b.c.c.   iron (below 910 °C), chromium, molybdenum, sodium, niobium, vanadium

The only familiar metal not appearing in this list is tin, which crystallises at ordinary temperatures in the body-centred tetragonal form (see Table 2.2 and Fig. 2.16).

5.1 ALLOTROPY

Crystallographically speaking, there is not a great deal of difference between the three main metallic crystal structures, and therefore there is often relatively little difference in the internal energy associated with different structures for any one metal. As a result, several metals can exist in more than one crystal structure, a pattern of behaviour known as *allotropy*. Iron, cobalt, tin, titanium and zirconium, of the metals listed above, all show allotropy, the structure adopted being dependant on temperature. Iron is the most important example, since steel technology is based on its allotropic change (see Chapter 14). Tin affords the most interesting example, although it has no commercial significance, since it changes to a nonmetallic allotrope below 18 °C — an amorphous white powder. The reason for this lies in the atomic structure of tin, which possesses four electrons in its outermost shell. Below 18 °C all four valency electrons are equally bound to the nucleus and the element obeys the $8 - N$ rule, forming the tetrahedral diamond structure with typically non-metallic properties. Above 18 °C two of the four valency electrons become much more loosely bound than the other two, and the element becomes a divalent metal with two free electrons per atom. Fortunately for commercial users of tin, the metal has to be cooled considerably below 18 °C before the nonmetallic allotrope nucleates, and pure tin has been kept at −40 °C for three years with no sign of transformation.

144

5.2 POSSIBLE INTERACTIONS BETWEEN DIFFERENT METALLIC ATOMS

Having established the physical nature of pure metals, we must now consider how their structure and behaviour are likely to be affected by the presence of foreign atoms in the lattice. In pure metals the crystallographic arrangement of the atoms is important in determining their physical properties, but with alloys this is not the case. The basic crystal structures of the component metals still play some part, but the relationship is more remote because alloys frequently show a mixture of grains of different crystal structures (usually referred to as phases), and it may be that the predominating phase has a crystal structure differing from that of any of the parent metals. The mechanical properties of alloys also depend on the size, shape, distribution and proportion of different grains present, which in turn depend to a large extent on the method of manufacture and shaping of the alloy.

Engineering makes little use of high-purity metals, except for electrical purposes. There are two simple reasons for this: a pure metal is almost invariably mechanically inferior to an alloy, even a very dilute alloy, based on that metal, whereas its electrical conductivity is superior; also the cost of metals rises very steeply with their degree of purity. Hence, almost all metals are used in the form of alloys with anything from one to six other metals added in varying amounts. For the sake of simplicity the following treatment is based entirely on binary alloys, since these exhibit all the general features of alloying and do not require the highly complex diagrams which ternary and higher alloy systems need for detailed presentation. Furthermore, the great majority of commercial alloys may usually be considered as approximate binary alloys with small additions, rather than true ternary or quaternary alloys.

Let us consider a binary alloy system made up from metals A and B, and restrict the discussion for the moment to the solid state. There are various ways in which atoms A and B may interact, and these may be classified according to the degree of attraction between like and unlike atoms. (The reasons for these attractions are considered below.) If the attractive forces between like atoms are written as AA and BB, and between unlike atoms as AB, the following possibilities present themselves:

Case 1: $AA \gg AB$, $BB \gg AB$. Here we have mutual insolubility, since there is a strong tendency for the atoms to separate out into regions of pure A and pure B. Absolute mutual insolubility is impossible on thermodynamic grounds, but a few systems approach this state and, of these, some are employed in engineering, e.g. copper—lead alloys for bearings and silver—tungsten alloys for switch gear. (Properly speaking the word 'alloy' is a misnomer in this instance, although commonly used in the context; since there is no interaction between the metals, the result is really a mixture.)

Case 2: $AA \simeq BB \simeq AB$. This represents the commonest form of

145

interaction, in which there is a random distribution of A and B atoms throughout the alloy, irrespective of the amount of solution occurring. The crystal structure of the alloys need not necessarily be the same as that of pure A or B; copper—zinc alloys (brasses) and copper—tin alloys (bronzes) are common examples.

Case 3: AB > AA, AB > BB. In this case, there is a tendency for A atoms to surround themselves with B atoms and vice versa. Hence a definite pattern of A and B results throughout the lattice, known as an 'ordered' solid solution or 'superlattice'. Superlattices confer considerably lower resistivities and higher specific heats, hardnesses and strengths than disordered alloys, but are comparatively rare; they have only been observed where the ratio of parent atoms is either 1 : 1 ot 3 : 1. Examples are given by the copper—gold system (based on Cu_3Au) and the copper—zinc system (based on Cu—Zn, brass).

Case 4: AB ≫ AA, AB ≫ BB. When very strong attraction occurs between unlike atoms in an alloy system, it is usually the result of a definite chemical affinity between the atoms, and the result is the formation of an intermetallic compound. Such compounds occur fairly frequently in binary alloy systems, but are not made use of except as minor constituents of alloys since their general physical properties are largely nonmetallic. The formation of iron carbide in steels, Fe_3C, otherwise known as *cementite*, is the most important example of compound in metallurgy — although, of course, it is not really an intermetallic compound. A schematic representation of cases 1—3 is shown in Fig. 5.6; no generalised picture of case 4 can

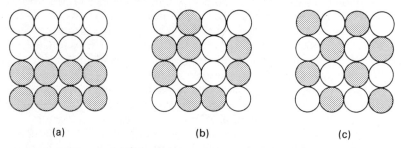

(a)   (b)   (c)

Figure 5.6   Atomic interaction in alloys: (a) mutual insolubility, (b) random solid solution, (c) ordered solid solution or *superlattice*.

be shown, since the bonding varies widely from compound to compound. Of these various combinations, the first, third and fourth are of relatively slight importance. Case 2, however, must be considered in more detail, since it represents the conditions which apply to most of the alloys employed in engineering.

5.3   SOLID SOLUTIONS

In all instances where alloying occurs, i.e. in all cases mentioned above except the first, the solute atoms may either replace certain solvent atoms in the crystal lattice (*substitutional alloys*), or fit between the solvent atoms (*interstitial alloys*). Both categories are shown schematically in Fig. 5.7. Of the two categories, substitutional

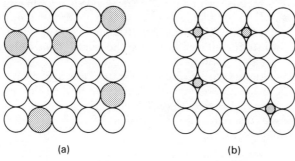

(a)   (b)

Figure 5.7   Solid solutions: (a) substitutional, (b) interstitial.

alloys form the great majority of industrial alloys. Only the non-metallic atoms hydrogen, boron, carbon, nitrogen and oxygen form interstitial alloys; metals, where they alloy at all, form substitutional alloys. The iron—carbon system, the basis of all steels, is the only really important interstitial system, although the iron—boron and iron—nitrogen systems have increased in importance in recent years. However, as regards general alloying behaviour, interstitial and substitutional alloys may be treated exactly alike, since that depends largely on the internal stresses produced by alloying, irrespective of the mechanism by which the internal stresses are produced. Therefore, in the ensuing discussion, no differentiation is made between the two types.

A metal will only dissolve in another metal when the internal strain generated by the introduction of unlike atoms into the lattice is not too great. Inspection of Fig. 5.7 will show that the relative sizes of solvent and solute atoms must play a large part in determining the amount of strain produced. Obviously for low degrees of internal strain it is necessary for the different atoms in a substitutional solid solution to be as nearly the same size as possible; on the other hand, for an interstitial solid solution, the solute atoms must be considerably smaller than the solvent atoms. (Hence the comparative rarity of interstitial alloys, and their restriction to combinations with the five nonmetals mentioned above.) These criteria have been formulated more precisely in the first of the Hume-Rothery principles of alloying, which states that for substitutional alloying the radius of the solute atom must not vary by more than 15% from that of the solvent atom if appreciable alloying is to take place. For interstitial alloying, the solute atomic radius must not *exceed* 59% of the solvent radius. It should be noted that the mere fact of two metals fulfilling these criteria does not necessarily guarantee that alloying will occur; there may be other factors which are unfavourable. However, if the atoms lie outside the permissible size ratios it is certain that alloying will *not* occur.

147

The second of the Hume-Rothery principles states that the tendency to form alloys, i.e. the range of solid solution, decreases as the difference in valency between the two alloying elements increases. At the same time, the tendency to form intermetallic compounds increases. For metals with large differences in valency, this is really another way of stating case 4 above, and may be deduced on thermodynamic grounds; where the valency difference is insufficient for the formation of intermetallic compounds, the effect is due to changes in the electron-atom ratio in the alloy.

The third Hume-Rothery principle is rather less universal; it states that if metal A has a lower valency than metal B, the solubility of B in A is probably higher than that of A in B. The principle holds reasonably well where metal A is univalent but becomes uncertain for higher valencies of A.

5.4 VARIATION OF STRUCTURE WITH ALLOY COMPOSITION

Let us now consider the various possible changes in structure as the composition of a binary alloy system varies from 100% A to 100% B. Unless otherwise stated, the temperature of the alloys under discussion may be taken as 0 °C, and compositions are conventionally expressed as weight rather than atomic percentages. The atomic percentage P_A of metal A in a binary alloy AB is related to the weight percentage by the formula

$$P_a = \frac{x/a_A}{x/a_A + y/a_B} \times 100$$

where x and y are the weight percentages of A and B respectively in the alloy, and a_A, a_B are the atomic weights of A and B. Structures in an alloy system having the same crystal structure as the parent metals are known as *primary* structures; thus, a primary solid solution of B in A has the crystal structure of pure A, whereas a primary solid solution of A in B has the crystal structure of pure B. If B is said to have a maximum primary solubility of 30% in A, then it means that A will dissolve up to 30% of B before the crystal structure of the alloy differs from pure A; the lattice constants change over this range, but not the crystal structure. Since no two elements have exactly the same atomic radius or lattice spacing, every atom of B in the crystal lattice of A introduces a region of asymmetry and therefore of strain. If the two metals have the same crystal structure and are sufficiently alike as regards atomic radius and electronic configuration, it is possible that the strains from the introduction of B may not be sufficient to cause the primary crystal structure to break down. Under these conditions (which are comparatively rare), a smooth change of lattice constant takes place from pure A to pure B. The internal strains present in the intermediate compositions usually result in the alloys being harder and stronger than the pure metals (see Section

148

16.2); other properties, such as ductility and conductivity, drop to a minimum below their values for the pure metals.

Much more frequent, however, is the case where there is only limited primary solid solubility of B in A and A in B, as must always occur when the crystal structure of A differs from that of B. As the proportion of B in A is increased, the stress within the crystal lattice of A is increased. Up to the composition represented by the point X (see Fig. 5.8) this stress is accommodated by adjustments in the lattice constants of A, but at X the stress becomes critical and any

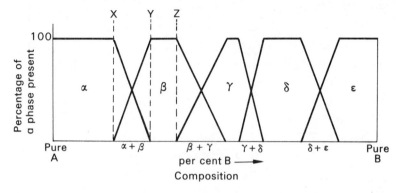

Figure 5.8 Proportions of phases present for various alloy compositions of the alloy system AB.

further increase in B results in a progressive breakdown of the primary lattice. The point X therefore marks the limit of primary solubility of B in A, which is usually referred to as the α solid solution or α phase region. With continuing increase in B, small islands of the β phase appear in the alloy and the structure passes through a two-phase region in which the proportion of α decreases and the proportion of β increases until at Y the entire structure of the alloy consists of β phase. The β phase may then take up a certain amount of B (all alloy phases have variable compositions, as opposed to intermetallic compounds, which have fixed compositions) until it reaches saturation by B at point Z. Beyond Z, another two-phase region occurs, the breakdown of the β being accompanied by the appearance of a γ phase and the whole procedure is repeated. Figure 5.8 shows the structures passing through a δ phase to an ϵ phase (which is in fact the structure of pure B), but many alloy systems are more simple and show only α and β phases. It is important to note that the chemical compositions of phases vary only in single-phase regions. In any of the two-phase regions in Fig. 5.8 the left-hand phase is saturated with B whilst the right-hand phase is saturated with A, and neither will change in composition while the alloy lies between (say) X and Y. This is illustrated in Fig. 5.9.

It is sometimes important to know the exact proportion of phases in a two-phase region, as opposed to the overall proportions of A and B. This may be done by means of the *lever rule*, an example of which

149

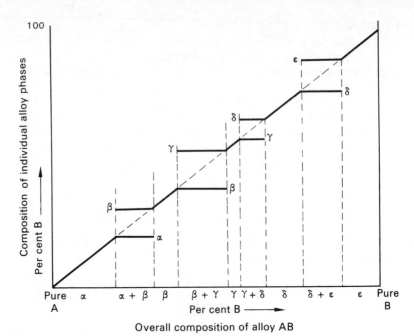

Figure 5.9   Variation in phase compositions for the alloy system in Fig. 5.8.

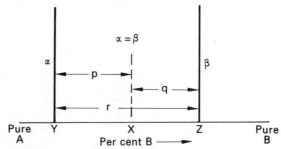

Figure 5.10   The lever rule.

is shown in Fig. 5.10. If we wish to determine the percentage of α phase in an alloy of composition X, then

$$\%\alpha = \frac{q}{r} \times 100$$

similarly

$$\%\beta = \frac{p}{r} \times 100$$

or, more simply,

$$\%\beta = 100 - \%\alpha$$

Taking X as 55% A, 45% B in Fig. 5.10, and the limits Y and Z of α and β solid solutions as 86% A and 28% A respectively, we have

$$r = 58 \quad p = 31 \quad q = 27$$

Hence % α phase in $X = \frac{27}{58} \times 100 = 46.6\%$

and therefore %β phase in $X = 53.4\%$.

150

5.5 PHASE DIAGRAMS

The structure of all metals also varies with the temperature. In the majority of cases, the only changes occur at the melting point and the boiling point. Certain metals, however, exhibit allotropy and then there may be one or more phase changes in the solid state before the melting point is reached. If, in a binary system, the variation in phase with composition is plotted horizontally against the variation in phase with temperature vertically, the result is the *phase diagram* of the alloy system. From the phase diagram, the phases present and their relative proportions may be determined for any combination of temperature and composition. Such diagrams are of great importance in determining suitable heat treatments for alloys. Theoretically, perhaps, the lower limit of temperature scale in phase diagrams should be taken as 0 K (-273 °C), but in practice the lower limit of accuracy may be considerably higher.

The term *'equilibrium diagram'* is frequently used as an alternative to 'phase diagram'; it is a revealing alternative, since these diagrams are limited to *equilibrium conditions only*. That is to say, the phases will only be present as plotted when the alloy is in a state of complete thermal and chemical equilibrium. Hitherto, in explaining the reasons for the existence of intermediate phases, we have been concerned only with composition changes in alloys; on the other hand, transformations in alloys during their heat treatment and fabrication are far more commonly confined to heating and cooling effects at *constant composition*. However, the validity of a phase diagram depends on the heating and cooling rates employed on the alloy concerned; the faster the heating or cooling rate, the greater the divergence — either temporary or permanent — from equilibrium. In many practical processes the exact attainment of equilibrium is uneconomic because of the extremely slow cooling rates required, even if it were desirable (which does not follow by any means). Herein lies the essential paradox involved in the use of equilibrium diagrams: they are used as a guide to the formation of structures which can only be attained under conditions in which the diagram by its essential nature must be inaccurate. Thus, the interpretation of phase diagrams demands the knowledge both of the ideal equilibrium structures and the degree of divergence from these under varying rates of heating and cooling. For the most part the divergence is fairly small, particularly at high temperatures where atomic movements (upon which any phase change must depend) are relatively rapid and equilibrium quickly attained, even for fast cooling rates. Under these conditions the phase diagram is adequate (with some reservations) as a general guide to alloy structures. However, at low temperatures, atomic mobilities become increasingly sluggish or even negligible, making true equilibrium extremely difficult to establish, especially in alloys with high melting points. Under these conditions phases may be formed or retained which are not predictable from the phase diagram, and in fact phase boundaries are usually sketched in very tentatively at low temperatures. The term 'low temperature' may mean anything up to 800 °C in this instance, and few diagrams give positive boundaries below 200 °C, except in the case of very low melting point metals. Fast cooling rates may also lead to nonequilibrium structures by minimising the time for atomic rearrangements to occur. The quenching of alloys may completely suppress certain phases — the quenching of steels (Chapter 16) affords a good example of this — and the phase diagram is useless for predicting the results of this sort of treatment. Bearing in mind these limitations, the study of phase diagrams may be divided conveniently into three stages: initially we shall examine the most important of the various patterns of phase diagram which exist; then the transformations which occur under equilibrium or approximately equilibrium conditions will be considered; and finally we shall study nonequilibrium transformations.

The study of phase diagrams is made easier by the fact that they are made up of standard 'units' or 'patterns' which are individually quite simple. Once these basic 'patterns' are mastered and the transformations which occur in them understood, any phase diagram may be easily interpreted. There are some six basic patterns, and all phase diagrams, however complex they appear at first sight, may be split up into sections which each contain one pattern. They may be classified according to the relative mutual solubility of the two metals in the solid and liquid states, and the stability of the intermediate phases.

Pattern 1. A and B are completely immiscible in the solid state and only partially miscible in the liquid state. This pattern corresponds to case 1 of Section 5.2 and is typified by the iron-lead system; the appropriate phase diagram is shown in Fig. 5.11. This

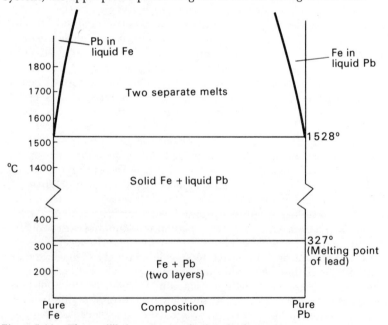

Figure 5.11   The equilibrium diagram for iron-lead.

152

type of interaction — or, rather, lack of interaction — is rare in that there is virtually no solubility below the melting point of the more refractory metal. Several alloy systems show insolubility in the solid state, but they mostly show some degree of solubility above the melting point of the more fusible metal. Copper—lead, used extensively as a bearing alloy, is just such a system, but for the most part systems of this nature have little use in civil engineering.

Pattern 2. A and B are completely miscible in both the solid and liquid states. This is the opposite of pattern 1, and is typified by the copper—nickel system, Fig. 5.12. This pattern is important, less

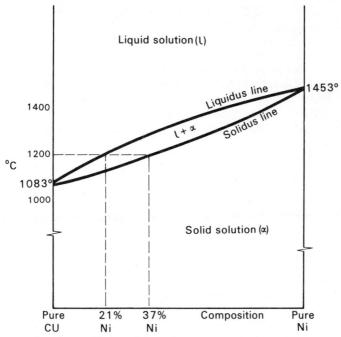

Figure 5.12 The equilibrium diagram for copper-nickel.

because of the alloy systems which exhibit it (it is relatively infrequent) than because it illustrates in its simplest form an extremely important feature, which applies to all systems where there is some mutual solid solubility. The melting point of the intermediate compositions, instead of lying on a straight line joining the melting points of pure A and pure B — as might perhaps be expected — takes the complex form of a double line enclosing a roughly lens-shaped region within which the solid and liquid phases exist in equilibrium. This means that at 1 200 °C, for example, liquid of composition 21% Ni/79% Cu is in equilibrium with solid of 37% Ni/63% Cu. The upper line, which marks the temperature at which solidification commences in any alloy, is known as the *liquidus line*; the lower line, which marks the temperature at which solidification is complete, is known as the *solidus line*. The explanation for this splitting of the melting point into liquidus and solidus lines is quite simple, but the details lie outside the scope of this book. Briefly, it can be shown that, for temperature-composition points which lie between the liquidus and solidus lines, the free energy of the alloy is at a minimum when it consists of a mixture of liquid and solid phases rather than a single homogeneous liquid or solid phase. The temperature difference between the liquidus and solidus points for any alloy is known as the *freezing range* of this alloy. When within this range, the alloy goes through a 'mushy' stage, which is turned to good use in the wiping of joints with plumber's solder. The existence of a freezing range in alloys also has a pronounced and important effect on the micro-structure of the solidified alloy when equilibrium cooling rates are slightly exceeded (see below).

Pattern 3. A and B are completely miscible in the liquid state and completely insoluble in the solid state. This in fact is a hypothetical case, since all binary alloys which are completely miscible in the liquid state show a definite mutual solid solubility, even it it is very small. For the purpose of discussion, the aluminium-silicon system approximates to these conditions, and is shown in a simplified form in Fig. 5.13. There are two important differences from pattern 2: the

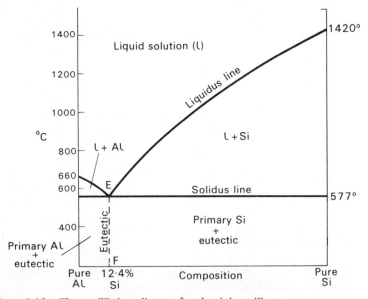

Figure 5.13 The equilibrium diagram for aluminium-silicon.

solidus line becomes horizontal, and the liquidus line becomes V-shaped. The two lines meet at the triple point E, known as the *eutectic point*, at which in the present example solid aluminium, solid silicon and liquid aluminium-silicon alloy containing 12.4% Si are all in equilibrium. The eutectic reaction is defined as the isothermal reversible reaction of a liquid that forms two different solid phases on cooling; the eutectic composition is the alloy composition that

154

freezes at constant temperature, undergoing the eutectic reaction completely. A eutectic structure (normally considered as one phase) is the finely divided mixture in fixed proportions of two phases formed by a eutectic reaction as a liquid alloy solidifies. Eutectic-type reactions can occur entirely in the solid state, when a solid phase breaks down to give a eutectic mixture of two further phases; transformations of this nature are referred to as *eutectoid* reactions, the best known example being the breakdown of austenite to form the pearlite eutectoid structure in steels (Fig. 5.14).

It will be seen from Fig. 5.13 that the eutectic is the one alloy composition with no freezing range, i.e. on melting and freezing it behaves like a pure metal rather than an alloy. Although eutectics are not defined by a vertical line in phase diagrams, as are pure metals and intermetallic compounds, they behave structurally in a rather similar way. They show a definite individual pattern of structure (see Figs. 5.14(a)–(e)) and always have an exact and characteristic composition; the constituents may be either pure metals or solid solutions, or intermetallic compounds. In the two examples cited, the Al–Si eutectic is made up of primary aluminium phase and pure silicon, and the steel eutectoid known as *pearlite* is a mixture of iron containing a small amount of dissolved carbon (ferrite) and iron carbide (cementite). A more detailed account of the structure of eutectics is given below, when we consider the actual process of formation. The proportion of primary constituents to eutectic varies in the same way as the phases in a two phase system: Fig. 5.15 shows the relationship. Thus the structures in the solid region of the aluminium–silicon diagram consist of crystals of primary Al and eutectic to the left of the line *EF*, and of primary silicon and eutectic to the right of *EF*. Within these regions the relative proportions may be determined by the use of the lever rule.

Pattern 4. A and B are completely miscible in the liquid state and show limited solubility in the solid state. By far the commonest and the most important of the basic types, it represents intermediate behaviour between Pattern 2 and Pattern 3. The lead–tin system (Fig. 5.16) will serve as an example here; it is used for soft solders and, with small additions of antimony and copper, in the white bearing metals. Basically the diagram is very similar to the previous one, except that there is an area at each end of the diagram marking the regions of primary solubility of lead in tin and tin in lead. Consequently the two-phase region and the eutectic are not made up of pure metals, but of saturated α and β solid solutions whose compositions are given by the points A (=1.9% Sn) and B (=0.3% Pb).

Pattern 5. A and B form an intermetallic compound $A_x B_y$ at some intermediate composition in the phase diagram. The presence of an intermetallic compound, denoted by a vertical line in the phase diagram, divides the diagram into separate sections which may then be treated independently as the $A–A_x B_y$ system and the $A_x B_y–B$ system respectively. For example, the phase diagram used for steels, normally referred to as the 'iron–carbon diagram', is really the iron–iron carbide diagram; carbon in the elemental form does not

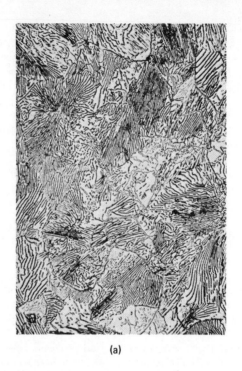

(a)

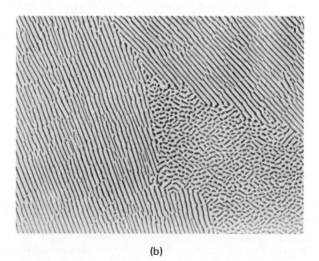

(b)

Figure 5.14 (a) Relatively coarse pearlite in a plain carbon steel (X 200).
(b) Cu-Hg eutectic (x 250). (c) Al-Si eutectic (x 150). (d) Zn-MgZn₂ eutectic
(x 500). (e) Cd-Bi eutectic (x 500). ((a) is from A. R. Bailey (1967) *The Structure
and Strength of Metals*, by permission of Metallurgical Services Laboratories Ltd.
and (b–e) are from G. A. Chadwick (1972) *Metallography of Phase Transformations*, by permission of Newnes-Butterworths).

156

(c)

(d)

(e)

157

normally occur in steels at all, but only in the combined form as iron carbide, Fe_3C.

Intermetallic compounds differ from the normal alloy phases in several respects. They have a single melting point and an exact composition, in which respects they resemble pure metals or eutectics rather than solid solutions. However, they differ in that the bonding between atoms is usually ionic or covalent rather than metallic, which confers on them high melting points (frequently higher than either parent metal) and predominantly nonmetallic physical properties, arising from the lack of free electrons in their structure.

Pattern 6. A phase in the system AB becomes unstable on heating and decomposes below its melting point. Figure 5.17 shows an imaginary example in which the β phase decomposes above a temperature T_p to give a mixture of α phase and liquid. Conversely, on cooling, α phase and liquid alloy react together to form β phase. The proportions of the reactants and their compositions are always fixed,

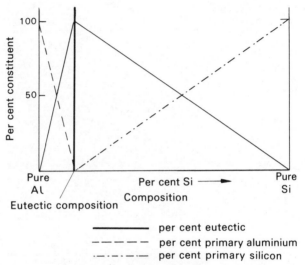

—————— per cent eutectic
— — — — — per cent primary aluminium
·—·—·—·— per cent primary silicon

Figure 5.15   This figure is approximate in that (like Fig. 5.13) it assumes zero mutual solubility of Al and Si in the solid state.

and may be read directly from the phase diagram, or calculated by means of the lever rule. The point P in Fig. 5.17 is known as the *peritectic point.* In the example given T_p corresponds to the solidus temperature over the composition range QP so that one of the reactants is liquid, but this need not necessarily be so; the reaction might occur entirely in the solid state so that (for example) the β phase might break up into a mixture of α and γ phases. Under these conditions the reaction is known as a perite*ctoid*, analogous to the eutectoid mentioned under pattern 5. The peritectic is really an inverted eutectic: in the peritectic a single phase breaks up into two different phases on heating, whilst in the eutectic the single phase breaks up into two on cooling. Peritectics occur frequently in alloy

158

systems (the copper–zinc (brass) alloys and copper–tin (bronze) alloys are good examples) but are less important than eutectics from the point of industrial processes.

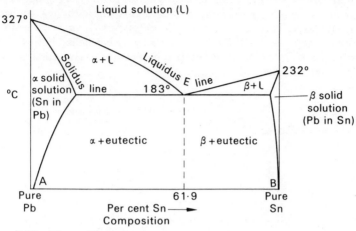

Figure 5.16   The equilibrium diagram for lead-tin.

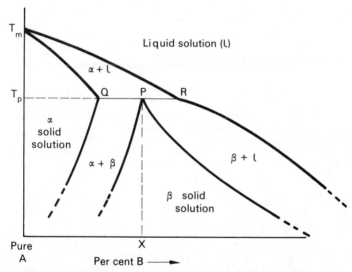

Figure 5.17   Hypothetical equilibrium diagram showing a *peritectic*.

All phase diagrams likely to be met with are made up from combinations of the above six patterns; a good example is the very important iron–carbon diagram, shown in Fig. 5.18. It will be seen that this includes a peritectic, two eutectics, a region (austenite) of appreciable solid solubility, and a region (ferrite) of near-zero solubility. However, each pattern may be considered separately; if this is kept in mind, the interpretation of quite complex diagrams becomes relatively simple. There are, in addition, two further simplifying factors. Not all phase changes need to be considered when studying

159

the final structure of an alloy; for example, the δ phase in Fig. 5.18 may be ignored during the heat treatment of steels. Also, the great majority of engineering alloys are based on the α or the α + β phase regions in the appropriate diagrams. This is true of the copper—zinc, copper—tin, and copper—aluminium systems where the maximum normal alloy additions to the copper are 40%, 15% and 10% respectively; in each case the limit of useful alloying is reached before the complex regions of the diagram. It is even more true of the iron—

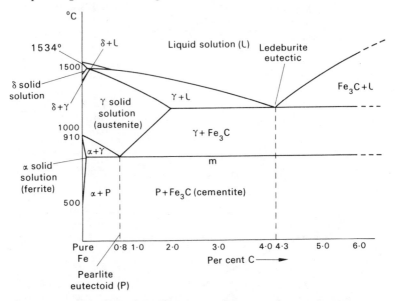

Figure 5.18   The conventional equilibrium diagram for Fe-C. It should be noted that, although composition is given in terms of percentage carbon, the carbon only appears as a separate phase in the form of iron carbide, Fe_3C (cementite). The true iron-carbon diagram, in which graphite separates out as a phase, follows the above diagram closely, but only under conditions of very slow cooling which are not normally met with in practice.

carbon diagram, which is seldom taken above 6% carbon. The reason for this lies in the crystal structure of the intermediate phases. In general, the successive phases that metals pass through with increasing alloy content become more complex in crystal structure, reaching a maximum of complexity roughly in the middle regions of the diagram. Thus, in Fig. 5.8, α and ε would be simple metal structures, β and δ more complex (though not necessarily equally so), and γ the most complex. This is not universal, but it is broadly true of the majority of systems. The complexity of the crystal structure (γ brass, for example, contains 52 atoms to the unit cell, as compared with 2 atoms for pure copper), together with the fact that atoms of different size are intermingled, means that the structure is more irregular on an atomic scale than the pure metal or a primary solid solution. This, in turn, results in a loss of the ductility and resilience typical of pure metals and simple alloys and causes complex phases to become relatively hard and brittle. In small amounts these phases may

160

considerably strengthen an alloy (see Section 16.2), but in larger quantities they are merely likely to produce an embrittling effect and make the alloy unworkable. The same argument applies to the presence of intermetallic compounds.

5.6   PHASE TRANSFORMATIONS I: NUCLEATION PROCESSES

Hitherto, this chapter has shown that intermediate phases and phase mixtures may exist in alloy systems, and has given on a very rough qualitative basis the principles which make the existence of these phases necessary. However, if we are to relate the mechanical properties of an alloy to its internal structure, a knowledge that the alloy contains (say) 70% α phase and 30% β phase is not enough. We must know the size, shape and distribution of the phases present and therefore we want to know *how* the phases came into existence. Phase transformations under equilibrium or near-equilibrium conditions occur by a mechanism known as *nucleation and growth*; there is a second category of transformation known as a *shear* transformation but, although important, it is of relatively rare occurrence and is primarily applicable to nonequilibrium conditions.

The nucleation and growth process is best approached by studying initially the simplest example: that of a phase transformation in a pure metal. As its name implies, it consists of two stages: the formation of a nucleus, and its subsequent growth. Let us now consider in more detail the first stage of the process, the formation of the nucleus.

It will be remembered from Section 2.7.4 that all systems tend to a state of minimum free energy at equilibrium and that a physical or chemical reaction will proceed only if the total free energy of the system is lowered as a result; the magnitude of the decrease in free energy is a measure of the driving force of the reaction. If the free energies of two possible phases in a metal, α and β, are denoted by F_α and F_β respectively and β is the more stable phase at high temperatures, their relationship with temperature may be shown schematically (Fig. 5.19) where T_c is the critical or transformation temperature. We may then say that

$$\Delta F = F_\alpha - F_\beta \qquad (5.1)$$

where ΔF is the free energy change during the reaction, and at first sight the transformation should commence and the α phase nucleate as soon as ΔF becomes negative. It can also be seen from Fig. 5.19 that the negative value of ΔF, i.e. the energy available to drive the reaction $\beta \rightarrow \alpha$ to the right, increases as the temperature drops below T_c. There are, however, two phenomena accompanying the formation of a nucleus formation. These are the energy required to form an interface between the new phase and the old, E_a, and the strain energy caused by the formation of a foreign nucleus in the parent matrix, E_b. Since the energy for these two changes can only be supplied by the free energy change in the structural transformation and the total free energy U of the system must decrease, it follows that for a stable nucleus to form

$$\Delta U = E_a + E_b - \Delta F < 0 \qquad (5.2)$$

where ΔU is the change in the free energy of the system. This immediately shows why supercooling occurs during phase transformation in metals, since at the critical temperature

$$F_\alpha = F_\beta$$

and therefore $\qquad\qquad \Delta F = 0$

Thus, at the exact transformation temperature, there is no energy available for the formation of an interface or for absorption as strain energy due to dimensional changes.

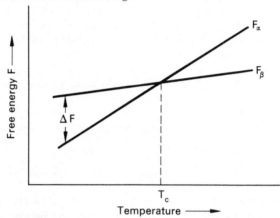

Figure 5.19   The variation in free energy F of two phases α and β with temperature.

If we consider the case of a pure metal solidifying from the liquid state, the conditions are simplified since

$$E_b = 0$$

(there can be no strain set up in a parent matrix of liquid), and therefore the criterion for the formation of a stable nucleus becomes

$$U = E_a - \Delta F < 0$$

The criterion for growth of the nucleus rests on exactly the same thermodynamic basis: it will occur only if the free energy of the system is lowered as a result. This requires that the nucleus must exceed a critical radius r_c for growth to be favoured; if $r < r_c$, then the nucleus will redissolve.

If we assume that the radius of the nucleus is initially spherical, we can put

$$\Delta F = \tfrac{4}{3}\pi r^3 E_v \qquad (5.3)$$

where r is the radius of the nucleus and E_v is the energy associated with the formation of a unit volume of nucleus.

162

The surface energy term E_a may also be related to the radius

$$E_a = 4\pi r^2 \gamma \tag{5.4}$$

where γ is the specific energy of the interface.

Since the condition for growth is that dU/dr be negative, we can say that the critical condition, i.e. when growth and redissolution are exactly balanced, is given by

$$\frac{dU}{dr} = \frac{d}{dr}\left(4\pi r_c^2 \gamma - \tfrac{4}{3}\pi r_c^3 E_v\right) = 0 \tag{5.5}$$

where r_c may be termed the critical radius of the nucleus.

From Equation 5.5 we have

$$8\pi r_c \gamma - 4\pi r_c^2 E_v = 0$$

and hence

$$r_c = \frac{2\gamma}{E_v} \tag{5.6}$$

(This is shown graphically in Fig. 5.20.)

For solid-state transformations the same principles apply, although there is then the additional volume-related strain energy term E_b to oppose the transformation and therefore increase the critical radius size. Also it is not justifiable to assume a spherical nucleus; nuclei in solid-state transformations are more frequently disc shaped, which gives a minimal strain energy term. However, the form of Equation 5.6 is independent of nucleus shape, since the volume must be related to the linear dimensions of the nucleus by a cubic function and the surface area by a square function.

The value of E_v (from Equation 5.3) is directly related to temperature; at the transformation temperature T_c, $E_v = 0$, but as the temperature falls E_v increases rapidly. On the other hand, the value of γ in Equation 5.4 may be taken as roughly constant over the normal temperature range of transformation. Hence, from Equation 5.6, it can be seen that r_c is infinite at the exact transformation temperature but diminishes rapidly as the temperature falls. Now the exact formation of an embryo nucleus depends on the random movement of atoms into positions approximating to the structure of the new phase. The chances of this occurring for a small number of atoms are reasonably good, especially in the case of a metal solidifying, since atoms in the liquid state are highly mobile. However, the chance of forming a particular nucleus becomes rapidly smaller as the nucleus increases in size; in other words, small nuclei are very much more probable than large ones. Thus, as the metal cools below a transformation temperature, nothing happens until the size of the critical nucleus (see Fig. 5.20(b)) has diminished sufficiently for it to form on the grounds of statistical probability. For any normal cooling rate, a large number of nuclei form nearly simultaneously at this point; they will grow rapidly and, since the larger they grow the greater is ΔF compared with E_a, the excess energy given out (the latent heat of transformation) may be sufficient to cause the temperature of the

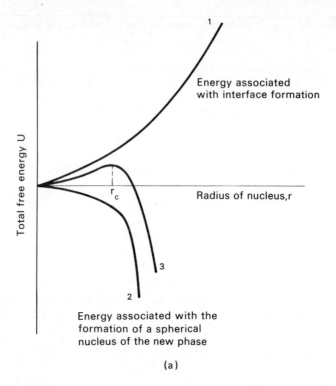

Energy associated
with interface formation

Total free energy U

Radius of nucleus, r

r_c

3

2

Energy associated with the
formation of a spherical
nucleus of the new phase

(a)

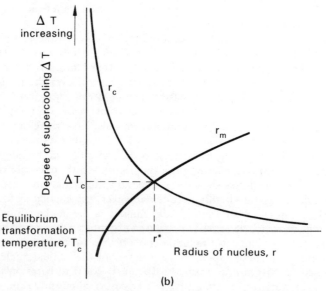

ΔT
increasing

Degree of supercooling ΔT

r_c

r_m

ΔT_c

Equilibrium
transformation
temperature, T_c

r^*

Radius of nucleus, r

(b)

Figure 5.20 (a) Energy barrier to nucleation which can only be overcome by the formation of a nucleus of radius $r > r_c$. 1, E_a; 2, ΔF; 3, $E_a - \Delta F$. (b) The critical nuclear radius, r_c, and the maximum radius, r_m, of a nucleus that is statistically probable, plotted against the degree of supercooling, ΔT. Nuclei of radius r^* will form at temperature $T_c - \Delta T_c$.

164

metal to rise again to the true transformation temperature until the transformation is complete. In the case of a metal solidifying the latent heat of fusion is relatively large, and so the temperature of the metal rises after supercooling to the true melting point; the result is the well known cooling curve obtained when a metal freezes (Fig. 5.21). In the case of a transformation entirely in the solid state (e.g. the transformation from austenite to ferrite in pure iron), the position

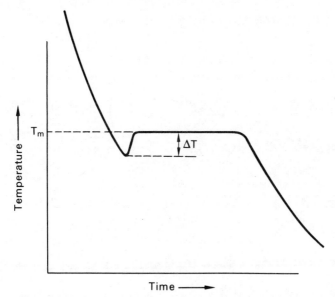

Figure 5.21 Cooling curve for a pure metal. T_m = melting point; ΔT = degree of supercooling.

is slightly different, and may lead to much higher degrees of super-cooling. In the first place, the difference in free energy between two solid metal phases is much smaller than that between a solid and a liquid phase, i.e. $\Delta F_{\text{solid} \rightarrow \text{solid}} \ll \Delta F_{\text{liquid} \rightarrow \text{solid}}$; hence there is less energy available for the interfacial and strain energy terms for any given degree of supercooling and, as a result, the temperature of the metal seldom rises to the true transformation temperature after the transformation has commenced. Secondly, the strain energy term E_b must be taken into account, which increases the size of the critical nucleus for any given degree of supercooling. Finally, atomic mobilities are much lower in solids than in liquids, and so the probability factor restricts nuclei to a much smaller size. In combination, these three factors are sometimes sufficient to suppress phase changes completely at low temperatures, and therefore in many cases only approximate phase boundaries can be determined in alloy systems below about 200 °C. Thus, for practical purposes, the temperature of solid—solid transformations varies considerably, depending on whether the alloy is being heated or cooled through the transformation temperature. On cooling the transformation always occurs at an appreciably lower temperature than on heating (when it is usually very close to the

165

theoretical temperature), the difference being sometimes as much as 40–50 °C, and depending in magnitude on the rates of heating and cooling. For this reason, the position of phase boundaries in phase diagrams may depend on whether the alloy is being heated or cooled, and the two cases are often differentiated by the use of the subscripts 'c' for heating (French 'chauffage') and 'r' for cooling (French 'refroidissement').

5.7   GRAIN SIZE AND SHAPE IN METALS

In deriving Equation 5.6 a spherical nucleus was assumed; this is reasonable in the case of a liquid–solid transformation since the interface area is a minimum for any given volume of nucleus.

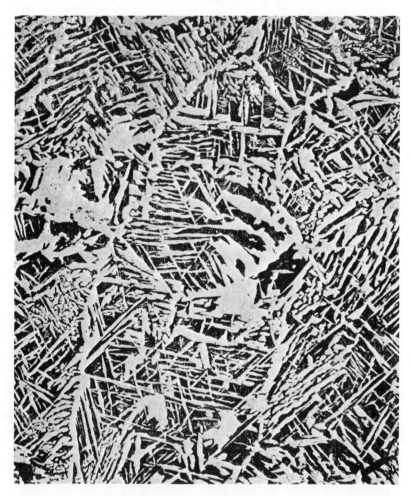

Figure 5.22   Widmanstatten structure of ferrite plates and pearlite in a hypo-eutectoid steel. (From R.H. Greaves and H. Wrighton (1939) *Practical Microscopical Metallography*, by permission of Chapman & Hall Ltd.)

However, in a solid—solid transformation the strain energy term E_b becomes appreciable, and is frequently more important than the interfacial energy term E_a; thus the shapes of nuclei, which are such as to keep the energy term $(E_a + E_b)$ associated with their formation to a minimum, vary for different transformations and are seldom spheroidal, their shape depending to some extent on the positions in the parent lattice at which they form. In fact, E_b is minimal for disc-shaped nuclei, intermediate for needle-shaped nuclei, and is at a maximum for spheroids. Hence, in transformations where the strain energy involved is significant, the tendency is towards disc-like grains of precipitate. The minimum strain energy criterion also results in the nuclei growing in preferential directions (along certain crystallographic planes) in the parent matrix, such that there is the best possible fit between the atoms in the new phase and those in the adjacent parent

Figure 5.23 Ingot of silicon-iron alloy. Macro-etching. (From C. H. Desch (1948) *Metallography*, by permission of Longman Group Ltd.)

167

phase. Hence new phases grow in characteristically shaped crystals in definite patterns; an example is shown in Fig. 5.22. It shows lens-shaped crystals (the section of a disc appears lens-shaped) growing in typical patterns; these patterns correspond to growth along the close-packed planes of the parent matrix.

Control of the grain size of metals is an important feature of manufacture, since the grain size has a considerable effect on the toughness of the product. Variation of the cooling rate when the metal is cast is one method of controlling grain size; this is because the degree of supercooling is related to the cooling rate. A very slow rate of cooling increases the chance for a few large nuclei to form in the liquid just below the melting point. Solidification is then based on few nuclei, and therefore, since each nucleus results in one grain, the grain size of the solid is large. If the liquid is cooled faster, the degree of super-cooling is greater; the critical nucleus size becomes smaller, allowing more nuclei to form, and therefore the grain size is reduced. Hence we arrive at the important general relationship that fast cooling, e.g. chill casting, gives a fine grain structure, whilst slow cooling gives a coarse grain structure. This principle is admirably illustrated in the grain structure of castings (Fig. 5.23); at the outside, where the metal is chilled by the mould wall, a fine grain structure results. Further in, the grains are larger and columnar in shape, and in the centre of the casting the grains are large and equiaxed. Since the grain size varies with the number of nuclei present in the molten metal, it can also be refined by the addition of solid impurities to the melt which are able to act as 'artificial' nuclei. (Not all solids will perform this function.) This process is normally carried out when casting steel ingots, by the addition of a small amount of aluminium powder to the melt; the powder is immediately oxidised to alumina, and the dispersed alumina particles act as nuclei and give a fine grained steel that is known as 'killed' steel, and which has superior mechanical properties by virtue of this.

5.8   PHASE TRANSFORMATIONS II: GROWTH OF PRODUCT

We must now consider the method of growth of nuclei in a liquid—solid transformation, of which there are two possible mechanisms. If the solid is cooler than the liquid, i.e. if the liquid is losing heat through the solid during solidification, the solid—liquid interface tends to be smooth. This state of affairs corresponds to the heat distribution shown in Fig. 5.24(a). Any lumps of solid projecting out into the liquid tend to grow more slowly than the overall inter-face, since heat is conducted away from them more slowly than from neighbouring regions. The crystallographic nature of the surface is usually a series of platelets growing edgewise, the surface of the platelets being parallel to the close-packed planes of the solid crystal. Hence, viewed from above, the surface would appear as a series of terraces. However, it will be seen from Fig. 5.24(a) that this mode of solidification involves no supercooling; the 'nucleus' for growth is

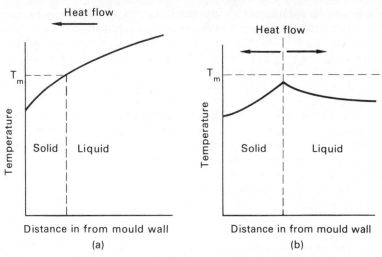

Figure 5.24 Modes of solidification. (a) No supercooling: smooth liquid-solid interface. Very slow cooling only. (b) The liquid has supercooled appreciably, so that the solid-liquid interface is the hottest part. Dendritic crystal growth into the melt ahead of the general mass of solid. This is the usual mode of solidification.

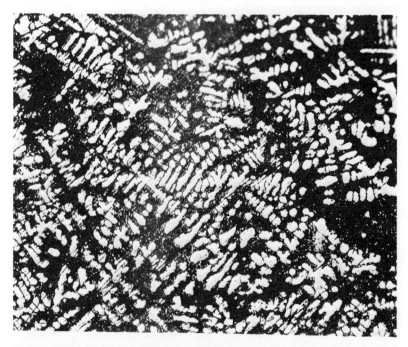

Figure 5.25 Dendrites of primary aluminium in a chill-cast aluminium-silicon alloy. The dark background is Al-Si eutectic (see Fig. 5.14(c)). (× 200)

169

present in the form of the solid surface. The conditions apply to very slow directional cooling of a melt (as when a long tube full of molten metal is very slowly withdrawn endways from a furnace), but are not commonly met with in practice. The normal mechanism, as explained above, always involves supercooling and since the solid is heated by the latent heat of solidification the solid is actually at a higher temperature than the surrounding liquid. Hence the temperature distribution is as shown in Fig. 5.24(b), and during the process of solidification the solid loses some of its heat through the liquid. Once formed, the growth of a stable nucleus is controlled by two factors: it will grow in the directions in which heat can be most easily lost, and it will grow by the deposition of atoms from the melt onto certain preferential lattice sites, so that the nucleus grows by the process of laying down layer upon layer of certain crystallographic planes. The first factor results in growth taking place most rapidly at the points where the nucleus has the greatest local surface area to volume ratio, since this is the condition under which heat can be most easily lost. Hence, in contrast to the first mode of solidification, any protuberances tend to grow preferentially out into the melt, and the result is the very common type of crystal growth which is shown occurring in an aluminium—silicon alloy in Fig. 5.25. Each individual growth is known as a *dendrite* (from the greek 'δενδρος', a fern), and the overall structure is referred to as *dendritic*. As solidification proceeds, the dendrites grow until they meet other dendrites, and the liquid metal remaining between the branches of the dendrites solidifies in the same crystal orientation as the surrounding dendrite. This process is illustrated schematically in Fig. 5.26. Hence dendrites are usually only visible in pure metals when the remaining liquid is poured away from a partially solidified ingot, and occasionally on the surface

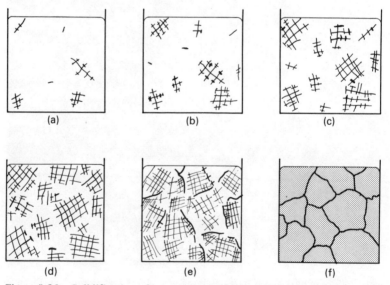

(a) (b) (c)

(d) (e) (f)

Figure 5.26 Solidification of a metal by dendritic crystal growth (schematic).

170

where they may be revealed by volume changes occurring during the solidification process, as for example on the surface of galvanized steel sheet. A polished microsection of a pure metal (Fig. 5.27) reveals nothing of the dendritic skeleton of the grains, since there is nothing to distinguish the metal of the dendrites from the metal in their interstices as both have the same crystal structure and orientation. Cast alloys, however, frequently show a dendritic structure in microsection, since their mode of solidification results in concentration differences between the dendrites and their interstices which may then be revealed by etching (see Section 5.11 and Fig. 5.33).

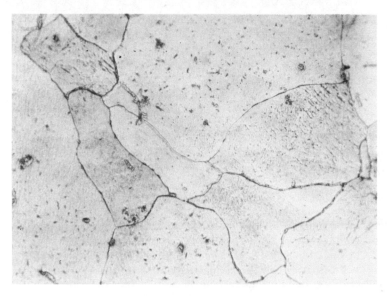

Figure 5.27   The microstructure of Armco iron.

5.9   GRAIN BOUNDARIES

As will be shown in Chapter 12, the boundaries between the individual grains of a metal have a considerable effect on its mechanical properties; the nature of these boundaries has therefore been of great interest to metallurgists and has been the subject of much speculation. There must be some degree of misfit between two contiguous but differently orientated grains; the problem is: how wide is the region of misfit? According to the views held at the present time, there are two general degrees of misfit. One is associated with considerable orientation differences, and is the sort normally seen in the grain boundaries of a metal; the other is concerned with extremely slight differences, normally invisible under the microscope, but playing some part in the deformation mechanisms of metals. Wide angle boundaries between metal grains consist of a disordered transition region of atoms which are attached to neither of the grains making

171

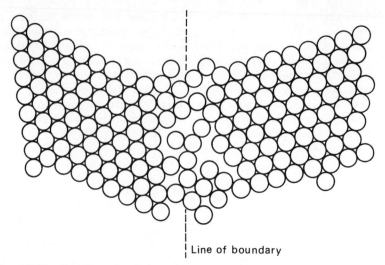

Line of boundary

Figure 5.28 Structure of grain boundary.

up the boundary, as shown in Fig. 5.28. This region is only of the order of 2–3 atoms thick, and its physical properties depend to some extent on the relative orientation of the grains. The atoms in this region, lacking the stabilising influence of the regular crystalline lattice within the grains, have a higher free energy than the atoms in the grains, and therefore grain boundaries show a surface tension typical of solid–solid interfaces. Hence grain boundaries tend to act as regions of nucleation for solid-state transformations (see Fig. 16.12) since the strain energy and the interface energy required for nucleus formation are already partially supplied. Also, the probability of a stable nucleus forming is greater since the atoms are more

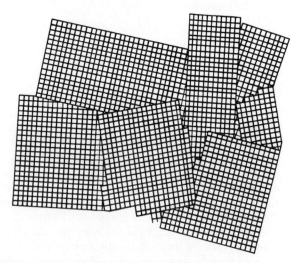

Figure 5.29 Mosaic structure of metal grain (schematic). The angles between the regions have been exaggerated for the sake of clarity.

172

mobile in the disordered region. Because of their relative instability, the atoms in grain boundaries are generally attacked more readily by chemicals. In this way we get the characteristic revealing of grain boundaries under reflected light by etching. The etchant is usually very mildly corrosive to the metal, sufficient to detach atoms from the grain boundaries but insufficient to attack the grains themselves. As a result, the grain boundaries after etching are a system of little channels on the surface of the metal, which appear dark (because of the angle of the sides) under reflected light.

The other type of orientation difference mentioned above occurs within grains, giving rise to what is known as mosaic structure. It is not revealed under the microscope by normal etching techniques. Differences in orientation between the regions are very small — of the order of 0.1° — so that the misfit involved is very slight. Any single grain of metal consists normally of a large number of these subgrains or crystallites made up as shown in Fig. 5.29. Their formation is thought to be due to the fact that a dendrite cannot keep a perfect orientation relationship between its separate branches, becoming slightly twisted or bent during its growth because of thermal fluctuations or convection currents in the liquid metal. Also, the last metal to solidify between the branches of the dendrites is probably under some constraint from the surrounding solid metal, which may lead to a slightly different orientation from the rest of the grain, because of localised stresses being set up in the vicinity.

Once formed in a metal, grains are stable in size and shape for most metals at ordinary temperatures. However, they may be altered by deforming the metal or by heating and cooling through a phase change, when grains of the new phase form by the process of nucleation and growth. Furthermore, if strained grains are heated, at a certain temperature depending on the particular metal concerned, new strain-free grains (which have a lower free energy) nucleate and grow through the entire metal. This process is known as *recrystallisation* (see Chapter 12 for further discussion).

5.10 TRANSFORMATIONS IN ALLOY SYSTEMS

5.10.1 Equilibrium Conditions

We must now consider how the presence of an alloying element affects the pattern of transformation discussed above. In many ways, transformations in alloy systems are very similar; supercooling, nucleation, grain growth and grain size are all governed by the same principles that apply to pure metals, although the possible presence of two physically different phases in equilibrium in the alloy may cause some modifications in behaviour. The salient difference stems from the way in which an alloy cools across a two-phase region in an equilibrium diagram, and the effect that it has on the distribution of the alloying element. A pure metal must always be homogeneous from the point of view of chemical composition; an alloy is often chemically inhomogeneous even in a single-phase region because of

173

the exact mechanism of transformation to that phase. These inhomogeneities are partly dependent on the rate of cooling and may have an important effect on the mechanical properties of the alloy. Let us examine first the solidification of an alloy AB and the changes that occur throughout its freezing range, assuming that in both the liquid and the solid state B is in complete solution in A, and that equilibrium is maintained at all stages of the process. The appropriate portion of a phase diagram is shown in Fig. 5.30.

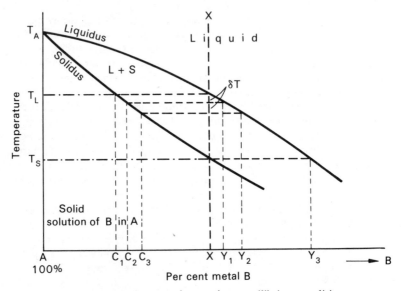

Figure 5.30 Cooling through a two-phase region: equilibrium conditions.

Consider an alloy of composition X in the liquid state. At any temperature above T_L it is a homogeneous liquid, and will remain so when cooled until the temperature T_L is reached, i.e. the liquidus temperature for this particular alloy. At this temperature the phase diagram indicates that solid of composition C_1 exists in equilibrium with liquid of composition X. Hence at T_L the first solid to separate is of composition C_1, and as a result the composition of the remaining liquid shifts towards B, to some composition (say) Y_1. An alternative approach is to say that below temperature T_L, liquid of composition X becomes supersaturated with solid of composition C_1. This view of the problem is useful when considering the solidification of eutectics (see below). (We are ignoring the possibility of supercooling here, since it does not affect the general principles involved.) The remaining liquid, of composition Y_1, has a slightly lower liquidus temperature than the original alloy, and at a slightly lower temperature (say $T_L - \delta T$) is in equilibrium with solid of composition C_2. Since we are considering an 'ideal' solidification under equilibrium conditions (unattainable in practice), the solidified alloy must at all times be in internal equilibrium, i.e. homogeneous, and therefore C_2 represents the overall composition of the solid at temperature $T_L - \delta T$ and *not*

174

the composition of the solid actually deposited at that temperature. (This point is important, because under practical cooling rates, when equilibrium is departed from, the solid alloy ceases to be homogeneous and the phase diagram ceases to give an accurate representation of the cooling mechanism.) The second precipitation causes in turn a further shift of liquid composition towards B, to Y_2, and therefore, after another increment of temperature drop δT, more solid alloy is deposited such that the *overall* composition of all the solid deposited is C_3. In practice, of course (whether equilibrium conditions are maintained or not), the steps involved in the sequence precipitate—composition shift—temperature drop—precipitate ... become vanishingly small, and the composition of the liquid simply follows the liquidus line continuously as the temperature falls, whilst the *overall* composition of the solid also follows a smooth curve — the solidus line under equilibrium conditions. This process continues until the last remaining drops of liquid have the composition Y_3; as these solidify (at temperature T_s) they finally bring back the overall composition of the solid to X, after which the alloy will cool through the solid range without any further change, unless some other phase boundary is encountered.

5.10.2 Nonequilibrium Conditions

Let us now see how the above mechanism must be modified to allow for normal rates of cooling, when equilibrium is not maintained throughout the phase change (see Fig. 5.31). The initial precipitation of solid from the liquid alloy takes place exactly as in the ideal case, solid of composition C_1 being deposited whilst the liquid composition

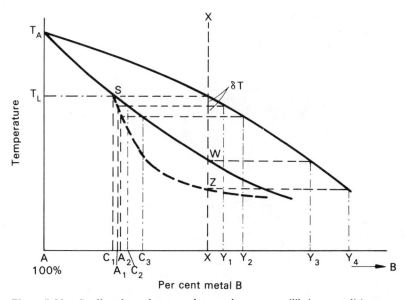

Figure 5.31　Cooling through a two-phase region: nonequilibrium conditions.

175

moves towards B. However, with normal rates of cooling diffusion in the solid alloy cannot take place fast enough to maintain a uniform composition throughout the solid *as well as* fulfilling the conditions imposed by the solidus line (i.e. a steady shift in overall composition from the first formed crystals of composition C_1 back to the basic composition X). As a result, if we again consider solidification as a stepwise process for the moment, the composition of the dendrites of solid deposited is such that the outermost layer contact with the liquid alloy, *and that layer only*, is of the composition predicted by the phase diagram to be in equilibrium with the liquid at that particular temperature. Thus, the initial solid (deposited at temperature T_L) has composition C_1; the next layer (deposited at $T_L - \delta T$) has composition C_2; the next (deposited at $T_L - 2\delta T$) has composition C_3, and so on. The resultant structure is shown diagrammatically in Fig. 5.32; in practice, of course, steps do not occur in the composition, which changes smoothly from the inside to the outside of the dendrite arms. Also, a certain amount of diffusion takes place, which lessens the shift in composition somewhat from that shown but never obliterates it.

The upshot of this mechanism is that since only the outermost layer of solid follows the solidus line, the *overall* composition of the total solid deposited at any stage must lie to the left of it. In other words, at temperature $T_L - \delta T$ the overall composition of the solid is the mean of C_1 and C_2 (shown as A_1 in Fig. 5.31) instead of the equilibrium value C_2, and at $T_L - 2\delta T$ the overall composition is the mean of C_1, C_2 and C_3 (A_2 in diagram), instead of C_3. Thus the actual solidus line followed is given by the line SA_1A_2Z instead of SW; solidification is completed not at W, but at Z, and the final solid deposited has the composition Y_4 which is appreciably richer in B than the final solid (Y_3) under equilibrium conditions.

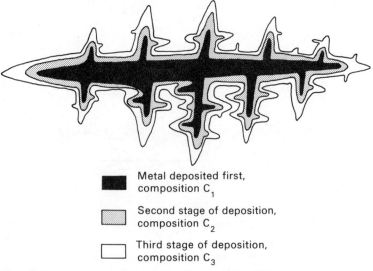

Metal deposited first, composition C_1

Second stage of deposition, composition C_2

Third stage of deposition, composition C_3

Figure 5.32 Schematic representation of a cored dendrite.

176

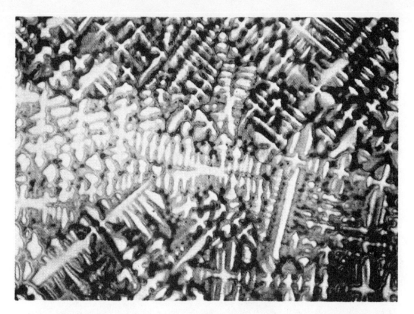

Figure 5.33 Cored structure in 70/30 Cu-Zn alloy (x 180). (From A.R. Bailey (1972) *The Role of Microstructure in Metals*, by permission of Metallurgical Services Laboratories Ltd.)

Hence, the final structure of the alloy is not homogeneous, but (in the example discussed) consists of dendrites which vary in composition from C_1 in their cores to Y_4 in their interstices. Brasses and bronzes show this effect very noticeably in castings; an example is shown in Fig. 5.33. This type of structure is known as a *cored* structure, and is important in that it may have a considerable effect on the mechanical properties of a casting. In extreme cases, the final liquid to solidify may become so enriched in metal B that its composition moves out of the α range into the $\alpha + \beta$ range, causing small

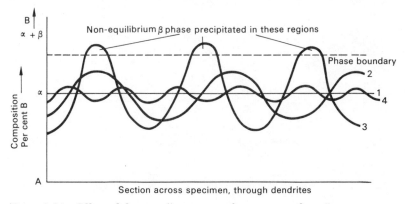

Figure 5.34 Effect of three cooling rates on the structure of an alloy.
(1) Uniform composition, produced by ideal equilibrium conditions, or annealing.
(2) Normal nonequilibrium cooling rates. (3) Faster cooling rate, resulting in precipitation of β phase. (4) Still faster rates: coring being partially suppressed.

177

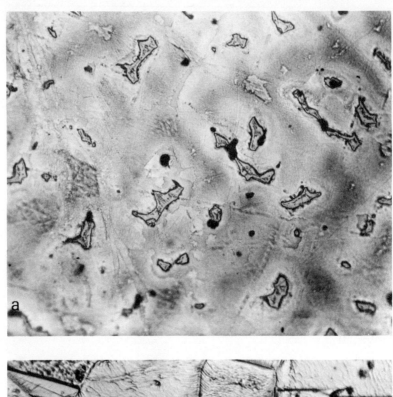

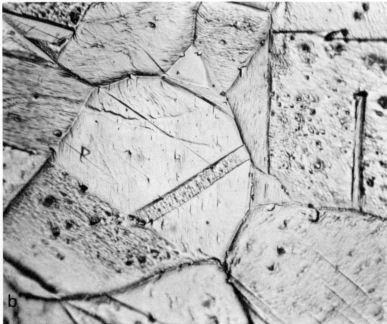

Figure 5.35 (a) Separation of nonequilibrium δ phase due to coring in a cast tin-bronze. (b) The same alloy after annealing; the δ phase has redissolved leaving only grains of the α phase. (x 400)

grains of the β phase to form in the alloy, even though the overall composition suggests that the alloy should be entirely α phase (see Fig. 5.34). If the β phase in these cases is hard or brittle, the alloy may be made unsuitable for hot or cold working processes, and coring must be eliminated before working can proceed. This applies in particular to the tin bronzes and more rarely in brasses, but is a possibility which must be borne in mind in any single-phase casting where the overall composition lies near to a two-phase region.

Coring may be eliminated by subsequent annealing of the alloy, which allows the various regions to interdiffuse and level out the concentration gradients. In practice, annealing is usually combined with plastic working (where there is no risk of embrittlement from the coring, see above), since the working helps to break down the structure and makes for shorter annealing times to equilibrate the composition. If the metal is cast to its final form and working is there-fore impossible, coring can only be eliminated by controlling the cooling rate of the casting, or by long annealing — either of which may have attendant disadvantages. Figure 5.35(a) shows the occur-rence of coring with the separation of a nonequilibrium phase, whilst Fig. 5.35(b) demonstrates the result of annealing on the cored structure.

The degree of coring (i.e. the composition difference between the cores of the dendrites and the interstices) which occurs in an alloy depends fundamentally on the freezing range of the alloy and also on the rate of cooling through the two-phase region. Large freezing ranges usually mean a considerable difference between the composi-tions of liquid and solid in equilibrium with each other during solidi-fication, which in turn means a large difference in composition between the first- and last-formed solids. Cooling rates affect coring by governing the time available for diffusion within the alloy, since both the development of coring and its elimination must depend on diffusion. Very slow cooling rates permit an approach to true equilib-rium conditions; therefore the concentration gradients which develop during the deposition of solid are partially annulled by the diffusion

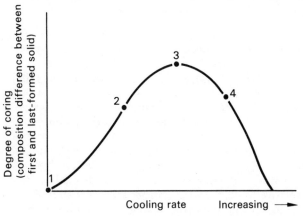

Figure 5.36   The effect of cooling rate on degree of coring.

of atoms in the solid state, and there is relatively little coring. Faster cooling rates suppress diffusion in the solid state (partially or completely), and therefore coring occurs. Still faster rates of cooling, however, tend to suppress coring by preventing diffusion in the liquid as well as the solid state. Thus there is a qualitative relationship between coring and cooling rate, which may be depicted roughly as in Fig. 5.36.

5.11 SHEAR TRANSFORMATIONS

The departures from equilibrium discussed in the preceeding section are sufficient to cause structural abnormalities after a transformation, but the essential *nature* of the transformation remains unimpaired. It is still a process of nucleation and growth. However, if the cooling rate becomes sufficiently rapid, transformations of this sort become impossible in the solid state because the individual atoms do not have time to diffuse into their new crystallographic habit. In this situation there are two possible consequences: either the transformation may be totally suppressed and the high-temperature phase retained in a metastable state down to room temperature, or a transformation may take place that does not require the diffusion of atoms in order to attain a new structure. Such a diffusionless transformation takes place by a process of cooperative shear of the atoms, and is known generally as a *shear transformation* or a *martensitic* transformation. Figure 5.37 demonstrates the principle of the transformation; note that it

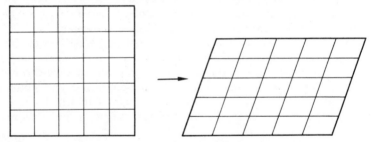

Figure 5.37 Shear transformation of a crystal lattice.

does not preclude movement of the atoms, but that it occurs by very small relative movements of less than one interatomic spacing.

Rapid cooling is not the only cause of shear transformations; they may also result from the shock loading of metals (particularly at low temperatures), or as a consequence of annealing processes. When formed in this way, they have the *same* structure as the parent matrix, but it is reflected at an angle to the parent structure across a particular crystallographic plane known as the *twin plane*. Figure 5.38 illustrates in two-dimensional form the movements involved in a simple twinning operation; it will be seen that the individual movements of atoms relative to neighbouring planes are less than one interatomic distance, but that the actual distance moved is proportional to the distance of the atom from the twin plane, and can be

180

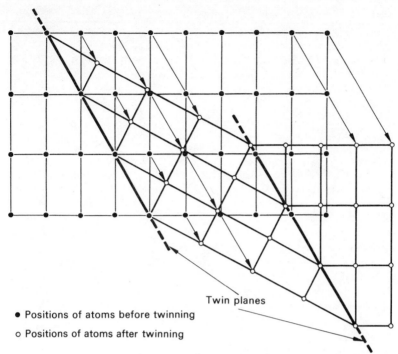

Twin planes

● Positions of atoms before twinning

○ Positions of atoms after twinning

Figure 5.38 Movements of atoms when a twin band forms across a crystal lattice. (For clarity, only a few of the movements are shown arrowed.)

very much more than one interatomic distance. A twin is thus a mirror image of the parent structure across the twin plane.

Because there is no actual crystalline misfit across the twin plane — merely a change in orientation — twin boundaries have a surface energy which is roughly only $\frac{1}{20}$ of the average grain boundary energy, and it is this that leads to the formation of copious annealing twins in a number of alloys with f.c.c. structures. If a recrystallised grain is growing out into a work-hardened structure, it will form boundaries with the surrounding unannealed grains. The energy of these boundaries varies according to the angular relations between the new and old grains and there will always be some cases where a high grain boundary energy can be reduced if the active growing grain rotates to form a twin over that particular section of boundary. Whether or not a twin will form there depends on whether the energy required to form a twin is more than offset by the lowered boundary energy, so that the energy of the system as a whole is lowered by twinning. The number of twins will depend on the number of suitably oriented grains which the growing grain meets, and therefore increases as the grain size of the worked sample decreases. Metals with f.c.c. and, to a lesser extent, hexagonal structures exhibit this form of twinning; it is particularly common in copper, α-phase copper alloys, and austenitic steels (see Fig. 5.39). It is not observed in b.c.c. metals such as iron.

Twins forming as a result of stress on an alloy (*deformation twins*) are much narrower than annealing twins, and the rotation associated

181

Figure 5.39 Grains of α brass showing twinning. (× 400)

with twinning normally increases plasticity since it serves to bring slip planes into a more favourable orientation to the applied stress. This category of twin does not occur in f.c.c. structures, but will appear in b.c.c. crystals (e.g. α iron) or hexagonal structures (e.g. zinc or cadmium). Twinning planes are specific for a particular structure; thus b.c.c. metals twin on the $\{112\}$ planes, whilst hexagonal metals twin on the $\{10\bar{1}2\}$ planes.

Figure 5.40 Kind bands in an aluminium crystal deformed 17.5% in tension (× 100). (From R.W.K. Honeycombe (1971) *The Plastic Deformation of Metals,* by permission of Edward Arnold.)

182

Because of the associated change of *shape* when a deformation twin forms, a considerable shock wave is induced in the alloy, and some plastic deformation results in the surrounding matrix. The curious rustling and clicking which may be heard when a bar of tin or zinc is bent in the hand (the 'tin cry') is due to the shock waves accompanying the formation of thousands of twin bands, whilst *accommodation kinks* (Fig. 5.40) will show on a polished surface that undergoes twinning. The intersection of two deformation twins may result in crack nucleation because the metal is unable to accommodate both movements adequately by plastic deformation, and it is possible that the mechanism may play some part in the brittle failure of steels under impact loading at low temperatures.

Shear transformations resulting in crystallographic change are much rarer than nucleation and growth processes. However, one example is so important that it has lent its name to this entire category: the *martensite* transformation in steels. The technological importance of this transformation will be considered in Chapter 16, but for the present it is sufficient to know that martensite is the transformation product when austenite is cooled below the eutectoid temperature at a rate which suppresses its normal breakdown to pearlite. In these circumstances the ferrite (the normal form of iron at room temperature) forms in a violently distorted state because it may contain up to 40 times the maximum carbon content that it can dissolve. The transformation pattern is shown in Fig. 5.41(a) which depicts two unit

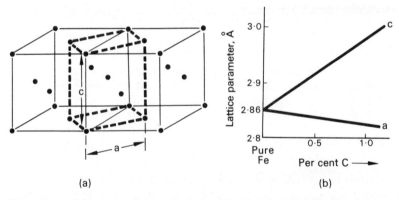

(a)   (b)

Figure 5.41 (a) Body-centred tetragonal cell (dashed outline), showing its derivation from the f.c.c. austenite structure. (b) Variation in the lattice parameters of b.c.t. martensite with carbon content.

cells of the f.c.c. structure of austenite. It will be seen, however, that this structure could also be considered as a body-centred tetragonal structure, the unit cell of which is given a dashed outline, with an axial ratio $c/a = 1.414$. Thus $(110)_{fcc} \rightarrow (100)_{bct}$ and $(100)_{fcc} \rightarrow (110)_{bct}$. By an appropriate contraction in the c direction and expansion in the a and b directions this new lattice could then transform by shear to the b.c.c. structure of ferrite; however, this is prevented in steels by the presence of the excess carbon atoms. These

183

are located in the centre of the f.c.c. cells, which means that on transformation they will be located preferentially in the vertical edges of the new body-centred structure, thus preventing it from contracting to the b.c.c. structure, and retaining the body-centred tetragonal habit. None the less, some contraction *towards* a cubic structure does occur; the amount, as might be expected, depends on the quantity of carbon present. Figure 5.41(b) shows the relation between carbon content and axial ratio. However, crystallographic studies have shown that the martensite crystals rotate and twin on the (112) planes at the same time that the lattice constants adjust themselves, the twinning serving to reduce the strain energy associated with the transformation. As with twins the formation of martensite is accompanied by accommodation kinks; the polished surface of a steel which undergoes quenching is considerably distorted so that the martensite needles become visible without need of etching.

It may be convenient to conclude this section by summarising the principal differences between shear transformation and nucleation-and-growth processes. (See Table 5.1.)

Table 5.1

Nucleation and growth	Shear
Dependent on atomic diffusion and therefore time dependent.	No dependence on diffusion; relative atomic movements are always less than one interatomic spacing.
Rate of transformation depends on degree of supercooling, but will proceed to completion at constant temperature.	Transformation propagates at speed of sound, but is temperature dependent, i.e. for transformation to be complete cooling must *continue* below the temperature at which the transformation commences.
Transformation will result in the redistribution of alloying elements (except in recrystallisation and solidification of pure metals).	Transformation produce has the same composition as the parent phase.
No macroscopic changes of shape are involved.	Transformation involves macroscopic changes of shape and the formation of accommodation kinks.
All crystal planes changed.	One crystal plane (the *twin* plane or *habit* plane) remains unchanged as a 'hinge' for the transformation.

5.12 DISTRIBUTION AND EFFECT OF IMPURITIES

At this stage it is convenient to consider the distribution of impurities throughout a metal, as these may under certain conditions have a drastic effect on the properties of the metal. Thus, as little as 0.002% bismuth in copper reduces the electrical conductivity nearly to zero and also embrittles it, whereas the presence of 0.05% oxygen improves the general mechanical properties, and yet both these elements are virtually insoluble in solid copper. The reason for this remarkable difference in behaviour lies in the melting points of the impurities (in the case of oxygen, the impurity we are concerned with is really cuprous oxide, Cu_2O), which in turn determine their distribution throughout the specimen. In general terms, an impurity with a higher melting point than the main metal alloy (e.g. cuprous oxide in copper) forms a dispersion of particles in the melt which either act as artificial nuclei for the alloy to solidify on, or are pushed to the grain boundaries by the advancing front of solidified metal where they remain as a series of discreet crystals (Fig. 5.42(a)). An impurity

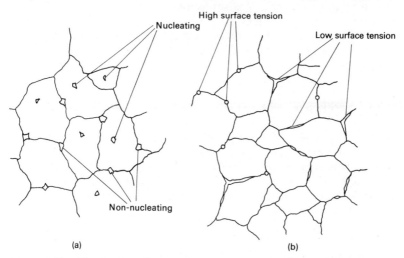

Figure 5.42 Distribution of minor impurities in alloys. (a) Impurities with melting points above the liquidus temperature. (b) Impurities with melting points below the solidus temperature.

with a lower melting point (e.g. bismuth in copper) is also pushed to the grain boundaries of the alloy, but since it is still liquid after complete solidification of the alloy proper has taken place, it tends to form a thin grain boundary film. The amount of grain boundary area which a small percentage of impurity can cover depends on the interfacial surface tension between impurity and alloy; high values of surface tension will condense the impurity into droplets which remain at the junction of these grains, whilst a low surface tension allows the impurity to spread out in a thin film and perhaps completely permeate the grain boundaries (Fig. 5.42(b)). The result is simple: in the case of the low melting point impurity, depending in

185

some degree on the surface tension of the impurity/alloy interface, the strength of the metal depends no longer upon itself but upon the strength of the impurity 'cement' which is now holding the grains together. In the examples cited above, the thin film of intergranular bismuth serves both to weaken the copper mechanically and to act as an electrical insulating film round each copper grain; the cuprous oxide, however, by remaining as discreet particles, has no such damaging effect on the electrical properties since there is an ample electrical path between the inclusions of cuprous oxide. Mechanically speaking the oxide, in small amounts at least, may even strengthen the metal by stabilising the grain boundaries. A further obvious danger of the low melting point impurity is that the melting point of the metal or alloy is effectively reduced to that of the impurity. e.g., in the above example, the melting point of copper is reduced from 1 083 °C to 271 °C (the melting point of bismuth). Further examples of the very damaging effect of low-melting impurities are afforded by sulphur in nickel and steels, phosphorous in steels, and lead in gold. Where the impurity cannot readily be removed from the metal, it is sometimes rendered innocuous by converting it from a low-melting-point to a high-melting-point impurity. Thus the addition of manganese to steels causes any sulphur present to form manganese sulphide (high m.p.t.) instead of iron sulphide (low m.p.t.), with the result that the harmful nature of the sulphur is effectively eliminated.

5.13   EUTECTIC TRANSFORMATIONS

The remaining way in which an alloy system differs in phase changes from a pure metal occurs when the alloy undergoes a eutectic or peritectic transformation, the eutectic being of more frequent occurrence and far greater importance. Let us consider the formation of a eutectic structure; Fig. 5.43 shows a typical example with some

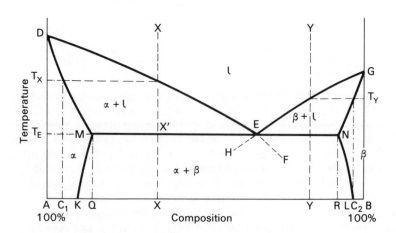

Figure 5.43   Mechanism of formation of eutectic structure.

186

degree of mutual solid solubility. On cooling the alloy X from some temperature in the liquid state, solid of composition C_1 separates out at temperature T_X, and on further cooling the composition of the liquid moves along DE and the overall composition of the solid along DM until the temperature has fallen to T_E, when the remaining liquid has the composition E. Up to this point the solidification process for the alloy is exactly similar to that discussed above for single phase alloys; if the cooling rate is such that equilibrium is departed from (for the time being we are assuming that it is not), then supercooling and coring will also occur in a similar manner. The behaviour of an alloy Y on cooling is the same as that of X, except of course that solid rich in B separates and the liquid and solid follow GE and GN respectively until T_E is reached. Thus, at temperature T_E the liquid remaining in any alloy between compositions M and N has composition E. (If the alloy composition lies outside MN, then it will solidify entirely as α or β solid solution, and there will be no eutectic present at all — unless coring results in the formation of some, see Section 5.10.2.) Remember that during the first stage of solidification of alloys X and Y, at no point can the state of the liquid drop below the liquidus line. If it were to, then in effect the liquid would become supersaturated in α (or β) and precipitate the one or the other until the liquidus line were regained. But after the remaining liquid has reached the eutectic point, neither alloy can avoid being supersaturated with respect to at least one of the α or β solid solutions, since if the liquid follows the line EF (remaining in equilibrium with α) it must drop below EH and become supersaturated with respect to β. Similarly, if the liquid follows the line EH, it must drop below EF, becoming supersaturated with respect to α. Hence solidification of the eutectic may be regarded as follows. The liquid follows EF a very small way below the eutectic temperature and deposits crystals of β phase. In so doing, it becomes supersaturated with α and therefore deposits crystals of α, thus moving across to the line EH. In so doing, it becomes supersaturated with β and deposits crystals of β, thus moving back to EF . . . and so on. Thus eutectic solidification consists of the alternate (in practice, simultaneous) deposition of crystals of α and β at a temperature just below T_E until solidification is complete. This simultaneous deposition explains why eutectics show finely divided structures, and also why a pure eutectic liquid has no freezing range, but solidifies at one temperature like a pure metal. (In theory, after solidification the α phase portion of the eutectic should reject some β, and the β should reject some α because of the slope of the lines MK and NL. However, this is to all intents and purposes negligible in common eutectics, and the phase changes involved may never take place if the diffusion of atoms becomes sluggish at low temperatures.)

The effect of increasing cooling rates upon eutectic transformations, i.e. of increasing divergence from equilibrium conditions, is to give an apparent spreading of the precise eutectic composition to a eutectic 'zone of composition'. In the initial stage of solidification, where primary α or β solid solution is deposited, some degree of supercooling and coring is inevitable. The effect of the coring is to increase the amount of eutectic liquid present for any given alloy at temperature T_E (since the solidus lines DM and GN are shifted to the left and right respectively by nonequilibrium conditions). In addition, alloys near the eutectic composition may sometimes be supercooled by rapid cooling into the region below HEF before solidification commences. In this region the conditions for eutectic solidification hold good, since the alloy is supersaturated with respect to both primary constituents. Therefore the greater the degree of supercooling, the greater the spread of composition over which completely eutectic structures may be obtained. However, very fast cooling rates may suppress the formation of a eutectic altogether by preventing the necessary diffusion of the atoms. Suppressions of this nature are naturally easier in the case of a solid—solid transformation than a solidification; the classic example of this is the quenching of steels to prevent the formation of the eutectoid known as pearlite, which is one of the normal equilibrium constituents of steels.

From the above considerations of eutectic transformations we may deduce the general pattern of structure produced in alloys by this mechanism. Depending on the proportion of primary phase (say α) to eutectic (= 100 X'E/ME in Fig. 5.43), the structure will vary from isolated grains of α surrounded by eutectic to an almost complete grain structure of α with a film of eutectic around the grain boundaries. If the eutectic involves solidification from the liquid state, the α phase will probably be in dendritic form with eutectic occurring in the interstices; if it forms from the solid, however, dendrites of the primary α phase are unlikely since atom mobilities are too low for them to form under these conditions. In the case of a eutectoid formation, it is possible that the primary phase, being the first to separate, will nucleate preferentially at the grain boundaries of the parent phase thus taking up a grain boundary pattern whilst the eutectoid forms at the centre of the grains. This is an inversion of conditions in liquid—solid transformations where the last phase to solidify — i.e. the eutectic — is invariably pushed to the grain boundaries. Thus, even if only a small amount of eutectic forms in an alloy, it probably has a disproportionate effect on the general strength and mechanical properties since it is always the last part to solidify. Examples of eutectic and eutectoid structures are shown in Fig. 5.14.

CHAPTER 6

Environmental Volume Change in Concrete

6.1 TWO KINDS OF DEFORMATION

The engineering designer has a first responsibility to ensure that his structure will carry the imposed loads; understandably, this leads him to give highest priority to the analysis of the stresses and deformation in the structure under the external forces. In many instances this is not sufficient, because concrete, like virtually all other materials, expands as a result of a temperature rise, and, like most porous solids, shrinks as a result of a drop in humidity. Since concrete is an isotropic material, both of these deformations are volumetric in that the strains resulting from a change of temperature or humidity are the same in all directions. This is in contrast to timber, which suffers both kinds of deformation, but, because of its anisotropic nature, the magnitudes are quite different in the three principal directions (longitudinal, radial and tangential — see Chapter 7). Metals are the simplest of the three materials in that they are isotropic and normally do not respond to change of humidity. Either deformation can cause both kinds of undesirable structural effect:

(a) direct change in dimensions causing movements of one part of the engineering structure relative to another part, for example in the thermal expansion of bridge spans;
(b) imposed external deformations on restrained or indeterminate structures, causing induced stresses, and the possibility of failure. A simple example is the cracking of concrete in a reinforced member in which the shrinkage of the concrete is restrained by the reinforcing steel.

The vulnerability of the particular structure depends on the magnitude of these environmental deformations, and it is the purpose of this chapter to explain their occurrence, to discuss the factors that influence them, and to show how their magnitudes may be calculated.

189

6.2 SHRINKAGE

6.2.1 Shrinkage Tests

Although shrinkage is a volumetric deformation it is normal to measure the changes that occur in length. This means that the apparatus can be very simple; the typical robust shrinkage frame shown in Fig. 6.1 is that suggested in the British Standard (1970).

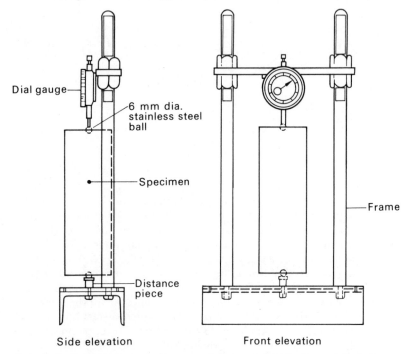

Side elevation   Front elevation

Figure 6.1 Typical apparatus for tests of drying shrinkage. (From BS 1881 (1970) Part 5, reproduced by permission of BSI, 2 Park Street, London W1A 2BS, from whom complete copies can be obtained.)

The specimen has stainless steel balls cemented to each end, and in the course of measurement these locate in conical seatings, one on the baseplate of the frame, and the other at the end of the stem of the measuring device. In this case a dial gauge is shown, but an alternative, particularly for automatic recording, is a linear displacement transducer. The frame is usually susceptible to temperature change, and an invar setting bar is provided so that corrections can be made for thermal expansion.

Most shrinkage is related to the loss of water from the concrete, and the rate of loss is governed by the rate of moisture transfer from the interior of the specimen to the surface. This process, as will be seen in Chapter 8, is slow, and, at normal temperatures, shrinkage continues for months rather than hours or days. Logically, it takes longer for a larger specimen to reach a given mean moisture content than for a smaller one, and it follows that large and small specimens

190

exposed to the same drying environment for the same length of time exhibit very different shrinkages. Even if both specimens are allowed to reach equilibrium, the shrinkage of the smaller is likely to be distinctly greater than that of the larger. The experimenter is thus faced with the difficulties that the test may be of extended duration and that different sizes of specimen give different answers. To overcome these problems the British Standard specifies a standard cross-sectional area of specimen of 75 mm x 75 mm, and adopts a system of oven drying to speed the process.

The shrinkages of supposedly identical specimens show a considerable variability, coefficients of variation between 5% and 10% being quite normal. The usual good practice of concrete laboratory work should be followed and the results from at least three, and preferably more, specimens should be averaged when determining shrinkage.

6.2.2 Drying Shrinkage and Hardened Cement Paste

A representative sample of concrete, including the coarse aggregate, necessarily suffers from the effects of size as mentioned in the last

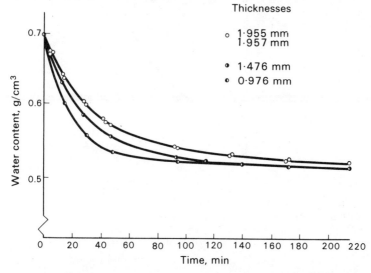

Figure 6.2 Drying rates for thin h.c.p. specimens. (From R. A. Helmuth and D. M. Turk (1967) *Journal of the Portland Cement Association Research & Development Laboratories*, 9, by permission of the Portland Cement Association (USA).)

section. A sample of h.c.p. can be much smaller, and its size is determined more by the practical consideration of handling than by the question of how small it can be without becoming unrepresentative of the material as a whole. It is a reasonable proposition to cast, handle and instrument cement paste samples down to 1 mm in thickness, at which the influence of size is greatly reduced, though, as discussed in Section 6.2.6.2, the potential carbonation shrinkage is greatly enhanced. For such specimens the process of drying is accelerated in comparison with the normal concrete testpiece, and,

191

as can be seen in Fig. 6.2, equilibrium is approached after only a few hours rather than the months needed for concrete. This figure is taken from a most rewarding investigation by Helmuth and Turk (1967), which is one of only a few carried out on thin h.c.p. specimens (with appropriate precautions against carbonation). However, the effects of size are so important and the results of shrinkage tests on large testpieces can be so misleading, that all the examples quoted in this section are taken from those few investigations that have been performed on thin specimens.

The observed volume change is the result of the interaction between the movement of moisture to or from the h.c.p. and the solid skeletal structure of the material. The structure is always in a

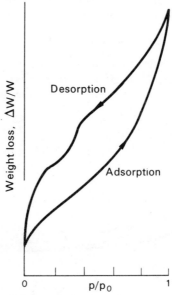

Figure 6.3 Weight change isotherm for h.c.p. (water/cement ratio 0.5). (From R. F. Feldman (1969) *Proceedings of the 5th International Symposium on the Chemistry of Cement, Vol. III*, by permission of the Cement Association of Japan.)

state of internal volumetric stress, the moisture movement changes the level of stress, and the consequent volume change is observed as shrinkage. Likely physical processes causing the internal stresses are described in the next section, and it is sufficient in this discussion of the general characteristics of shrinkage merely to accept that the solid structure of the paste is subjected to internal volumetric stresses that vary in sympathy with the changes in the moisture state.

Drying of h.c.p. at constant temperature yields a continuous curve of relative weight loss, $\Delta W/W$, against decreasing relative water vapour pressure (RWVP) p/p_0. This isotherm, as shown in Fig. 6.3, has an adsorption branch on rewetting a previously dried sample, followed by a desorption branch on re-drying. The experimental method consists of drying (or wetting) by exposure to each of a series of relative vapour pressures, and weighing when equilibrium is reached in each case. Overall the process is reversible, the water loss on drying being the same as the gain on wetting, but between the extremes very different equilibrium water contents are found for adsorption and desorption at a given value of p/p_0. Thus there is a large hysteresis loop. The corresponding plot of shrinkage, ϵ_s, against RWVP is given in Fig. 6.4 and, although the shape is different from

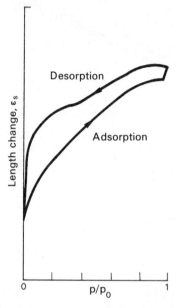

Figure 6.4 Length change isotherm for h.c.p. (water/cement ratio 0.5). (From R. F. Feldman (1969) *Proceedings of the 5th International Symposium on the Chemistry of Cement, Vol. III*, by permission of the Cement Association of Japan.)

that of Fig. 6.3, there is again a large hysteresis loop and an overall reversibility of shrinkage between the extreme vapour pressures. However, it must be added that if very strong drying is followed by a further cycle of rewetting to a high vapour pressure and re-drying, the weight change is still reversible but additional shrinkage occurs that is irreversible.

In engineering practice concrete begins its life in the saturated state, and it is the first drying shrinkage that is most important. This is not covered by the results quoted so far, and it is found that the shrinkage on first drying is distinctly greater than that on redrying; it follows that a substantial part of the first drying shrinkage is ir-reversible. This is shown in Fig. 6.5, in which the successive equilibrium relative humidities are indicated by the digits. It seems that reversibility sets in after first drying to a relative humidity of 50% (i.e. RWVP of 0.5). In this case drying was not strong, the lowest RH being 11%, and the cycles of shrinkage (after first drying) are reversible.

If the h.c.p. is never dried but remains immersed in water, the

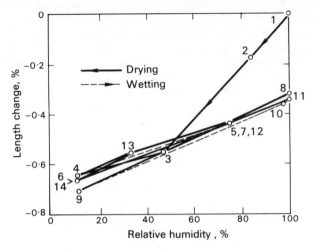

Figure 6.5 Length change of h.c.p. during first drying, followed by cycles of wetting and drying (water/cement ratio 0.6). (From R. A. Helmuth and D. M. Turk (1967) *Journal of the Portland Cement Association Research & Development Laboratories,* 9, by permission of the Portland Cement Association (USA).)

moisture content increases. Firstly, water enters the paste merely to make up the water taken up in hydration, and to keep the material saturated. Secondly, additional water is drawn into the hydrate structure with a consequent increase in volume. This swelling is similar to that observed in some clays such as montmorillonite, and can be described as resulting from the swelling pressure of a gel. In h.c.p. the expansion is resisted by the skeletal structure so that the swelling is only a fraction of the drying shrinkage; the restraint imposed by the structure accounts for the classification of h.c.p. as a limited swelling gel.

In principle, the stronger the structure of the h.c.p. the less it will respond to the forces of swelling or shrinkage. This is borne out by the results given in Fig. 6.6, in which decreasing structural strength is represented by increasing total porosity. The interesting observation is that the reversible shrinkage is *not* influenced by porosity, and the overall trend shown on first drying is entirely due to the irreversible shrinkage.

Pastes of different water/cement ratios were used to provide the variations of porosity shown in Fig. 6.6. Changes in porosity also result from the progress of hydration of pastes with the same water/cement ratio, but the effect of shrinkage is not so simple. The porosity diminishes as the degree of hydration increases, and this should mean that the shrinkage of an h.c.p. cured under water for a long time will be distinctly less than that of a young paste. However, the unhydrated cement grains also provide restraint to the shrinkage forces, and their volume decreases as hydration goes on, thus allowing an increase in shrinkage with time. Furthermore, it can be argued that the more mature paste contains more water of the kind that causes shrinkage, so that drying of such a paste results in higher internal

194

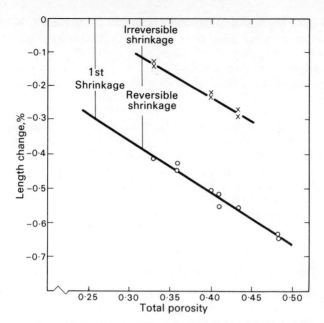

Figure 6.6 Reversible and irreversible shrinkage of h.c.p. after drying at 47% RH (From R. A. Helmuth and D. M. Turk (1967) *Journal of the Portland Cement Association Research & Development Laboratories*, 9, by permission of the Portland Cement Association (USA).)

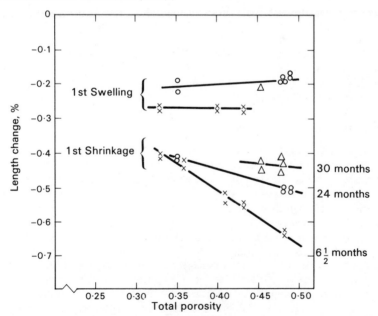

Figure 6.7 The effect of ageing under prolonged moist curing on the reversible and irreversible shrinkages of h.c.p. Curing times of 6½, 24 and 30 months. (From R. A. Helmuth and D. M. Turk (1967) *Journal of the Portland Cement Association Research & Development Laboratories*, 9, by permission of the Portland Cement Association (USA).)

stresses and a greater shrinkage. This is another example, of the sort so common in concrete technology, in which it is hard to deduce the overall effect of opposing influences.

Hydration is virtually complete after say 6 months, and porosity and the proportion of combined water do not change significantly from then on. It is to be expected that the shrinkages of pastes cured for longer periods would not be too different, but this is not the case. This is demonstrated in Fig. 6.7, which shows that prolonged wet curing results in a substantial reduction in the first shrinkage (and in the reversible shrinkage), while the irreversible shrinkage remains much the same. After some 30 months of curing it seems that a situation is attained in which neither the reversible nor the irreversible shrinkage is dependent on the porosity. The explanation lies in an ageing process, which may be regarded as a continuing stabilisation of the solid structure, perhaps related to a chemical change in which the silicates convert to polymeric forms.

6.2.3 The Physical Processes of Drying Shrinkage

Various physical processes have been put forward to account for the characteristics of shrinkage described in the last section. None has been directly observed in h.c.p. itself, but they have established scientific respectability, based on the experimental evidence from materials of a simpler and more regular structure. It is not unreasonable to postulate that they also operate in h.c.p. which is known at least to have similar structural forms, albeit of a more disordered nature. On the other hand the greatly distorted geometry of h.c.p. prevents the application of formulae derived from the appropriate theoretical treatments in other than the most approximate way. There is no real hope of predicting sorption isotherms for even a single physical process, let alone the combined effect of the several processes which overlap and interact in h.c.p.

6.2.3.1 *Capillary Tension*

With drying, capillary (or hydrostatic) tension builds up in the free water which is beyond the influence of the surface forces of the hydration solids. The free water exists both in the capillary pores between aggregations of hydration products, and also in the larger gel pores, and it follows that it is contained in a range of sizes of pores with a variety of different geometries. The pores may be isolated but after first drying are more usually interconnected in a random and tortuous manner.

For the sake of illustration consider the cylindrical pore shown in Fig. 6.8(a), which is full of water and exposed only at the open ends to the atmosphere, initially containing saturated water vapour. As drying occurs the RWVP drops and the radius of the menisci at the open ends of the pore decrease to maintain hygrometric equilibrium between the water and the vapour outside. This also give rise to an equilibrating pressure difference between the phases on either side of the surface. This is expressed by the Laplace equation, giving the

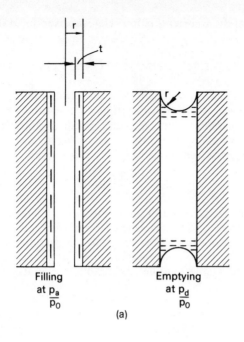

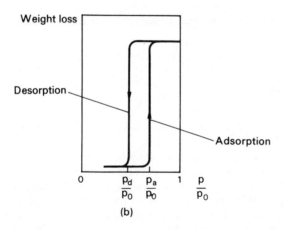

Figure 6.8 Capillary condensation in a cylindrical pore: (a) cylindrical pore, (b) isotherm for a cylindrical pore. (Reproduced, with permission, from T. Z. Harmathy (1967) Technical Paper No. 242, Division of Building Research, National Research Council of Canada.)

pressure difference, Δp,

$$\Delta p = \gamma \left(\frac{1}{r_1} + \frac{1}{r_2} \right) \tag{6.1}$$

where r_1 and r_2 are the radii of the meniscus (here $r_1 = r_2 = r$) and γ is the surface tension of the interface.

In addition, the Kelvin equation relates the radii and the RWVP, p/p_0. Thus

$$\ln \frac{p}{p_0} = \frac{-\gamma(Mv_w)}{RT}\left(\frac{1}{r_1} + \frac{1}{r_2}\right) \qquad (6.2)$$

where M is the molecular weight of water (18.2 g/mol), and v_w is the specific volume of capillary water, thus making (Mv_w) the molal volume of the capillary water. T is the absolute temperature and R is the gas constant.

Then, from Equations 6.1 and 6.2:

$$\Delta p = -\frac{RT}{Mv_w}\ln\frac{p}{p_0} \qquad (6.3)$$

Δp is the hydrostatic tension in the water which acts on the pore surface. It induces a compressive reaction in the solid around the pore, which in turn causes a corresponding contraction in volume of the solid. The contraction is observed as shrinkage, and its magnitude depends on the relative volume of the solid and on its rigidity.

Returning to Fig. 6.8(a), as soon as the radius of the meniscus falls below that of the cross-section of the pore, equilibrium is no longer possible, and the pore rapidly empties. This is shown on the desorption branch in Fig. 6.8(b), in which weight loss is plotted against RWVP (p/p_0). During the earlier stages of drying the capillary tension acts on a constant area of inner pore surface, and the force increases in step with the tension. However, when the pore empties the force falls to zero, not because the tension disappears, but because the area on which it acts becomes zero. The implication is that, with the disappearance of the capillary tensile force, the solid structure is relieved of its compressive force, and the earlier shrinkage is recovered.

On rewetting, a layer of adsorbed water of thickness t forms on inner surface of the pore, followed by an overlying condensate of capillary water with radii $r = r_1 - t$ and $r_2 = \infty$. The radii are again those needed for hygrometric equilibrium, and any further increase in RWVP leads to a lack of stability and the pore fills immediately, giving the adsorption branch of Fig. 6.8(b). At the same time the capillary tension and shrinkage are regenerated, only to disappear when the RWVP reaches saturation once more. This simple model demonstrates the essential features of the capillary tension process. In particular, it shows that hysteresis occurs, and that, assuming no structural changes during the drying and wetting, the material returns to its original state on completion of the whole cycle. Other geometries and pore sizes can be shown to lead to similar adsorption and desorption curves, though with different sizes of hysteresis loop and between different RWVP's. Their integrated effect could quite conceivably look like the continuous h.c.p. sorption curves shown in the earlier figures for the upper range of vapour pressures.

When the RWVP drops to 0.4—0.45 the capillary tension exceeds the breaking stress of water of 1 100 atmospheres, so that there can be no contribution to shrinkage from this source at the lower vapour pressures.

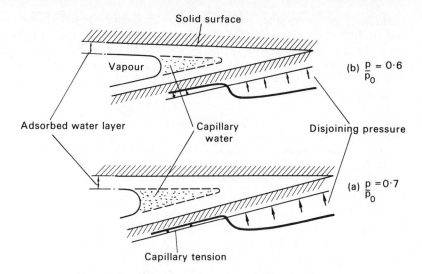

Solid surface

Vapour

(b) $\dfrac{p}{p_0} = 0\cdot6$

Adsorbed water layer

Capillary
water

Disjoining pressure

(a) $\dfrac{p}{p_0} = 0\cdot7$

Capillary tension

Figure 6.9 Water forces in a gel pore at two different relative water vapour pressures (p/p_0). (From Z. P. Bazant (1972) *Cement and Concrete Research*, **2**, 1–16, by permission of Pergamon Press.)

6.2.3.2 *Surface Forces*

Figure 6.9 shows a typical gel pore, narrowing from a wider section containing free water in contact with vapour, to a much narrower space between the solid layers in which all the water is under the influence of the surface forces. The two layers are prevented from moving apart by an interparticle bond or intersolid connection of the sort shown in Fig. 3.5 (see Section 3.2.3). Here the pore is shown in two different states of moisture equilibrium (a) with the vapour at a higher RWVP, with a large radius on the free water meniscus, and (b) at a lower RWVP with a consequently smaller radius needed to preserve the equilibrium.

The adsorbed water forms a layer on the solid surface with a maximum average thickness of five molecules (13 Å) at saturation. For equilibrium with the RWVP the thickness decreases with the RWVP and is down to 1 molecule (2.63 Å) at about 11% RH. Within the narrower part of the pore the gap is insufficiently wide for the full equilibrium thickness to be absorbed, and this is referred to as the area of hindered adsorption. As can be seen, its area diminishes as drying progresses.

There are two possible explanations of the effect of the adsorbed water layer on volume change, and it should be noted that there is no agreed view as to the relative significance of these two alternatives.

(i) *Surface Tension.* The adsorbed water can be regarded as part of the outer layers the solid which carry a surface tension. This sets up a compressive stress system in the inner part of the solid, and a reduction in surface energy, that is in surface tension, leads to a reduction in the compressive stress and an expansion of the solid

199

volume. This occurs when additional water molecules are adsorbed on to the surface when the RWVP rises. The change of surface energy is expressed mathematically by the Gibbs adsorption equation in which a constant surface area a undergoes a decrease in surface energy, $\Delta\gamma$.

Thus,

$$a\Delta\gamma = -RT \int nd\left(\frac{p}{p_0}\right) \tag{6.3}$$

where n is the number of water molecules adsorbed. The change in volumetric strain is then taken to be proportional to $\Delta\gamma$, and it is to be expected that the process will operate at all stages of drying.

(ii) *Disjoining Pressure.* The concept of disjoining pressure derives from the assumption that physical and chemical bonds on the solid surfaces within the areas of hindered adsorption are not satisfied, so that there is a tendency for water molecules to be drawn in from the areas outside. This is equivalent to a pressure gradient on the water in the direction into the narrow spaces. If equilibrium exists it follows that this gradient is neutralised by a reverse pressure exerted by the walls of the solid. Conversely the walls suffer a pressure tending to force them apart, that is to disjoin them. This is the disjoining pressure, the integrated effect of which is to apply a tensile force to the intersolid connection. As drying occurs so both the disjoining pressure and the area on which it acts decrease, and so does the tension in the bond, with a consequent contraction of volume. Thus shrinkage occurs by means of a relief of disjoining pressure. Disjoining pressures have been measured in thin sheet structures, and it seems that they should not be discounted in h.c.p. Although they should occur at all humidities their effects on shrinkage are likely to be most significant at the intermediate and higher humidities, above say 40% RH.

6.2.3.3 *Interlayer Penetration*

The shrinkage processes under the previous two headings were associated with the free water and the adsorbed water. This heading relates to the third kind of evaporable water, namely that held within the hydrate solid, between the molecular layers, that is the inter-layer water (see Section 3.2.4). As the h.c.p. dries, the interlayer water escapes under the drive of the hygrometrically-induced energy gradient. Its intimate contact with the solid and the deviousness of its path to the open air suggest that hard drying under a steep energy gradient is required to move it, and that it is associated with the lower reaches of the sorption isotherm. While this is probably broadly true, the evidence of changes in lattice spacings at higher humidities in other similar materials suggests that at least some inter-layer water will be lost from 80% RH down to 50%.

The process of movement of interlayer water has not received the same degree of detailed scientific attention as capillary tension and the surfaces forces, so that the behaviour can only be summed up in

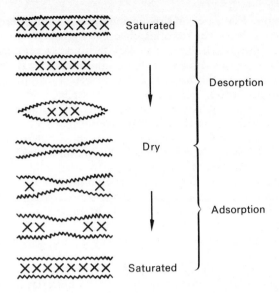

Figure 6.10 Schematic representation of the movement of water from and to an interlayer space. (From R. F. Feldman (1969) *Proceedings of the 5th International Symposium on the Chemistry of Cement, Vol. III,* by permission of the Cement Association of Japan.)

a qualitative way, and demonstrated by means of the simple conceptual model of Fig. 6.10. It is to be hoped that this does not encourage an undervaluation of its importance, for it is likely that it contributes at least as much to shrinkage as the other two processes. Figure 6.10 shows the number and positions of molecules of interlayer water in a typical interlayer space at various vapour pressures common to the adsorption and desorption branches of an isotherm. The model starts at saturation and passes to complete aridity before returning again to saturation. Initially as drying occurs water molecules diffuse slowly from the open ends of the layer, and the presence of the remaining molecules in the centre of the space prevents more than a small volume change. It is only when the last of the molecules move out as the dry state is approached that the solid layers are drawn together and the majority of the shrinkage takes place. The changes in the other direction on rewetting follow the same pattern. Initially molecules penetrate slowly into the outer parts of interlayer space, without causing any great expansion, and it is only when saturation is approached that the centre of the space refills and volume movement becomes more pronounced. Clearly, this cycle reproduces the sort of hysteresis observed for both weight change and shrinkage in h.c.p.

All three physical processes are essentially reversible in that the water removed in drying is returnable on rewetting. However, this depends on the all-important assumption that there is no change in the structure during the humidity cycle, and this is hardly possible, especially during first drying. Thus:

(a) The capillary pores are initially more or less isolated, but the movement of water on first drying opens up the interconnections between them. The resulting changes in geometry are likely to reduce the area on which capillary tension can act during humidity cycles and with it the magnitude of the shrinkage attributable to this process. Since the effect of capillary tension disappears both at saturation and at complete aridity, the formation of capillary interconnections will not alter the overall shrinkage in the complete humidity cycle.

(b) When adsorbed water moves out of the h.c.p. the solid surfaces come somewhat nearer together, the complete closing of the gap between them being resisted by the intersolid connections. Nevertheless some new interparticle bonds will probably form, providing some degree of irreversible consolidation of the structure. Again this will occur mainly on first drying, the surface area of the solid particles will be permanently reduced and with it the effects of both surface tension and disjoining pressure.

(c) The process of consolidation in (6) above leads to an aggregation of layers actually producing an increase in interlayer space (and of interlayer water). However, the original layers are drawn much closer together as the interlayer water dries out, and, in much the same manner as irrecoverable creep, irreversible shrinkage can occur through structural adjustment, either by the formation of new bonds in the collapsed layers, or by slippage between the layers forming a consolidated and more stable structure.

6.2.4   Size Effects in Drying

So far in this chapter consideration has been confined to thin h.c.p. specimens which dry and shrink quickly without the occurrence of significant internal moisture gradients. In larger specimens of h.c.p. and in all concretes, size effects become important. Drying is no longer a rapid process; in a saturated specimen placed in a drier environment, moisture moves gradually to the surface so that, in general, the outside is drier than the inside. That is, a moisture gradient is set up from the interior of the specimen to the outside surface; the shape of the gradient, and the mass transfer equations for its calculation are given in some detail in Chapter 8. If shrinkage is assumed to be directly proportional to moisture loss (a not unreasonable assumption for the conditions of normal atmospheric drying) the same equations can be employed to determine the gradients of free shrinkage. Such a gradient is shown diagrammatically in Fig. 6.11, which refers to the radial gradients in a cylinder. The free shrinkage would only be observed if successive ring elements of the specimen were free to move relative to one another, that is if they were discontinuous. Of course, the material is actually continuous and the larger shrinkages of the outside of the specimen are restrained by the more dormant interior. As shown in Fig. 16.11(b) and (c), the restraint in a long cylinder is so great that the ends of a middle section

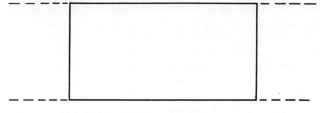

(a) Section of long cylinder

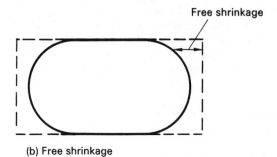

Free shrinkage

(b) Free shrinkage

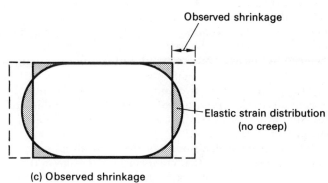

Observed shrinkage

Elastic strain distribution
(no creep)

(c) Observed shrinkage

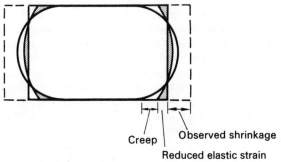

Creep | Observed shrinkage

Reduced elastic strain

(d) The effect of creep

Figure 6.11   Internal strain distributions in a long cylinder drying radially.

203

are plane so that the observed strains on the longitudinal axis and on the surface are identical. This means that the end section carries a distribution of induced stress. If the material is elastic, the difference in strain between the free shrinkage and actual overall uniform strain is the elastic strain distribution associated with the induced stresses (see Fig. 16.11(c)), and the relative positions of the two strain distributions are governed by the need for equilibrium; the integration of the induced stress over the whole section must yield a zero total force. If the material, like concrete, also creeps there will be some relief of the induced stress, giving the strain distribution of Fig. 6.11(d). The implication of this sequence is that measured so-called shrinkage is, in reality, a combination of free shrinkage and elastic and creep strains, the relative proportions of which depends on the moisture gradients. This should be kept in mind throughout the remainder of the shrinkage part of this chapter, which deals exclusively with larger specimens.

6.2.5 Shrinkage of Concrete

6.2.5.1 *The Aggregate*

The various trends introduced previously on the shrinkage of h.c.p. apply also to concrete. It is true that the presence of the aggregate can modify the moisture content of the h.c.p., but this usually has but a secondary effect on the shrinkage; or, there may be some small degree of cracking in the paste because of the local tensile stresses induced by the aggregate restraint, but this does nothing drastic to the form of the relationship between shrinkage and such variables as size of specimen or relative humidity. Rather, the inclusion of aggregate modifies the magnitude of the volumetric strains, and this will usually mean that the shrinkage of concrete is a fraction of that of the paste that it contains.

The normal mechanism is that the h.c.p. shrinks around the non-shrinking aggregate so that compensating tensile (in the h.c.p.) and compressive (in the aggregate) stresses are built up, with the result that the overall volumetric strain is greatly reduced. If, as has been suggested, the h.c.p. cracks, then the stresses are relieved and the volumetric strain is even smaller. The consideration of the concrete as a two-phase material is again appropriate (see also Section 3.1) and two properties are sufficient to quantify the aggregate restraint. They are the ratio of the bulk moduli of aggregate and paste, K_a/K_p, and the volume concentration of the aggregate, g. These property values can be fed into the relevant equation to give the shrinkage of the concrete, ϵ_{sc}, in terms of the shrinkage of the h.c.p., ϵ_{sp}. One such expression derived theoretically for the volumetric behaviour of particles in a shrinking matrix is:

$$\epsilon_{sc} = \epsilon_{sp} - \frac{(\epsilon_{sp} - \epsilon_{sa})2(K_a/K_p)g}{1 + (K_a/K_p) + g(K_a/K_p - 1)} \tag{6.4}$$

The computed linear strains given by this formula are plotted in Fig. 6.12. Other, simpler expressions have been developed and shown

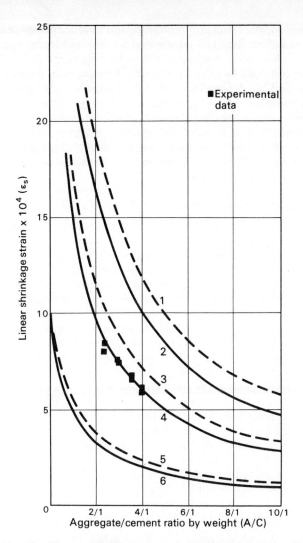

Figure 6.12 The effects on shrinkage of the proportion of aggregate in concrete, and of the relative stiffness of aggregate and h.c.p. Curve 1, $\epsilon_{sp} = 50 \times 10^{-4}$, $K_a/K_p = 4$; Curve 2, $\epsilon_{sp} = 50 \times 10^{-4}$, $K_a/K_p \rightarrow \infty$; Curve 3, $\epsilon_{sp} = 30$, $K_a/K_p = 4$; Curve 4, $\epsilon_{sp} = 30$, $K_a/K_p \rightarrow \infty$; Curve 5, $\epsilon_{sp} = 10$, $K_a/K_p = 4$; Curve 6, $\epsilon_{sp} = 10$, $K_a/K_p \rightarrow \infty$. (From D. W. Hobbs (1971) *Materiaux et Constructions*, **4**, by permission of RILEM.)

to be successful predictors, and Equation 6.4 is quoted here less for its success in estimating shrinkage of real materials than for its admirably succinct statement of the effect of the aggregate. As can also be seen in Fig. 6.12, the shrinkage of a concrete (made with normal hard aggregates) is reduced:

(a) greatly by increasing the volume concentration of the aggregate;
(b) slightly by increasing the rigidity of the aggregate.

The equation also shows that if the aggregate has a shrinkage ϵ_{sa},

205

and some do, the restraint diminishes, and large shrinkages can result.

The equations based on concrete as a two-phase material are likely to be used but rarely because of the lack of starting data. The volume concentration can be calculated readily enough from the mix proportions and the densities of the constituents, but the values of K_a and K_p and above all of ϵ_{sp} itself are not often known. The usefulness of the approach embodied in the equation lies in the realisation that it brings of the way in which the aggregate changes the behaviour of the material, and hence an awareness of the situations in design when shrinkage may be a problem, and an indication of some possible means of solving that problem.

6.2.5.2 *Estimation of Shrinkage in Practice*

Over the years many engineering tests have been performed to measure the shrinkage of concrete. The techniques employed were and are similar to those described in Section 6.2.1, and the strains measured represent the volume change in a certain size and shape of concrete specimen existing in the unloaded state; it is the apparent (so-called) shrinkage rather than the free shrinkage. Furthermore, while the results reflect the findings for h.c.p. given in the earlier part of this chapter, they are expressed in terms of engineering variables, water/cement ratio rather than porosity, or relative humidity rather than moisture content. The combined wisdom of a large number of such tests has been assembled in the form of the prediction charts (see Fig. 6.13) given in the International Recommendations for the Design and Construction of Concrete Structures (CEB 1970).

Design engineers are referred to these charts by the British Code of Practice, CP 110; The Structural Use of Concrete.

On the chart (a) the shrinkage ϵ_c is related to the relative humidity in which the concrete is stored. At 100% RH with no drying the shrinkage is taken to be zero, rising to over 400 microstrain at 40% RH. The chart gives no indication of the swelling that occurs when the concrete is stored under water.

ϵ_c is modified by three further coefficients:

k_b depending on the mix (chart b), expressed by the water/cement ratio and the cement content (rather than porosity and volume concentration of aggregate);

k_t showing the development of shrinkage with time from the start of drying (chart c), on the assumption that the relative humidity of storage remains constant;

k_e for the reduction in shrinkage with the size of member (chart d). This is expressed in terms of the effective thickness, e_m, where

$$e_m = \frac{\text{area of section}}{\text{semi perimeter in contact with the atmosphere}}$$

e_m is effectively a volume/surface ratio which ties in sensibly with the principles of mass transfer.

The charts for k_b and k_t are also used in the estimation of creep (see Section 10.3.7).

206

The final estimated shrinkage of the concrete, ϵ_{cs}, is

$$\epsilon_{cs} = \epsilon_c k_b k_e k_t \tag{6.5}$$

This method of estimation is for Portland cements and, as shown in the next section, there is a considerable variation in the shrinkages between the different types of Portland cement, and even between cements of the same type from different sources. Thus even if the practical conditions are exactly in accordance with those of the charts there will be a range of observed shrinkages, and the estimated values can be no more than the expected mean of such observations. In fact, the concrete in a real structure will be subjected to rather different conditions:

(a) The humidity will not be constant but will vary in an erratic manner, including, for external structures, the effects of rainfall.

(b) The temperature will also vary from the assumed mean of 20 °C, with consequent changes in the degree of hydration and the rate of moisture movement.

(c) The stiffness of the aggregate may be exceptional. The possibility of a soft aggregate is recognised in the Code by the suggestion that the shrinkage of lightweight concrete is between one and two times that of a concrete made with normal aggregate. The recognition of the importance of the stiffness is commendable, but the vagueness of the suggested numerical adjustment hardly induces confidence in the estimation procedure.

(d) The early curing may be inadequate so that there are high losses of moisture in early days, including the period immediately after casting and before the concrete sets (see plastic shrinkage, Section 6.2.6.4). Estimates of shrinkage based on a later starting date may then be greatly in error.

The combination of these uncertainties must lead to a healthy scepticism about the accuracy of the answer. This does not mean that it is valueless, but rather that the engineering calculations should be based on a statistical view of shrinkage, in much the same way as strength (see Section 13.7.2.4).

6.2.6 Shrinkage and Chemistry

6.2.6.1 *Composition of Cement*

In Chapter 3 the composition of cement was given in terms of the proportions of the main constituents, C_3A, C_2S, C_3S and C_4AH.

It is logical to refer to these proportions when examining the effects of composition on the properties of the h.c.p. This was done for the shrinkage of no less than 199 different Portland cements in the USA (Blaine, 1968), most of which could be classified as ordinary Portland cement but with some which were sulphate resisting or rapid hardening and a few which were low-heat cements. The shrinkages after eight weeks of concrete testpieces of 8 in by 6 in section

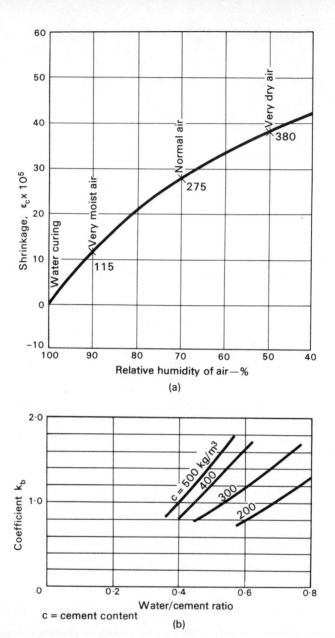

(a)

(b)

c = cement content

made with the various cements were compared and it was found that they lay between 150 and 420 microstrain, with a mean of 300. As all other variables were kept constant this remarkably large range of results can be attributed to the differences in the cements. An analysis of statistical significance showed that a number of factors could be influential. Changes in the C_3A content and in the sulphate content (SO_3, from the addition of gypsum) were found to be responsible for the greatest variations in shrinkage, but other variables

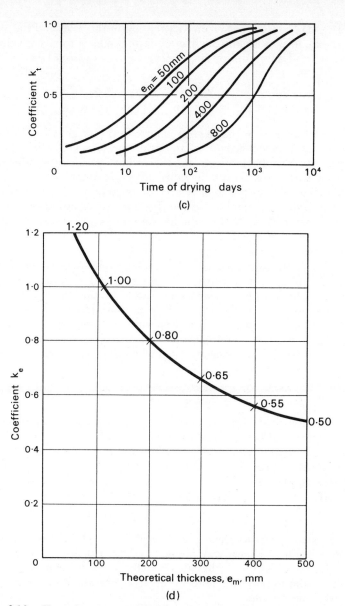

Figure 6.13   Charts for estimating the magnitude of shrinkage for use in design. (From *International Recommendations for the Design and Construction of Concrete Structures* (1970), CEB/FIP.)

such as the alkalis, the fineness of the cement and the air content were also significant. The final picture is far from clear-cut, and it is quite impossible to rank cements in their order of shrinkage. The conclusions to be drawn from the figures are that the composition of the cement does matter for shrinkage, and that even in a smaller country like Britain estimated values should be viewed with a proper caution.

The SO_3 content has received attention in the past, and the gypsum addition is controlled at an optimum level to give maximum strength and minimum shrinkage. Typical figures from Lea (1970) are:

SO_3 content %	1.06	2.23	3.21	3.97
shrinkage %	0.142	0.132	0.099	0.101

These illustrate both the great change in shrinkage with SO_3 content and the occurrence of an optimum value. Unfortunately, some cases have been demonstrated in which no such optimum was reached within the acceptable range of SO_3 contents, and any small variation in SO_3 would then be accompanied by a significant variation in shrinkage. The relevance of the C_3A content to shrinkage and the known effect of the SO_3 on the hydration of C_3A suggest that it is the SO_3–C_3A interaction that is responsible for the influence of the SO_3 on shrinkage.

The sensitivity of shrinkage to chemical composition points to the possible unhappy performance of concretes with chemical admixtures. Fortunately, most have been shown not to produce any alarming volume movements. Apart from gypsum, the exception is the accelerator, calcium chloride, $CaCl_2$. Experiments have indicated that an increase in the $CaCl_2$ content leads to higher shrinkage, and in one extreme case the shrinkage doubled when the $CaCl_2$ content was raised from 0 to 2%.

6.2.6.2 *Carbonation Shrinkage*

Carbonation shrinkage differs from drying shrinkage in that its cause is chemical, and has nothing to do with loss of water from the h.c.p. Carbon dioxide in combination with water, as carbonic acid, attacks the hydration products, and even the low concentration of CO_2 in the atmosphere is sufficient to have an appreciable effect. The most important reaction is that with calcium hydroxide ($Ca(OH)_2$). Thus

$$CO_2 + Ca(OH)_2 = CaCO_3 + H_2O$$

Water is released, and there is an increase in weight of the h.c.p. The volume decreases because, it is suggested, the calcium hydroxide is dissolved and calcium carbonate crystallises out in the pores. This view is supported by the observation of a decrease in permeability of a carbonated paste.

Fortunately the penetration of the carbonic acid is extremely slow so that its effect is encountered only on the outer layers, say up to 50 mm thick, of concrete members, and the phenomenon is associated with surface shrinkage such as crazing. Carbonation cannot occur if the carbonic acid ingredients do not infiltrate, and this is the case if the concrete remains saturated so that the carbon dioxide cannot enter, or if it is very dry when there is a paucity of water. For the intermediate situation of drying at 50% RH, carbonation shrinkage reaches a maximum and is of the same order as drying shrinkage.

210

6.2.6.3 *Expanding Cements*

Shrinkage is a nuisance to engineers and the idea of a cement whose shrinkage is suppressed or counteracted is attractive. Even better, a cement which expands on hydration presents the engineer with the pleasing prospect of automatically prestressing any included steel. The two functions are recognised individually in the terms shrinkage-compensation and chemical prestressing.

These desirable properties can be achieved by the inclusion of an expansive additive, which usually makes up about a tenth of the total cement. Several versions have been developed by somewhat different manufacturing processes, but the basis of their action is common in the early formation of hydrated calcium sulphoaluminate (ettringite) in a controlled manner.

The volume movements are indicated in Fig. 6.14 in comparison with the movement of an h.c.p. without additive. The expansion

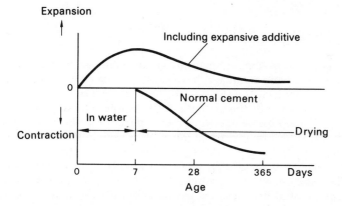

Figure 6.14 Shrinkage compensation using expanding cement.

occurs soon after mixing and is completed in a few days. The later drying shrinkage occurs unabated, and, for a shrinkage-compensating function finally produces a zero volume change. With higher proportions of additive, the expansion is greater and is far from cancelled by the drying shrinkage, thus performing the chemical prestressing function.

There are problems in practice in that it is difficult to ensure that the necessary chemical and curing conditions obtain in order to achieve the predetermined volume change. If the concrete dries too soon, the full expansion may not be reached, and overdosage may cause catastrophic disintegration of the concrete. Thus, while the potentialities of the additives are exciting, their use demands understanding and careful control.

6.2.6.4 *Plastic Shrinkage*

The adjective 'plastic' does not here carry its definition from solid mechanics, but refers rather to concrete in its fresh state when, in the

211

first few hours of its life, it has not set and has no structural strength. It has been found in practice that quite wide cracks can develop at this stage at or near to the surface. There is no doubt that cracking of this sort can occur when the surface is subjected to rapid drying, either in hot climates or under a high wind velocity. In these circumstances large shrinkages do occur through the loss of water. It has also been observed that cracks sometimes occur when the surface is not subjected to rapid drying; indeed it may be covered with a water layer. It is suggested that this is associated with bleeding and settlement, in which, in less cohesive mixes, the water separates and rises to the surface, and the solids sink. The tendency to crack is encouraged by the presence of reinforcing steel or of other impediments to the settlement process. It is also possible that thermal gradients caused by the heat of hydration contribute to the early formation of cracks, and this view is supported by the evidence of cracking in deep foundations made from pumped concrete.

6.3   THERMAL EXPANSION

The deformations and temperature gradients in concrete are analogous to those of drying shrinkage; both cause volume change, the former resulting from heat transfer and the latter from moisture transfer. However, in spite of this similarity of principle, a difference in treatment is often possible in engineering applications. This is because thermal equilibrium is reached much more rapidly than moisture equilibrium, and it is possible to consider thermal expansion in a simple manner through an instantaneous coefficient of expansion.

The measurement of thermal deformation is also simpler than the measurement of drying shrinkage, because the time-dependent effects can be ignored merely by waiting for a short period (a few hours at most) for thermal equilibrium before taking readings. The experimental method is thus merely a question of recording a linear dimension at each of several different temperatures; this apparent simplicity is, in practice, somewhat diminished by the necessity of allowing for any possible thermal deformation of the measuring apparatus itself. In more sophisticated research work, the simplicity disappears as time dependence cannot be ignored. Not only must the time to reach thermal equilibrium be considered, but also the time dependence of the thermal deformation itself. Unlike many other materials, h.c.p. and concrete do not respond instantaneously to a change of temperature and reasons for their delayed movement are given in the next section.

6.3.1   Response of Hardened Cement Paste to Temperature Change

Careful experimental work on the thermal deformation of h.c.p. has shown not only that there is time dependence but also that there is a reversal of sign; an immediate expansion is followed by a time-dependent contraction (and vice versa). Bazant (1970) described

212

three appropriate physical processes to account for these observations, and these are stated below in relation to the structural concepts of Chapter 3.

6.3.1.1 *Pure Thermal Dilatation*

This consists mainly of normal thermal deformation, taking place immediately, and representing the response of the molecular structure to the heat input corresponding to a given increase of temperature. The structure consists of various components which respond somewhat differently to temperature, and may interact with each other in a time-dependent manner. It is suggested here that the water in the smaller spaces has a larger coefficient of expansion than the solid around the space, and this must lead to equilibriating compressive forces (i.e. pressure) in the water and tension forces in the solid. With time, the water moves under the influence of the pressure which dissipates as does the tension in the solid. The resulting time-dependent contraction is sketched in Fig. 6.15(a) which indicates the trends for different relative water vapour pressures. It can be seen that if the h.c.p. is dry ($p/p_0 = 0$) there can be no water pressure and the thermal deformation occurs immediately.

6.3.1.2 *Thermal Shrinkage*

An increase in temperature causes water to move to positions where it is less firmly bonded. This is equivalent to drying, and shrinkage therefore occurs. Alternatively this may be stated in terms of free energy. For the adsorbed water, the temperature increase causes a smaller decrease in the free energy of the hindered adsorbed water than in the thicker unhindered waters, thus giving rise to an energy gradient and a movement of water out of the hindered region. The effect, a contraction, is shown in Fig. 6.15(b).

6.3.1.3 *Hygrometric Dilatation*

In this third process attention is directed at the larger pores in the h.c.p. If no exchange of moisture with the atmosphere occurs, then an increase in temperature causes a corresponding increase in pressure in the water vapour and, it is suggested, a decrease in the surface tension of the free water. This reduces the reactive compressive forces in the surrounding solid with a consequent expansion, observed immediately the temperature is raised. As shown in Fig. 6.15(c), this is followed by a time-dependent expansion, which is attributed to a higher relative humidity in the water vapour resulting from the temperature rise. Hygrometric equilibrium is restored slowly by movement of water into the smaller spaces and layers, with an accompanying swelling. There is no effect both when the h.c.p. is dry and when it is saturated, because in neither case can there be a change in vapour pressure.

The total effect of the three processes is sketched in Fig. 6.16 and the trends shown are in agreement with experimental observations.

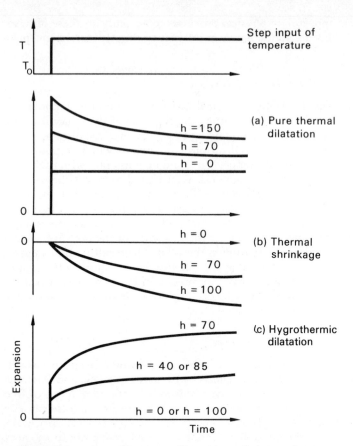

Figure 6.15 Components of response of h.c.p. to a step input of temperature. For various relative humidities (h in %). (From Z. P. Bazant (1970) *Nuclear Engineering and Design*, 14, 308, by permission of North-Holland Publishing Company.)

The coefficient of thermal expansion, measured by normal methods should be enhanced at intermediate water states which offer scope for temperature-induced change in vapour pressure. As shown in Fig. 6.17, this sort of behaviour has been observed.

6.3.2 Thermal Expansion of Concrete

Concrete members are massive in comparison with the small h.c.p. specimens used in the research mentioned in the last section, and where necessary temperature distributions can be found in a similar manner to those for shrinkage using a diffusion type equation (see Section 8.3).

In practice structurally important temperature gradients arise in, for example, the dispersal of the heat of hydration in massive foundations, or in the response of a pavement to the continuously varying diurnal temperatures. It is in the more common problems of overall movements of bridge decks or building frames that time dependence

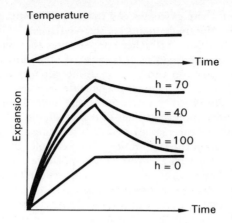

Figure 6.16 Overall strain caused by a temperature increase. (From Z. P. Bazant (1970) *Nuclear Engineering and Design,* 14, 308, by permission of North-Holland Publishing Company.)

can be ignored, and the dimensional changes calculated as an immediate result of the temperature variations. In contrast, shrinkage movements in these structures develop very gradually over months or even years as previously described.

The temperature in the above applications lies within the range 0–60 °C. This is fortunate in that the behaviour of the concrete is

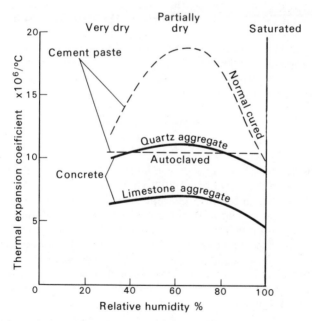

Figure 6.17 Effect of dryness on the thermal expansion of h.c.p. and concrete. (From S. L. Meyers (1950) *Proceedings of the Highway Research Board,* 30, 193, by permission of the Transportation Research Board, National Academy of Sciences.)

215

stable between these extremes and the thermal response can then be represented by the single quantity, the coefficient of linear expansion α, which is a constant. At higher temperatures the value changes and the situation becomes confused because of breakdown of the material through decomposition and because of incompatibility between the thermal coefficients of h.c.p. and aggregate (see Section 8.3.1). The remainder of this chapter is devoted to the lower temperature range and concentrates upon the values of α that should be used in analysis.

The presence of the aggregate has a restricting action on the phenomena introduced in the last section. For example, the enhancement of thermal deformation at intermediate moisture states is not nearly so pronounced in concrete as it is in h.c.p., as can be seen in Fig. 6.17. The time dependence can also be neglected so that α is not only a constant (for a given concrete) but also characterises a movement following instantaneously in step with change of temperature. In fact, things are even better for the engineer in that the water/cement ratio has little effect on α, and, except for very young concrete, nor does the age. This means that the only mix variables that matter concern the aggregate.

6.3.2.1 *Effect of the Aggregate*

In general, the coefficient of expansion of the h.c.p. is greater than that of the aggregate, and the combined effect, in concrete, can be found from an appropriate law of mixtures. Once again the comparison with shrinkage is apt, the aggregate acting as a restraining influence on the movement of the paste. The relevant variables are the volume concentration (g) and relative stiffness of the aggregate (K_a/K_p) and an equation like that for shrinkage (see Equation 6.4) is again appropriate. Thus the coefficient of expansion for the concrete, α_c, can be related to those for paste and aggregate, α_p and α_a.

$$\alpha_c = \alpha_p - \frac{(\alpha_p - \alpha_a) \cdot g \cdot 2\left(\dfrac{K_a}{K_p}\right)}{1 + \left(\dfrac{K_a}{K_p}\right) + g\left(\dfrac{K_a}{K_p} - 1\right)} \qquad (6.6)$$

The influence of the aggregate volume is shown in Fig. 6.18, in which a typical value for α_p of $20 \times 10^{-6}/^\circ C$ is assumed. The two dotted lines then approximately represent the limits of the range of normal aggregates. The vertical lines define the usual volume concentrations, and it is immediately apparent that most concretes have coefficients of thermal expansion close to those of the aggregates that they include. As a rough guide it is the siliceous aggregates such as quartzite and sandstone which have the higher values of α_a, while limestones have the lowest; large variations can be expected within any one rock type.

6.3.2.2 *Code of Practice*

For the purposes of design the British Code of Practice, CP 110 (1973), follows the simplest method of suggesting a single value for

216

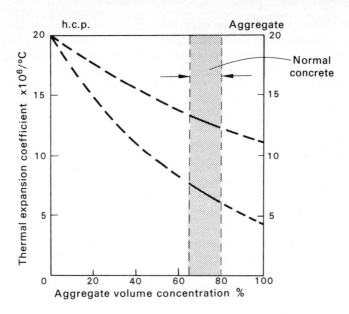

Figure 6.18   The effect on the coefficient of thermal expansion of concrete of the volume concentration of aggregate. (From R. D. Browne (1972) *Concrete*, 6, 51.)

the coefficient of linear expansion for all normal concretes — and to make it even easier that number is the nice round one of $10 \times 10^{-6}/°C$. The Recommendations of the CEB (1970) go rather further and give a value of $8 \times 10^{-6}/°C$ for lightweight concrete together with a statement of the ranges likely for both categories.

6.4   REFERENCES

Bazant, Z. P. (1970) Delayed thermal dilatations of cement paste and concrete due to mass transport, *Nuclear Engineering and Design*, 14, 308.

Bazant, Z. P. (1972) Thermodynamics of hindered adsorption and its implications for hardened cement paste and concrete, *Cement and Concrete Research*, 2, 1.

Blaine, R. L. (1968) A statistical study of the shrinkage of neat cements and concretes, *International Colloquium on the shrinkage of hydraulic concretes*, RILEM/Cembureau, Madrid, Paper I—M.

BS 1881 (1970) *Methods of testing concrete for other than strength, Part 5, Methods of testing concrete*, British Standards Institution, London.

CP 110 (1973) *The Structural use of Concrete, Part 1*, British Standards Institution, London.

Browne, R. D. (1972) Thermal movement of concrete, Current practice sheet 3PC/06/1, *Concrete*, 6, 51.

International recommendations for the design and construction of concrete structures (1970), CEB/FIP.

Feldman, R. F. (1969) Sorption and length-change scanning isotherms of methanol and water on hydrated Portland cement, *Proceedings of the Fifth International Symposium on the Chemistry of Cement*, **III**, 53, The Cement Association of Japan.

Harmathy, T. Z. (1967) Moisture sorption of building materials, *Technical Paper No. 242 of the Division of Building Research*, National Research Council of Canada, Ottawa.

Helmuth, R. A. and Turk, D. H. (1967) The reversible and irreversible drying shrinkage of hardened Portland cement and tricalcium silicate pastes, *Journal of the Portland Cement Association R and D Laboratories*, 9.

Hobbs, D. W. (1971) The dependence of the bulk modulus, Young's modulus, creep, shrinkage and thermal expansion of concrete upon aggregate volume concentration, *Matériaux et Constructions*, 4, 107.

Lea, F. M. (1970) *The chemistry of cement and concrete*, Arnold, London.

Meyers, S. L. (1950) Thermal expansion characteristics of hardened cement paste and concrete, *Proceedings of the Highway Research Board*, 30, 193.

CHAPTER 7

Movement in Timber

7.1 INTRODUCTION

Timber in an unstressed state may undergo dimensional changes following variations in its moisture content and/or temperature. The magnitude and consequently the significance of such changes in the dimensions of timber is much greater in the case of alterations in moisture content compared with temperature. Although thermal movement will be discussed, the greater emphasis in this chapter is placed on the influence of changing moisture content.

7.2 MOISTURE IN TIMBER

Timber is hygroscopic, that is it will absorb moisture from the atmosphere if it is dry and correspondingly yield moisture to the atmosphere when wet, thereby attaining a moisture content which is in equilibrium with the water vapour conditions of the surrounding atmosphere. Thus for any combination of relative humidity and temperature of the atmosphere there is a corresponding moisture content of the timber such that there will be no inward or outward diffusion of water vapour: this moisture content is referred to as the *equilibrium moisture content.*

The fundamental relationships between moisture content and atmospheric conditions have been determined experimentally and the average equilibrium moisture content values are shown graphically in Fig. 7.1. A timber in an atmosphere of 20 °C and 22% relative humidity will have a moisture content of 6% (see below), while the same timber if moved to an atmosphere of 40 °C and 64% relative humidity will double its moisture content. It should be emphasised that the curves in Fig. 7.1 are average values and that slight variations in moisture content will occur due to differences between timbers or to the previous history of the timber.

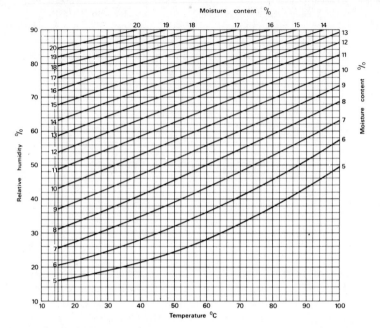

Figure 7.1 Chart showing the relationship between the moisture content of wood and the temperature and relative humidity of the surrounding air; approximate curves based on values obtained during drying from green condition. (From the Princes Risborough Laboratory, Building Research Establishment. © Crown copyright.)

7.2.1 Moisture Content

It is customary to express the moisture content of timber in terms of its oven-dry weight using the equation

$$\mu = 100 \frac{W_m - W_0}{W_0} \tag{7.1}$$

where W_m is the weight of wet timber, W_0 is the weight of timber after oven drying at $105\,°C$ and μ is the moisture content (%).

The expression of moisture content of timber on a dry weight basis is in contrast to the procedure adopted for other materials where moisture content is expressed in terms of the wet weight of the material.

Determination of moisture content in timber is usually carried out using the basic gravimetric technique, though it should be noted that at least a dozen different methods have been recorded in the literature. Suffice it here to mention only two of these alternatives. First, where the timber contains volatile extractives which would normally be lost during oven drying, thereby resulting in erroneous moisture content values, it is customary to use a distillation process, heating the timber in the presence of a water-immiscible liquid such as toluene, and collecting the condensed water vapour in a calibrated trap. Second, where ease and speed of operation are preferred to

220

extreme accuracy, moisture contents are assessed using electric
moisture meters: these may be either DC in operation, measuring the
change in resistivity of wet timber compared with dry, or AC in
operation, determining the increase in dielectric constant and loss
tangent that occurs with higher moisture contents. These meters
require calibrating and different scales are present for different
groups of timbers.

7.2.2 Dimensional Changes

In timber it is customary to distinguish between those changes that
occur when green timber is dried to very low moisture contents, and
those that arise in timber of low moisture content due to seasonal or
daily changes in the relative humidity of the surrounding atmosphere.
The former changes are called *shrinkage* while the latter are known
as *movement*.

7.2.2.1 Shrinkage

In the living tree, water is to be found not only in the cell cavity but
also within the cell wall. Consequently the moisture content of green
wood (newly felled) is high, usually varying from about 60% to nearly
200% depending on the location of the timber in the tree and the
season of the year. However, seasonal variation is slight compared to
the differences that occur within a tree between the sapwood and
heartwood regions. The degree of variation is illustrated for a number
of softwoods and hardwoods in Table 7.1: within the former group

Table 7.1 Average Green Moisture Contents
of the Sapwood and Heartwood

Botanical name	Commercial name	Moisture content (%)	
		Heartwood	Sapwood
Hardwoods			
Betula lutea	Yellow birch	64	68
Fagus grandifolia	American beech	58	79
Ulmus americana	American elm	92	84
Softwoods			
Pseudotsuga menziesii	Douglas fir	40	116
Tsuga heterophylla	Western hemlock	93	167
Picea sitchensis	Sitka spruce	50	131

the sapwood may contain twice the percentage of moisture to be
found in the corresponding heartwood, while in the hardwoods this
difference is appreciably smaller or even absent. Nevertheless, in
general terms, the weight of water present in newly felled timber is
approximately equal to the weight of cell wall material of the timber.
    Green timber will yield moisture to the atmosphere with conse-
quent changes in its dimensions: at moisture contents above 20%

221

many timbers, especially their sapwood, are susceptible to attack by fungi: the strength and stiffness of green wood is considerably lower than for the same timber when dry. For all these reasons it is necessary to dry or *season* timber following felling of the tree and prior to its use in service.

Drying or seasoning of timber can be carried out in the open, preferably with a top cover. However, it will be appreciated from the previous discussion on equilibrium moisture contents that the minimum moisture content that can be achieved is determined by the lowest relative humidity of the summer period. In this country it is seldom possible to achieve moisture contents less than 16% by air-seasoning. The planks of timber are separated in rows by stickers which permit air currents to pass through the pile; nevertheless it may take from two to ten years to air-season timber.

The process of seasoning may be accelerated artificially by placing the stacked timber in a drying kiln, basically a large chamber in which the temperature and humidity can be controlled and altered throughout the drying process: the control may be carried out manually or programmed automatically. Humidification is sometimes required in order to keep the humidity of the surrounding air at a desired level when insufficient moisture is coming out of the timber; it is frequently required towards the end of the drying run and is achieved either by the admission of small quantities of live steam or by the use of water atomisers or low-pressure steam evaporators. Various designs of kilns are used and these are reviewed in detail by Pratt (1974).

Drying of timber in a kiln can be accomplished in 2–5 days, the optimum rate of drying varying widely from one timber to the next. Following many years of experimentation kiln schedules have been published representing a compromise between time on the one hand and degree of degrade, splitting and twisting on the other (Pratt, 1974). Most timber is now seasoned by kilning: little air drying is carried out.

As previously mentioned in the introduction to this chapter, water in green or freshly felled timber is present both in the cell cavity and within the cell wall. During the seasoning process, irrespective of whether this is by air or within a kiln, water is first removed from within the cell cavity: this holds true down to moisture contents of about 27–30%. Since the water in the cell cavities is free, not being chemically bonded to any part of the timber, it can readily be appreciated that its removal will have no effect on the strength or dimensions of the timber. The lack of variation of the former parameter over the moisture content range of 110–27% is illustrated in Fig. 7.2.

However, at moisture contents below 27–30% water is no longer present in the cell cavity but is restricted to the cell wall where it is chemically bonded (hydrogen bonding) to the matrix constituents, to the hydroxyl groups of the cellulose molecules in the noncrystalline regions and to the surface of the crystallites. The uptake of water by the lignin component is considerably lower than that by either the hemicellulose or the amorphous cellulose: water may be present as a

222

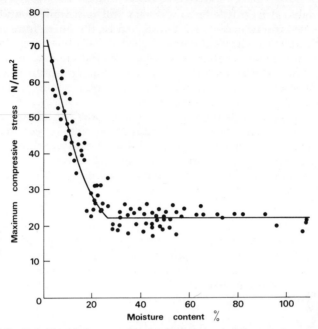

Figure 7.2 Relationship between longitudinal compressive strength and moisture content. (From the Princes Risborough Laboratory, Building Research Establishment. © Crown copyright.)

monomolecular layer though frequently up to six layers can be present. Water cannot penetrate the crystalline cellulose since the hygroscopic hydroxyl groups are mutually satisfied by the formulation of both intra- and intermolecular bonds within the crystalline region as described in Chapter 4. This view is confirmed from X-ray analyses which indicate no change of state as the timber gains or loses moisture.

However, the percentage of noncrystalline material in the cell wall varies between 8 and 33% and the influence of this fraction of cell wall material as it changes moisture content on the behaviour of the total cell wall is very significant. The removal of water from these areas within the cell wall results first in increased strength and secondly in marked shrinkage. Both changes can be accounted for in terms of drying out of the water-reactive matrix, thereby causing the microfibrils to come into closer proximity, with a commensurate increase in interfibrillar bonding and decrease in overall dimensions. Such changes are reversible, or almost completely so.

Fibre Saturation Point.   The increase in strength on drying is clearly indicated in Fig. 7.2, from which it will be noted that there is a threefold increase in strength as the moisture content of the timber is reduced from about 27% to zero. The moisture content corresponding to the inflexion in the graph is termed the *fibre saturation point,* where in theory there is no free water in the cell cavities while the walls are holding the maximum amount of bound water. In practice

223

this rarely exists; a little free water may still exist while bound water is removed from the cell wall. Consequently, the fibre saturation 'point', while a convenient concept, should really be regarded as a 'range', in moisture contents over which the transition occurs.

The fibre saturation point therefore corresponds to the moisture content of the timber when placed in a relative humidity of 100% and it is generally found that the moisture content of hardwoods at this level are from 1 to 2% higher than for softwoods. At least nine different methods of determining fibre saturation point are recorded in the literature. These range from extrapolation of moisture content absorption isotherms to unit relative vapour pressure, to the determination of the inflection point in plotting the logarithm of electrical conductivity against moisture content.

Anisotropy in Shrinkage. The reduction in dimensions of the timber, technically known as *shrinkage*, can be considerable but, owing to the complex structure of the material, the degree of shrinkage is different on the three principal axes: in other words timber is anisotropic in its water relationships. The variation in degree of shrinkage that occurs between different timbers, and, more important, the variation among the different axes is illustrated in Table 7.2. It

Table 7.2   Shrinkage (%) on Drying from
Green to 12% Moisture Content

| Botanical name | Commercial name | Transverse | | Longi-tudinal |
		Tangen-tial	Radial	
Chlorophora excelsa	Iroko	2.0	1.5	<0.1
Tectona grandis	Teak	2.5	1.5	<0.1
Pinus strobus	Yellow pine	3.5	1.5	<0.1
Picea abies	Whitewood	4.0	2.0	<0.1
Pinus sylvestris	Redwood	4.5	3.0	<0.1
Tsuga heterophylla	Western hemlock	5.0	3.0	<0.1
Quercus robur	European oak	7.5	4.0	<0.1
Fagus sylvatica	European beech	9.5	4.5	<0.1

should be noted that the values quoted in the table represent shrinkage on drying from the green state (i.e. >27%) to 12% moisture content, a level which is of considerable practical significance since at 12% moisture content timber is in equilibrium with an atmosphere having a relative humidity of 60% and a temperature of 20 °C; these conditions would be found in buildings having regular but intermittent heating.

From Table 7.2 it will be observed that shrinkage ranges from 0.1% to 10%, i.e. a 100-fold range. Longitudinal shrinkage, it will be noted, is always an order of magnitude less than transverse, while in the transverse plane radial shrinkage is usually some 60–70% of the corresponding tangential figure.

The anisotropy between longitudinal and transverse shrinkage amounting to 40:1 is due in part to the arrangement of cells in timber and in part to the particular orientation of the microfibrils in the middle layer of the secondary cell wall (S_2).

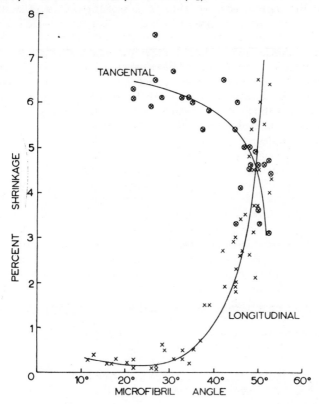

Figure 7.3   The relationship between longitudinal and tangential shrinkage and microfibril angle of the S_2 layer in *Pinus jeffreyi*. (From B. A. Meylan (1968) *Forest Products Journal*, 18 (4), by permission of the Forest Products Research Society.)

Since the microfibrils of the S_2 layer of the cell wall are inclined at an angle of about 15° to the vertical, the removal of water from the matrix and the consequent movement closer together of the microfibrils will result in a horizontal component of the movement considerably greater than the corresponding longitudinal component. This, of course, is apparent in practice (see Table 7.2) though the relationship between shrinkage and microfibrillar angle is nonlinear (Fig. 7.3). The variation in shrinkage with angle for tangential shrinkage is, as might be expected, almost the mirror image of that for longitudinal shrinkage. The cross-over point, when the degree of anisotropy is zero, occurs experimentally at a microfibrillar angle of about 48°. Although microfibrillar angle is of supreme importance, it is apparent that the relationship is complex and that other factors, as yet unknown, are playing additional though minor roles.

225

Various models have been used to account for shrinkage in terms of microfibrillar angles. These generally consider the cell wall to consist of an amorphous hygroscopic matrix in which are embedded parallel crystalline microfibrils which restrain swelling or shrinking of the matrix. Early models considered part of the wall as a flat sheet consisting only of an S_2 layer in which microfibrillar angle has a constant value. Later models have treated the cell wall as two equal-thickness layers having microfibrillar angles of equal and opposite sense, and these two-ply models have been developed extensively over the years to take into account the layered structured of the cell wall, differences in structure between radial and tangential walls, and variations in wall thickness. Although models such as these are still relatively crude in simulating the anatomical and chemical properties of wood, the degree of agreement between calculated and experimental values is usually very good.

The influence of microfibrillar angle on degree of longitudinal and transverse shrinkage described for normal wood is supported by evidence derived from experimental work on *compression wood*, one of the forms of reaction wood described in Chapter 4. Compression wood is characterised by possessing a middle layer to the cell wall, the microfibrillar angle of which can be as high as $45°$ though $20-30°$ is more usual. The longitudinal shrinkage is much higher and the transverse shrinkage correspondingly lower than in normal wood and it has been demonstrated that the values for compression wood can be accommodated on the shrinkage/angle curve for normal wood.

Differences in the degree of transverse shrinkage between tangential and radial planes (Table 7.2) can be explained in terms of: first, the restricting effect of the rays on the radial plane; second, the increased thickness of the middle lamella on the tangential plane compared with the radial; third, the difference in degree of lignification between the radial and tangential cell walls; fourth, the small difference in microfibrillar angle between the two walls; and fifth, the alternation of earlywood and latewood in the radial plane, which, because of the greater shrinkage of latewood, induces the weaker earlywood to shrink more tangentially than it would do if isolated. Considerable controversy reigns as to whether all five factors are actually involved and their relative significance.

Volumetric shrinkage, s_v, is slightly less than the sums of the three directional components and is given by:

$$s_v = 100[1 - (1 - 0.01s_l)(1 - 0.01s_r)(1 - 0.01s_t)] \qquad (7.2)$$

where the shrinkages are in percentages. This simplifies to

$$s_v = s_l + s_r + s_t - 0.01s_r s_t$$

and subsequently to

$$s_v = s_r + s_t \qquad (7.3)$$

as greater approximations become acceptable.

Sorption and Diffusion. Timber, as already noted, assumes with the passage of time a moisture content which is in equilibrium with the relative vapour pressure of the atmosphere. This process of water sorption is typical of solids with a complex capillary structure and this phenomenon has already been mentioned for concrete. The similarity in behaviour between timber and concrete with regard to moisture relationships is further illustrated by the presence of S-shaped isotherms when moisture content is plotted against relative vapour pressure. Both materials have isotherms which differ according to whether the moisture content is reducing (desorption) or increasing (adsorption) thereby producing a hysteresis loop (Fig. 7.4).

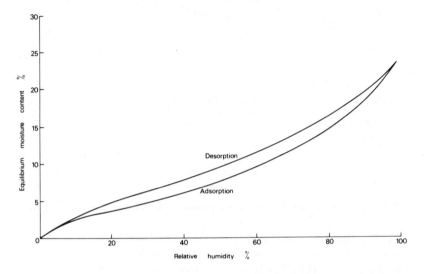

Figure 7.4 Hysteresis loop resulting from the average adsorption and desorption isotherms for six species of timber at 40 °C. (From the Princes Risborough Laboratory, Building Research Establishment. © Crown copyright.)

The hysteresis coefficient, A/D, is defined as the ratio of the equilibrium moisture content for adsorption to that for desorption for any given relative vapour pressure. For timber this varies from 0.8 to 0.9 depending on species of timber and on ambient temperature. The effect of increasing temperature on reducing equilibrium moisture content of timber is illustrated in Fig. 7.5.

The hysteresis effect in sorption was at first explained in terms of differences in contact angle of the advancing and receding water front within the cell cavities which are considered to be capillaries. Later the hysteresis effect was related to the behaviour of the hydroxyl groups in both the cellulose and lignin: in drying these groups are thought to satisfy each other and on rewetting some continue to do so and are not available for water adsorption. According to a third hypothesis plasticity is considered to be the principal cause of hysteresis and is related to irreversible inelastic exchanges of

227

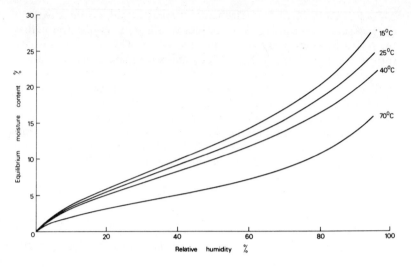

Figure 7.5 Average desorption isotherms for six species at four temperatures. (From the Princes Risborough Laboratory, Building Research Establishment. © Crown copyright.)

hydroxyl groups between neighbouring cellulose molecules (Barkas, 1949).

Practical Significance. In order to avoid shrinkage of timber after fabrication, it is essential that it is dried down to a moisture content which is in equilibrium with the relative humidity of the atmosphere in which the article is to be located. A certain latitude can be tolerated in the case of timber frames and roof trusses, but in the production of furniture, window frames, flooring and sports goods it is essential that the timber is seasoned to the expected equilibrium conditions, namely 12% for regular intermittent heating and 10% in buildings with central heating, otherwise shrinkage in service will occur with loosening of joints, crazing of paint films, and buckling and delamination of laminates. An indication of the moisture content of timber used in different environments is presented in Fig. 7.6.

7.2.2.2 *Movement*

So far only those dimensional changes associated with the initial reduction in moisture content have been considered. However, dimensional changes, albeit smaller in extent, can also occur in seasoned or dried wood due to changes in the relative humidity of the atmosphere. Such changes certainly occur on a seasonal basis and frequently also on a daily basis. Since these changes in humidity are usually fairly small, inducing only slight changes in the moisture content of the timber, and since a considerable delay occurs in the diffusion of water vapour into the centre of a piece of timber it follows that these

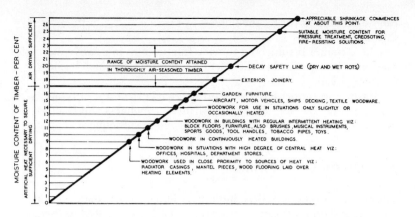

Figure 7.6 Equilibrium moisture content of timber in various environments. The figures for different species vary, and the chart shows only average values. (From the Princes Risborough Laboratory, Building Research Establishment. © Crown copyright.)

dimensional changes in seasoned timber are small, considerably smaller than those for shrinkage.

To quantify such movements for different timbers, dimensional changes are recorded over an arbitrary range of relative humidities. In the UK the standard procedure is to condition the timber in a chamber at 90% relative humidity and 25 °C, to measure its dimensions and to transfer it to a chamber at 60% relative humidity, allowing it to come to equilibrium before remeasuring it; the corresponding change in moisture content is from 21% to 12%. Movement values in the tangential and radial planes for those timbers listed in Table 7.2 are presented in Table 7.3. The timbers are recorded in the same

Table 7.3   Movement (%) on Transferring Timber from
90% Relative Humidity to 60%

Botanical name	Commercial name	Tangential	Radial
Chlorophora excelsa	Iroko	1.0	0.5
Tectona grandis	Teak	1.2	0.7
Pinus strobus	Yellow pine	1.8	0.9
Picea abies	Whitewood	1.5	0.7
Pinus sylvestris	Redwood	2.2	1.0
Tsuga heterophylla	Western hemlock	1.9	0.9
Quercus robur	European oak	2.5	1.5
Fagus sylvatica	European beech	3.2	1.7

order, thus illustrating that although a broad relationship holds between values of shrinkage and movement, individual timbers can behave differently over the reduced range of moisture contents associated with movement. Since movement in the longitudinal plane is so small, it is generally ignored. Anisotropy within the transverse plane

229

can be accounted for by the same set of variables that influence shrinkage.

Where timber is subjected to wide fluctuations in relative humidity care must be exercised in the selection of a species which has low movement values.

Moisture in timber has a very pronounced effect not only on its strength (Fig. 7.2) but also on its stiffness, toughness and fracture morphology; these aspects will be discussed in subsequent chapters.

7.3 THERMAL MOVEMENT

Timber, like other materials, undergoes dimensional changes commensurate with increasing temperature. This is attributed to the increasing distances between the molecules as they increase the magnitude of their oscillations with increasing temperature. Such movement is usually quantified for practical purposes as the coefficient of linear expansion and values for certain timbers are listed in Table 7.4. Although differences occur between species these appear

Table 7.4   Coefficient of Linear Thermal Expansion of Various Woods and Other Materials Per Degree Centigrade

		Coefficient of thermal expansion $\times 10^{-6}$	
		Longitudinal	Transverse
Picea abies	Whitewood	5.41	34.1
Pinus strobus	Yellow pine	4.00	72.7
Quercus robur	European oak	4.92	54.4
GRP, 60/40, unidirectional		10.0	10.0
CFRP, 60/40, unidirectional		10.0	−1.00
Mild steel		12.6	
Duralumin		22.5	
Nylon 6/6		125	
Polypropylene		110	

to be smaller than those occurring for shrinkage and movement. The coefficient for transverse expansion is an order of magnitude greater than that in the longitudinal direction. This degree of anisotropy (10:1) can be related to the ratio of length to breadth dimensions of the crystalline regions within the cell wall.

The expansion of timber with increasing temperature appears to be linear over a wide temperature range: the slight differences in expansion which occur between the radial and tangential planes are usually ignored and the coefficients are averaged to give a transverse value as recorded in Table 7.4. For comparative purposes, the coefficients of

230

linear thermal expansion for glass- and carbon-reinforced plastic, two metals and two plastics are also listed. Even the transverse expansion of timber is considerably less than that for the plastics.

The dimensional changes of timber caused by differences in temperature are small when compared to changes in dimensions resulting from the uptake or loss of moisture. For most practical purposes thermal expansion or contraction can be safely ignored over the range of temperatures in which timber is generally employed.

7.4 REFERENCES

Barkass, W. W. (1949) *The Swelling of Wood under Stress*, HMSO, 103 pp.

Meylan, B. A. (1968) Cause of high longitudinal shrinkage in wood, *Forest Products Journal*, 18 (4), 75–78.

Pratt, G. H. (1974) *Timber Drying Manual*, HMSO, 152 pp.

CHAPTER 8

Flow of Moisture and Heat in Concrete

8.1 SCOPE

Flow is one of the topics in which there is a marked similarity in behaviour between concrete and timber. This arises from their porous structures, and it is evident both in the same mathematical relationships and in the similar influence of variables such as moisture content and temperature. The emphases in this chapter and in Chapter 9 on timber are complementary and it would be advantageous to read *both* chapters in order to gain the full picture of flow in *either* material.

In this chapter the characteristics of flow of greatest importance in concrete technology are discussed. They are developed from a brief treatment of fundamentals, and a broader description of the topic of flow is given in Chapter 9. As pointed out there, flow is the passage through the material of one or more of a variety of entities which, for concrete, include moisture, carbon dioxide, heat, electric current and chloride ions. Here only moisture is treated in any detail, and a brief reference is made to heat.

From the practical standpoint the interest in flow in timber is generated by the need to impregnate with preservatives. In contrast, flow in concrete has been studied primarily because of the potentially damaging gradients of temperature and moisture content that arise as a result of heat loss or drying. Although both these applications come under the heading of flow they differ in that the driving force behind the transport processes are not the same. This is recognised by referring to:

(a) permeability, in which the fluid is forced through the material in response to an imposed absolute pressure. Thus the impregnants are forced into timber, or water flows through the concrete of a dam under the pressure of the water retained on the upstream surface.

(b) diffusion (because it is expressed by a diffusion-type equation), in which the fluid is driven by the action of a chemical potential

(or moisture potential). This includes the combined effects of gradients of:
(1) temperature,
(2) solute concentration giving an osmotic pressure,
(3) moisture content, giving rise to surface (adsorption) forces and vapour/water interface forces in the capillaries (see Section 6.2.3).

In concrete (and timber) drying occurs under the influence of the third of these gradients, which may in some cases be assisted by an imposed thermal gradient. It is also possible for the different gradients to act against each other giving rise to the possibility of water moving 'uphill' against the moisture content gradient.

Permeability and diffusion are really different aspects of the same phenomenon. Permeability is characterisd by the imposed pressure and usually by the material being in a saturated state. Diffusion, on the other hand, is caused by a chemical potential and normally takes place in a material that is partially dry. This means that although the governing equations for the two processes stem from the same kind of physical laws (as shown in the next section), the working quantities and mathematical forms differ somewhat.

8.2 MOISTURE MOVEMENT

8.2.1 Basic Relationships

The equations of permeability and diffusion are based on two relationships:

(a) between the velocity of flow (of the water) and the driving force;
(b) expressing the continuity of movement of the water through the material.

The flow rates in concrete are very low and, under an imposed pressure head, this means that laminar rather than turbulent flow occurs. The effects of the chemical potential are usually negligible and hence Darcy's law is applicable. The velocity, v_x, considering flow in the x direction only, is given by

$$v_x = ki_x = -k \frac{\partial h}{\partial x} \tag{8.1}$$

where i_x is the hydraulic gradient (h being the head of water, so that i_x is dimensionless), and k is the coefficient of permeability, expressed in m/s. Similar linear relationships are used for the diffusion processes (see Section 8.2.3).

Concrete shrinks as the moisture content reduces (see Section 6.2.2), but the volume change is relatively small and it is reasonable to assume that it does not affect the volume of the pores through which moisture movement occurs. It is also reasonable to assume that water is incompressible. Then consider the element of material

233

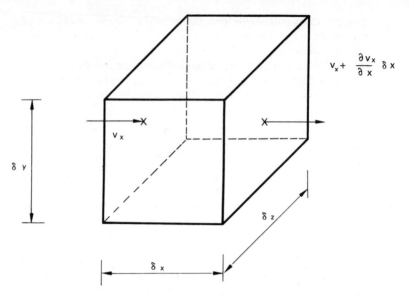

Figure 8.1 The continuity of unidirectional flow through an element of material.

$\delta_x \times \delta_y \times \delta_z$ through which flow occurs in the x direction (see Fig. 8.1). The velocity of flow into the element is v_x and the change in the volume of water contained by the element is δV in a time δt. An additional volume change of δS can also occur where δS is the change in a source or sink volume representing the creation (or loss) of moisture within the volume. The loss is caused by water used in hydration (fixation-drying), and is a common occurrence; the creation is less likely but could result from carbonation of the hardened cement paste. δS is only of importance in young concretes, and it can usually be omitted from the analysis.

Then $\delta V - \delta S$ = volume of water entering the element $-$ volume of water leaving the element,

$$= \left[v_x \delta y \delta z - \left(v_x + \frac{\partial v_x}{\partial x} \cdot \delta x \right) \delta y \delta z \right] \delta t$$

or
$$\frac{\partial v_x}{\partial x} = -\frac{\left(\frac{\delta V - \delta S}{\delta x \delta y \delta z} \right)}{\delta t} = \frac{\partial c_v}{\partial t} + \frac{\partial s}{\partial t} \tag{8.2}$$

where c_v is the moisture concentration (by volume), and s is the unit sink or source volume. Equation 8.1 is the general equation for the velocity of flow. Equation 8.2 is the general equation for the continuity of flow. The application of these equations to the particular cases of permeability and diffusion is given in the following sections.

8.2.2 Permeability

The lower the permeability of the concrete the greater is its resistance to the ingress of aggressive agents. Since, in general, the low permeability is associated with low porosity, it is also associated with high strength. The combination of low permeability and high strength implies good durability, and permeability is one of the few properties of concrete that provides a measure of durability. Its relation to durability is discussed in more detail in Chapter 17, while the details of its measurement, analytical treatment and characteristics are given in this section.

8.2.2.1 *Measurement*

The measurement of permeability in the laboratory is simple in principle as can be seen in the sketch of Powers' apparatus for h.c.p. shown in Fig. 8.2. The specimen is in the form of a truncated cone

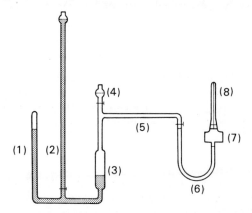

Figure 8.2 Schematic drawing of a permeability apparatus. 1, closed end manometer; 2, Hg pressure column; 3, Hg–H$_2$O reservoir; 4, H$_2$O supply reservoir; 5, pressure manifold; 6, rubber hose connection; 7, permeability cell; 8, capillary flow tube. (From T. C. Powers, L. E. Copeland, J. C. Hayes and H. M. Mann (1954) *Proceedings of the American Concrete Institute,* **51**, by permission of the American Concrete Institute.)

(to assist sealing of the sides by the force of the permeating liquid), and for most of the tests reported a constant pressure of about three atmospheres was imposed. This imposed pressure was indicated by the closed-end manometer, and the flow was recorded by the movement of the meniscus in the outlet capillary flow tube.

The permeability is determined from the rate of flow under steady-state conditions, and since h.c.p. is relative impermeable it took at least three days for the flow rate to become constant, and, in some cases, the required period was as much as four weeks. Permeability is very susceptible to variations in the experimental conditions, and great care had to be taken to prepare representative specimens, to ensure that they were cured in a saturated state, and to boil any dissolved air out of the permeating water so that there was no danger of it being released within the specimen. An additional complication arose because the chemical potential set up by variations in alkali concentration (osmotic pressure) was not insignificant in comparison with the imposed pressure head. A correction was applied by regarding the head loss through the specimen, Δh, as the sum of the hydraulic head and the head corresponding to the osmotic pressure.

Permeability tests of this kind are hardly suitable for everyday testing, especially as they cannot be performed on concrete in situ in the engineering structure. Of more practical interest is the ISAT (initial surface absorption test) which is more directly related to reality of durability in that it is a measure of the penetration of water into the concrete surface. It is described in Section 17.2.1, and penetration is discussed in Section 8.2.4.

8.2.2.2 *Expressions for Permeability*

For the test method described in the previous section, Darcy's law can be written

$$\frac{dq}{dt} \cdot \frac{1}{A} = k \frac{\Delta h}{L} \tag{8.3}$$

where q is the volume of flow, A is the area and L is the length of the specimen, so that Δh is the head loss over the distance L.

The permeability can often be generalised by introducing the specific permeability coefficient K which is given by

$$K = \frac{k\eta}{\gamma_w} \tag{8.4}$$

where η is the viscosity of water and γ_w is the unit weight of water. K has units of m^2 and is theoretically applicable to all permeating liquids. It is dependent only on the characteristics of the skeletal structure of the material and can thus be expected to be a function of parameters such as porosity and pore size. However, as Powers pointed out, the use of K depends on the viscosity of water being independent of the pore structure, and for h.c.p. with its very small flow channels this is not the case. Hence the permeability of h.c.p. and concrete is better investigated in terms of k rather than K. It is not possible to make an assessment of the usefulness for h.c.p. and concrete of the structural representations embodied in the Poiseuille equation for flow in capillaries or the further extensions in the Kozeny equation (see Section 9.2.2.1).

8.2.2.3 *Factors Influencing Permeability*

The effect of the hydration of the calcium silicates is to infill the initial skeletal structure of the hydrated cement paste (see Section 3.2.2), thus inevitably blocking some of the flow channels. Hence it is to be expected that permeability will drop with the progress of hydration and this is confirmed by the figures given in Table 8.1. The remarkable sensitivity of permeability is well demonstrated by the reduction by a factor of 1/100 between the ages of 6 and 24 days.

Hydration also has the effect of reducing porosity and there is a

236

Table 8.1   The Effect of Hydration
on Permeability

From T. C. Powers et al. (1954) Proceedings of the American Concrete Institute, 51, 285 by permission of the American Concrete Institute.

Age	Permeability coefficient, k (cm/s)
fresh	2×10^{-4}
5 days	4×10^{-8}
6 days	1×10^{-8}
8 days	4×10^{-9}
13 days	5×10^{-10}
24 days	1×10^{-10}
ultimate	0.6×10^{-10} (calculated)

direct relationship between porosity and permeability. In h.c.p. the gel pores are very small and tortuously connected so that a paste is virtually impermeable. The capillaries offer much less resistance and it is thought that in normal pastes water movement occurs almost wholly within the capillary system. The relation between coefficient of permeability and capillary porosity is shown in Fig. 8.3, and again

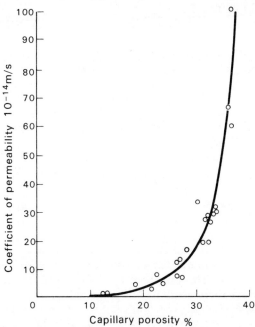

Figure 8.3   Relation between permeability and capillary porosity of hardened cement paste. (From T. C. Powers (1958) *Journal of the American Ceramic Society*, 41, 1–6, by permission of the American Ceramic Society.)

the sensitivity of permeability to changes in porosity is to be seen, especially at the higher level of porosity of pastes (about 20–25%) such as would be found in normal concretes.

The slightest crack (or air void), if correctly oriented, provides a low-resistance bypass for the permeating fluid, and the permeability rises dramatically if shrinkage cracks form. Even if no such cracks are observed, drying has the effect of giving increased permeability when the material is subsequently resaturated. This can be attributed to the water movement on drying forming connecting links between capillary spaces that were isolated in the undried h.c.p., thus providing a more ready passage under a subsequent applied pressure.

The presence of an impermeable aggregate has a dual effect. Firstly, it reduces the area of flow and increases the flow path length, thus tending to reduce the permeability of the concrete in comparison with that of the h.c.p. Secondly, entrapped air and water-filled voids (for instance, see water gain, Section 3.5.2) are more likely to occur in concrete than in h.c.p. and this, in contrast to the first effect, tends to increase the permeability of the concrete relative to that of the paste. The porosity of normal strong natural aggregates is about 3% in comparison with, say, 30% for h.c.p. and it might be thought that the assumption of impermeability made above is indeed reasonable for aggregates. In fact, although their porosities are so much smaller, their proportions of larger pores are much higher than those of cement pastes, so that aggregates have much the same permeabilities as h.c.p. Some typical figures are given in Table 8.2. There is thus no certainty

Table 8.2   Typical Permeabilities of Natural Rocks

From H. Woods (1968) *Durability of Concrete Construction*, ACI Monograph No. 4, by permission of the American Concrete Institute.

Rock type	Permeability of rocks (cm/s)	Water/cement ratio of mature cement paste for same permeability
Dense trap	0.003×10^{-10}	0.38
Quartz diorite	0.01×10^{-10}	0.42
Marble 1	0.03×10^{-10}	0.48
Marble 2	0.08×10^{-10}	0.66
Granite 1	7×10^{-10}	0.70
Sandstone	20×10^{-10}	0.71
Granite 2	20×10^{-10}	0.71

about how the permeability of a concrete will compare with that of the paste that it contains but there is little support for the intuitive feeling that the concrete will be less permeable.

The data available on permeability of h.c.p. and concrete are few, and most of those quoted above come from a single investigation by Powers and his coworkers. This scarcity can be attributed to the difficulties of making the measurements, and also to the limited need (in the past) for such information.

8.2.2.4   The Continuity Condition

If the concrete remains saturated throughout its life, as may reasonably be assumed for a mass concrete dam, the continuity condition (see Section 8.2.1) is simplified, because there is no change in the volume of water within each element. Thus in Equation 8.2

$$\delta V = \delta c_v = 0$$

and hence the continuity condition becomes

$$\frac{\partial v_x}{\partial x} = 0$$

and from Darcy's law (Equation 8.1)

$$\frac{\partial}{\partial x}\left(k\,\frac{\partial h}{\partial x}\right) = 0$$

Assuming that k is constant,

$$k\,\frac{\partial^2 h}{\partial x^2} = 0 \tag{8.5}$$

Equation 8.5 corresponds to the equation for the potential function governing flow in soils and used in flow net analyses. This is no surprise as the problem of the determination of the distributions of pressure and flow in a mass concrete structure is exactly analogous to the flow net analyses for uniform permeable soils.

8.2.3   Diffusion

The chemical potential driving the diffusion transfer of moisture includes several different energy sources, as listed earlier in Section 8.1. The most important is the gradient of moisture content which is associated with a gradient of internal relative humidity resulting in surface and capillary forces tending to drive the moisture down the concentration gradient. It is natural, therefore, to express the driving force in terms of the concentration, and in similar fashion to Darcy's law to quote the flux (mass flow per unit area), Q, as a linear function of the gradient. Thus, for one dimension,

$$Q = -D_g\,\frac{\partial c_m}{\partial x} \tag{8.6}$$

where D_g is the moisture diffusivity (or moisture conductivity) and has units of m^2/s, and c_m is the moisture concentration by mass.

Substituting in the continuity equation (after Equation 8.2) yields the governing equation for diffusion, commonly referred to as Fick's second law.

$$\frac{\partial c_m}{\partial t} = \frac{\partial\left(D_g\,\frac{\partial c_m}{\partial x}\right)}{\partial x^2} \tag{8.7}$$

239

If k is constant this becomes

$$\frac{\partial c_m}{\partial t} = D_g \frac{\partial^2 c_m}{\partial x^2} \tag{8.8}$$

The justification for assuming D_g constant is examined in the next section.

8.2.3.1 Moisture Conductivity, D_g

The rate of diffusion is affected for thermodynamic reasons by an overall change of temperature, and for mechanical reasons by an overall change of pore structure caused by the progress of hydration. Such changes are important but in many practical circumstances they are relatively insignificant and can be neglected. D_g could then be taken as constant with respect to time. Much more serious is the possible variation of D_g with the moisture concentration c_m, which, if important, means that the more difficult Equation 8.7 must be solved rather than the relatively simple Equation 8.8. The dependence of D_g on c_m turns on the actual mechanisms providing the driving forces , and a general view on the modes of transport in a typical constricted capillary pore is given below with reference to Fig. 8.4.

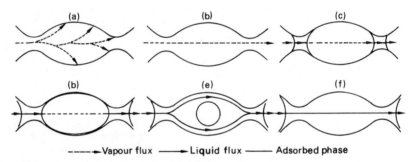

Figure 8.4 Modes of moisture transport through a typical pore. (From D. A. Rose (1965) *RILEM Bulletin*, 29, by permission of RILEM.)

The effect of increasing internal relative humidity is considered in six stages, a to f. At very low RH the moisture is in the vapour state and the concentration increases as the water is adsorbed onto the dry surfaces. As the RH rises so the adsorbed layers build up, and direct vapour movement through the pore (in the manner of an inert gas) becomes possible. When the RH reaches about 50%, water condenses in the constricted part of the pore, forming a meniscus of a radius to suit the prevailing RH. The presence of the condensed water zones shortens the path for vapour transfer, thus increasing the rate of movement. The condensed water zones extend with rising RH and the flow through the pore is augmented by transfer in the adsorbed films. In the later stages straightforward liquid flows occurs, initially in the unsaturated state and finally in the saturated.

It is not immediately possible to translate these qualitative ideas into mathematical theories for the estimation of moisture concentration, but experimental work does confirm the increases to be expected in the moisture conductivity as the RH goes up. Empirical forms of the relation between D_g and c_m have been suggested (Kasperkiewicz, 1972; and Pihlajavaara, 1965, for instance) but these are too uncertain to be used with confidence in practice. A comparison between predicted moisture distributions using constant D_g and concentration-dependent D_g is given later, but it is evident that the use of a constant D_g is normally satisfactory in practice, and it is in any case a more sensible choice than a concentration-dependent D_g of doubtful validity. In most engineering applications it is the strains induced by the moisture gradient that matter and this problem of differential shrinkage is dealt with in Chapter 6.

8.2.3.2 *Solution of the Diffusion Equation*

There are advantages in the adoption of nondimensional forms of the variables. The concentration of moisture becomes

$$u = \frac{c_m - c_0}{c_0 - c_e}$$

where c_m is the current concentration, c_0 is its initial value, and c_e the value for equilibrium.

At the start of drying, u is unity, and at final equilibrium it drops to zero; it will be used in all further discussion.

x can be replaced by $X = x/L$ where L is a standard dimension such as the thickness of a plate. t can be replaced by the Fourier number, $F_0 = kT/L^2$.

The nondimensional form of the diffusion equation is then

$$\frac{\partial u}{\partial F_0} = \frac{\partial^2 u}{\partial X^2} \tag{8.9}$$

This form is ideal for the graphical presentation of the general result, and this is shown in Fig. 8.8, which is discussed in some detail later. It is not used in the development here as the meaning of the variables previously introduced is more immediately obvious.

The drying of porous solids by evaporation of the water from the surface can be divided into three distinct stages as shown in Fig. 8.5. In the first stage, that of constant rate, the surface remains completely wet, and the rate of evaporation is the same as that from a liquid surface, being independent of the moisture content of the material. The drying rate is influenced by the velocity of the air around the solid. The second stage is the first falling rate period during which the rate of moisture loss declines rapidly as the surface dries out. The air velocity can still be expected to affect the rate. Finally, the second falling rate period occurs in which the surface remains virtually dry and water is evaporated as soon as it reaches the surface from the interior. The rate of moisture loss is then largely controlled by the process

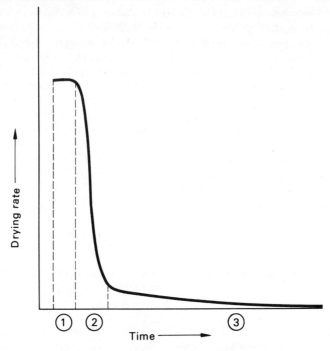

Figure 8.5 Drying rates for concrete. (From S. E. Pihlajavaara (1963) *Notes on the Drying of Concrete,* by permission of the Technical Research Centre of Finland.)

of diffusion, and it declines much more slowly than during the first falling rate period. Air velocity is no longer an influence.

In concrete the first two stages take place rapidly and have very little to do with long-term drying associated with shrinkage.

Two boundary conditions are relevant to the interface between the solid and the surrounding environment. The first is a convection type in which the rate of flow is proportional to the moisture gradient across the interface. It is similar to the Newton Law of Cooling. Thus

$$-D_g \left(\frac{\partial u}{\partial x}\right)_b = \mu \left(u_b - u_e\right) \tag{8.10}$$

where suffix b refers to the boundary, and u_e is the equilibrium value ($=0$) of moisture content. μ is the transfer coefficient representing the surface characteristics. This condition applies when water exists on the surface or arrives there more rapidly than the evaporation processes can remove it; it is relevant to the constant rate period and the first falling rate period, but less so to the second falling rate period.

The second boundary condition states that the moisture content on the surface has the value for equilibrium with the surrounding atmosphere. That is,

$$u_b = u_e \qquad (=0) \tag{8.11}$$

242

This applies when water arriving by mass transfer at the surface is not hindered in its escape by the surface. It is relevant to the majority of the drying time of normal concretes, and it is a good engineering decision to adopt it for the normal calculation of the distribution of free drying shrinkage.

Finally, the boundary condition at a sealed face is often required, Here the moisture gradient is zero, so that

$$\left(\frac{\partial u}{\partial x}\right)_b = 0 \tag{8.12}$$

The diffusion equation, with constant D_g and no sink on source terms is

$$\frac{\partial u}{\partial t} = D_g \frac{\partial^2 u}{\partial x^2} \tag{8.13}$$

For simple geometries and boundary conditions it can be solved in closed form (see Crank, 1957, for example). The expressions are involved and it is preferable to have a method of solution that is of greater generality without losing the essential simplicity of Equation 8.13, even if there is some loss of precision. This wider scope is provided by numerical solutions of the equation written using finite difference approximations to each of the terms. Various forms and methods of solution of the finite difference equations are possible and the simplest (but by no means the least effective) is given below.

The case is considered of a large plate sealed on one face and open to the drying atmosphere on the other; moisture then moves across

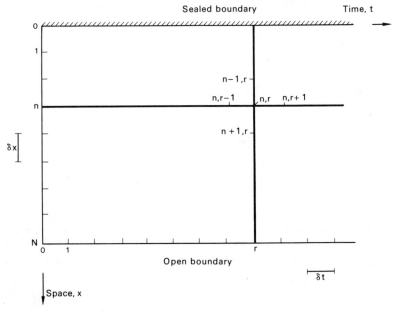

Figure 8.6   The space-time network for solving the diffusion equation for uni-dimensional drying.

243

the thickness of the plate towards the open face. The development of the distribution of moisture is traced through the values of the moisture concentration, u, at the node points of the space—time chart shown in Fig. 8.6. The space dimension is divided into increments, δx, with nodes running from 0 at the sealed boundary to N at the open boundary. Time is similarly divided into intervals, δt, starting at the left-hand side and increasing indefinitely to the right. Each of the nodes has space—time coordinates, given typically as n, r, and the moisture concentration at that node is then $u_{n,r}$. We seek to establish the values of u along the space axis at each of a succession of times, starting with the known values of u at time zero. Thus, typically, $u_{n,r+1}$ at space node n and time $r + 1$ is found from known values of u at time r and at various space nodes $n - 1, n, n + 1 \ldots$.

One of the difficulties with finite difference solutions is that they can become unstable and give absurd answers. Sometimes it is not obvious that the answers are absurd, and the methods of solution should always be checked, preferably against the answers given by closed form solutions. Sometimes the stability is affected by the form of finite difference representation chosen. This can be true of the left-hand side of Equation 8.3, and it is best written in terms of the forward difference.

$$\left(\frac{\partial u}{\partial t}\right)_{n,r} = \frac{u_{n,r+1} - u_{n,r}}{\delta t} \tag{8.14}$$

Similarly,

$$\left(\frac{\partial^2 u}{\partial x^2}\right)_{n,r} = \frac{1}{(\delta x)^2}\left(u_{n+1,r} - 2u_{n,r} + u_{n-1,r}\right)$$

which leads to

$$\frac{u_{n,r+1} - u_{n,r}}{\delta t} = \frac{D_g}{(\delta x)^2}\left(u_{n+1,r} - 2u_{n,r} + u_{n-1,r}\right) \tag{8.15}$$

which can be rearranged to give the expression for the unknown $u_{n,r+1}$

$$u_{n,r+1} = u_{n,r}\left(1 - 2\lambda\right) + \lambda\left(u_{n-1,r} + u_{n+1,r}\right)$$

in which

$$\lambda = \frac{D_g \delta t}{(\delta x)^2}$$

For stability $\lambda \leqslant \frac{1}{2}$. The value of k is known because it is a property of the concrete and δx is chosen to provide a suitable number of nodes (a minimum of eight is likely to be satisfactory). Hence the limit of δt can be found. If $\lambda = \frac{1}{2}$, a particularly simple form of the equation emerges,

$$u_{n,r+1} = \frac{1}{2}\left(u_{n-1,r} + u_{n+1,r}\right) \tag{8.16}$$

This can be solved graphically, and is referred to as the Schmidt—Binder method. It is illustrated in Fig. 8.7, which shows that a u_n

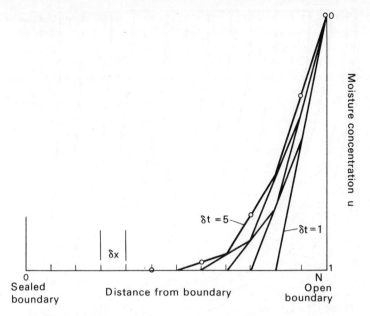

Figure 8.7 Simple graphical solution for the uni-dimensional drying of concrete.

value at time $r + 1$ is found simply by averaging the values of u_{n-1} and u_{n+1} at time r.

Initially the solid is placed in a saturated state in the drying atmosphere at time zero. All points within the solid have the same moisture concentration of unity. That is,

$$u_{0,0} = u_{1,0} \cdots u_{n,0} = 1$$

On the open boundary the concentration falls immediately to the equilibrium value, and remains there indefinitely. Thus

$$u_{N,0} = u_{N,1} = u_{N,r} = 0$$

For the sealed boundary a fictional point, $-1,r$, outside the boundary is contemplated. Then,

$$\left(\frac{\partial u}{\partial x}\right)_b = \frac{u_{-1,r} - u_{1,r}}{2\delta x} = 0$$

from which

$$u_{-1,r} = u_{1,r}$$

and in setting up the equation for the point on the boundary $0,r$ the point outside the boundary $u_{-1,r}$ is substituted by the known value within the boundary $u_{1,r}$.

8.2.3.3 *Computation of Moisture Distribution*

Pihlajavaara (1969) proposed the following expression for the moisture conductivity:

$$D_g = D_{ge} (1 + Bu^n) \tag{8.17}$$

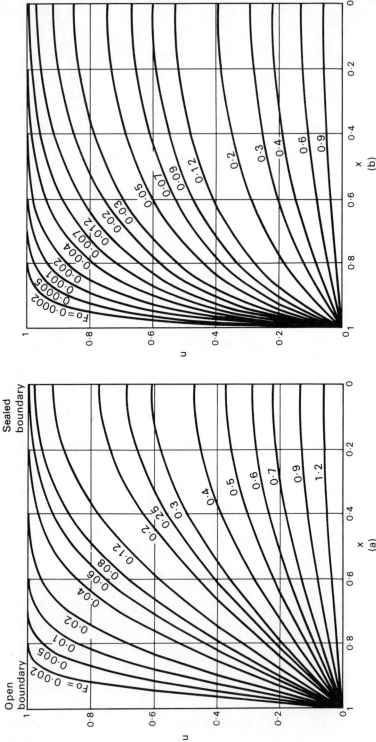

Figure 8.8 Computed moisture distributions in concrete: (a) with moisture conductivity constant at final equilibrium value, i.e. $D_g = D_{ge}$; (b) with D a function of moisture concentration, u, i.e. $D_g = D_{ge}(1+10u^2)$. (From S. E. Pihlajavaara (1969) *Publication 153, An Approximate Solution of the Quasi-Linear Diffusion Problem*, by permission of the Technical Research Centre of Finland.)

in which D_{ge} is the equilibrium value which applies on the boundary throughout the drying period (it is the lowest value of D_g), u is the (nondimensional) moisture concentration and B and n are constants.

He solved the case of the slab drying from one large face using a variety of combinations of B and n. Two typical charts are given in Fig. 8.8, (a) for constant D_g equal to the minimum (equilibrium) value D_{ge} and (b) for $n = 2$ and $B = 10$. The moisture-dependent D_g reduces the diffusivity of the outer layers of the body nearest to the drying atmosphere, so that the shape of the moisture distributions tends towards a plateau in the interior of the body. Also, in further comparison with the constant D_g case, the average diffusivity is higher so that the rate of loss is greater for the case of moisture-dependent D_g. This is apparent on the charts when comparing curves for equal Fourier numbers (representing equal times, other conditions being identical). If, for the case of constant moisture conductivity, D_g were taken equal to the initial predrying value rather than to that at equilibrium, the comparative shapes of the two charts would not be affected, but the comparative rates of loss would be reversed, the greater loss of moisture-dependent D_g. The curves of Fig. 8.8 are computed but such experimental evidence as exists confirms the general trends shown in chart (b).

Equation 8.17 can hardly be considered for practical use because the constants B and n are not known. The alternative is to adopt a mean constant value of D_g and to use the simple analysis embodied in chart (a). Some appropriate values of D_g are given in Table 8.3.

Table 8.3   Suggested Values for Moisture
Conductivity, D_g (m^2/s), for Use in
Engineering Analysis

Wet refers to drying at RH $> 50\%$. From
S.E. Pihlajavaara (1965) *Estimation of the
Drying of Concrete,* Proceedings of RILEM/
CIB conference on moisture problems in
buildings, by permission of the Technical
Research Centre of Finland.

	Strength class				
Low		Middle		High	
Wet	Dry	Wet	Dry	Wet	Dry
10^{-8}	10^{-9}	10^{-10}	10^{-11}	10^{-11}	10^{-12}

8.2.3.4   *Effect of Temperature*

The presence of a temperature gradient adds a second component to the chemical potential, and this can be incorporated as an equivalent moisture gradient. Thus the flux, Q, becomes

$$Q = -D_g \left(\frac{\partial c_m}{\partial x} + a_t \frac{\partial T}{\partial x} \right) \qquad (8.18)$$

where a_t is a constant converting temperature gradient to equivalent moisture gradient.

The effect of a temperature gradient is given in the experimental results of Fig. 8.9, which shows the moisture (and shrinkage) distributions for a concrete specimen 1500 mm long, drying virtually

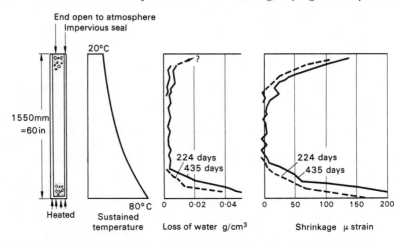

Figure 8.9 The effects of a temperature gradient on the moisture distribution in concrete, in comparison with drying from an open surface. (From A. D. Ross, J. M. Illston and G. L. England (1968) Short and long-term deformations of concrete as influenced by its physical structure and state, in *Conference on the Structure of Concrete*, by permission of the Cement and Concrete Association.)

normally from the top and heated at the bottom. It can be seen that after 450 days drying has penetrated only some 300 mm at the top, thus demonstrating the very slow rates of moisture movement that normally obtain in concrete. At the bottom the thermal gradient has driven the water up the specimen, but again the movement is very slow, The centre section of the specimen gained water, if anything, and it is unlikely that it would ever be affected by either of the drying gradients.

8.2.4   Penetration

The phenomena of permeability and diffusion have been described in the previous two sections. Permeability is characterised by the movement of moisture in a saturated material under an imposed hydraulic pressure, while diffusion occurs in a partially dry material under the drive of a chemical potential, usually deriving from gradients of internal relative humidity. Penetration (of moisture) combines features of both permeability and diffusion in that it refers to the ingress of water into a partially dry material, so that the part penetrated becomes saturated. The driving force certainly includes the effects of a humidity gradient which may be supplemented or indeed overwhelmed by an imposed hydraulic pressure.

A simple analysis of the effect of an imposed pressure has been presented by Valenta (1970). The penetrating water front is assumed to be sharply defined so that its depth of penetration is represented by a unique position, x, which can be determined. Thus it is assumed that the material is saturated to a depth x, beyond which there is no change from the initial partially dry state. The applied pressure head, h, acts at the surface of the material and declines linearly over the depth x to the water front where it is fully dissipated. Then by Darcy's law the velocity of penetration, v_x, is given by

$$v_x = \frac{dx}{dt} = k\frac{h}{x} \qquad (8.19)$$

Integrating and setting $x = 0$ when $t = 0$,

$$x = \sqrt{(2kht)} \qquad (8.20)$$

Levitt (1971) tackled the case of the internal humidity gradient in much the same way by considering the pore structure as a system of capillaries so that with reference to Equation 8.19 the head becomes the appropriate capillary rise, which, in accordance with the Poiseuille equation, is proportional to the flow velocity. Then, as before,

$$x \propto t^{\frac{1}{2}}$$

In practice the power of t has been found to vary over the range 0.3 to 0.7. The low value implies a faster flow rate than that predicted and can be explained as a reduction in the capillary resistance caused by a process of flushing out by the water as it penetrates. Conversely, the higher values can be attributed to a silting up. Back substitution of t for x in Equation 8.19 gives

$$\frac{dx}{dt} = bt^{-\frac{1}{2}} \qquad (8.21)$$

where b is a constant.

Levitt developed his initial surface absorption test to record flow rate at given times after exposure to penetration, thus providing an empirical but comparative measure of the resistance to flow of a material. This is of significance in durability and the test is described in Section 17.2.1.

8.3   HEAT MOVEMENT

Heat transfer is analogous to moisture transfer and the principles are embodied in Fourier's law, which is of the same form as Fick's law. Thus

$$\frac{\partial \left(D_h \frac{\partial T}{\partial x} \right)}{\partial x} = \frac{\partial T}{\partial t} \qquad (8.22)$$

where T is temperature and D_h is the coefficient of thermal diffusivity in m²/s. D_h is given by

$$D_h = \frac{K_h}{\rho c} \qquad (8.23)$$

where K_h is the coefficient of thermal conductivity in kcal/m s °C, ρ is the density in kg/m³ and c is the specific heat in kcal/kg °C. For engineering purposes K_h and ρ can be taken as independent of temperature, but c increases linearly with temperature and it follows that D_h must correspondingly decrease with temperature. Taking D_h outside the partial differential in Equation 8.22 introduces a significant degree of approximation.

The solution of Equation 8.22 yields distributions of temperature in space and time resulting from the transfer of heat through the material in accordance with the appropriate boundary conditions (see Section 8.2.3.2). The rate at which the steady-state distribution of temperature is approached depends on D_h, which has an order of magnitude of about 1×10^{-6} m²/s. This is much greater than the corresponding value for moisture movement and it follows that transient temperature gradients will not persist in the way that transient moisture gradients do, and only in special circumstances will the temperature distributions in concrete assume importance in design. Among such special circumstances are problems caused by the heat of hydration in mass concrete structures such as deep foundations, and the effects of the changing temperature of the atmosphere on concrete pavements. The increasingly important consideration of thermal insulation is essentially a steady-state application in which it is the heat loss that matters rather than the temperature gradient itself; nevertheless any proper analysis of the loss of heat through a wall must include the effects of diffusivity on the response to the changing temperatures at the boundaries.

8.3.1 Thermal Conductivity

Under steady-state conditions the heat flow is controlled by the thermal conductivity of the material and, as can be seen from the comparison between materials in Section 9.2.3, concrete does not conduct heat readily; although its thermal conductivity is about three times that of water and 6 times that of timber it is only 1/200 that of copper. The composition of the concrete greatly affects the value of the conductivity and it is advantageous to regard it as a three-phase system — water, dry h.c.p. and aggregate. The influence of the water is to raise the conductivity by a factor of roughly two between the dry and the wet states. In the two-phase material composed of water and h.c.p., that is moist h.c.p., the most important variable for thermal conductivity is the moisture content by volume. If it is held constant then the more porous paste has the lower conductivity. That is, for constant moisture content, the h.c.p. with the highest water/cement ratio has the lowest conductivity.

The effect of the aggregate can be predicted by the use of a law of mixtures applied to the two-phase material composed of moist h.c.p. and aggregate (Campbell-Allen, 1963). Usually normal aggregates have higher thermal conductivities than paste while lightweight aggregates have lower conductivities than paste. The lower the density of the aggregate, the lower the thermal conductivity. Thus, for a concrete to have the lowest thermal conductivity, it should have the maximum content of the lowest density aggregate, and the highest water/cement ratio, and it should be dried out. Typical figures for K_h are 2 kcal/m s °C for a wet normal-weight concrete down to 0.4 kcal/ m s °C for a lightweight concrete. The lightweight concretes with low thermal conductivity have a commensurately low strength, but this is no great disadvantage for facings or blocks for the building industry, where lightness and low thermal conductivity are the more important requirements.

8.4   REFERENCES

Campbell-Allen, D. and Thorne, C. P. (1963). The thermal conductivity of concrete, *Magazine of Concrete Research*, 15, No. 43, 39.
Crank, J. (1957) *The Mathematics of Diffusion*, Oxford University Press.
Kasperkiewicz, J. (1972) Some aspects of water diffusion process in Concrete, *Materiaux et Constructions*, 5, 209.
Levitt, M. (1971) The ISAT — a non-destructive test for the durability of concrete, *British Journal of Non-Destructive Testing*, 13.
Pihlajavaara, S. E. (1963) *Notes on the drying of concrete*, Report of the State Institute for Technical Research, Helsinki, Finland.
Pihlajavaara, S. E. (1965) *Estimation of the Drying of Concrete*, Proceedings of RILEM/CIB Conference on Moisture Problems in Buildings, Otaniemi, Finland.
Pihlajavaara, S. E. (1969) *An approximate solution of a quasi-linear diffusion problem*, Report of the State Institute for Technical Research, Helsinki, Finland.
Powers, T. C., Copeland, L. E., Hayes, J. C. and Mann, H. M. (1954) Permeability of Portland cement paste, *Proceedings of the American Concrete Institute*, 51, 285.
Powers, T. C. (1958) Structure and physical properties of hardened Portland cement paste, *Journal of the American Ceramic Society*, 41, 577.
Rose, D. A. (1965) Water movement in unsaturated porous materials, RILEM Bulletin, No. 29, 119.
Ross, A. D., Illston, J. M. and England, G. L. (1968) Short and long-term deformations of concrete as influenced by its physical structure and state, *The Structure of Concrete*, Proceedings of an International Conference, Cement and Concrete Association, London.
Valenta, O. (1970) The permeability and durability of Concrete in aggressive conditions, Proceedings of the Dixieme Congres des Grands Barrages, Montreal, Canada.

Woods, H. (1968) *Durability of Concrete Construction*, ACI Mono-graph No. 4, American Concrete Institute and the Iowa State University Press.

CHAPTER 9

Flow in Timber

9.1 INTRODUCTION

In timber the term *flow* is associated almost exclusively with the impregnation of timber with artificial preservatives and, while such a viewpoint can be justified in terms of the practical significance of impregnation, it is nonetheless a somewhat restricted interpretation of the term. Certainly flow is synonymous with passage of liquids, but the term is also applicable to the passage of gases, thermal energy, or even electrical energy, and this wider interpretation is applied in this chapter.

It is convenient when discussing flow to think of it in terms of being either constant or variable, with respect to either time or location within the specimen; flow under the former conditions is referred to as *steady-state flow*, whereas when flow is time and space dependent it is referred to as *unsteady-state flow*, and these are treated separately below.

9.2 STEADY-STATE FLOW

One of the most interesting features of flow in timber in common with many other materials is that irrespective of whether one is concerned with liquid or gas flow, diffusion of moisture, or thermal and electrical conductivity, the same basic relationship holds, namely that the flux or rate of flow is proportional to the pressure gradient:

$$\frac{\text{Flux}}{\text{Gradient}} = K \tag{9.1}$$

where flux is rate of flow per unit cross-sectional area, gradient is pressure difference causing flow per unit length, and K is constant, dependent on form of flow, e.g. permeability, diffusion or conductivity. This relationship when relating to fluid flow is ascribed to Darcy: however, where thermal conductivity is involved the equation is referred to as Fourier's law. However, this is of secondary importance: the really significant point is that irrespective of whether it is fluid flow by way of the cell cavities, moisture diffusion through the cell wall, or thermal or electrical flow along the cell walls, the same general relationship relating rate of flow to pressure gradient applies. Each of these variants of flow will now be discussed, but because of the practical significance of fluid flow greater emphasis will be placed upon it.

9.2.1   Permeability

Permeability is simply the quantitative expression of the flow of fluids through a porous material. Of all the numerous physical and mechanical properties of timber, permeability is by far the most variable; when differences between timbers and differences between the principal directions within a timber are taken into consideration the range is of the order of 10^7.

Passing reference has already been made to the technical significance of permeability in timber utilisation. Not only is it important in the impregnation of artificial preservatives, fire retardants and stabilising chemicals, but it is also significant in the chemical removal of lignin in the manufacture of wood pulp and in the removal of *free* water during drying.

9.2.1.1   *Flow of Fluids*

Three types of flow can occur and, although all three are applicable to gases, only the first two are relevant to the flow of liquids:

(a) viscous or laminar flow;
(b) turbulent flow;
(c) molecular diffusion, i.e. slip or Knudsen flow.

Laminar flow occurs in capillaries where the rate of flow is relatively low and when the viscous forces of the fluid are overcome in shear thereby producing an even and smooth flow pattern. In laminar flow Darcy's law is directly applicable, but a more specific relationship for flow in capillaries is given by the Poiseuille equation, which for liquids is:

$$Q = \frac{\pi r^4 \, \Delta P}{8\eta l} \qquad (9.2)$$

where Q is the volume rate of flow, r is the capillary radius, ΔP is the pressure drop across capillary, l is the capillary length and η is the viscosity. For gas flow, the above equation has to be modified slightly to take into account the expansion of the gas along the pressure gradient. The amended equation is:

$$Q = \frac{\pi r^4 \, \Delta P}{8\eta l} \cdot \frac{\bar{P}}{P_0} \qquad (9.3)$$

where $\bar{P}$ is the mean gas pressure within capillary and P_0 is the pressure

254

of gas where Q was measured. That is

$$Q \propto \frac{\Delta P}{l}$$

or flow is proportional to the pressure gradient, which conforms with the basic relationship.

Turbulent flow occurs as the rate of flow increases significantly causing eddies to form and disrupting the smooth laminar flow. So far, turbulent flow has eluded mathematical description because of its complexity, but it has been shown experimentally that

$$Q \propto \sqrt{\Delta P} \qquad (9.4)$$

which means that Darcy's law is not upheld. However, in timber it is unlikely that turbulent flow occurs, because of the relatively low fluid velocities attained and the large diameter of the 'capillaries'.

Molecular diffusion is restricted to gases, becoming significant when the gas pressure in a capillary is lowered to such an extent that the mean free path of the gas becomes approximately equal to the diameter of the capillary; under these conditions the flow rate under a pressure gradient exceeds that predicted by Poiseuille's law. It is thought that this extra component of flow is the result of specular reflection occurring between the gas molecules and the capillary wall, in addition to the intermolecular collisions responsible for the laminar or viscous forces. This process can be interpreted as a sliding or slipping layer of molecules immediately adjacent to the capillary wall and is known as *slip flow*.

Eventually a state is reached where further reduction of pressure results in the virtual cessation of molecular collisions and consequently the absence of viscous forces. Gases then progress entirely by specular reflection from the capillary walls, a process known as *Knudsen flow*. Both slip and Knudsen flow are diffusion processes being controlled by the concentration gradient of the molecules concerned.

Slip flow along a capillary of circular section is given by:

$$Q_s = \frac{2\pi r^3 \delta_1 \overline{V}}{3l} \qquad (9.5)$$

where δ_1 is a factor depending on the fraction of molecules undergoing diffuse reflection upon collision with capillary wall, and $\overline{V}$ is the molecular mean thermal velocity. Slip flow occurs in wood cells at pressures equal to or less than atmospheric; gas flow with slip in timber will be given by the combination of Equations 9.3 and 9.5, thus:

$$\frac{QP_0}{\Delta P} = \frac{\pi r^4 \overline{P}}{8\eta l} + \frac{2\pi r^3 \delta_1 \overline{V}}{3l} \qquad (9.6)$$

In the Knudsen region where the viscous forces are virtually absent flow is given by:

$$\frac{QP_0}{\Delta P} = \frac{2\pi r^3 \delta_0 \overline{V}}{3l} \qquad (9.7)$$

255

where δ_0 is a factor depending on the fraction of molecules undergoing diffuse reflection at the capillary wall.

At the present time no mathematical treatment of the transition from Equation 9.7 to Equation 9.6 is available. Equations 9.2, 9.3 and 9.6 are applicable only for long tubes and a correction factor is necessary for flow in short tubes such as occur in timber; the correction involves the increase in the length term by an amount dependent on the ratio r/l.

Such theory has been modified for flow through tubes differing from the perfect cylindrical capillaries considered above. Thus the Kozeny Carman Equation has been applied to flow through tracheid lumina of rectangular cross section (Petty and Puritch, 1970), while theory due to Hanks and Weissberg has been applied to flow through pit membrane pores which are considerably wider than they are long (Petty, 1974). In many cases, use of such theory involves considerable approximation since pores in wood are seldom entirely regular.

For liquids all this theory can, however, be summarised by Darcy's Law:

$$K_p = -\frac{F\eta}{\Delta P} \tag{9.8}$$

where K_p is the liquid viscous permeability constant, F is the flux (V/tA). ΔP is the pressure gradient and η is the viscosity of the liquid. This expression can be rewritten in terms of volume flow per unit time (Q) as

$$Q = \frac{K_p A \, \Delta P}{\eta l} \tag{9.9}$$

where A is the cross-sectional area of the capillary. It may be shown theoretically that this law remains applicable for a number of capillaries in parallel, or even for a heterogeneous porous medium in which different types of capillary are combined in series, providing that A is then taken as the area of the medium normal to flow.

Due to the compressible nature of gases, Equation 9.9 for gas flow becomes

$$Q = \frac{K_{pg} A \, \Delta P}{\eta l} \cdot \frac{\bar{P}}{P} \tag{9.10}$$

where $\bar{P}$ is the mean pressure in the specimen, P is the pressure at which Q is measured, K_{pg} is the gas viscous permeability constant and η is the viscosity of the gas. At low mean gas pressures, a slip flow term must clearly be added to Equation 9.10. By analogy with Equation 9.5 this takes the form

$$Q_s = \frac{\delta K_s \bar{V} \Delta P A}{P l} \tag{9.11}$$

where K_s is the slip constant, δ is a dimensionless coefficient $\simeq 1$, and $\bar{V}$ is the mean thermal velocity of gas. $\bar{V} = \sqrt{(8RT/\pi M)}$, where R is the universal gas constant and M is the molecular weight of the gas.

Where this slip flow term makes no significant contribution to total flow (i.e. at high mean gas pressures) Equation 9.10 will be equally applicable to homogeneous and heterogeneous porous media, just as Equation 9.9 may be used to describe the permeability of such media to liquids. However, when the contribution of slip flow to total flow is significant, Equation 9.10, even with the above slip term, can only be applied to homogeneous media, or to a hetero-geneous collection of capillaries conducting in parallel. Darcy's Law will not be upheld at low mean gas pressures for a heterogeneous porous medium composed of different types of capillary *in series*: the gaseous conductivity of such a medium will vary nonlinearly with mean gas pressure at low mean gas pressures.

9.2.1.2 *Flow Paths in Timber*

Softwoods. Because of their simpler structure and their greater economic significance much more attention has been paid to flow in softwood timbers than in the hardwood timbers. It will be recalled from Chapter 4 that both tracheids and parenchyma cells have closed ends and that movement of liquids and gases must be by way of the pits in the cell wall. Three types of pit are present. The first is the bordered pit (Fig. 4.10) which is almost entirely restricted to the radial walls of the tracheids, tending to be located towards the ends of the cells. The second type of pit is the ray or semi-bordered pit which interconnects the vertical tracheid with the horizontal ray parenchyma cell, while the third type is the simple pit between adjacent parenchyma cells.

For very many years it was firmly believed that since the diameter of the pit opening or of the openings between the margo strands was very much less than the diameter of the cell cavity, and since perme-ability is proportional to a power function of the capillary radius (Equation 9.6), the bordered pits would be the limiting factor con-trolling longitudinal flow. However, it has recently been demonstrated that this concept is falacious and that at least 40% of the total resist-ance to longitudinal flow in *Abies grandis* sapwood that has been specially dried to ensure that the torus remains in its natural position can be accounted for by the resistance of the cell cavity (Petty and Puritch, 1970).

Both longitudinal and tangential flowpaths in softwoods are pre-dominantly by way of the bordered pits as illustrated in Fig. 9.1, while the horizontally aligned ray cells constitute the principal path-way for radial flow, though it has been suggested that very fine capil-laries within the cell wall may contribute slightly to radial flow. The rates of radial flow are found to vary very widely between species.

It is not surprising to find that the different pathways to flow in the three principal axes result in anisotropy in permeability. Perme-ability values quoted in the literature illustrate that for most timbers longitudinal permeability is about 10^4 times the transverse perme-ability; mathematical modelling of longitudinal and tangential flow supports a degree of anisotropy of this order. Since both longitudinal and tangential flow in softwoods are associated with bordered pits, a

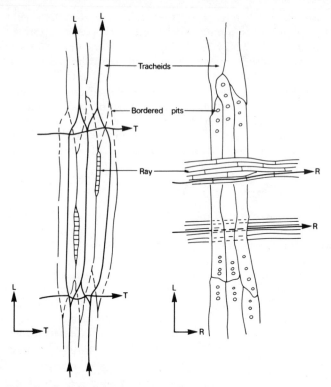

Figure 9.1 On the left, a representation of the cellular structure of a softwood in a longitudinal-tangential plane illustrating the significance of the bordered pits in both longitudinal and tangential flow; on the right, softwood timber in the longitudinal-radial plane, indicating the role of the ray cells in defining the principal pathway for radial flow. (From the Princes Risborough Laboratory, Building Research Establishment. © Crown copyright.)

good correlation is to be expected between them; radial permeability is only poorly correlated with that in either of the other two directions and is frequently found to be greater than tangential permeability.

Permeability is not only directionally dependent, but is also found to vary with moisture content and between earlywood and latewood (Fig. 9.2). In *green* timber the torus of the bordered pit is usually located in a central position and flow can be at a maximum (Fig. 9.3(a)). Since the earlywood cells possess larger and more frequent bordered pits, the flow through the earlywood is considerably greater than that through the latewood. However, on drying, the torus of the earlywood cells becomes aspirated (Fig. 9.3(b)), owing, it is thought, to the tension stresses set up by the retreating water meniscus (Hart and Thomas, 1967): in this process the margo strands obviously undergo very considerable extension (Petty (1972) has suggested that this could be as high as 8% in the earlywood) and the torus is rigidly held in a displaced position by strong hydrogen bonding.

This displacement of the torus effectively seals the pit and reduces

258

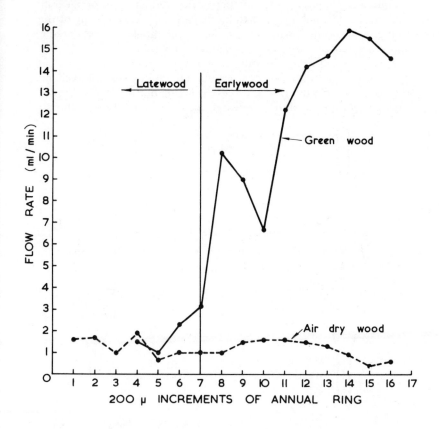

SAPWOOD

HEARTWOOD

Figure 9.2 The variation in rate of longitudinal flow through samples of green and dry earlywood and latewood of Scots pine sapwood and heartwood. (From Banks (1968). © Crown copyright.)

the level of permeability of dry earlywood to a value similar to that of green latewood. In the latewood, only about half the pits become aspirated and consequently the percentage reduction in permeability with drying is much lower than in the earlywood. Rewetting of the timber causes only a partial reduction in the number of aspirated pits and it appears that aspiration is mainly irreversible.

It is possible to prevent aspiration of the bordered pits by replacing the water in the timber by acetone and ethanol by solvent exchange, thus reducing the interfacial forces occurring during drying. While this is a useful procedure in research for the examination of the torus

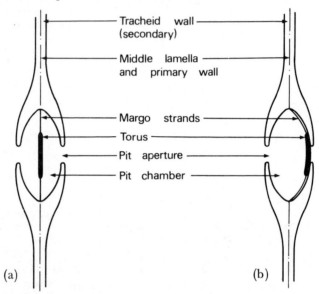

Tracheid   wall
(secondary)

Middle   lamella
and   primary   wall

Margo   strands

Torus

Pit   aperture

Pit   chamber

(a)   (b)

Figure 9.3   Cross-section of a bordered pit in the sapwood of a softwood timber: (a) in timber in the green condition with the torus in the 'normal' position; and (b) in timber in the dried state with the torus in an aspirated position. (From the Princes Risborough Laboratory, Building Research Establishment. © Crown copyright.)

in an unaspirated state, it cannot be applied commercially because of the cost.

Quite apart from the fact that many earlywood pits are aspirated in the heartwood of softwoods, the permeability of the heartwood is usually appreciably lower than that of the sapwood due to the deposition of encrusting materials over the torus and margo strands and also within the ray cells (Fig. 9.2).

Hardwoods. The longitudinal permeability is usually high in the sapwood of hardwoods. This is because these timbers possess vessels, the ends of which have been either completely or partially dissolved away. Radial flow is again by way of the rays, while tangential flow is more complicated, relying on the presence of pits interconnecting adjacent vessels, fibres and vertical parenchyma. Transverse flow rates are usually much lower than in the softwoods, but somewhat surprisingly a good correlation exists between tangential and radial permeability; this is due in part to the very low permeability of the rays in hardwoods.

Since the effects of bordered pit aspiration, so dominant in controlling the permeability of softwoods, are absent in hardwoods, the influence of drying on the level of permeability in hardwoods is very much less than is the case with softwoods.

Permeability is highest in the outer sapwood, decreasing inwards and reducing markedly with the onset of heartwood formation as the cells become blocked either by the deposition of gums or resins or, as happens in certain timbers, by the ingrowth into the vessels of the cell wall material of neighbouring cells, a process known as the formation of *tyloses*.

9.2.1.3 *Timber and the Laws of Flow*

Initial research on liquid permeability indicated that Darcy's Law did not appear to be valid since flow rate decreased with time. Further investigation has revealed that the initial results were an artefact caused by the presence of air and impurities in the impregnating liquid. By de-aeration and filtration of their liquid a number of workers have since been able to achieve steady-state flow, and to find that in very general terms Darcy's Law was upheld in timber.

A second requirement of the 'improved' Darcy's Law is that rate of flow is inversely related to the viscosity of the liquid. Once again the early work on flow was bedeviled by a lack of appreciation of all the factors involved and it was not until all variables occurring within the experimental period were eliminated that this second requirement was satisfied.

Gas, because of its lower viscosity and the ease with which steady flow rates can be obtained, is a most attractive fluid for permeability studies. It has been shown in previous sections that timber is a complex medium composed of different types of capillary in series, and that with such a medium deviations from Darcy's Law for gases are to be expected at low mean gas pressures, due to the presence of slip flow; such deviations have been observed by a number of investigators. At higher mean gas pressures, an approximately linear relationship between conductivity and mean pressure is expected and this, too, has been observed experimentally. However, at even higher mean gas pressures, flow rate is sometimes less than proportional to the applied pressure differential due, it is thought, to the onset of turbulence. Darcy's Law may thus appear to be valid only in the middle range of mean gas pressures.

In a study of gas flow through *Abies grandis*, Smith and Banks (1971) have verified experimentally that the contribution of laminar flow to total flow is inversely proportional to the viscosity of the gas, as is expected from Equation 9.10, and that the contribution of slip flow to total flow is inversely related to the square root of the molecular weight of the gas, as is expected from Equation 9.11.

Flow rate has also been found to decrease with increasing specimen length at a rate faster than that predicted by Darcy, at least up to 20 mm, but thereafter to conform with Darcy's Law. This phenomenon is explained in terms of the increased probability of encounter-

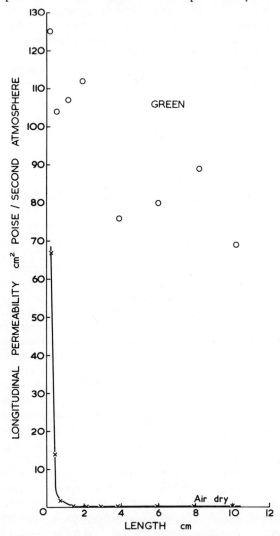

Figure 9.4 The marked decrease in longitudinal permeability of air dry samples of Norway spruce sapwood of increasing length. Permeability of the green samples with respect to length varies randomly about a mean value. (From Banks (1970). © Crown copyright.)

ing a cell with all the pits aspirated as specimen length is increased; therefore, permeability of dried timber will decrease with increasing specimen length until a length is reached where all the easy pit-flow paths are blocked and flow is controlled by 'residual' flow paths whose probability tends to zero (Fig. 9.4). The residual flow paths are assumed to be the unaspirated bordered pits of the latewood or, exceptionally, some of the semi-bordered ray pits (Banks, 1970; Bramhall, 1971).

Bramhall (1971), working on Douglas fir, incorporated a decay function in the Darcy equation to account for the random occurrence of aspirated pits and the reduced number of conducting tracheids with increasing depth of penetration, thus

$$Q = \frac{K_p A e^{-bl} \Delta P}{\eta l} \qquad (9.12)$$

where Q is the volume flow rate, K_p is the permeability constant, A is the cross-sectional area, e is the base of natural logs, b is the positive constant, ΔP is the pressure differential, η is the fluid viscosity and l is the specimen length. This model, which differs from the Darcy Equation (9.9) by the inclusion of the exponential term, gave good agreement with experimental data.

In theory, it should be a simple matter to relate the laminar flow of gases to that of liquids through a given medium. In practice, this is difficult with a heterogeneous medium such as timber because of the presence of slip flow of the gases, and the resultant apparent departure from Darcy's Law at low mean gas pressures. Experimental and analytical approaches to the solution of this problem have been moderately successful (Comstock, 1967; Smith and Banks, 1971).

To summarise, too little is known about flow in hardwoods to make any firm conclusions, but for softwoods it may be concluded that Darcy's law is only partially applicable; certainly the rate of flow in the laminar region is proportional to the applied pressure for both liquids and gases, but the rate of flow is not always inversely related to specimen length, nor is it proportional to pressure at the higher pressure levels: for gases, however, the Darcy equation can be made more appropriate by the incorporation of a slip-flow term as in Equation 9.11.

Looking specifically at the flow of moisture in timber the conclusions put forward by Bateman, Hohf and Stamm (1939) still hold good at the present time. They visualised that flow occurred in three ways: (a) as *free*, liquid water in the cell cavities above the fibre saturation point (see Section 7.2.2.1); (b) as *bound* water within the cell walls below the fibre saturation point; and (c) as water *vapour* both above and below the fibre saturation point.

The movement of the free water is caused by capillary forces and the flow can be described in terms of the Poiseuille equation (Equation 9.2). However, since the molecules of water adjacent to the capillary walls are bound to the capillary by chemisorption, a term must be introduced in the equation to account for the disturbance of flow. Flow is produced by differences in tension due to surface forces in the menisci within the capillaries; the basic mathematical relationship for the tension force in a capillary has already been presented in Chapter 8 for concrete and need not be repeated here. However, the very variable dimensions of both the cell cavities and the pores in the pit membranes render almost impossible any

263

numerical evaluation of capillary movement of moisture in timber; this is most unfortunate in view of the practical significance of moisture movement in the drying of timber.

The significance of pit aspiration in reducing capillary flow or permeability has already been discussed; one practical manifestation of the existence of capillary forces occurs when water is removed too rapidly in drying. When the capillary tension exceeds the compression strength of the timber perpendicular to the grain the cells collapse, resulting in a corrugated surface to the timber.

Below the fibre saturation point bound water moves through the cell walls and this process of moisture diffusion is discussed in the following section. Because of the capillary structure of timber, vapour pressures are set up and vapour can pass through timber both above and below the fibre saturation point: the flow of vapour is usually regarded as being of secondary importance to that of both bound and free water.

9.2.2 Moisture Diffusion

Flow of water below the fibre saturation point embraces both the diffusion of water vapour through the void structure comprising the cell cavities and pit membrane pores and the diffusion of bound water through the cell walls. Diffusion is another manifestation of flow, conforming with the general relationship between flux and pressure. Thus it is possible to express diffusion of moisture in timber in terms of Fick's first law, which states that the flux of moisture diffusion is directly proportional to the gradient of moisture concentration; as such, it is analogous to the Darcy law on flow of fluids through porous media.

The total flux F of moisture diffusion through a plane surface under steady-state conditions is equal to the sum of the flux of the bound water component F_b and that of the vapour component F_v

$$F = F_b + F_v \tag{9.13}$$

The flux of the bound water can be written in terms of Fick's first law, thus:

$$F_b = \frac{dm}{dt} = -K_b \frac{du}{dx} \tag{9.14}$$

where dm/dt ($= m/tA$) is the flux (rate of mass transfer), du/dx is the moisture concentration gradient in the x direction, and K_b is the bound water moisture conductivity coefficient.

Similarly the flux of the vapour movement can be expressed as

$$F_v = \frac{dm}{dt} = -\frac{K_v}{u} \cdot \frac{dh}{dx} \tag{9.15}$$

where dh/dx is the vapour pressure gradient in the x direction, K_v is the moisture conductivity of the water vapour and

$$u = \frac{\text{resistance of gross wood to vapour movement}}{\text{resistance offered by still air of same dimensions}}$$

264

The vapour component of the total flux is usually much less than that for the bound water. The rate of diffusion of water vapour through timber at moisture contents below the fibre saturation point has been shown to yield coefficients similar to those for the diffusion of carbon dioxide, provided corrections are made for differences in molecular weight between the gases. This means that water vapour must follow the same pathway through timber as does carbon dioxide and implies that diffusion of water vapour through the cell walls is negligible in comparison to that through the cell cavities and pits (Tarkow and Stamm, 1960).

Much more attention has been devoted to the study of bound water diffusion. In the literature on this subject readers will find that the conductivity coefficient is frequently referred to as a *diffusion coefficient*; this is quite erroneous as these are separate though related parameters. The diffusion coefficient can be obtained from the conductivity coefficient thus:

$$D_b = \frac{100K_p}{G\rho} \tag{9.16}$$

where D_b is the diffusion coefficient for bound water in gross timber, G is the specific gravity of timber, and ρ is the density of water.

The most important factors affecting the diffusion coefficient of water in timber are temperature, moisture content and density of the timber. Thus Stamm (1959) has shown that the bound water diffusion coefficient of the cell wall substance increases with temperature approximately in proportion to the increase in the saturated vapour pressure of water, and increases exponentially with increasing moisture content at constant temperature. The diffusion coefficient has also been shown to decrease with increasing density and to differ according to the method of determination at high moisture contents. Consequently, the diffusion coefficient for timber cannot be regarded as an absolute value: rather it should be treated as an *apparent diffusion coefficient*.

The evidence to which the bound water diffusion theories relate has recently been re-examined and it has been proposed that the driving force is not a moisture gradient, but a vapour pressure gradient (Bramhall, 1976). Bound water diffusion occurs when water molecules bound to their sorption sites receive energy in excess of the bonding energy, thereby allowing them to move to new sites. At any one time the number of molecules with excess energy is proportional to the vapour pressure of the water in the timber at that moisture content and temperature. The rate of diffusion is proportional to the concentration gradient of the migrating molecules, which in turn is proportional to the vapour pressure gradient. Fick's law can be expressed therefore as:

$$F = -D_p \cdot \frac{dP}{dx} \tag{9.17}$$

where F is the moisture flux, and D_p is the diffusion coefficient for bound water under a vapour pressure gradient of dP/dx.

The values of moisture content and vapour pressure are not inter-changeable, but in the lower half of the hygroscopic moisture range under isothermal conditions the two are approximately proportional. Most of the experiments reported in the literature are within this moisture range and considerable controversy now reigns as to whether the driving force should be a vapour pressure gradient rather than the moisture content gradient. The position is further confused since the proponents of a moisture content driving force believe it is a function of temperature and moisture content as previously dis-cussed, while the proponents of the vapour pressure concept believe it to be dependent on moisture content but independent of tempera-ture.

The diffusion coefficient is also dependent on the grain direction: the ratio of longitudinal to transverse coefficients is approximately 2.5.

9.2.3 Thermal and Electrical Conductivity

The basic law for flow of thermal energy is ascribed to Fourier and when described mathematically as

$$K_h = \frac{Hl}{tA\,\Delta T} \tag{9.18}$$

(where K_h is the thermal conductivity, H is the quantity of heat, t is time, A is the cross-sectional area, l is the length and ΔT is the temperature differential) is analogous to that of Darcy (Equation 9.9) for fluid flow.

Compared with permeability, where the Darcy equation was shown to be only partially valid for timber, thermal flow is explained adequately by the Fourier equation, provided the boundary con-ditions are defined clearly.

Thermal conductivity will increase slightly with increased moisture content, especially when calculated on a volume-fraction-of-cell-wall basis; however, it appears that conductivity of the cell wall substance is independent of moisture content. Conductivity is influenced con-siderably by the density of the timber, i.e. by the volume-fraction-of-cell-wall substance, and various empirical and linear relations between conductivity and density have been established. Conductivity will also vary with timber orientation due to its anisotropic structure: the longitudinal thermal conductivity is about 2.5 times the transverse conductivity.

Compared with the metals the thermal conductivity of timber is extremely low, though it is generally about 1–8 times higher than that of insulating materials (Table 9.1). The transverse value for timber is about one quarter that for brick thereby explaining the lower heating requirements of timber houses compared with the traditional brick house.

Electrical conductivity once again conforms to the basic relation-ship between flux and pressure gradient, but it is much more sensitive to moisture content than is the case with thermal conductivity: as moisture content increases from zero to fibre saturation point the electrical conductivity increases by at least 10^{10} times. Conductivity

266

Table 9.1 Thermal Conductivity of Timber and Other Materials

Material	K_h (W m^{-1}K^{-1})
Copper	400
Aluminium	201
Concrete	1.5
Glass	1.1
Brick wall	1.0
Water	0.59
TIMBER parallel to grain	0.38
TIMBER perpendicular to grain	0.15
Polyisoprene rubber	0.15
Plaster	0.13
Cork (baked slab)	0.05
Glass wool	0.04
Polystyrene, cellular	0.035

is also temperature dependent, each 10 °C rise in temperature doubling the electrical conductivity.

Electrical conductivity, or rather its reciprocal the *resistivity*, is a most useful measure of the moisture content of timber and mention has already been made in Chapter 7 of the use of electrical resistance meters for the rapid determination of moisture content below the fibre saturation point.

9.3 UNSTEADY-STATE FLOW

As recorded above, steady-state flow in timber has received considerable attention but, unfortunately, those processes which depend on the movement of fluids into or out of timber are concerned with unsteady-state flow which is usually most complex in analysis, necessitating many simplifying assumptions and consequently receiving but little attention. Unsteady-state flow occurs when the rate of flow and the pressure gradient are varying in space and time, as actually occurs during the drying of timber or its impregnation with chemical solutions. The same broad similarities in behaviour between fluids, moisture vapour, and heat that are seen with steady-state flow are again recognisable in unsteady-state flow.

9.3.1 Unsteady-State Flow of Fluids

The unsteady-state equation can be derived from the steady-state equation by differentiating with respect to time. For gaseous flow this becomes:

$$\frac{\Delta V}{\Delta t} = -\frac{K_g A (P_2{}^2 - P_1{}^2)}{2\Delta x P_2} \qquad (9.19)$$

where P_2 is the pressure at distance x, P_1 is the pressure at distance $x + \Delta x$, and K_g is the superficial gas permeability. Taking the loss in the mass of the gas into consideration the equation may be developed and written in partial derivative form as

$$\frac{\partial P^2}{\partial t} = D_p \frac{\partial^2 P^2}{\partial x^2} \qquad (9.20)$$

where D_p is the diffusion coefficient for hydrodynamic flow based on average pressure, averaged over space and time, which equals $K_g \bar{P}/p$, where K_g is the superficial gas permeability, p is the porosity, and hence $p/\bar{P}$ is the volume of gas required to raise the pressure by 1 atmosphere (Siau, 1971).

There is almost complete absence of analytical solutions for this diffusion equation when it is used to describe the flow of compressible fluids. This results from the very complex variation that occurs in the diffusion coefficient with changing pressure.

Early investigations on unsteady-state flow resolved the problem by using a constant value for the diffusion coefficient. Such an assumption is not valid for gas flow since the coefficient varies systematically with pressure, which in turn is dependent on time and location. While these early results have to be interpreted with caution, they did indicate the very marked anisotropy in the values of the diffusion coefficient with a ratio of longitudinal to transverse of over 10 000.

Recently, the problem has been resolved by the use of a dimensionless, pressure-dependent dynamic flow coefficient ($\equiv$ diffusion coefficient) and the equation solved for particular boundary conditions (Sebastian et al., 1973). By replacement of the nonlinear differential equation by a finite-difference equation and solution by computer, both the viscous and the slip flow parameters for longitudinal flow have been determined. Despite the simplified mathematical model that was adopted, the numerical solution of the partial differential equation was in reasonable agreement with the experimental data.

Such studies of unsteady-state fluid flow through timber have, to date, assumed that timber is a homogeneous porous medium and that Darcy's Law may be used as a basis for unsteady-state flow theory. It was stated previously that this assumption is tenable for liquid flow, but incorrect for gas flow through timber at low mean gas pressures. The implications of this for unsteady-state gas flow theory have apparently not been investigated. Unsteady-state flow theory is certainly an area which demands much more attention.

9.3.2 Unsteady-State Moisture Diffusion

As in the case of fluid flow, the unsteady-state equation for moisture diffusion can be derived by differentiation from the steady-state equation. In the partial derivative form it becomes:

$$\frac{\partial M}{\partial t} = D_m \frac{\partial^2 M}{\partial x^2} \qquad (9.21)$$

where D_m is the average diffusion coefficient for moisture movement. This equation is referred to as Fick's second law of diffusion. The diffusion coefficient can be derived experimentally assuming that it is constant. However, once again this assumption is invalid since it is now known that the diffusion coefficient is concentration dependent. In order to circumvent the otherwise extremely complex solution of the equation, average values for the diffusion coefficient for selected moisture content ranges have been used.

As in the case of steady-state flow, Bramhall (1976) has indicated that the driving mechanism for unsteady-state diffusion is a vapour pressure gradient rather than a moisture content gradient, thus:

$$\frac{\partial M}{\partial t} = D_p \frac{\partial^2 P}{\partial x^2} \qquad (9.22)$$

where D_p is the diffusion coefficient for bound water under a vapour pressure gradient. Values for the diffusion coefficient vary with direction: at a moisture content of 25% the ratio of longitudinal to transverse diffusion coefficients is less than 4, while at 5% moisture content the ratio lies between 50 and 100. Not only moisture content, but also temperature (except in the case of D_p) and timber density are known to affect the coefficient, thereby resulting in a wider range of values than occurs with either gaseous or thermal diffusion coefficients.

9.3.3 Unsteady-State Thermal Conductivity

The equivalent equation in partial derivative form for thermal flow is:

$$\frac{\partial T}{\partial t} = D_h \frac{\partial^2 T}{\partial x^2} \qquad (9.23)$$

where D_h is the thermal diffusion coefficient. ($D_h = K_h/\kappa\rho$, where K_h is the thermal conductivity of gross timber, and $\kappa\rho$ is the quantity of heat required to increase the temperature of a unit cube of timber by 1 °C.)

Once again the diffusion coefficient is not constant, being dependent on the thermal conductivity, density and moisture content. The ratio of the coefficients in the longitudinal and transverse directions is about 2.5.

The work done so far on unsteady-state flow has been informative, but caution must be exercised in any attempt to accept the findings in absolute terms since many of the assumptions on which the analysis of unsteady-state flow is based have been rendered untenable by recent findings under steady-state conditions.

Although it has been possible to obtain good correlations between theory and experimental results of both steady- and unsteady-state flow when carried out under laboratory conditions, generally rather disappointing relations have been found under practical conditions as, for example, in the artificial seasoning of timber, the migration of moisture in buildings, or the impregnation of timber with artificial preservatives. Generally the rates of flow obtained are very much lower than theory would indicate and unfortunately the reasons for the discrepancy are not understood. Many of the coefficients used in timber, particularly the diffusion coefficients, have been derived empirically; flow in timber is only one of many aspects of material performance where theory and practice have yet to be reconciled.

9.4 REFERENCES

Banks, W. B. (1968) A technique for measuring the lateral permeability of wood, *Journal of the Institute of Wood Science*, 4 (2), 35–41.

Banks, W. B. (1970) Some factors affecting the permeability of Scots pine and Norway spruce, *Journal of the Institute of Wood Science*, 5 (1), 10–17.

Bateman, E., Hohf, J. P. and Stamm, A. J. (1939) Unidirectional drying of wood, *Ind. Eng. Chem.*, 31, 1150–1154.

Bramhall, G. (1971) The validity of Darcy's Law in the axial penetration of wood, *Wood Science and Technology*, 5, 121–134.

Bramhall, G. (1976) Fick's Laws and bound-water diffusion, *Wood Science*, 8 (3), 153–161.

Comstock, G. L. (1967) Longitudinal permeability of wood to gases and nonswelling liquids, *Forest Products Journal*, 17 (10), 41–46.

Hart, C. A. and Thomas, R. J. (1967) Mechanism of bordered pit aspiration as caused by capillarity, *Forest Products Journal*, 17 (11), 61–68.

Petty, J. A. (1972) The aspiration of bordered pits in conifer wood, *Proceedings of the Royal Society, London*, B 181, 395–406.

Petty, J. A. (1974) Laminar flow of fluids through short capillaries in conifer wood, *Wood Science and Technology*, 8 (4), 275–282.

Petty, J. A. and Puritch, G. S. (1970) The effects of drying on the structure and permeability of the wood of *Abies grandis*, *Wood Science and Technology*, 4 (2), 140–154.

Sebastian, L. P., Siau, J. F. and Skaar, C. (1973) Unsteady-state axial flow of gas in wood, *Wood Science*, 6 (2), 167–174.

Siau, J. F. (1971) *Flow in Wood*. Syracuse University Press, pp. 99, 100.

Smith, D. N. R. and Banks, W. B. (1971) The mechanism of flow of gases through coniferous wood, *Proceedings of the Royal Society, London*, B 177, 197–223.

Stamm, A. J. (1959) Bound water diffusion into wood in the fiber direction, *Forest Products Journal*, 9 (1), 27–32.

Tarkow, H. and Stamm, A. J. (1960) Diffusion through the air filled capillaries of softwoods: I, Carbon dioxide; II, Water vapour, *Forest Products Journal*, 10, 247–250 and 323–324.

CHAPTER 10

Response of Concrete to Stress

10.1 STRESS-STRAIN RELATIONS

This chapter is concerned with the strains that occur in concrete in conjunction with stress. Sometimes the stress arises from an imposed strain but, more often, it is a matter of the concrete deforming under an applied load, and it is most convenient to consider the strain response to an applied stress.

The structure of h.c.p. is sufficiently disordered to ensure that stress-strain relations will not be simple. There are three ways in which this is apparent.

(a) The strain is not reversible; that is, the loading and unloading curves do not coincide.
(b) The strain is time-dependent; that is, it continues to increase under a constant sustained stress.
(c) The magnitudes of the strains are affected by many influences both intrinsic (such as the concrete mix proportions) and extrinsic (such as the relative humidity of storage).

These effects are considered in this chapter and a first view of the stress-strain-time relations is given in Fig. 10.1, which shows what is observed when a uniaxial compressive stress is applied to a concrete specimen. It is applied at time t_1 and then sustained, unchanged, until time t_2, when it is removed. There is an immediate strain which for lower levels of stress is approximately proportional to the applied stress. It is thus convenient to treat it as an elastic strain, and an appropriate value of Young's modulus, E_c, can be found as discussed in the next section. With time the concrete continues to contract, partly because of volume changes caused by shrinkage or thermal movement, but additionally because of the applied stress. The strain, additional to the volume change, is creep and, as can be seen, it proceeds at an ever-decreasing rate. When the stress is removed there is an immediate strain recovery which is often less than

271

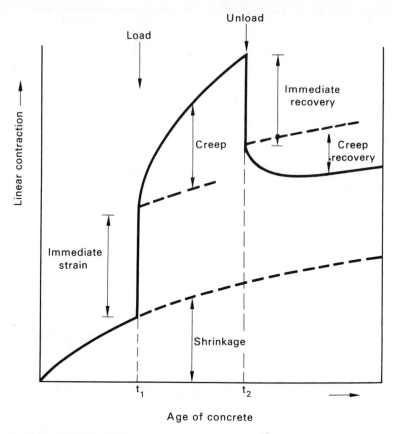

Figure 10.1 The response of concrete to stress for applied compressive stress in a constant drying environment.

the immediate strain on loading. It is followed by a time-dependent creep recovery which is less than the preceding creep, and which, unlike creep, reaches completion in due course.

For the lower levels of stress, i.e. at working level, of the kind that are not usually exceeded in concrete structures, both creep and creep recovery are, like the immediate strains, approximately proportional to the applied stress, σ. Thus, creep, ϵ_c, is given by

$$\epsilon_c = \sigma \cdot c_c \tag{10.1}$$

where c_c is the creep per unit stress, and is referred to as specific creep. Equation 10.1 is the constitutive equation relating time-dependent uniaxial stress and strain, and it corresponds to the simple elastic constitutive law relating stress and elastic strain ϵ_e.

$$\epsilon_e = \sigma/E_c \tag{10.2}$$

In the following sections the characteristics of the constitutive co-efficients, E_c and c_c, will be presented and discussed.

272

10.2   ELASTIC MODULUS

10.2.1   Test Methods

There are various ways of determining the elastic modulus, all of which can be classified as either static or dynamic. The static methods cover the possibilities of applying direct stress, compressive or tensile, or shear stress, and of measuring deflections or strains. The dynamic methods include the various modes of resonant vibration — longitudinal, flexural or torsional — and the determination of ultrasonic pulse velocities. Two methods, one of each kind, are described below; they are the two most widely used and both have the seal of approval given by their appearance in British Standard 1881: 'Methods of Testing Concrete' (1970).

10.2.1.1   *The Static Test*

This is the most obvious form of test in which the variables, stress and strain, are measured in the simplest possible manner, to give the static Young's modulus of elasticity, E_{cs}. A specimen of concrete, normally a cylinder, is loaded axially in compression and the contraction of the concrete in the direction of the load is measured after each of a series of load increments. Stress and strain are found by dividing load and contraction by the cross-sectional area and gauge length respectively.

A typical observation is shown in the sketch of Fig. 10.2, which gives the results for a test in which two successive cycles of load were applied, with a maximum stress of about 65% of the concrete strength. There are distinctive hysteresis loops, and a residual strain (OX) after unloading. In addition the stress-strain relation is at no stage truly linear, and it is clear that the use of the elastic modulus is an approximation to the real behaviour of concrete. The non-linearity becomes more marked if:

(a)   the load is applied more slowly;
(b)   the stress is increased to higher levels.

(a) This allows a larger creep strain to occur during the test so that the slope of the curve diminishes at all stress levels. The effect can obviously be reduced by more rapid loading, and it is also found that the effect is less in second or later cycles of loading to the same level. This can be seen in Fig. 10.2 in the greatly diminished area of the second hysteresis loop compared to the first. It is a recognised experimental technique to exercise a specimen through several loading cycles before recording the strains from which the elastic modulus is deduced. As shown earlier the creep rate diminishes with time, and it is noticeable that there is no great difference between the stress-strain curves of specimens loaded in more than 2 and less than 15 minutes.

(b) Failure in concrete is preceded by a gradual proliferation and propagation of microcracks starting at quite low stresses and making a significant and increasing contribution to the curvature of the

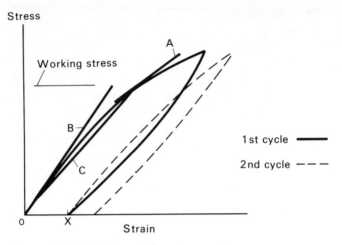

Figure 10.2 Typical short-term stress-strain curves for concrete.

stress-strain curve as the stress rises above working level. This is discussed in detail elsewhere (see Section 13.4.2) in connection with the complete stress-strain curve and the failure of concrete.

The Young's modulus is given by the slope of the stress-strain relation, and there are two possible choices:

(a) the tangent modulus, A, which has a value which reduces greatly as the stress increases. As a general measure of the elastic properties of the material, the initial tangent modulus, B, drawn to the curve as its origin, is the best choice.

(b) the secant modulus, C, given by the slope of the chord drawn from the origin. This too reduces with stress, but to a lesser extent than the tangent modulus, and it has the advantages of being the best linear approximation to the actual stress-strain curve.

At first sight it appears that Young's modulus for concrete should be quoted with reference to a particular rate of loading and a particular stress level, but it emerges that careful definition and a well considered experimental technique can lead to the deduction of values that enable good predictions to be made of the immediate strains in concrete after loading. To sum up, the secant modulus corresponding to a stress of, say, 40% of the concrete strength should be found, in a test in which the specimen is exercised for several cycles before recording the strains during a loading lasting about 5 min.

10.2.1.2 The Dynamic Test

The centrepiece of the apparatus is shown in Fig. 10.3. A variable frequency oscillator drives the electromagnetic exciter unit placed in contact with one end of the concrete specimen. A piezoelectric pickup is attached to the other end, the vibrations are fed to an

274

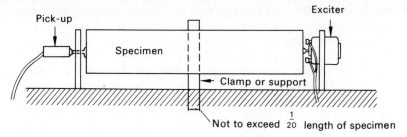

Figure 10.3 Arrangement of specimen for determination of dynamic modulus of elasticity by an electrodynamic method. (From BS 1881 (1970) Part 5, reproduced by permission of BSI, 2 Park Street, London W1A 2BS, from whom complete copies can be obtained.)

amplifier, and the amplitude is displayed. The specimen is supported at the midpoint so that the frequency can be varied until resonance is achieved in the fundamental mode of longitudinal vibration. This can be identified as giving maximum amplitude, and a check should be made that it is not the resonance at the first harmonic that has been located. The frequency is recorded and the dynamic modulus, E_{cd}, in MN/m^2 is given by:

$$E_{cd} = 4n^2 L^2 \rho \times 10^{-12} \tag{10.3}$$

where L is the length of the specimen in mm, n is the frequency (of the fundamental mode of longitudinal vibration) in hertz, and ρ is the density of the concrete in kg/m^3.

The dynamic test produces good reproducible results, and it has the additional advantage that it is nondestructive. The force applied to the specimen in setting it in vibration is negligible and rapid, so that the causes of nonlinearity in the static test do not arise. It is to be expected that the dynamic modulus should be of the same magnitude as the initial tangent modulus, and it has been shown that this is indeed not far from the truth.

It also follows that the dynamic modulus will be greater than the static, and a typical linear relation between them is given in the BS Code of Practice 110: 'The Structural Use of Concrete' (1972).

$$E_{cs} = 1.25 E_{cd} - 19 \qquad \text{in } kN/mm^2 \tag{10.4}$$

The dynamic test is most valuable as a laboratory method for checking the quality of concrete. This can be a matter of monitoring the changes that occur with time, or of identifying the weaker specimens in a batch. The static test is more appropriate to the estimation of the immediate strain in loaded concrete, and it is the more likely to agree with the first observed strain in a creep test in which the process of loading may take some minutes.

10.2.2 Elasticity and Hardened Cement Paste

10.2.2.1 *Porosity*

When a load is applied to h.c.p. the forces are carried internally by

275

the solid structure, made up of both the larger crystalline hydration products and the calcium silicate gel. The solid skeleton is interspersed with a large volume of gel and capillary pores (see Section 3.2.3) with a wide range of sizes. It is conceivable that water within the pores initially carries some of the load, and hence comes under pressure. The gradual dissipation of the pressure in the water would lead to time-dependent strain which would be observed as creep, and the elastic (immediate) strain for wet concrete would differ from that of dry concrete. However, the evidence indicates that this effect, although it may be present, does not result in more than a small part of the load being diverted from the solid material. It follows that the elastic response of the h.c.p. depends on the rigidity of the skeletal structure.

There is an analogy here with the performance under load of a framed building, in which the displacements depend on the stiffness, size and configuration of the members. This is helpful in thinking about the strains in h.c.p. but the structure of h.c.p. is too complicated for any sort of realistic mathematical treatment. It seems reasonable to assume that the stiffnesses of the various solid hydration products and of the unhydrated cement grains incorporated in the structure are not too dissimilar, and that the configuration does not change with the development of hydration. The difference between one h.c.p. and another then reduces to differences in size of member; that is, to different densities of structure. This implies that the elastic modulus of h.c.p. (E_p) will rise with the increasing content of solid material, so that it is to be expected that E_p will increase with both the degree of hydration and with the lowering of the water/cement ratio of the mix. This expectation is confirmed by the experimental results shown in Fig. 10.4, in which the development of hydration is represented by increase in age.

The two separate influences can be incorporated in a single, more fundamental quantity — porosity — which may be either capillary porosity or total porosity (see Section 3.2.5.3). Figure 10.5 shows the relation between E_p and capillary porosity, p_c and this can be expressed

$$E_p = E_g(1 - p_c)^3 \tag{10.5}$$

where E_g is the elastic modulus of cement gel. From Fig. 10.5, E_g is the value of E_p when p_c is zero. That is, E_g has the value 32 kN/mm^2.

10.2.2.2 Environment

Drying removes water from the larger pores which, according to the argument above, may be load-bearing, with a consequent reduction of elastic modulus with decreasing moisture content of the h.c.p. More significantly, water is also removed by drying from the finer pores and from between the layers of the solid material. This water is bound relatively strongly and it can be regarded as a part of the solid material, contributing to its stiffness. The loss of this water is a second cause of a reduction in modulus with drying. Credence to

276

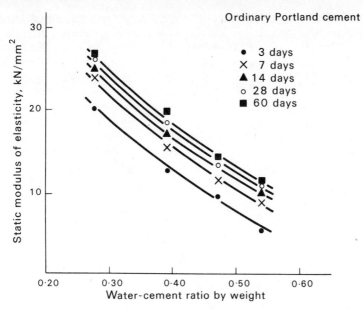

Figure 10.4 The effect of water/cement ratio and age on the static modulus of elasticity of saturated cement paste. (From T. J. Hirsch (1962) *Proceedings of the American Concrete Institute,* **59,** by permission of the American Concrete Institute.)

these ideas is given by Fig. 10.6, in which compliance is plotted against shrinkage. Compliance is elastic strain per unit stress, allowance being made for the nonlinearity of the stress-strain relation, and the shrinkage is that occurring in predrying the h.c.p. before testing. Since shrinkage is related directly to moisture loss, increasing shrinkage on this axis is a measure of decreasing moisture content.

Temperature, too, affects the elastic modulus of h.c.p., higher temperatures giving lower elastic moduli. This is shown, for concretes, in Fig. 10.7, and a similar trend has been found for h.c.p.

The temperature dependence may well be related to the increased mobility of the moisture at higher temperatures, giving some loss of stiffness in the solid structure. This reasoning assumes that there is no difference in the degree of hydration between the h.c.p.s being compared; if the temperature is raised some time before the elastic strains are measured the acceleration of hydration caused by the higher temperature has the reverse effect — the greater degree of hydration at higher temperature increases the stiffness, and hence the elastic modulus.

10.2.3   Influence of the Aggregate

Concrete is a multiphase material with a number of constituents including unhydrated cement, cement gel, water, air, sand and stones. The geometries of these various constituents are complicated in themselves, and the situation becomes confused indeed when they

277

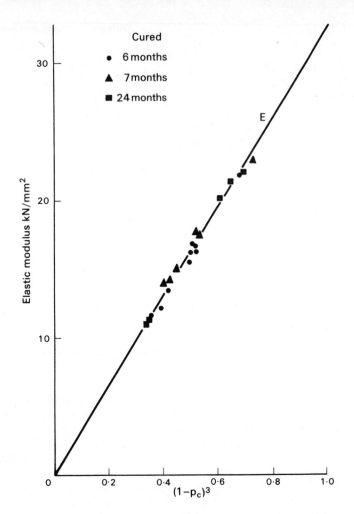

Figure 10. 5 Elastic moduli of hardened Portland cement pastes of various capillary porosities, p_c, cured 6, 7 and 24 months. (From R. A. Helmuth and D. M. Turk (1966) Elastic moduli of hardened cement paste and tricalcium silicate pastes, in *Symposium on Structure of Portland Cement Paste and Concrete*, by permission of the Transportation Research Board, National Academy of Sciences,)

are mixed together. It is not possible to analyse the real material to determine the forces and deformation of each constituent, but concrete can be treated as a two-phase material, consisting only of h.c.p. and aggregate (sand plus stones). It is reasonable to assume that that the interfacial effects are negligible and hence the elastic modulus of the concrete can be found from a model incorporating:

(1)   a suitable geometrical arrangement of the phases;
(2)   the property values representing the phases. If a sufficiently simple view is taken these can be reduced to a mere three:
        elastic modulus of the aggregate, E_a

278

elastic modulus of the h.c.p., E_p
volume concentration of the aggregate, g.

In general, the elastic modulus for the concrete, E_c, is given by

$$E_c = f(E_a, E_p, g) \qquad (10.6)$$

where the form of the function f depends on the particular geometrical configuration adopted. Three possible arrangements are

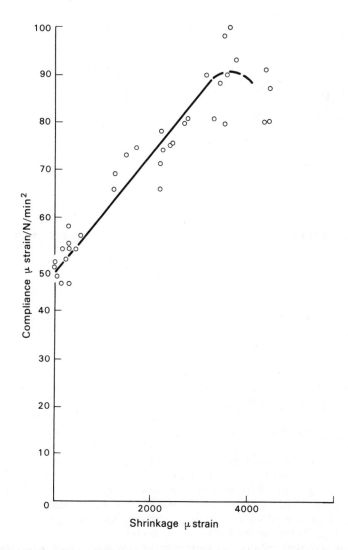

Figure 10.6   Effect of drying upon 'compliance' h.c.p. Water/cement ratio = 0.47. Data derived from $\epsilon_{el} = a\sigma^{1.07}$ where ϵ_{el} = elastic strain ($\times 10^{-6}$), σ = stress (N/mm^2) and a = compliance. (From L. J. Parrott (1973) *Magazine of Concrete Research*, **25**, 19.)

279

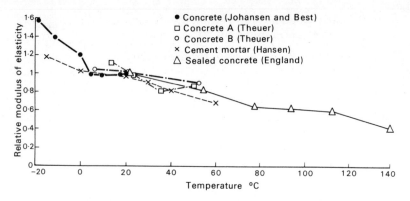

Figure 10.7 Effect of temperature on modulus of elasticity. (From A. D. Ross, J. M. Illston and G. L. England (1968) Short and long-term deformations of concrete as influenced by its physical structure and state, *Proceedings of an International Conference on the Structure of Concrete,* by permission of the Cement and Concrete Association.)

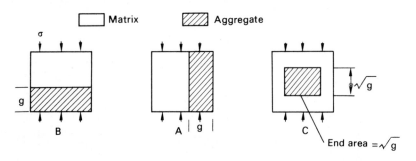

Figure 10.8 Simple two-phase models for concrete. (From T. C. Hansen (1960) *Proceedings of the Swedish Cement and Concrete Research Institute,* No. 31 and U. J. Counto (1964) *Magazine of Concrete Research,* 16.)

shown in Fig. 10.8. All the models consist of unit cubes, but while Aand B have the phases arranged as adjacent layers, model C has the aggregate set within the h.c.p. such that the height of the aggregate block equals its base area. Models A and B differ in that under the applied uniaxial compressive stress, σ, model A suffers uniform strains in the two phases, while the phases of model B carry the same stress. It can be shown theoretically that these two models represent upper and lower bounds to the predicted modulus of elasticity for concrete as a two-phase material. Model C is treated as three separate layers, two of h.c.p. alone and the third of h.c.p. and aggregate in parallel (as in model A); it is intuitively more satisfactory in that it does actually have some resemblance to concrete.

The actual distributions of stress and strain within concrete show great variations, but the models are concerned with the prediction of average behaviour; that is, for behaviour of a volume of concrete greater than the representative cell.

280

In the analysis of the models three further assumptions are made:

(a) The applied uniaxial compressive stress remains uniaxial and compressive throughout the model.
(b) The effects of lateral continuity between layers can be ignored.
(c) Any local bond failure or crushing does not contribute to the deformation.

The derivation of the form of Equation 10.6 for a model A is given below; those for models B and C can be found in a similar manner, and the final results only are quoted.

Model A

Strain and Compatibility. The strain in the concrete, ϵ_c, is equal to the strain in both the aggregate, ϵ_a, and the h.c.p., ϵ_p, i.e.

$$\epsilon_c = \epsilon_a = \epsilon_p \tag{10.7}$$

Equilibrium. The total force applied to the model is given by applied stress acting on unit area. Thus

$$\sigma \cdot 1 = \sigma_a \cdot g + \sigma_p (1 - g) \tag{10.9}$$

Constitutive Relation. Both constituent materials are elastic, and so is the concrete.

$$\sigma = \epsilon_c E_c \qquad \sigma_a = \epsilon_a E_a \qquad \sigma_p = \epsilon_p E_p \tag{10.9}$$

Substituting in Equation 10.8 from Equation 10.9

$$\epsilon_c E_c = \epsilon_a E_a g + \epsilon_p E_p (1 - g)$$

and hence, from Equation 10.7

$$E_c = E_a g + E_p (1 - g) \tag{10.10}$$

Model B

$$\frac{1}{E_c} = \frac{g}{E_a} + \frac{(1 - g)}{E_p} \tag{10.11}$$

Model C

$$\frac{1}{E_c} = \frac{(1 - \sqrt{g})}{E_p} + \frac{1}{\left[\frac{1 - \sqrt{g}}{\sqrt{g}}\right] E_p + E_a} \tag{10.12}$$

For a given h.c.p. the modulus of the concrete depends on the stiffness of the aggregate and its volume concentration. The effect of stiffness is shown in nondimensional form in Fig. 10.9, in which the predictions of the three models are plotted and compared with some typical experimental results. The lower bound is the more nearly correct when the aggregate is stiffer than the h.c.p. $[(E_a/E_p) > 1]$. This is the zone in which most natural aggregates lie, while light-weight aggregates are less stiff than the h.c.p. $[(E_a/E_p) < 1]$ and thus

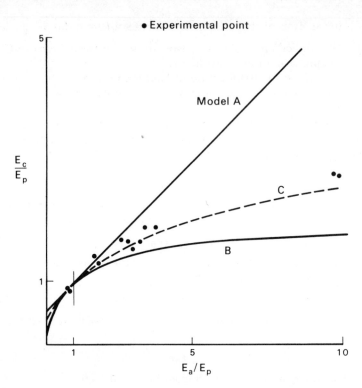

Figure 10.9 Prediction of elastic modulus by various two-phase models. Volume concentration of aggregate = 50%.

the elastic moduli of lightweight concretes are considerably lower than those of normal concrete. In the zone of softer aggregates it is the upper bound that is nearer to the experimental observations, but the predictions of model C are undoubtedly the best and they apply for the full range of aggregate stiffnesses.

The effect of changing the volume concentration, g, is shown in Fig. 10.10, in which typical values of E_a and E_p have been chosen.

10.2.4 Elastic Modulus and Poisson's Ratio in Practice

In the preceding sections various factors that influence the magnitude of the elastic modulus of concrete have been introduced and discussed. They include:

for the h.c.p.
the mix — the degree of hydration and the water/cement ratio
the state — the moisture content, the temperature
for the aggregate
the stiffness, the volume concentration.

Ideally all these quantities should be considered when deciding on the value of the elastic modulus of the concrete in a real structure, but there are two constraints.

282

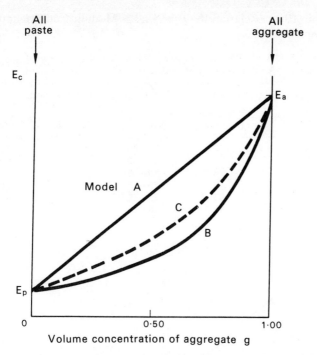

Figure 10.10 Variation in elastic modulus of concrete with volume concentration of aggregate by simple two-phase models.

Firstly, although it is easy enough to demonstrate the importance of the various quantities, this is not the same as establishing usable relationships which are universally applicable. Until such relationships are found and proved, the engineering profession will not, and should not, attempt to calculate elastic moduli on this desirable basis.

Secondly, the vital data are not always, or even often, available. For example, although the two-phase models have been tested and found reliable (and thus satisfy the first constraint) the moduli of the aggregate and the h.c.p. must be known in order to deduce the modulus of the concrete. Not only may these not be known but, at least for some aggregates, it is not easy to suggest how they should be found.

In these circumstances the fundamental approach has limited applicability and an alternative relationship must be available. That between elastic modulus and strength is the most favoured, and it has the overwhelming advantage that strength is the one quantity that is virtually always known by the engineer. A reasonable correlation exists between elastic modulus and strength, and this is presented in numerical form in Table 10.1, taken from the Code of Practice CP 110 (1972).

In principle this is an unsatisfactory approach since it introduces a variable, strength, which does not truly represent the full range of

Table 10.1　Relation Between Elastic Modulus and Strength of Concrete

From CP 110 (1972), reproduced by permission of BSI, 2 Park Street, London W1A 2BS, from whom complete copies can be obtained.

Compressive strength f_{cu} (N/mm^2)	Static modulus E_{cs}		Dynamic modulus E_{cd}	
	Mean value (kN/mm^2)	Typical range (kN/mm^2)	Mean value (kN/mm^2)	Typical range (kN/mm^2)
20	25	21–29	35	31–39
25	26	22–30	36	32–40
30	28	23–33	38	33–43
40	31	26–36	40	35–45
50	34	28–40	42	36–48
60	36	30–42	44	38–50

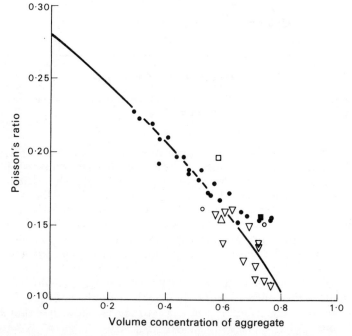

Figure 10.11　Relation between Poisson's ratio and volumetric content of aggregate (saturated concretes made with gravel aggregate, tested at 30 days). (From M. Anson and K. Newman (1966) *Magazine of Concrete Research,* **18,** 125.)

factors that influence the elastic modulus. Thus the properties of the aggregate do not have a great effect on strength, though, as has been seen, they are very important to elastic modulus. Similarly, a reduction in moisture content causes a decrease in modulus, but has the opposite influence on strength. These are deficiencies in the strength-elastic modulus relation, but, on the other hand, the effect of porosity (degree of hydration and water/cement ratio) is similar for both quantities. It can be concluded that the present practical method must be accepted as the best that can be done, but that engineers should be aware of its obvious faults.

Poisson's ratio, ν_e, is found by measuring lateral strains, as well as longitudinal ones, in a static test of the kind described earlier. For h.c.p. it is much the same for different water/cement ratios, but reduces with moisture content; typical figures are Poisson's ratio of 0.25 for a saturated h.c.p. and 0.17 for an h.c.p. dried out at 20%.

Addition of aggregate again modifies the behaviour, as shown in Fig. 10.11. There are few situations in which such variations in the value of ν_e have engineering significance, and the use of a constant value of 0.20 is suggested.

10.3   CREEP

10.3.1   Test Methods

A creep test is essentially simple, consisting merely of applying a constant sustained load to a concrete specimen, and measuring the resulting strains. In practice, there are considerable problems in relation to both the loading and the strain measurement (Illston and Pomeroy, 1975).

10.3.1.1   *Loading*

A variety of imposed loading systems have been developed, including direct tension, flexure, torsion, multiaxial compression and, above all, uniaxial compression. The requirements for loading are:

(1)   to apply the load as rapidly as possible,
(2)   to maintain it reasonably constant for the duration of the test,
(3)   to measure it accurately,
(4)   to induce a uniform state of uniaxial compressive stress over the measurement zone of the specimen.

Three devices that have proved successful are shown in sketch form in Fig. 10.12. They employ:

(a)   the nutcracker principle, in which the lever can be adjusted to impose a high stress on the specimen with a comparatively low applied load. Dead weights have apparent advantages in speed of loading and long term constancy, but the difficulties of handling large metal weights should not be overlooked.
(b)   colinear spring loading, in which the applied load will decline somewhat as the concrete creeps, so that topping up may be necessary. The contraction in length of the calibrated spring is the measure of the load.
(c)   fluid pressure, which is transferred to the specimen through a diaphragm and piston. The piston follows the contraction of

the specimen so that there is no drop off in load; there may, however, be some difficulty in maintaining a constant fluid pressure over a long period. The speed of loading is greatest in this case, and the device is the most complicated and expensive of the three.

There are possible refinements, and all three rigs usually incorporate ball-seatings at the ends of the specimens to improve the uniformity of the applied stress. The aspect ratio of the specimens is at least 2 : 1 so as to give a central zone for strain measurement which is free of end effects.

10.3.1.2 *Strain Measurement*

The requirements for strain measurement are:

(1) to ensure that the strain gauges are stable with time,
(2) to measure the strains with sufficient precision;
(3) differences of 5 microstrain or less should be measurable, to ensure that the strain is the true average of the response of the concrete.

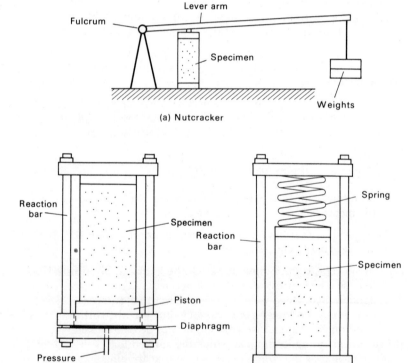

Figure 10.12 Loading systems for creep testing.

286

To eliminate the effects of any eccentricity of the load at least two strains are recorded on each specimen. In fact, the rigs discussed under (b) and (c) above have three or four reaction bars, and it is normal to operate on three or four longitudinal gauge lengths spaced regularly around the circumference of the specimen. The most common form of gauge is the demountable mechanical type which locates in specially drilled metal studs cast into the surface of the concrete. The variability of concrete is such that, for research work, tests are performed on groups of at least three specimens from the same batch of concrete.

The first measured strain is regarded as the elastic strain, and this will be in error if the loading takes too long. It is found that the rate of creep between 2 and 10 min after the start of loading is not great, and it is a reasonable compromise between the ideal of elasticity and the reality of testing to suggest that the recording of the first strains should be completed during this period.

10.3.2 Creep-Time Relations

The longer the concrete is under load the smaller becomes the rate of creep. This feature of the relation between creep and time under load can be represented mathematically by a number of different expressions which are useful but entirely empirical. Their relative merits lie in how well they fit the experimental curves and how easily their constants may be determined. Two relations that give a good overall fit to the longtime creep curve are demonstrated in Fig. 10.13.

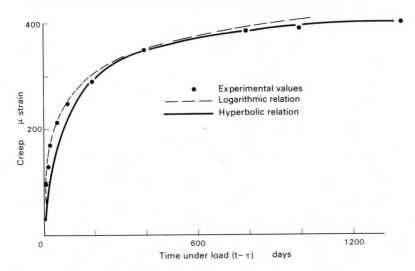

Figure 10.13 Expressions for creep-time curves.

The first is the hyperbolic function;

$$\epsilon_c = \frac{(t - \tau)}{a + b(t - \tau)} \tag{10.13}$$

287

where t is the age of the concrete, τ is the age of loading, so that $(t - \tau)$ is the time under load. The great advantage of this equation is that it can be rearranged in the form

$$\frac{t - \tau}{\epsilon_c} = a + b(t - \tau) \tag{10.14}$$

so that if, for a given creep test, $(t - \tau)/\epsilon_c$ is plotted against $(t - \tau)$ the best straight line through the plotted points yields the constants a (the intercept on the $(t - \tau)/\epsilon_c$ axis) and b (the slope). This expression indicates that creep has a limiting value, $\epsilon_{c\infty} = 1/b$. While this is helpful in extrapolating the results of a short creep test, it is likely to underestimate longtime creep, because tests show that creep is still occurring, albeit very slowly, even after 30 years under load.

The second is the logarithmic relation

$$c = A + B \log(1 + t - \tau) \tag{10.15}$$

Here the plot of creep against the logarithm of time under load is linear, and it is again possible to find the constants A and B with minimal difficulty. This relation does not set a limit to creep and it is likely to overestimate the longtime magnitudes.

If, for the same concrete, specimens are loaded at different ages, a family of creep curves is obtained, which is referred to as a creep surface (see Fig. 10.14). While, in general, the curves forming the surface have the same geometrical shape, the creep at any given time under load $(t - \tau)$ is a function of the age of loading τ. Then, remembering the linear relation between creep and stress, and emphasising the dependence of creep on both τ and $t - \tau$, creep can be written in the form

$$\epsilon_c = \sigma c_c(\tau, t - \tau) \tag{10.16}$$

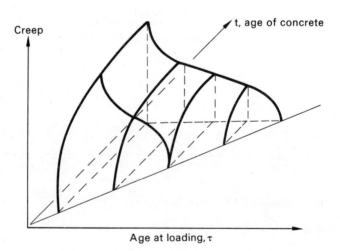

Figure 10.14   Creep surface.

288

10.3.3 Creep under Varying Stress

The expressions of the last section and the curves of the creep surface apply to concrete under a constant stress. When the stress is a function of time it is reasonable to expect concrete to obey the laws of linear viscoelasticity so that the principle of superposition can be invoked. A series of stress increments can be considered individually and their effects superposed to yield the total creep strain.

Considering n increments, giving stress increments $\Delta\sigma_1$ applied at τ_1, $\Delta\sigma_2$ at τ_2, up to $\Delta\sigma_n$ at τ_n, then the total creep at some later age t is given by the sum of the creeps for the individual stress increments applied for their differing times under load $t - \tau_1, t - \tau_2, \ldots, t - \tau_n$. Thus

$$\epsilon_c(t) = \Sigma_1^n \Delta\sigma_i \cdot c_{ci} \tag{10.17a}$$

or in integral form

$$\epsilon_c(t) = \int_{\tau_1}^{t} c_c(\tau, t - \tau) \cdot \frac{\partial\sigma(\tau)}{\partial\tau} \cdot d\tau \tag{10.17b}$$

This integral is usually solved numerically, though particular versions of the expression for c_c enable direct solutions to be found. Four possibilities are discussed below, two of a simple kind and two more complicated.

10.3.3.1 Simple Methods

(a) *Rate of Creep.* In this method the rate of creep is considered to be independent of the stress history. In other words a big simplification is introduced by taking rate of creep independent of the age of loading. Specific creep can then be written $c_c(t - \tau_1)$ for a first age of loading τ_1. It follows that the creep during an increment of time for which the specific creep is Δc_c is given by $\Delta\epsilon_c = \sigma \cdot \Delta c_c$, and the creep history can easily be found by summing a sequence of such increments. The integral form can be deduced from Equation 10.17(b). Integration by parts of that equation gives

$$\epsilon_c(t) = [\sigma(\tau) \cdot c_c(\tau, t - \tau)]_{\tau_1}^{t} - \int_{\tau_1}^{t} \sigma(\tau) \cdot \frac{\partial c_c(\tau, t - \tau)\, d\tau}{\partial\tau} \tag{10.18}$$

Insertion of the limits in the bracketed term reduces it to zero, since $\sigma(\tau) = 0$ when $\tau = \tau_1$, $c_c(\tau, t - \tau) = 0$ when $\tau = t$.

Further, the single variable, time under load, $t - \tau_1$ suffices. From

$$\frac{\partial c_c}{\partial\tau} = -\frac{\partial c_c}{\partial(t - \tau_1)} \qquad \sigma(\tau) = \sigma(t - \tau_1) \qquad d\tau = d(t - \tau_1)$$

the final form of the integral is:

$$\epsilon_c(t) = \int_{\tau_1}^{t} \sigma(t - \tau_1) \cdot \frac{dc_c(t - \tau_1)}{d(t - \tau_1)} \cdot d(t - \tau_1) \tag{10.19}$$

289

The rate of creep method requires only the single creep curve for the age of loading, τ_1, and not the series of curves making up the creep surface. It is also a ready source of direct solutions to important structural engineering problems.

(b) *Effective Modulus.* This method adopts an even greater degree of simplification in that the integration of Equation 10.17(b) is taken to be just the product of the stress at the reference age t and the specific creep for the first age of loading, τ_1. That is,

$$\epsilon_c(t) = \sigma(t) \cdot c_c(t - \tau_1)$$

which is equivalent to assuming that the stress at age t has been sustained since loading age τ_1. The elastic strain too is a function of the stress at age t so further simplification is possible by defining an effective modulus, E'_c, as

$$E'_c = \frac{\text{stress}}{\text{total strain}} = \frac{\text{stress}}{\text{creep} + \text{elastic strain}}$$

that is,
$$E'_c = \frac{\sigma}{\sigma/E_c + \sigma \cdot c_c} = \frac{E_c}{1 + c_c \cdot E_c} \qquad (10.20)$$

Clearly, the longer the concrete is under load, the smaller E'_c becomes. It is sometimes referred to as the reduced modulus. As its name implies, the effective modulus method regards concrete as an (effectively) elastic material, and thus opens the door to the multitude of elastic solutions to engineering problems.

Both methods estimate creep exactly if the sustained stress remains constant, and in general both become more inaccurate as the variation in stress becomes greater. As demonstrated later, this is very evident when the stress is removed (see Fig. 10.15). The effective modulus gives an estimate in terms of the current stress which is zero, thus incorrectly indicating that the strain is also zero. The rate of creep method estimates the rate of creep in terms of the current stress, and thus predicts (incorrectly) zero recovery after unloading.

10.3.3.2   *More Complicated Methods*

For some applications, such as mass concrete dams and nuclear pressure vessels, the simple methods of the last section are too imprecise, and they are supplanted by more sophisticated methods, which are characterised by their ability to predict creep recovery.

Virgin Superposition. This method applies directly the principle of superposition as described in Equation 10.17(a) and (b). Numerical integration is normally employed and the stress history is approximated by a step function thus providing a sequence of stress increments. The creep at any age is then given by the sum of the responses of virgin concrete to the individual increments. This is demonstrated in Fig. 10.15 for single increments of (a) increasing stress, and (b) decreasing stress. Here two creep curves are needed, and for general application the method of virgin superposition requires data to represent a creep surface. It should also be noted that the method somewhat overestimates the recovery of creep, which implies that concrete does not obey the Boltzmann principle of superposition perfectly. For comparison the predictions of recovery by rate of creep and effective modulus are also shown in Fig. 10.15.

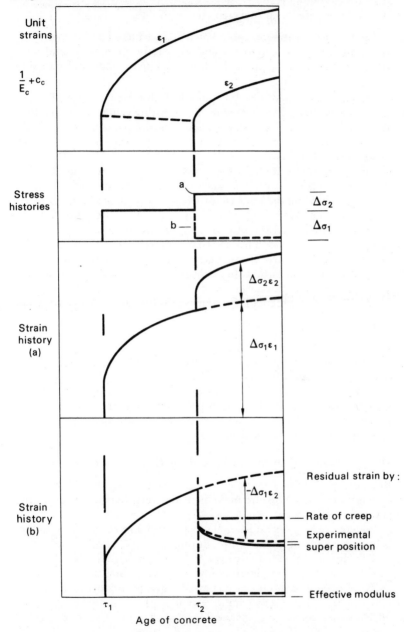

Figure 10.15 Creep under variable stress. The method of virgin superposition.

Rate of Flow. Creep is considered as the sum of reversible and irreversible components, sometimes referred to as delayed elastic strain and flow, respectively. Both are proportional to stress and they have the following characteristics:

(a) irrecoverable creep (flow);
(b) recoverable creep (delayed elastic strain).

(a) The rate of irrecoverable creep is considered as independent of stress history, so that the treatment is exactly the same as that for creep in the rate of creep method; there is just a single curve of flow against age.

(b) This component develops rapidly following a change of stress, and soon reaches a limiting value. The limiting value is not greatly affected by age, moisture state or temperature (over the normal working range of 0–80 °C) and for practical purposes it can be taken as equal to 0.4 x elastic strain.

The recoverable strain is observed as a creep recovery and it is determined experimentally by measuring the time-dependent recovery following the removal of stress. Consequently, the prediction of strain in Fig. 10.15 is exact, in contrast to the other three methods. Conversely, the rate of flow method underestimates the creep of concrete loaded for the first time at an age later than the origin of the flow curve itself. There is scope for further simplification of the method in that the recoverable creep may be considered as occurring instantaneously and be treated as an additional elastic strain. That is, the immediate strain, ϵ_e, following a change of stress is

$$\epsilon_e = 1.4\sigma/E_c \qquad (10.21)$$

Both virgin superposition and rate of flow provide better predictions than rate of creep and effective modulus, but which is the better of the two depends on the particular stress history. Rate of flow has the advantage of requiring fewer data and simpler numerical processing. In addition it offers much better prospects of closed-form solutions to practical problems.

10.3.4   Basic and Drying Creep and the Causes of Creep

Basic creep is defined as the creep that occurs under conditions in which there is no moisture movement between the concrete and the ambient environment.

Drying creep is the additional creep occurring when the concrete is drying, or taking up moisture, during the period under load.

This division of creep into two components results from the evidence of many experimental investigations which show that the creep of drying concrete is substantially greater than that of the same concrete sealed in the wet state. A typical plot is shown in Fig. 10.16, which records the classical tests started by Davis in the 1920's and pursued for 30 years. (Troxell *et al.*, 1958). While the effect of drying is very significant, it is equally important that creep still

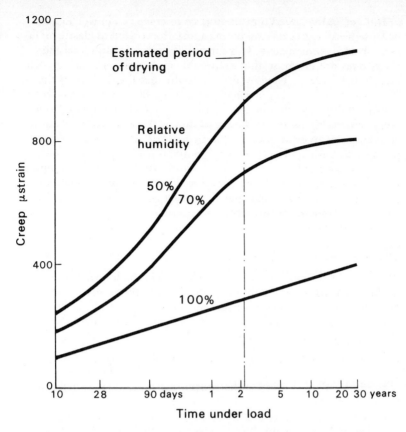

Figure 10.16 Creep cured in fog for 28 days then loaded and stored at different relative himidities. (From G. E. Troxell, J. M. Raphael and R. E. Davies (1958) *ASTM Proceedings*, **58** required by permission of the American Society for Testing and Materials.)

occurs when there is *no* moisture movement to or from the atmosphere. These two findings must be explained in any attempt to identify the causes of creep. As a phenomenon it has obvious similarities to shrinkage and it seems likely that the causes are related, *but* creep is still observed when drying shrinkage is not.

The physical and chemical happenings that are associated with creep are on a molecular scale, and there is no convincing direct evidence of what actually goes on; so the explanation of creep has, perforce, consisted of interpreting engineering level observations in terms of likely physical and chemical phenomena. This is the way that the creep mechanism game is played, and many of the research workers in this field have indulged themselves and enjoyed the game. It may well be that recounting some of the moves may be instructive to the reader, and the bones of some of the arguments are given below.

293

It is generally agreed that the source of creep lies in the h.c.p. because most aggregates are inert and can be treated as elastic inclusions in a creeping matrix. It is further agreed that completely dry h.c.p. does not creep, so that evaporable moisture is a vital ingredient, even in basic creep. There is also no doubt that the density of the paste structure is of importance in much the same way as it is for elastic modulus, so that creep and porosity are closely related, and creep and elastic modulus change in step as the water/cement ratio and degree of hydration vary. The characteristics of creep are complex; it has basic and drying components, it is partly recoverable, partly irrecoverable, it occurs under tensile stress as well as compressive, and so on. It seems inconceivable that this variety of behaviour can be attributed to a single physical process, and some or all of the following mechanisms probably contribute to the overall strain that is observed as creep.

10.3.4.1 *Moisture Diffusion*

The concept is that the load causes changes in the internal stresses in the h.c.p. structure, with the result that the thermodynamic equilibrium is upset, and moisture moves down the induced free energy gradient. This implies a move from smaller to larger pores, and it can happen at several different levels:

(a) in capillary water as a rapid and reversible pressure drop;
(b) in adsorbed water in moving more gradually from zones of hindered adsorption — this movement should be reversible and the resulting strain should be recoverable;
(c) in interlayer water in diffusing very slowly out into the gel pores. It is thought that the spaces vacated by the interlayer water close up and bonds develop between the solid layers; this would mean that at least some of the creep would be irrecoverable.

Although, unlike shrinkage, drying creep does not cause a measurable weight loss, the movement of the water displaced by the load to the atmosphere is taken to be the explanation of the extra creep on concurrent drying. Prevention of moisture exchange by sealing is unlikely to suppress this mechanism, as there are always voids in concrete which can accommodate a purely internal diffusion of moisture.

10.3.4.2 *Structural Adjustment*

Stress concentrations arise in the structure of h.c.p. because of its heterogeneous nature, and the mechanisms included here are all concerned with a consolidation of the structure at points of high stress. They have it in common that this takes place without any reduction in strength; indeed there may even be some gain in strength.

(a) Viscous flow, in which adjacent particles slide past one another under the action of shear stresses. The decrease in volume

294

accompanying creep implies that there must, in some sense, be a process of flow of material into empty pores.

(b) Recrystallisation, in which the effect of high stress is to dissolve solid material, to consolidate the structure and to reprecipitate the material in a region of lower stress.

(c) Bond breakage, in which primary bonds are ruptured locally, allowing a collapse of the structure, followed by a reconnection of the bonds to restore or improve the local strength.

The effect of concurrent moisture movement, or of a temperature gradient, is taken to disturb the molecular pattern, so as to encourage a greater structural adjustment. All the mechanisms are essentially irreversible, and the gradual progression from the metastable state towards stability is in accord with the observed decreasing rate of creep.

10.3.4.3 *Microcracking*

The failure of concrete is initiated by the formation of microcracks in the h.c.p. and at the interface between the h.c.p. and the aggregate. Although creep, as defined here, is at a level of stress below that at which significant propagation of microcracks occurs, some degree of microcracking is evident in a creeping concrete or h.c.p. This mechanism is not an adjustment of structure because it is accompanied by a local increase in volume, and a local loss of strength. Clearly, the contribution of microcracking is greater at higher stresses, and possibly under tensile stress. In a drying concrete the existence of a moisture gradient would undoubtedly enhance the stress levels producing more microcracking and greater creep.

10.3.4.4 *Delayed Elastic Strain*

If material that is creeping is connected in parallel with inert material, delayed elastic strain will occur. That is, the stress in the creeping material declines as load is transferred to the inert material, which deforms elastically as its stress gradually increases. Removal of the load allows the process to act in reverse so that the inert material finally returns to its initial unstressed condition. Thus delayed elastic strain is fully recoverable. It is observed in h.c.p. and can be attributed to any of the mechanisms discussed above if they function within and in connection with the skeletal structure of the h.c.p. The aggregate, too, acts as a parallel inert material and it has a considerable influence on both the irreversible and reversible components.

Delayed elastic action is well represented by the Kelvin rheological model, shown in Fig. 10.17(a). The inert material is represented by a spring which is connected parallel with a dashpot for the creeping material. The spring deforms in a linear elastic manner, and the dashpot is characterised by a linear relation between the force on the piston and its rate of movement.

The similarity in behaviour between concrete and timber is pointed out in Chapter 11 (see Section 11.4) and some similarity of mole-

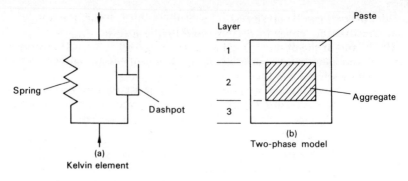

Figure 10.17   Delayed elastic strain and the effect of the aggregate.

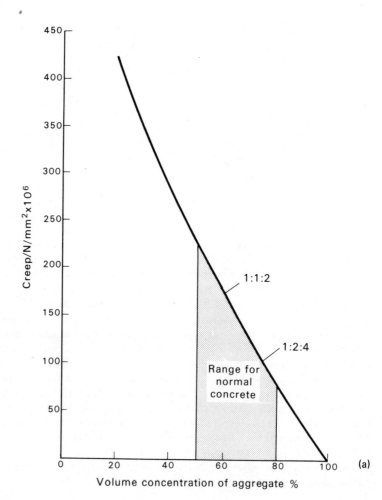

cular process can be presumed. This can be found in the attribution
of irrecoverable creep to the breaking and remaking of molecular
bonds, and in the occurrence of delayed elastic interactions between
active and inert components. Creep in metals differs in that it
becomes a serious consideration only under special circumstances
such as abnormally high temperatures or abnormally high stresses.
It is associated with the time-dependent movement of dislocations
(see Section 12.7).

10.3.5   Effect of the Aggregate

The influence of the aggregate is demonstrated with reference to
model C of Fig. 10.8, which is shown again in Fig. 10.17(b). The
h.c.p. creeps under a constant sustained stress, but the aggregate is
purely elastic. Considering the model in three layers, as shown: layers
1 and 3 deform elastically when the stress is applied, and then creep
without restraint when it is sustained; layer 2 also deforms elastically,
and continues to strain with time as load is transferred from the
creeping h.c.p. to the elastic aggregate. Movement stops when the

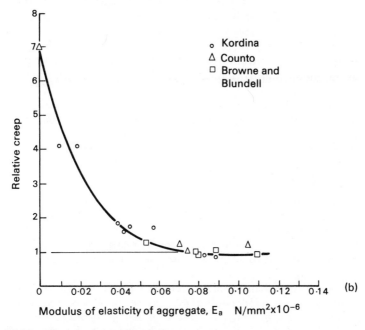

Figure 10.18   The influence of the aggregate on the creep of concrete. (a)
Volume concentration of aggregate. (b) Relative stiffness of aggregate. Creep
assumed as unity for $E_{agg} = 6.9 \times 10^4$ N/mm². Coarse aggregate content = 45—55%
by volume.

$$\text{Relative creep} = \frac{\text{creep}}{\text{creep at } E_{agg} = 6.9 \times 10^4 \text{ N/mm}^2}$$

(From C. S. Technical Report No. 101 (1973) *The Creep of Structural Concrete*,
by permission of the Concrete Society.)

stress in the h.c.p. reduces to zero. On removal of the applied stress the whole strain is recoverable, and this layer behaves like the Kelvin model.

If the aggregate is a normal natural stone, of greater stiffness than the h.c.p., its presence will restrain the creep, and, as for elastic modulus, the relevant variables are the volume concentration and the stiffness relative to that of the h.c.p. Thus there is a similar basis for the curves for creep in Fig. 10.18 and those for elastic modulus in Figs. 10.9 and 10.10.

10.3.6 Further Environmental Effects

10.3.6.1 *Size*

Drying of concrete is a slow process (see Section 8.2.3.3), and the inside of a block of concrete remains wet long after the outer layers have dried. If the behaviours of two blocks of different sizes are compared, the larger takes longer to lose the moisture from its core, and it thus has the higher average moisture content. This has two effects. Firstly, moisture is available for the continuation of hydration for a longer period in the larger block, and, secondly, the rate of moisture movement, is on average, slower. Both of these effects reduce the rate of creep relative to that of the smaller block and this can be seen in Fig. 10.19. The rate of drying is so slow in the more massive concrete structures that they are far from hygral equilibrium at the end of their structural life, and, correspondingly, their creeps are always far less than those of smaller members.

The actual rate of drying depends on both the size and the shape of the concrete block, and this double dependence is conveniently expressed by the single quantity, volume/surface ratio, or as a theoretical thickness defined as the area/the semi-perimeter.

10.3.6.2 *Temperature*

Temperature has a double effect on creep, just as it has on elastic modulus. If the temperature of the concrete is raised as part of the curing regime, that is *before* it is loaded, the degree of maturity will be increased, the structure will become relatively denser, and the creep will be less than that of a similar concrete stored at lower temperature.

On the other hand, a higher temperature during the period under load causes an enhanced rate of creep; this aspect is of particular interest within the range 0—100 °C, in which the moisture in the concrete is in the liquid state. It can be argued that the rates of all the mechanisms discussed in the last section will be significantly increased by a rise in temperature, and so it is no surprise to find that the rate of creep does indeed go up steeply with temperature, as shown in Fig. 10.20. Creep can be observed at temperatures down to −10°C, and the rate of increase with temperature is approximately constant between 0° and 70 °C. At higher temperatures the rate of increase is less, and some investigators have recorded creep rates at

90 °C+ that are lower than those at, say, 80 °C.

The temperature dependence of chemical reactions is expressed by the Arrhenius equation:

$$\text{rate} = \text{constant} \times e^{-Q/RT}$$

where Q is the activation energy, R is the gas constant, and T is the absolute temperature. This equation can be applied generally to chemical or physical processes that are thermally activated, so that rate of creep, $\dot{\epsilon}_c$, can be expressed:

$$\dot{\epsilon}_c = F(\sigma, M, T) \exp \frac{-Q(\sigma, M, T)}{RT} \qquad (10.22)$$

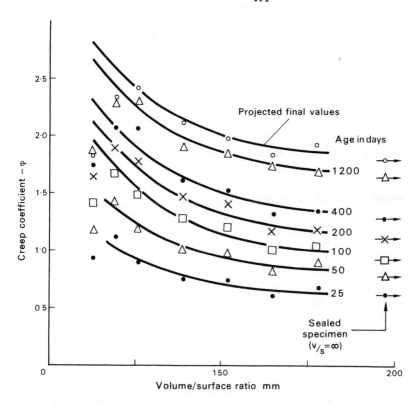

Figure 10. 19   Variation of creep coefficient with volume/surface ratio at different ages. Gravel aggregate concrete cylinders loaded at 8 days stored at 50% RH. (From T. C. Hansen and A. H. Mattock (1966) *Proceedings of the American Concrete Institute*, **63**, by permission of the American Concrete Institute.)

As has been seen, the rate of creep is a function of age, mix and state of the concrete, and this dependence is recognised by the parameter M which characterises the molecular structure, including the changes that take place under load. F and Q are also functions of the applied stress σ, and, although T also appears as one of the variables, neither F nor Q is very sensitive to it. It follows that, for a given

299

stress and concrete, the rate of creep is a function of the absolute temperature; it is possible to deduce the activation energy for particular ages, but there is no obvious way in which such quantitative results can throw light on the problems of creep mechanisms. The value of the activation energy approach lies more in the possible application of the concepts of rate process theory to creep of concrete.

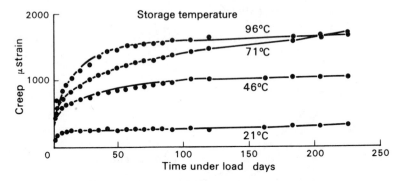

Figure 10.20   Creep of concrete at different temperatures. (From K. W. Nasser and A. M. Neville (1967) *Proceedings of the American Concrete Institute*, 64, by permission of the American Concrete Institute.)

10.3.7   Creep in Practice

The presentation of data for general use in practice is always a matter of compromise. On the one hand the information must be in a succinct form that is easily understood by practitioners, who are not specialists in concrete technology, and who need to abstract their numbers quickly and accurately. On the other hand, any vital conditions which affect the application of the data must be made known. Creep data are among the most difficult to present, and the existing method, given in the CEB International Recommendations for the design and construction of concrete structures (1970) can be and is criticised both because it is complicated to use and because it fails to include important factors.

It is couched in terms of the creep coefficient, ϕ_c, which is related to specific creep through the elastic modulus.

$$c_c = \frac{\phi_c}{E_c} \qquad (10.23)$$

so that creep coefficient is the ratio of creep per unit stress, c_c, to elastic strain per unit stress, $1/E_c$.

ϕ_c is given as the product of five partial coefficients, each of which represents one or more influential factors on creep.

$$\phi_c = k_c k_d k_b k_e k_t \qquad (10.24)$$

where the values of the partial factors are taken from the charts shown in Fig. 10.21.

300

(a) k_c represents the effect on creep of the relative humidity in which the concrete is drying under load.

(b) k_b covers the mix, described here by means of water/cement ratio and cement content (which is an alternative to the volume concentration of aggregate). No mention is made of aggregate stiffness, except that lightweight concrete is recognised as a separate case.

(c) k_d is concerned with age of loading, and also shows the difference between the two most popular cements. A correction factor is suggested in the text to take account of the more rapid rate of hydration of concretes cured at higher temperatures.

(d) k_e treats size effects, using the theoretical thickness, mentioned above.

(e) k_t gives the development of creep with time under load.

The virgin superposition method is suggested for dealing with variable stress regimes.

Predictions of creep by this method are by no means exact. There have been comparatively few attempts to verify the efficacy of the method, so that it is foolish to attempt to assess it; however, it is probably not unreasonable to take upper and lower bounds on the estimates of ±20%. This scatter is certainly due, in part, to the choice of engineering variables, such as water/cement ratio and relative humidity, rather than variables more expressive of the structure and state of the material, such as porosity and moisture content. Furthermore some influential variables, and most obviously temperature, are neglected completely.

10.4 REFERENCES

Anson, M. and Newman, K. (1966) The effect of mix proportions and method of testing on Poisson's ratio for mortars and concretes, *Magazine of Concrete Research*, **18**, 115.

British Standard 1881 (1970) Methods of Testing Concrete, Part 5, Methods of Testing Hardened Concrete for Other than Strength, British Standards Institution.

British Standards Institution (1972) *The Structural Use of Concrete,* CP 110, Part 1.

Concrete Society Technical Paper No. 101 (1973) *The Creep of Structural Concrete,* 47 p.

Counto, U. J. (1964) The effect of the elastic modulus of the aggregate on the elastic modulus, creep and creep recovery of concrete, *Magazine of Concrete Research*, **16**, 129.

Hansen, T. C. (1960) Creep and stress relaxation of concrete, *Proceedings of the Swedish Cement and Concrete Research Institute,* No. 31.

Hansen, T. C. and Mattock, A. H. (1966). Influence of size and shape of member on the shrinkage and creep of concrete, *Proceedings of the American Concrete Institute*, **63**, 267.

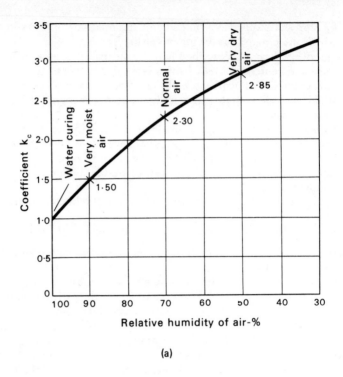

(a)

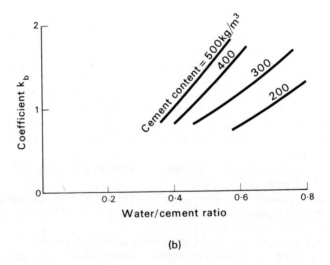

(b)

Figure 10.21 Charts for estimating the magnitude of creep for use in design. (From *International Recommendations for the Design and Construction of Concrete Structures* (1970) CEB/FIP.)

302

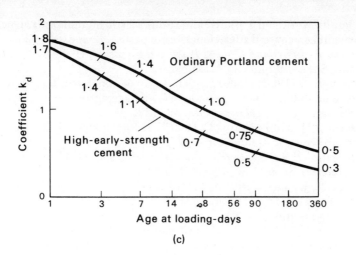

(c)

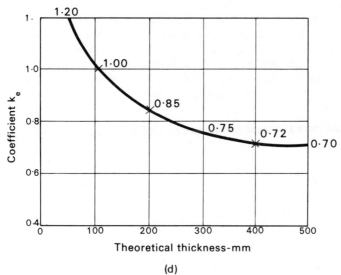

(d)

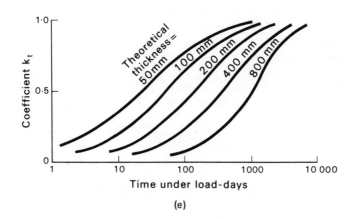

(e)

303

Helmuth, R. A. and Turk, D. M. (1966) Elastic moduli of hardened Portland cement and tricalcium silicate pastes, *Symposium on Structure of Portland Cement Paste and Concrete*, Highway Research Board, Special Report 90, Washington.

Hirsch, T. J. (1962) Modulus of elasticity of concrete affected by elastic moduli of cement paste matrix and aggregate, *Proceedings of the American Concrete Institute*, 59, 427.

Illston, J. M. and Pomeroy, C. D. (1975) Recommendations for a standard creep test, *Concrete*, 9, 24.

International Recommendations for the Design and Construction of Concrete Structures (1970) CEB/FIP.

Nasser, K. W. and Neville, A. M. (1967) Creep of old concrete at normal and elevated temperatures, *Proceedings of the American Concrete Institute*, 64, 97.

Parrott, L. J. (1973) The effect of moisture content on the elasticity of hardened cement paste, *Magazine of Concrete Research*, 25, 17.

Ross, A. D., Illston, J. M. and England, G. L. (1968) Short and long-term deformations of concrete as influenced by its physical structure and state, *Proceedings of an International Conference on the Structure of Concrete*, Cement and Concrete Association, London, p. 407.

Troxell, G. E., Raphael, J. M. and Davies, R. E. (1958) Long-time creep and shrinkage tests of plain and reinforced concrete, *Proceedings of the American Society for Testing Materials*, 58, 1101.

CHAPTER 11

Deformation in Timber

11.1 INTRODUCTION

This chapter is concerned with the type and magnitude of the deformation that results from the application of external stress. As in the case of both concrete and high polymers the stress-strain relationship is exceedingly complex, resulting from the facts that

(a) timber does not behave in a truly elastic mode, rather its behaviour is time-dependent, and

(b) the magnitude of the strain is influenced by a wide range of factors; some of these are property dependent, such as density of the timber, angle of the grain relative to direction of load application, angle of the microfibrils within the cell wall; others are environmentally dependent, such as temperature and relative humidity.

Under service conditions timber has to withstand an imposed load for many years, perhaps even centuries. When loaded, timber will deform and a generalised interpretation of the variation of deformation with time together with the various components of this deformation is illustrated in Fig. 11.1. On the application of a load at time zero an instantaneous (and reversible) deformation occurs which represents elastic behaviour. On maintaining the load to time t_1 the deformation increases, though the rate of increase is continually decreasing; this increase in deformation with time is termed *creep*. On removal of the load at time t_1 an instantaneous reduction in deformation occurs which is approximately equal in magnitude to the initial elastic deformation. With time, the remaining deformation will decrease at an ever-decreasing rate until at time t_2 no further reduction occurs. The creep that has occurred during stressing can be conveniently subdivided into a *reversible* component, which disappears with time and which can be regarded as *delayed elastic* behaviour, and an *irreversible* component which results from *plastic* or *viscous* flow. Therefore, timber on loading possesses three forms of deformation behaviour — elastic, delayed elastic and viscous. Like so many other materials timber can be treated neither as a truly elastic material where, by Hooke's law, stress is proportional to strain but independent of the rate of strain, nor as a truly viscous liquid where, according to Newton's law, stress is proportional to rate of strain but independent of strain itself. Where combinations of behaviour are encountered the material is said to be viscoelastic and timber, like many high polymers, is a viscoelastic material.

Having defined timber as such, the reader will no doubt be surprised to find that half of this chapter is devoted to the elastic

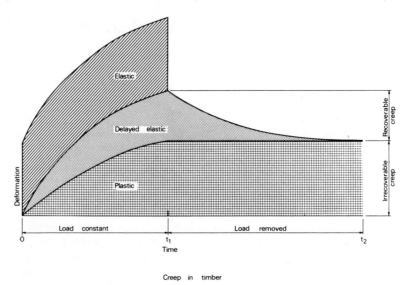

Creep in timber

Figure 11.1   The various elastic and plastic components of the deformation of timber under constant load. (From the Princes Risborough Laboratory, Building Research Establishment. © Crown copyright.)

behaviour of timber. It has already been discussed how part of the deformation can be described as elastic and the section below will indicate how at low levels of stressing and short periods of time there is considerable justification for treating the material as such. Perhaps the greatest incentive to this viewpoint is the fact that classical elasticity theory is well established and when applied to timber has been shown to work very well. The question of time in any stress analysis can be accommodated by the use of safety factors in design calculations.

Consequently, this chapter will deal first with elastic deformation as representing a very good approximation of what happens, while the second part will deal with viscoelastic deformation, which embraces both delayed elastic and irreversible deformation. Although technically more applicable to timber, viscoelasticity is certainly less well understood and developed in its application than is the case with elasticity theory.

306

11.2   ELASTIC DEFORMATION

When a sample of timber is loaded in tension, compression or bending the deformations obtained with increasing load are approximately proportional to the values of the applied load. Figure 11.2 illustrates

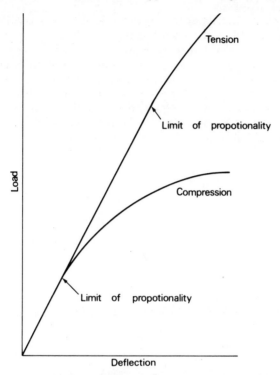

Figure 11.2   Load-deflection graphs for timber stressed in tension and compression parallel to the grain. The assumed *limit of proportionality* for each graph is indicated. (From the Princes Risborough Laboratory, Building Research Establishment. © Crown copyright.)

that this approximation is certainly truer of the experimental evidence in longitudinal tensile loading than in the case of longitudinal compression. In both modes of loading, the approximation appears to become a reality at the lower levels of loading. Thus it has become convenient to recognise a point of inflection on the load-deflection curve known as the *limit of proportionality*, below which the relationship between load and deformation is linear, and above which nonlinearity occurs. Generally the limit of proportionality in longitudinal tension is found to occur at about 60% of the ultimate load to failure while in longitudinal compression the limit is considerably lower, varying from 30% to 50% of the failure value.

At the lower levels of loading, therefore, where the straight-line relationship appears to be valid the material is said to be linearly elastic. Hence

$$\epsilon = \frac{\sigma}{E} \qquad\qquad (11.1)$$

307

where ϵ is the strain (change in dimension/original dimension), σ is the stress (load/cross-sectional area), and E is a constant, known as the modulus of elasticity. The modulus of elasticity, MOE, is also referred to in the literature as the elastic modulus, Young's modulus, or simply as stiffness.

The apparent linearity at the lower levels of loading is really an artefact introduced by the rate of testing and the lack of precision: it is found not only in timber but in many other fibre composites. It is true that at fast rates of loading, or testing on insensitive equipment, a very good approximation to a straight line occurs, but as the rate of loading decreases, or the testing equipment sensitivity increases, the load-deflection line assumes a curvilinear shape (Fig. 11.3). Such curves can be treated as linear by introducing a straight-line approximation, which can take the form of either a tangent or

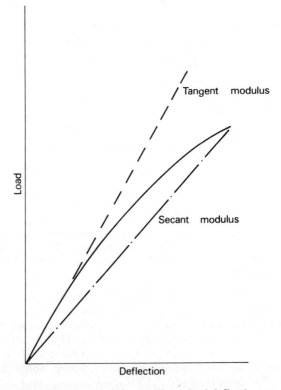

Figure 11.3   The approximation of a curvilinear load-deflection curve for timber stressed at low loading rates, by linear tangent or secant regressions. (From the Princes Risborough Laboratory, Building Research Establishment. © Crown copyright.)

secant. Traditionally for timber and wood fibre composites tangent lines have been used as linear approximations of load deflection curves.

Thus, while in theory it should be possible to obtain a true elastic response, in practice this is rarely the case, though the degree of divergence is frequently very low. It should be appreciated in passing that a curvilinear load-deflection curve must not be interpreted as an absence of true elastic behaviour. The material may still behave elastically, though not linearly elastically: the prime criterion for elastic behaviour is that the load-deflection curve is truly reversible, i.e. that no permanent deformation has occurred on release of the load.

11.2.1 Modulus of Elasticity

The modulus of elasticity or elastic modulus in the longitudinal direction is one of the principal elastic constants of our material. While the following test methods can be modified to measure elasticity in other planes they are described specifically for the determination of elasticity in the longitudinal direction.

11.2.1.1 *Test Methods*

A whole spectrum of test methods, many of them fairly simple in concept, exists for the determination of the elastic modulus: many of the methods are also applicable to the determination of one of the other sets of elastic constants, namely the modulus of rigidity or shear modulus. These methods can be conveniently treated in two groups, the first comprising *static* methods based on the application of direct stress and the measurement of resultant strains; and the second comprising *dynamic* methods, which are based on resonant vibration from flexural, torsional or ultrasonic pulse excitation. It is appropriate to examine a few of these techniques.

The determination of the elastic modulus from stress-strain curves has already been described and despite its sensitivity to rate of loading it remains one of the more common methods. Although frequently carried out in the bending mode, it can also be derived from compression, tension or shear tests: the values of the modulus in tensile, compressive and bending modes are approximately equal.

Where the tensile mode is adopted, waisted samples of timber must be used because of the very high longitudinal tensile strength. BS 373 provides dimensions for the standard, knot-free, straight-grained timber sample. Deformation under load is measured by an extensometer which is fitted to the waisted region and can be clearly seen in Fig. 11.4. The slope of the line on the load-deflection graph corrected for cross-sectional area of the sample and distance between the extensometer grips provides the elastic modulus.

In the three-point bending test, measurement of deflection by gauge or extensometer will provide the static modulus using the following equation:

$$E_s = \frac{Pl^3}{48I\delta} \tag{11.2}$$

where P is the load applied to the centre of the span, l is the distance between supports, I is the moment of inertia of the cross section about the neutral axis, and δ is the deflection at the centre of the span. This modulus includes a contribution from shear deflection; four-point bending provides a true modulus of bending between the two loading points.

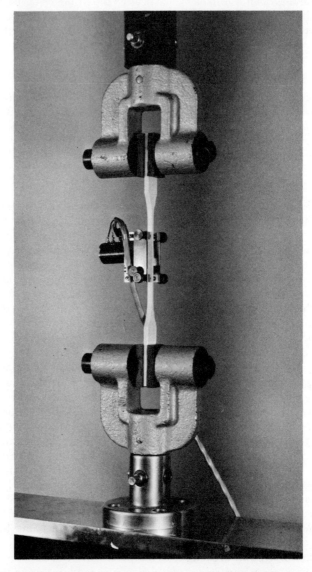

Figure 11.4 Measurement of elongation in the waisted region of a timber sample under tensile loading: changes in voltage across the transducer attached to the two pairs of grips provide a measure of the extension. (From the Princes Risborough Laboratory, Building Research Establishment. © Crown copyright.)

Perhaps the simplest method of determining E_s is the loading of a cantilever beam and the measurement of deflection by, for example, a cathotometer for each increment of load. The static elastic modulus

310

is then given by

$$E_s = \frac{Pl^3}{3I\delta}$$ (11.3)

where P and δ are at the free end, and l is the length of the cantilever.

The determination of the dynamic elastic modulus can be obtained by either longitudinal or flexural vibration; the latter is used more frequently and may take the form of a small unloaded beam to which are attached thin metal plates. Under the action of an oscillating electromagnetic impulse the beam vibrates. The response is measured as a function of the frequency and the dynamic elastic modulus is calculated from the specimen dimensions and resonant frequency.

The dynamic elastic modulus can be obtained very simply and accurately by clamping a strip of timber firmly to a bench in such a way that one end of the strip extends horizontally beyond the bench. A suitable weight is then attached to the free end of the strip and the period of the resulting vibrations determined (Hearmon, 1966). The dynamic modulus of elasticity is then given by

$$E_d = \frac{4\pi^2 l^3}{3IT^2}\left(M_t + \frac{33}{140}M\right)$$ (11.4)

where l is the free length of the strip, I is the moment of inertia about the neutral axis ($bd^3/12$ for a rectangular strip), T is the time of period, M_t is the mass of the free length of timber, and M is the load applied.

Ultrasonic techniques can also be employed to measure E_d. These are based on the delay time of a set of pulses sent out by one transducer and received by a second; elasticity is proportional to velocity of propagation, which of course will vary with grain direction.

The values of modulus obtained dynamically are generally only marginally greater than those obtained by static methods. The relationship obtained between E_d and E_s in one experiment using the same cantilever strips prepared from a number of species is presented in Fig. 11.5. Although the mean value of E_d is only about 3% higher than that for E_s, the differences between the readings are nevertheless significant at the 0.1% level. Thus for timber, like concrete, it would appear that the value of the elastic modulus is dependent on test method, though the difference appears smaller for timber than for concrete.

11.2.2 Modulus of Rigidity

Within the elastic range of the material, shearing stress is proportional to shearing strain. The constant relating these parameters is called the shear or rigidity modulus and is designated by the letter G. Thus

$$G = \frac{\tau}{\gamma}$$ (11.5)

where τ is the shearing stress and γ is the shearing strain.

311

Simple torsion tests are available to determine material properties under shearing conditions: these may be either static or dynamic in operation (Hearmon, 1966).

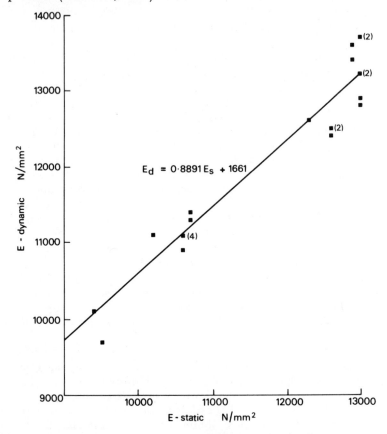

Figure 11.5   The relation between the modulus of elasticity obtained by dynamic and static methods on the same timber samples. (From the Princes Risborough Laboratory, Building Research Establishment. © Crown copyright.)

11.2.3   Poisson's Ratio

In general, when a body is subjected to a stress in one direction, the body will undergo a change in dimensions at right angles to the direction of stressing. The ratio of the contraction or extension to the applied strain is known as Poisson's ratio and for isotropic bodies is given as

$$\nu = -\frac{\epsilon_y}{\epsilon_x} \tag{11.6}$$

where ϵ_x, ϵ_y are strains in the x, y directions resulting from an applied stress in the x direction. (The minus sign indicates that, when ϵ_x is a tensile positive strain, ϵ_y is a compressive negative strain.) In timber,

312

because of its anisotropic behaviour and its treatment as a rhombic system, six Poisson's ratios occur.

11.2.4 Orthotropic Elasticity and Timber

The theory of elasticity is a well developed area of applied mathematics which is usually discussed in relation to isotropic materials. However, by suitable development it can be applied to materials possessing orthotropic symmetry. Providing the assumption is made that timber has orthotropic symmetry (and the justification for this assumption will be discussed later) then this form of elasticity can be applied to timber.

Let us start with the generalised stress condition and work towards the particular. If we imagine a cube of material (Fig. 11.6) with its

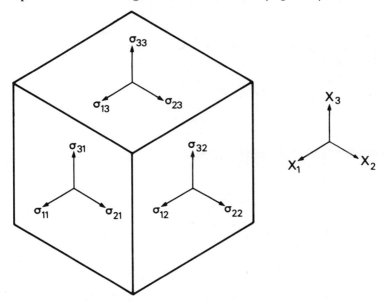

Figure 11.6 Stresses acting on a cube of timber. (From the Princes Risborough Laboratory, Building Research Establishment. © Crown copyright.)

edges lying along the coordinate axes, there will be a set of three mutually perpendicular stresses acting on each face. Stresses are labelled by the notation σ_{ij} where i refers to the direction of stress and j to the direction of the perpendicular to the face on which it acts; Fig. 11.6 is labelled accordingly. The shear stresses can be readily identified as those with $i \neq j$.

Now it is important to state that our cube will not rotate and for this to happen $\sigma_{ij} = \sigma_{ji}$, whereupon we reduce the number of components working on the cube from nine to six, three normal and three shear. Similarly it can be argued that there will be six strain components.

313

Since the handling of these double suffixes proves awkward at times it is customary to use only single suffixes, the conversion being set out below in matrix form.

$$\begin{bmatrix} \sigma_{11} & \sigma_{12} & \sigma_{13} \\ - & \sigma_{22} & \sigma_{23} \\ - & - & \sigma_{33} \end{bmatrix} = \begin{bmatrix} \sigma_1 & \sigma_6 & \sigma_5 \\ - & \sigma_2 & \sigma_4 \\ - & - & \sigma_3 \end{bmatrix}$$

and

$$\begin{bmatrix} \epsilon_{11} & 2\epsilon_{12} & 2\epsilon_{13} \\ - & \epsilon_{22} & 2\epsilon_{23} \\ - & - & \epsilon_{33} \end{bmatrix} = \begin{bmatrix} \epsilon_1 & \epsilon_6 & \epsilon_5 \\ - & \epsilon_2 & \epsilon_4 \\ - & - & \epsilon_3 \end{bmatrix} \tag{11.7}$$

Note the factor 2 which enters into the conversion of the shear strains: this does not affect the form of the law but proves to be mathematically convenient.

The relationship between stress and strain is expressed by the generalised Hooke's law, which states that each stress component is linearly related to all the strain components and *vice versa*. Now these six components of strain can be expressed in terms of the six products of the appropriate components of stress and elastic compliances (denoted by S), thus:

$$\epsilon_1 = S_{11}\sigma_1 + S_{12}\sigma_2 + S_{13}\sigma_3 + S_{14}\sigma_4 + S_{15}\sigma_5 + S_{16}\sigma_6$$
$$\epsilon_2 = S_{21}\sigma_1 + S_{22}\sigma_2 + S_{23}\sigma_3 + S_{24}\sigma_4 + S_{25}\sigma_5 + S_{26}\sigma_6$$
$$\epsilon_3 = S_{31}\sigma_1 + S_{32}\sigma_2 + S_{33}\sigma_3 + S_{34}\sigma_4 + S_{35}\sigma_5 + S_{36}\sigma_6$$
$$\epsilon_4 = S_{41}\sigma_1 + S_{42}\sigma_2 + S_{43}\sigma_3 + S_{44}\sigma_4 + S_{45}\sigma_5 + S_{46}\sigma_6$$
$$\epsilon_5 = S_{51}\sigma_1 + S_{52}\sigma_2 + S_{53}\sigma_3 + S_{54}\sigma_4 + S_{55}\sigma_5 + S_{56}\sigma_6$$
$$\epsilon_6 = S_{61}\sigma_1 + S_{62}\sigma_2 + S_{63}\sigma_3 + S_{64}\sigma_4 + S_{65}\sigma_5 + S_{66}\sigma_6$$

$$\tag{11.8}$$

Similarly the stresses could be expressed in terms of all the strain components and the appropriate elastic stiffnesses. Since there are six stress and six strain components, there are altogether 36 compliances (S) and 36 stiffnesses (C). However, due to thermodynamic considerations,

$$S_{ij} = S_{ji} \qquad \text{and} \qquad C_{ij} = C_{ji} \tag{11.9}$$

which reduces the number of compliances and stiffnesses each to 21 in the most general case. However, many materials possess a particular structure such that the number of independent constants is reduced: such is the case of orthotropic materials, which possess three mutually perpendicular planes of elastic symmetry and, most important, the principal axes are chosen to be in the directions perpendicular to the orthotropic planes.

314

The generalised Hooke's law for orthotropic materials becomes (in matrix form)

$$
\begin{bmatrix} \epsilon_1 \\ \epsilon_2 \\ \epsilon_3 \\ \epsilon_4 \\ \epsilon_5 \\ \epsilon_6 \end{bmatrix} =
\begin{bmatrix}
S_{11} & S_{12} & S_{13} & 0 & 0 & 0 \\
S_{21} & S_{22} & S_{23} & 0 & 0 & 0 \\
S_{31} & S_{32} & S_{33} & 0 & 0 & 0 \\
0 & 0 & 0 & S_{44} & 0 & 0 \\
0 & 0 & 0 & 0 & S_{55} & 0 \\
0 & 0 & 0 & 0 & 0 & S_{66}
\end{bmatrix}
\begin{bmatrix} \sigma_1 \\ \sigma_2 \\ \sigma_3 \\ \sigma_4 \\ \sigma_5 \\ \sigma_6 \end{bmatrix}
\tag{11.10}
$$

The compliance matrix is thus symmetrical, comprising nine independent compliance parameters. Now an orthotropic material is characterised by six elastic moduli, three of which are the ratios of normal stress to strain in the principal directions (E) and three are the ratio of shear stress to strain in the orthotropic planes (G). The relationship of these constants to the compliances listed above can be determined by taking each in turn and assuming that only a single stress is acting. Thus by applying stress σ_i and measuring ϵ_i the slope of the stress-strain diagram can be obtained. Thus

$$
\frac{\sigma_1}{\epsilon_1} = E_1 \quad \frac{\sigma_2}{\epsilon_2} = E_2 \quad \frac{\sigma_3}{\epsilon_3} = E_3 \quad \frac{\sigma_4}{\epsilon_4} = G_{44} \quad \frac{\sigma_5}{\epsilon_5} = G_{55} \quad \frac{\sigma_6}{\epsilon_6} = G_{66}
\tag{11.11}
$$

where E is the modulus of elasticity, and G is the modulus of rigidity. Now under these conditions

$$
\epsilon_1 = \sigma_1 S_{11} \qquad \epsilon_2 = \sigma_1 S_{21} \qquad \epsilon_3 = \sigma_1 S_{31} \tag{11.12}
$$

Substituting in Equation 11.11

$$
E_1 = \frac{1}{S_{11}} \tag{11.13}
$$

Similarly, it can be shown that

$$
E_2 = \frac{1}{S_{22}} \quad E_3 = \frac{1}{S_{33}} \quad G_{44} = \frac{1}{S_{44}} \quad G_{55} = \frac{1}{S_{55}} \quad G_{66} = \frac{1}{S_{66}}
\tag{11.14}
$$

As previously discussed, orthotropic materials are characterised by possessing six Poisson's ratios. However, three are linked with the modulus of elasticity and only three are truly independent.

Returning to the condition of the application of only a single stress, this will result in two induced strains and consequently two Poisson's ratios thus:

$$\frac{\epsilon_2}{\epsilon_1} = -\nu_{21} \quad \text{and} \quad \frac{\epsilon_3}{\epsilon_1} = -\nu_{31} \qquad (11.15)$$

Substituting in Equation 11.12 gives:

$$-\nu_{21} = \frac{S_{21}}{S_{11}} \quad \text{and} \quad -\nu_{31} = \frac{S_{31}}{S_{11}} \qquad (11.16)$$

Similarly it can be shown that:

$$-\nu_{12} = \frac{S_{12}}{S_{22}} \qquad -\nu_{32} = \frac{S_{32}}{S_{22}}$$

$$-\nu_{13} = \frac{S_{13}}{S_{33}} \qquad -\nu_{23} = \frac{S_{23}}{S_{33}} \qquad (11.17)$$

Substituting in Equations 11.13 and 11.14 gives:

$$S_{12} = \frac{-\nu_{12}}{E_2} \qquad S_{13} = \frac{-\nu_{13}}{E_3} \qquad S_{23} = \frac{-\nu_{23}}{E_3}$$

$$S_{21} = \frac{-\nu_{21}}{E_1} \qquad S_{31} = \frac{-\nu_{31}}{E_1} \qquad S_{32} = \frac{-\nu_{32}}{E_2} \qquad (11.18)$$

Using these definitions, the constitutive equation for timber as an assumed orthotropic elastic body becomes

$$
\begin{bmatrix} \epsilon_1 \\ \epsilon_2 \\ \epsilon_3 \\ \epsilon_4 \\ \epsilon_5 \\ \epsilon_6 \end{bmatrix}
=
\begin{bmatrix}
\dfrac{1}{E_1} & \dfrac{-\nu_{12}}{E_2} & \dfrac{-\nu_{13}}{E_3} & 0 & 0 & 0 \\[2mm]
\dfrac{-\nu_{21}}{E_1} & \dfrac{1}{E_2} & \dfrac{-\nu_{23}}{E_3} & 0 & 0 & 0 \\[2mm]
\dfrac{-\nu_{31}}{E_1} & \dfrac{-\nu_{32}}{E_2} & \dfrac{1}{E_3} & 0 & 0 & 0 \\[2mm]
0 & 0 & 0 & \dfrac{1}{G_{44}} & 0 & 0 \\[2mm]
0 & 0 & 0 & 0 & \dfrac{1}{G_{55}} & 0 \\[2mm]
0 & 0 & 0 & 0 & 0 & \dfrac{1}{G_{66}}
\end{bmatrix}
\begin{bmatrix} \sigma_1 \\ \sigma_2 \\ \sigma_3 \\ \sigma_4 \\ \sigma_5 \\ \sigma_6 \end{bmatrix}
\qquad (11.19)
$$

Reconverting the suffixes (Equation 11.7) and substituting letters in place of numbers, Equation 11.19 becomes

$$
\begin{bmatrix}
\epsilon_{LL} \\
\epsilon_{TT} \\
\epsilon_{RR} \\
2\epsilon_{TR} \\
2\epsilon_{LR} \\
2\epsilon_{LT}
\end{bmatrix}
=
\begin{bmatrix}
\dfrac{1}{E_L} & \dfrac{-\nu_{LT}}{E_T} & \dfrac{-\nu_{LR}}{E_R} & 0 & 0 & 0 \\
\dfrac{-\nu_{TL}}{E_L} & \dfrac{1}{E_T} & \dfrac{-\nu_{TR}}{E_R} & 0 & 0 & 0 \\
\dfrac{-\nu_{RL}}{E_L} & \dfrac{-\nu_{RT}}{E_T} & \dfrac{1}{E_R} & 0 & 0 & 0 \\
0 & 0 & 0 & \dfrac{1}{G_{TR}} & 0 & 0 \\
0 & 0 & 0 & 0 & \dfrac{1}{G_{LR}} & 0 \\
0 & 0 & 0 & 0 & 0 & \dfrac{1}{G_{LT}}
\end{bmatrix}
\begin{bmatrix}
\sigma_{LL} \\
\sigma_{TT} \\
\sigma_{RR} \\
\sigma_{TR} \\
\sigma_{LR} \\
\sigma_{LT}
\end{bmatrix}
\qquad (11.20)
$$

Orthotropic elasticity is applied only infrequently to the solution of three-dimensional problems, owing to the complexity of the solutions. Wherever possible two-dimensional approximations are made (known as plane systems), thereby reducing the number of constitutive equations with commensurate simplification of the solution. Plane stress and plane strain systems will nearly always result in three-dimensional states of strain and stress respectively. For example, taking the case of a plane stress situation (in the 1–2 plane) where the stress components σ_1, σ_2 and σ_6 are present and σ_3, σ_4 and σ_5 are zero, Equation 11.10 will yield strain components ϵ_1, ϵ_2, ϵ_3 and ϵ_6. Now ϵ_3, the strain normal to the plane of the applied stress, is very small and can usually be safely ignored: under these conditions the constitutive equation in matrix form can be written as:

$$
\begin{bmatrix}
\epsilon_1 \\
\epsilon_2 \\
\epsilon_6
\end{bmatrix}
=
\begin{bmatrix}
S_{11} & S_{12} & 0 \\
S_{21} & S_{22} & 0 \\
0 & 0 & S_{66}
\end{bmatrix}
\begin{bmatrix}
\sigma_1 \\
\sigma_2 \\
\sigma_6
\end{bmatrix}
\qquad (11.21)
$$

An important property of anisotropic materials is the change in the values of stiffness and compliance with orientation. The effect of rotation on these parameters is obtained by transforming the stresses or strains from the rotated to the principal axes on the grounds that the potential energy of the deformed body is independent of orientation. For an orthotropic material transformation gives the new compliances in terms of the original compliances and the direction cosines.

In applying the elements of orthotropic elasticity to wood and board materials made from wood the assumption is made that the three principal elasticity directions coincide with the longitudinal, radial and tangential directions in the tree. The assumption implies that the tangential faces are straight and not curved, and that the radial faces are parallel and not diverging. However, by dealing with small pieces of timber removed at some distance from the centre of the tree, the approximation of rhombic symmetry for a system possessing circular symmetry becomes more and more acceptable.

The nine independent constants required to specify the elastic behaviour of timber are the three moduli of elasticity, one in each of the L, R and T directions; the three moduli of rigidity, one in each of the principal planes LT, LR and TR; and three Poisson's ratios, namely ν_{RT}, ν_{LR}, ν_{TL}. These constants, together with the three dependent Poisson's ratios ν_{RL}, ν_{TR}, ν_{LT} are presented in Table 11.1 for a selection of hardwoods and softwoods.

The table illustrates the high degree of anisotropy present in timber. Comparison of E_L with either E_R or E_T, and G_{TR} with G_{LT} or G_{LR} will indicate a degree of anisotropy which can be as high as 60 : 1. Note should be taken that the values of ν_{TR} are frequently greater than 0.5.

11.2.5 Factors Influencing the Elastic Modulus

The stiffness of timber is influenced by many factors, some of them properties of the material while others are components of the environment.

11.2.5.1 *Grain Angle*

The significance of grain angle in influencing stiffness has already been described in the section dealing with elasticity theory. It will be recalled that the effect of rotation on the elastic constants can be obtained by transforming the stresses or strains from the rotated to the principal axes. Figure 11.7, in addition to illustrating the marked influence of grain angle on stiffness, shows the degree of fit between experimentally derived values and those obtained from transformation equations.

11.2.5.2 *Density*

Stiffness is related to density of the timber, a relationship which was apparent in Table 11.1 and which is confirmed by the plot of over two hundred species of timber contained in Bulletin 50 of the former Forest Products Research Laboratory (Fig. 11.8); the correlation coefficient was 0.88 for timber at 12% moisture content and 0.81 for green timber and the relation is curvilinear. A high correlation is to be expected, since density is a function of the ratio of cell wall thickness to cell diameter: consequently increasing density will result in increasing stiffness of the cell.

Owing to the variability in structure that exists between different timbers the relation between density and stiffness will be higher where only a single species is under investigation. Because of the reduced range in density the regression is usually linear.

318

Table 11.1   Elastic Constants of Certain Timbers

E is the modulus of elasticity in a direction indicated by the subscript.

G is the modulus of rigidity in a plane indicated by the subscript (in N/mm^2).

ν_{ij} is the Poisson's ratio for an extensional stress in j direction,

$$= \frac{\text{compressive strain in } i \text{ direction}}{\text{extensional strain in } j \text{ direction}}$$

From Hearmon (1948), but with different notation for the Poisson's ratios.

Species	Density (kg/m³)	Moisture content (%)	E_L	E_R	E_T	ν_{TR}	ν_{LR}	ν_{RT}	ν_{LT}	ν_{RL}	ν_{TL}	G_{LT}	G_{LR}	G_{TR}
Hardwoods														
Balsa	200	9	6 300	300	106	0.66	0.018	0.24	0.009	0.23	0.49	203	312	33
Khaya	440	11	10 200	1 130	510	0.60	0.033	0.26	0.032	0.30	0.64	600	900	210
Walnut	590	11	11 200	1 190	630	0.72	0.052	0.37	0.036	0.49	0.63	700	960	230
Birch	620	9	16 300	1 110	620	0.78	0.034	0.38	0.018	0.49	0.43	910	1 180	190
Ash	670	9	15 800	1 510	800	0.71	0.051	0.36	0.030	0.46	0.51	890	1 340	270
Beech	750	11	13 700	2 240	1 140	0.75	0.073	0.36	0.044	0.45	0.51	1 060	1 610	460
Softwoods														
Norway spruce	390	12	10 700	710	430	0.51	0.030	0.31	0.025	0.38	0.51	620	500	23
Sitka spruce	390	12	11 600	900	500	0.43	0.029	0.25	0.020	0.37	0.47	720	750	39
Scots pine	550	10	16 300	1 100	570	0.68	0.038	0.31	0.015	0.42	0.51	680	1 160	66
Douglas fir*	590	9	16 400	1 300	900	0.63	0.028	0.40	0.024	0.43	0.37	910	1 180	79

*listed in original as Oregon pine.

Similar relations with density have been recorded for the modulus of rigidity in certain species: in others, however, for example spruce, both the longitudinal–tangential and longitudinal–radial shear moduli have been found to be independent of density. Most investigators agree, however, that the Poisson's ratio are independent of density.

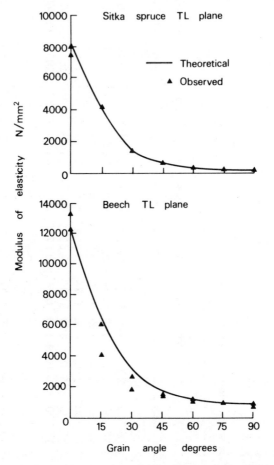

Figure 11.7 Effect of grain angle on the modulus of elasticity. (From the Princes Risborough Laboratory, Building Research Establishment. © Crown copyright.)

11.2.5.3 *Knots*

Since timber is anisotropic in behaviour, and since knots are characterised by the occurrence of distorted grain, it is not surprising to find that the presence of knots in timber results in a reduction in the stiffness. The relation is difficult to quantify since the effect of the knots will depend not only on their number and size but also on their distribution both along the length of the sample and across the faces. Dead knots, especially where the knot has fallen out, will result in larger reductions in stiffness than will green knots (see Chapter 4).

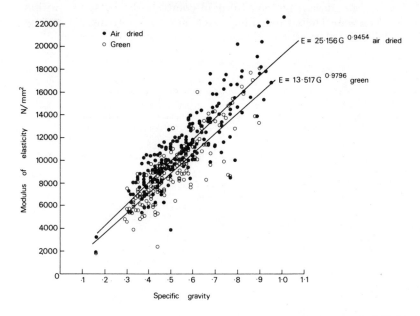

Figure 11.8 Effect of specific gravity on the modulus of elasticity for over 200 species of timber tested in the green and dry states. (From the Princes Risborough Laboratory, Building Research Establishment. © Crown copyright.)

11.2.5.4   *Ultrastructure*

Two components of the fine or chemical structure have a profound influence on both the elastic and rigidity moduli. The first relates to the existence of a matrix material with particular emphasis on the presence of lignin. In those plants devoid of lignin, e.g. the grasses, or in wood fibres which have been delignified, the stiffness of the cells is low and it would appear that lignin, apart from its hydrophilic protective role for the cellulosic crystallites, is responsible to a considerable extent for the high stiffness found in timber.

The significance of lignin in determining stiffness is not to imply that the cellulose fraction plays no part: on the contrary it has been shown that the angle at which the microfibrils are lying in the middle layer of the secondary cell wall, S_2, also plays a significant role in controlling stiffness (Fig. 11.9).

Over the last decade and a half a considerable number of mathematical models have been devised to relate stiffness to microfibrillar angle. The early ones were two-dimensional in approach, treating the cell wall as a planar slab of material, but over the years the models have become much more sophisticated, taking into account the existence of cell wall layers other than the S_2, the variation in microfibrillar angle between the radial and tangential walls and consequently the probability that they undergo different strains, and, lastly, the possibility of complete shear restraint within the cell wall. These three-dimensional models are frequently analysed using finite element techniques.

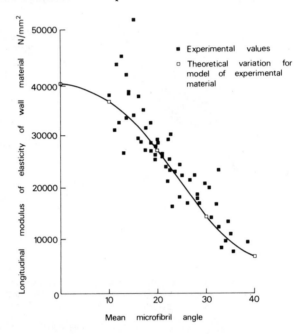

Figure 11.9 Effect of the mean microfibrillar angle of the cell wall on the longitudinal modulus of elasticity of the wall material in *Pinus radiata*. Calculated values from a mathematical model are also included. (From I. D. Cave (1968) *Wood Science and Technology*, 2, 268–278, by permission of Springer-Verlag.)

Recent modelling of timber behaviour in terms of structure uses the concept of an elastic fibre composite consisting of an inert fibre phase embedded in a water-reactive matrix. The constitutive relation is related to the overall stiffness of the composite, the volume fraction, stiffness and sorption characteristics of the matrix, and unlike previous models the equation can be applied not only to elasticity but also to shrinkage and even moisture induced creep (Cave, 1975).

Stiffness of a material is very dependent on the type and degree of chemical bonding within its structure and the abundance of covalent bonding in the longitudinal plane and hydrogen bonding in the transverse planes contributes considerably to the moderately high levels of stiffness characteristic of timber.

11.2.5.5 *Moisture Content*

The influence of moisture content on stiffness is similar though not quite so sensitive as that for strength briefly described in Chapter 7 and illustrated in Fig. 7.1. Early experiments by Carrington (1922), in which stiffness was measured on a specimen of Sitka spruce as it took up moisture from the dry state, clearly indicated a linear loss in stiffness as the moisture content increased to about 30%, corresponding to the fibre saturation point discussed in the previous chapter; further increase in moisture content has no influence in stiffness (Fig. 11.10). It will be noted from this figure that the longitudinal elastic modulus is less sensitive to moisture content than either the radial or tangential elastic moduli.

Carrington also measured the rigidity moduli and Poisson's ratios. While the former showed a similar trend to the elastic moduli, only one of the latter displayed the same trend while the three other Poisson's ratios that were measured showed an inverted relationship (Fig. 11.10).

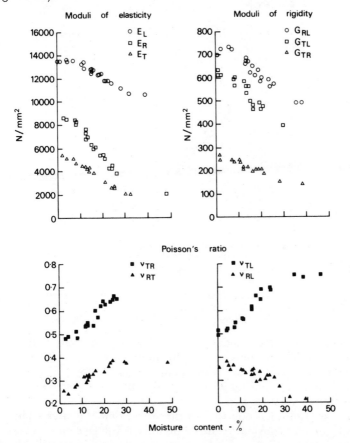

Figure 11.10 Effect of moisture content on the elastic constants of Sitka spruce. (From H. Carrington (1922) *Aeronautical Journal*, 24, 462, by permission of the Royal Aeronautical Society.)

Carrington's results for the variation in longitudinal moduli have been confirmed using the simple dynamic methods described earlier in this chapter. Measurement of the frequency of vibration was carried out at regular intervals as samples of Sitka spruce were dried from 70% to zero moisture content (Fig. 11.11).

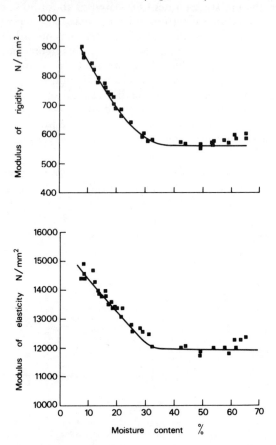

Figure 11.11   Effect of moisture content on the moduli of elasticity and rigidity in Sitka spruce. Both moduli were determined dynamically. (From the Princes Risborough Laboratory, Building Research Establishment. © Crown copyright.)

Confirmation of the reduction in modulus of elasticity with increasing moisture content is forthcoming from Fig. 11.8, in which the regression lines of elasticity against density for over 200 species of timber at 12% moisture content and in the green state are presented.

Barkas (1945) has pointed out that when timber is stressed in compression at constant relative humidity it will give up water to the atmosphere, and conversely under tensile stressing it will absorb moisture. The equilibrium strain, therefore, will be the sum of that produced elastically and that caused by moisture loss or gain. It is therefore necessary to distinguish between the elastic constants at constant humidity (E_h) and those measured at constant moisture content (E_m). Hearmon (1948) has indicated that the ratio of

324

E_h/E_m is 0.92 in the tangential direction, 0.95 in the radial direction and 1.0 in the longitudinal direction when spruce is stressed at 90% relative humidity. At 40% the ratio increases to 0.98 and 0.99 in the tangential and radial directions respectively.

11.2.5.6 *Temperature*

In timber, like most other materials, increasing temperature results in greater oscillatory movement of the molecules and an enlargement of the crystal lattice. These in turn affect the mechanical properties and the stiffness and strength of the material decreases.

Although the relationship between stiffness and temperature has been shown experimentally to be curvilinear, the degree of curvature is usually slight at low moisture contents (Fig. 11.12) and the relation is frequently treated as linear thus:

$$E_T = E_t[1 - a(T - t)] \qquad (11.22)$$

where E is the elastic modulus, T is a higher temperature, t is a lower temperature, and a is the temperature coefficient. The value a for longitudinal modulus has been shown to lie between 0.001 and 0.007 for low moisture contents.

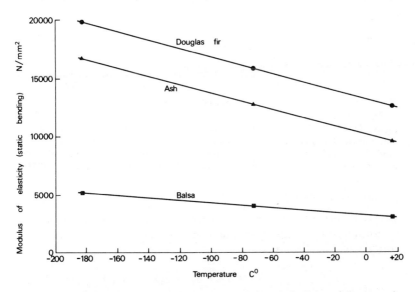

Figure 11.12 Effect of temperature on the modulus of elasticity of three species of timber. (From the Princes Risborough Laboratory, Building Research Establishment. © Crown copyright.)

At higher moisture contents the relationship of stiffness and temperature is markedly curvilinear and the interaction of moisture content and temperature in influencing stiffness is clearly shown in Fig. 11.13, which summarises the extensive work of Sulzberger (1947). At zero moisture content the reduction in stiffness between

325

−20 °C and +60 °C is only 6%: at 20% moisture content the reduction is 40%.

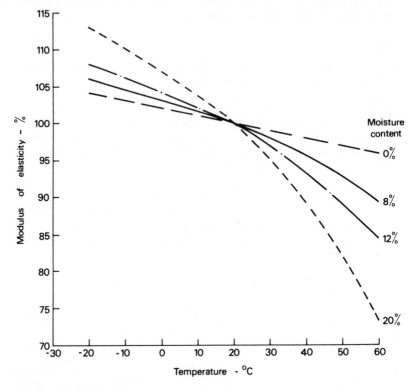

Figure 11.13 The interaction of temperature and moisture content on the modulus of elasticity. Results are averaged for six species of timber and the modulus at 20 °C and 0% moisture content is taken as unity. (From the Princes Risborough Laboratory, Building Research Establishment. © Crown copyright.)

11.2.6 Bulk Modulus

This is a measure of the cubic compressibility of timber under a uniform hydrostatic pressure and can be obtained from the elastic compliances by the equation:

$$K = \frac{1}{P} \cdot \frac{dV}{V} = S_{11} + S_{22} + S_{33} + 2(S_{23} + S_{31} + S_{12}) \qquad (11.23)$$

where K is the isothermic cubic compressibility, P is the uniform hydrostatic pressure, and dV/V is the relative change of volume.

From equation 11.23 it follows that

$$\nu_K = \frac{S_{31} + S_{12} + S_{23}}{S_{11} + S_{22} + S_{33}} \qquad (11.24)$$

where ν_K is the bulk Poisson's ratio, which for timber has been found to lie between 0.17 and 0.31.

326

11.3 VISCOELASTIC DEFORMATION

In the introduction to this chapter (Section 11.1) timber was described as being neither truly elastic in its behaviour nor truly viscous, but rather a combination of both states; such behaviour is usually described as viscoelastic and, in addition to timber, materials such as concrete, bitumen and the thermoplastics are also viscoelastic in their response to stress.

Viscoelasticity infers that the behaviour of the material is time dependent; at any instant in time under load its performance will be a function of its past history. Now if the time factor under load is reduced to zero, a state which we can picture in concept but never attain in practice, the material will behave truly elastically, and we have seen in Section 11.2 how timber can in fact be treated as an elastic material and how the principles of orthotropic elasticity can be applied. However, where stresses are applied for a period of time, viscoelastic behaviour will be experienced and, while it is possible to apply elasticity theory with a factor covering the increase in deformation with time, this procedure is at best only a first approximation.

In a material such as timber; time-dependent behaviour manifests itself in a number of ways of which the more common are *creep, relaxation, damping capacity*, and the dependence of strength on *duration of load*. When the load on a sample of timber is held constant for a period of time the increase in deformation over the initial instantaneous elastic deformation is called creep and Fig. 11.1 illustrates not only the increase in creep with time, but also the subdivision of creep into a reversible and an irreversible component of which more will be said in a later section.

Most timber structures carry a considerable dead load and the component members of these will undergo creep; the dip towards the centre of the ridge of the roof of very old buildings bears testament to the fact that timber does creep. However, compared to thermoplastics and bitumen the amount of creep in timber is appreciably lower. Although creep behaviour in timber has been recognised for many decades, considerably less research has been carried out on this property compared with the other mechanical properties: this is due to a large extent to the treatment of timber in structural analyses as an elastic material with correction factors for the time variable.

Viscoelastic behaviour is also apparent in the form of relaxation where the load necessary to maintain a constant deformation decreases with time; in timber utilisation this has limited practical significance and the area has attracted very little research. Damping capacity is a measure of the fractional energy converted to heat compared with that stored per cycle under the influence of mechanical vibrations; this ratio is time dependent. A further manifestation of viscoelastic behaviour is the apparent loss in strength of timber with increasing duration of load; this feature is discussed in detail in Section 14.7.12.2 and illustrated in Fig. 14.11.

11.3.1 Creep

11.3.1.1 *Test Methods*

Although a simple test in concept, the creep test necessitates the solution of many problems in relation to both loading and recording if high degrees of accuracy and reproducibility are to be achieved. Not only must the load be applied in as near instantaneous time as possible, but it must be held constant for the duration of the test: if dead weights are used their handling is a considerable task, though the quantity can be reduced by applying the load through a lever arm. The deformation of timber under load is affected by temperature and is particularly sensitive to changes in relative humidity; consequently creep tests must be performed in a controlled environment. Because of the small deflections obtained their measurement must be determined with a high degree of accuracy, at least of the order of 0.01 of a millimeter.

Deformation can be recorded manually employing dial gauges of the necessary accuracy, or automatically by using either strain gauges or potentiometers. Use of the former is restricted due to the stiffening effect which the adhesive induces on the surface of the timber and to uncertainties over their time stability. Linear potentiometers, though not without operational problems, are usually employed for test work on timber.

Duration of test can vary from as much as a few hours to a few years; greater accuracy in predicting future behaviour is obtained from the tests of longer duration.

Although creep in timber under tensile loading has been measured occasionally, much more attention has been given to creep in bending, due to the high proportion of structural timber that is loaded in this mode.

It is possible to quantify creep by a number of time-dependent parameters of which the two most common are *creep compliance* (known also as *specific creep*) and *relative creep* (known also as the *creep coefficient*); both parameters are a function of temperature.

Creep compliance is the ratio of increasing strain with time to the applied constant stress, i.e.

$$c_c(t, T) = \frac{\text{(varying) strain}}{\text{applied constant stress}} \qquad (11.25)$$

while relative creep is defined as the deflection at time t expressed in terms of the initial elastic deflection, i.e.

$$c_r(t, T) = \frac{\epsilon_t}{\epsilon_0} \qquad (11.26)$$

where ϵ_t is the deflection at time t, and ϵ_0 is the initial deflection.

Relative creep has also been defined as the change in compliance during the test expressed in terms of the original compliance.

328

11.3.1.2 Creep Relationships

In both timber and timber products such as plywood or chipboard the rate of deflection or creep slows down progressively with time (Fig. 11.14); the creep is frequently plotted against log time and the graph assumes an exponential shape. Results of creep tests can also be plotted as relative creep against log time or as creep compliance against stress as a percentage of the ultimate short-time stress.

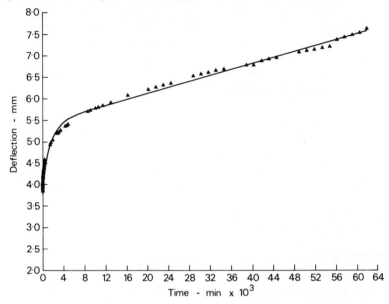

Figure 11.14 The increase in deformation with time of urea-formaldehyde-bonded chipboard: the regression line has been fitted to the experimental values using Equation 11.32. (From the Princes Risborough Laboratory, Building Research Establishment. © Crown copyright.)

In Section 11.2.4 it was shown that the degree of elasticity varied considerably between the horizontal and longitudinal planes. Creep, as one particular manifestation of viscoelastic behaviour, is also directionally dependent. In tensile stressing of longitudinal sections produced with the grain running at different angles it was found that relative creep was greater in the direction perpendicular to the grain than it was parallel to the grain. Furthermore, in those sections with the grain running parallel to the longitudinal axis, the creep compliance and relative creep across the grain was not only less than the parallel value but also negative (Fig. 11.15). Such anisotropic variation must result in at least some of the Poisson's ratios being time dependent and actual measurement of some of these ratios has confirmed this view (Schniewind and Barrett, 1972). Such evidence has important consequences; it means that in the stressing of timber, at least in tension parallel to the grain, an increase in volume occurs and therefore the deformation of timber with time appears to be quite different from those materials which deform at constant

329

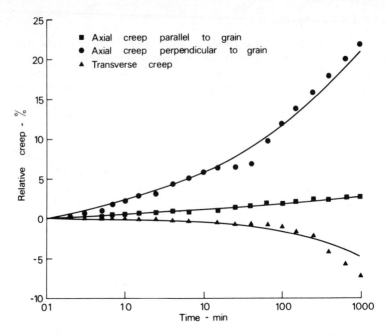

Figure 11.15 The variation in relative creep with time in samples of Douglas fir stressed at different angles to the grain. (From A. P. Schniewind and J. D. Barrett (1972) *Wood Science and Technology*, 6, 43—57, by permission of Springer-Verlag.)

volume. To these is frequently applied the theory of plasticity, but the application of this concept does not appear to be valid for timber.

When creep compliance is plotted against stress as a percentage of the ultimate short-term stress, the relation is linear over the lower half of the range in stress. For viscoelastic behaviour to be defined as linear the instantaneous, recoverable and nonrecoverable components of the deformation must vary directly with the applied stress. An alternative definition is that the ratio of stress to strain is a function of time (or frequency) alone and not of stress magnitude; such a relation appears as a straight isochronous graph.

The linear limit for the relation between creep and applied stress varies with mode of testing, with species of timber (Fig. 11.16), and with both temperature and moisture content. In tension parallel to the grain at constant temperature and moisture content timber has been found to behave as a linear viscoelastic material up to about 75% of the ultimate tensile strength, though some workers have found considerable variability and have indicated a range from 36% to 84%. In compression parallel to the grain the onset of non-linearity appears to occur at about 70%, though the level of actual stress will be much lower than in the case of tensile strength since the ultimate compression strength is only one third that of the tensile strength. In bending, nonlinearity seems to develop very much earlier at about 56—60% (Fig. 11.16); the actual stress levels will be very similar to those for compression.

330

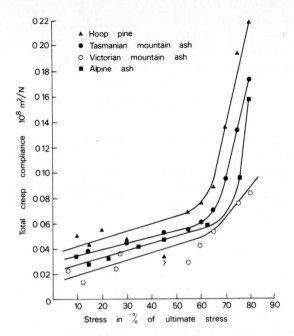

Figure 11.16 The relation of total creep compliance to stress as a percentage of the ultimate for four Australian species loaded in bending for 20 hours. (From R. S. T. Kingston and B. Budgen (1972) *Wood Science and Technology*, 6, 230–238, by permission of Springer-Verlag.

In both compression and bending the divergence from linearity is usually greater than in the case of tensile stressing; much of the increased deformation occurs in the nonrecoverable component of creep and is associated with progressive structural changes including the development of incipient failure (Section 14.8.1.2).

Increase not only in stress level, but also in temperature to a limited extent, and moisture content to a considerable degree, result in an earlier onset of nonlinearity and a more marked departure from it. For most practical situations, however, working stresses are only a small percentage of the ultimate, rarely approaching even 50%, and it can be safely assumed that timber, like concrete (see Chapter 10), will behave as a linear viscoelastic material.

11.3.1.3 *Principle of Superposition*

Since timber behaves as a linear viscoelastic material under conditions of normal temperature and humidity and at low to moderate levels of stressing, it is possible to apply the *Boltzmann's principle of superposition* to predict the response of timber to complex or extended loading sequences. This principle states that the creep occurring under a sequence of stress increments is taken as the super-posed sum of the responses to the individual increments. This can be expressed mathematically in a number of forms, one of which for linear materials is:

$$\epsilon_c(t) = \Sigma_1^n \, \Delta\sigma_i c_{ci} \qquad (11.27)$$

331

where n is the number of load increments, $\Delta\sigma_i$ is the stress increment, c_{ci} is the creep compliance for the individual stress increments
applied for differing times, $t - \tau_1, t - \tau_2, \ldots, t - \tau_n$, and $\epsilon_c(t)$ is the
total creep at time t; or in integrated form

$$\epsilon_c(t) = \int_{\tau_1}^{t} c_c(t - \tau) \cdot \frac{d\sigma}{d\tau}(\tau) \cdot d\tau \tag{11.28}$$

In experiments on timber the superposition principal was found to
be applicable even to high stresses in both shear and tension of dry
samples. However, at high moisture contents the limits of linear
behaviour appear to be considerably lower and it has been shown
that superposition is no longer applicable at stresses somewhat below
half the failing stress.

11.3.1.4 *Viscoelasticity: Constitutive Equations for Creep*

Provided that some limitations on the conditions for linear behaviour
are observed, it is possible to apply linear viscoelastic theory to the
long-term deformation of timber. The constitutive equation for a
linear anisotropic viscoelastic material can be written in terms of
the creep compliance; as in the treatment of timber as an elastic
solid it is again necessary to treat the circular symmetry of timber as
orthotropic. This reduces the 81 compliances of the general case to
36; thermodynamic arguments provide a further reduction to 21,
while coincidence between the reference axes and the axes of
symmetry leads to a final 9 independent compliance parameters. The
constitutive equation written in matrix form is therefore:

$$
\begin{bmatrix}
\epsilon_1(t) \\
\epsilon_2(t) \\
\epsilon_3(t) \\
\epsilon_4(t) \\
\epsilon_5(t) \\
\epsilon_6(t)
\end{bmatrix}
=
\begin{bmatrix}
S_{11}(t) & S_{12}(t) & S_{13}(t) & 0 & 0 & 0 \\
S_{21}(t) & S_{22}(t) & S_{23}(t) & 0 & 0 & 0 \\
S_{31}(t) & S_{32}(t) & S_{33}(t) & 0 & 0 & 0 \\
0 & 0 & 0 & S_{44}(t) & 0 & 0 \\
0 & 0 & 0 & 0 & S_{55}(t) & 0 \\
0 & 0 & 0 & 0 & 0 & S_{66}(t)
\end{bmatrix}
\begin{bmatrix}
\sigma_1 \\
\sigma_2 \\
\sigma_3 \\
\sigma_4 \\
\sigma_5 \\
\sigma_6
\end{bmatrix}
\tag{11.29}
$$

and, as such, is analogous to Equation 11.10 for the elastic behaviour
of orthotropic materials.

The nine independent components could be determined experimentally, but to date generally only isolated compliance components have been determined using uniaxial tests. Schniewind and
Barrett (1972), however, measured the principal components of the
creep compliance tensor in two planes, thereby permitting the solution of problems of generalised plane stress. In the case of plane
stress systems, where it is usual to ignore the very small strain component normal to the plane of applied stress, the constitutive

equations, for stress applied in the 1–2 plane, can be written in matrix form as:

$$
\begin{bmatrix} \epsilon_1(t) \\ \epsilon_2(t) \\ \epsilon_6(t) \end{bmatrix} = \begin{bmatrix} S_{11}(t) & S_{12}(t) & 0 \\ S_{21}(t) & S_{22}(t) & 0 \\ 0 & 0 & S_{66}(t) \end{bmatrix} \begin{bmatrix} \sigma_1 \\ \sigma_2 \\ \sigma_6 \end{bmatrix} \tag{11.30}
$$

Equation 11.30 for viscoelastic deformation is therefore analogous to Equation 11.21. for elastic deformation. The similarity is continued for just as it is possible to use transformation equations to obtain the value of strain at some angle to the grain when the timber is treated as deforming elastically, so it is possible to obtain data on creep at some angle to the grain using transformation equations written in terms of the creep compliances.

11.3.1.5 Mathematical Expressions of Creep

The relationship between creep and time has been expressed mathematically using a wide range of equations. It should be appreciated that such expressions are purely empirical, none of them possessing any sound theoretical basis. Their relative merits depend on how easily their constants can be determined and how well they fit the experimental results.

The most successful mathematical description for timber appears to be of the type

$$
\epsilon(t) = \epsilon_0 + at^m \tag{11.31}
$$

where $\epsilon(t)$ is the time-dependent strain, ϵ_0 is the initial deformation, a and m are constants ($m = 0.33$ for timber), and t is the elapsed time.

Creep behaviour in timber, like that of many other high polymers, has been interpreted with the aid of mechanical models comprising different combinations of springs and dashpots; the springs act as a mechanical analogue of the elastic component of deformation, while the dashpot simulates the viscous or flow component. When more than a single member of each type is used, these components can be combined in a wide variety of ways, though only one or two will be able to describe adequately the creep and relaxation behaviour of the material.

The simplest linear model which successfully simulates the time-dependent behaviour of timber is the four-element model illustrated in Fig. 11.17; the central part of the model will be recognised as the Kelvin element described in Chapter 10 dealing with deformation in concrete. To this unit has been added in series a second spring and dashpot. The strain at any time t under a constant load is given by the mathematical model

$$
\epsilon(t) = \frac{\sigma}{E_1} + \frac{\sigma}{E_2} \left(1 - e^{-t/\tau_2} \right) + \frac{\sigma t}{\eta_3} \tag{11.32}
$$

333

where $\epsilon(t)$ is the strain at time t, E_1 is the elasticity of spring 1, E_2 is the elasticity of spring 2, σ is the stress applied, η_3 is the viscosity of dashpot 3, and $\tau_2 = \eta_2/E_2 =$ viscosity of dashpot 2/elasticity of spring 2.

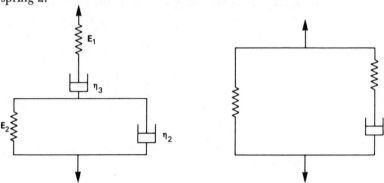

Figure 11.17 Mechanical analogues of the components of creep: the springs simulate elastic deformation and the dashpots viscous flow. The model on the left corresponds to Equation 11.32, while that on the right is known as the standard linear solid. (From the Princes Risborough Laboratory, Building Research Establishment. © Crown copyright.)

The first term on the right hand side represents the instantaneous deformation, while the second term describes the delayed elasticity and the third term the plastic flow component. Thus the first term describes the elastic behaviour while the combination of the second and third terms accounts for the viscoelastic or creep behaviour. The response of this particular model will be linear and it will obey the Boltzmann superposition principle.

The degree of fit between the behaviour predicted by the model and experimentally derived values can be exceedingly good: an example is illustrated in Fig. 11.14, where the degree of correlation between predicted and experimental results for creep in bonding of urea-formaldehyde chipboard beams was as high as 0.941.

The model known as the standard linear solid (Fig. 11.17) has also been used to describe creep behaviour in timber: the degree of fit is not very good, though this can be improved considerably if the linear elastic spring (parallel to the dashpot) and the linear dashpot are replaced by corresponding nonlinear elements to produce a nonlinear viscoelastic model (Ylinen, 1965).

It should be appreciated, however, that while models such as the one described above have been of value in the derivation of equations to describe the behaviour of timber, thereby allowing the prediction of performance under longer periods of time, they are of no assistance in interpreting viscoelastic behaviour in terms of the basic structure of the material.

11.3.1.6   *Reversible and Irreversible Components of Creep*

In timber and many of the high polymers creep under load can be subdivided into reversible and irreversible components: passing reference to this was made in Section 11.1 and the generalised relationship with time was depicted in Fig. 11.1. The relative proportions of these two components of total creep appears to be related to stress level and to prevailing conditions of temperature and moisture content.

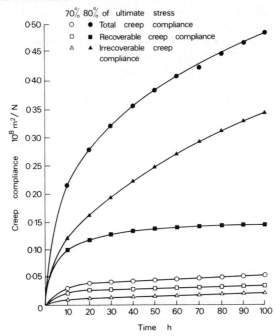

Figure 11.18 The relative proportions of the recoverable and irrecoverable creep compliance in samples of hoop pine (*Araucaria cunninghamii*) stressed in bending. (From R. S. T. Kingston and B. Budgen (1972) *Wood Science and Technology*, **6**, 230–238, by permission of Springer-Verlag.)

The influence of level of stress is clearly illustrated in Fig. 11.18, where the total compliance at 70% and 80% of the ultimate stress for hoop pine in compression is subdivided into the separate components: at 70% the irreversible creep compliance accounts for about 45% of the total creep compliance, while at 80% of the ultimate the irreversible creep compliance has increased appreciably to 70% of the total creep compliance at the longer periods of time, though not at the shorter durations. Increased moisture content and increased temperature will also result in an enlargement of the irreversible component of total creep.

Reversible creep is frequently referred to in the literature as delayed elastic or primary creep and is ascribed to either polymeric uncoiling or the existence of a creeping matrix. Owing to the close longitudinal association of the molecules of the various components in the amorphous regions it appears unlikely that uncoiling of the polymers under stress can account for much of the reversible component of creep.

335

The second explanation of reversible creep utilises the concept of time-dependent two-stage molecular motions of the cellulose, hemicellulose and the lignin constituents. The pattern of molecular motion for each component is dependent on that of the other constituents and it has been shown that the difference in directional movement of the lignin and nonlignin molecules results in considerable molecular interference such that stresses set up in loading can be transferred from one component (a creeping matrix) to another component (an attached, but noncreeping structure). It is postulated that the lignin network could act as an energy sink, maintaining and controlling the energy set up by stressing (Chow, 1973): this most recent explanation of reversible creep is certainly most attractive.

Irreversible creep, also referred to as viscous, plastic or secondary creep has been related to either time-dependent changes in the active number of hydrogen bonds, or to the loosening and subsequent remaking of hydrogen bonds as moisture diffuses through timber with the passage of time. Such diffusion can result directly from stressing; thus Barkas (1945) found that when timber was stressed in tension it gained in moisture content, and conversely when stressed in compression its moisture content was lowered. It is argued, though certainly not proven, that the movement of moisture by diffusion occurs in a series of steps from one absorption site to the next, necessitating the rupture and subsequent reformation of hydrogen bonds. The process is viewed as resulting in loss of stiffness and/or strength, possibly through slippage at the molecular level. Recently, however, it has been demonstrated that moisture movement, while affecting creep, can account for only part of the total creep obtained, and this explanation of creep at the molecular level warrants more investigation; certainly not all the observed phenomena support the hypothesis that creep is due to the breaking and remaking of hydrogen bonds under stress bias. At moderate to high levels of stressing, particularly in bending and compression parallel to the grain, the amount of irreversible creep is closely associated with the development of incipient failures: this point is discussed further in Section 14.8.1.2.

Attempts have been made to describe creep in terms of the fine structure of timber and it has been demonstrated that creep in the short term is highly correlated with the angle of the microfibrils in the S_2 layer of the cell wall, and inversely with the degree of crystallinity. However, such correlations do not necessarily prove any causal relation and it is possible to explain these correlations in terms of the presence or absence of moisture which would be closely associated with these particular variables.

11.3.1.7 *Environmental Effects on Rate of Creep*

Temperature. Common with many other materials, especially the high polymers, the effect of increasing temperature on timber under stress is to increase both the rate and the total amount of creep. Figure 11.19 illustrates a two and a half-fold increase in the amount

336

of creep as the temperature is raised from 20 °C to 54 °C; there is a marked increase in the irreversible component of creep at the higher temperatures. Cycling between low and high temperatures will induce in stressed timber a higher creep response than would occur if the temperature was held constant at the higher level.

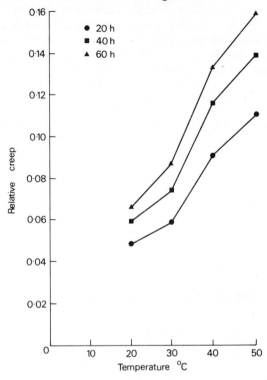

Figure 11.19 The effect of temperature on relative creep of samples of hoop pine (*Araucaria cunninghamii*) loaded in compression for 20, 40 and 60 hours. (From R. S. T. Kingston and B. Budgen (1972) *Wood Science and Technology*, 6, 230–238, by permission of Springer-Verlag.)

Research has indicated that the principle of time-temperature superposition which is widely used in creep investigations on plastics is not applicable to creep in timber or timber-based board materials.

Moisture Content: Steady State. The rate and amount of creep in timber of high moisture content is appreciably higher than that of dry timber: Clouser (1959) has shown that an increase in moisture content from 6% to 12% increases the deflection at any given stress level by 20%. It is interesting to note the occurrence of a similar increase in creep in nylon when in the wet condition.

Moisture Content: Unsteady State. If the moisture content of small timber beams under load is cycled from dry to wet and back to dry again the deformation will also follow a cyclic pattern; however, the recovery in each cycle is only partial and over a number of cycles the total amount of creep is very large: the greater the moisture differential in each cycle the higher the amount of creep (Hearmon and Paton, 1964). Figure 11.20 illustrates the deflection that occurs with time in matched samples loaded to $\frac{3}{8}$ ultimate short-term load where one is maintained in an atmosphere of 93% relative humidity, while the other is cycled between 0 and 93% relative humidity. After 14 complete cycles the latter sample had broken after increasing its initial deflection by 25 times; the former sample was still intact having increased in deflection by only twice its initial deflection. Failure of the first beam occurred, therefore, after only a short period of time and at a stress equivalent to only $\frac{3}{8}$ of its ultimate.

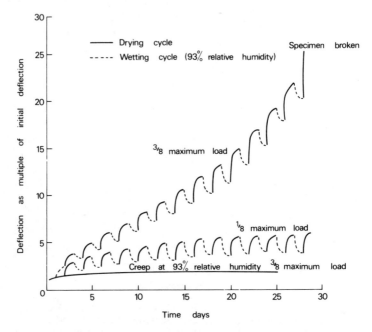

Figure 11.20   The effect of cyclic variations in moisture content on relative creep of samples of beech loaded to $\frac{1}{8}$ and $\frac{3}{8}$ of ultimate load. (From the Princes Risborough Laboratory, Building Research Establishment. © Crown copyright.)

It should be appreciated that creep increased during the drying cycle and decreased during the wetting cycle with the exception of the initial wetting when creep increased. Such creep behaviour has been recorded and confirmed only for timber. The negative deflection observed during absorption has not yet been explained, though the energy for the change is probably provided by the heat of absorption. The net change at the end of a complete cycle of wetting and drying is considered to be a redistribution of hydrogen bonds which manifests itself as an increase in deformation of the stressed sample.

The rate of moisture change has been found to affect the rate of creep, but not total creep; this appears to be proportional to the total change in moisture content.

Theory of Hydroviscoelasticity. The behaviour of timber under long-term load as a linear viscoelastic material assumes constant climatic conditions. In reality those never exist and, as described above, changes in moisture content of the timber have an appreciable effect on the rate and amount of creep.

Two approaches to the difficulty of including temperature and moisture content variation have been proposed: the first, and certainly the approach which is mathematically more developed, is the recently proposed theory of hydroviscoelastity which describes a functional relationship between deformation and stress, time, temperature and moisture content. It makes the important assumption that some component or components of the timber act as a memory with regard to those four variables. Thus:

$$\epsilon(t) = \sum_{\Upsilon=0}^{t} [\sigma(\Upsilon), \mu(\Upsilon), T(\Upsilon)] \qquad (11.33)$$

where σ is the stress, μ is the moisture content, T is the temperature, and Υ are the time co-ordinates. This equation is expanded in a Frechet series containing Kernel functions and so far has been tested against experimental evidence for plywood and wood veneer; reasonable fit between theory and practice has been obtained (Ranta-Mannus, 1973).

Mechanosorptive behaviour. The second approach to the handling of the effect of moisture content variation on creep is at the present time only at the descriptive state: it attempts to separate from the true viscoelastic creep certain deformations which are directly related to the interaction of moisture and mechanical stressing. Such an approach is almost the direct antithesis to the theory of hydroviscoelasticity.

It is argued that creep under cyclic humidity conditions is much more dependent on the magnitude of the moisture change (see Unsteady State) than on the duration of the process. Furthermore, deformation is not truly permanent since a large part of it is recoverable when the timber is taken through a moisture cycle after removal of load. Once again it is argued that some component is acting as a memory function. Hence it has been questioned whether increased creep under conditions of cyclic humidity is a true viscoelastic phenomenon or whether it should be treated as a mechanosorptive deformation (Grossman, 1976).

11.3.2 Relaxation

Another characteristic of viscoelastic materials is that they will relax with time; i.e. when a sample is stressed for a long period of time, the level of stress necessary to maintain a constant deflection will decrease with time. The process is usually quantified in terms of a relaxation modulus:

$$M_r(t, T) = \frac{\text{stress (varying)}}{\text{applied constant strain}} \qquad (11.34)$$

Although the relaxation modulus and the creep compliance are related it must be appreciated that it is only when time (t) is zero that they are exactly reciprocals. Studies of relaxation in timber have been carried out and the factors responsible for this form of time-dependent behaviour are identical with those causing creep. Non-linearity appears to develop early at low levels of initial strain, but fortunately the degree of nonlinearity is very small over a large range of initial values and timber can still be treated as a linear viscoelastic material.

11.4 COMPARISON WITH CONCRETE

Many readers will already be conscious of the similarity in behaviour under load of concrete and timber. Both composites display visco-elastic behaviour with reversible and irreversible components: both can be treated as linear viscoelastic and both obey Boltzmann's superposition principle. Creep in both is influenced by density (or porosity), by temperature and by moisture content. Both materials display instantaneous elastic deformation and in structural analysis both are treated as elastic materials, though strictly speaking they are viscoelastic.

Two fairly important differences occur: first, under cyclic humidity conditions creep in timber increases on the desorption cycle and decreases during adsorption; such behaviour has not been recorded in concrete. Secondly, due to the high degree of anisotropy of timber, orthotropic elasticity is applicable in stress analysis.

11.5 REFERENCES

Barkas, W. W. (1945) Swelling stresses in gels, *Special Report on Forest Products Research, London*, No 6.

Carrington, H. (1922) The elastic constants of spruce as affected by moisture content, *Aeronautical Journal*, 26, 462.

Cave, I. D. (1975) Wood substance as a water-reactive fibre-reinforced composite, *Journal of Microscopy*, 104(1), 47–52.

Chow, S. (1973) Molecular rheology of coniferous wood tissues, *Transactions of the Society of Rheology*, 17, 109–128.

Clouser, W. S. (1959) Creep of small wood beams under constant Bending load, *FPL Madison, Report*, 2150, 25 pp.

Grossman, P. U. A. (1976) Requirements for a model that exhibits mechanosorptive behaviour, *Wood Science and Technology*, 10, 163–168.

Hearmon, R. F. S. (1948) Elasticity of wood and plywood, *Special Report on Forest Products Research, London*, No 7, 87 pp.

Hearmon, R. F. S. (1966) Vibration testing of wood, *Forest Products Journal*, 16 (8), 29–40.

Hearmon, R. F. S. and Paton, J. M. (1964) Moisture content changes and creep in wood, *Forest Products Journal*, 14, 357–359.

Kingston, R. S. T. and Budgen, B. (1972) Some aspects of the rheological behaviour of wood, Part IV: Nonlinear behaviour at high stresses in bending and compression, *Wood Science and Technology*, 6, 230–238.

Ranta-Mannus, A. (1973) A theory for the creep of wood with application to birch and spruce plywood, *Technical Research Centre of Finland, Building Technology and Community Development*, Publication 4, pp. 35.

Schniewind, A. P. and Barrett, J. D. (1972) Wood as a linear ortho-tropic viscoelastic material, *Wood Science and Technology*, 6, 43–57.

Sulzberger, P. H. (1947) The effect of temperature on the strength of wood at various moisture contents in static bending, *Progress Report 7, Project TP 10–3, CSIRO (Aust) Div of Forest Products*.

Ylinen, A. (1965) Prediction of the time-dependent elastic and strength properties of wood by the aid of a general nonlinear visco-elastic rheological model, *Holz als Roh und Werk*, 5, 193–196.

CHAPTER 12

Deformation of Metals

12.1 ELASTIC DEFORMATION

Elastic deformation in solids may occur in two ways: the atom-to-atom bonds may be strained in tension, shear, or compression, or the angular relation between bonds may be changed to accommodate the applied stress. The great majority of solids, including metals, deform in the first way; the second is limited to long-chain molecules with many kinks in the chain, and is typified by the behaviour of the elastomer family of polymers, especially indiarubber. (The 'rubber principle' is, of course, used in metals in the form of helical springs.) Both modes of deformation are reversible; on removal of the stress, the solid regains its former dimensions, and the status quo is completely restored. Metals deform almost entirely by the stretching of the interatomic bonds; changes of bond angle occur, but the elastic effect is negligible by comparison.

We have seen in Section 2.5 that F, the force on the atoms in a crystal lattice, varies linearly with small displacements from r_0; the relationship is shown in Fig. 2.26. Since metals, in common with most crystalline solids, can only endure a limited degree of elastic deformation — normally of the order of 1% — it follows that the curve of Fig. 2.26 may be taken as a straight line over the elastic range, and therefore the elastic strain is proportional to stress, i.e. the metal obeys Hooke's Law. When the stress is purely tensile or compressive, the relatinship is given by

$$\sigma = E\epsilon \qquad (12.1)$$

where σ is the normal stress, ϵ the strain, and E the elastic modulus or Young's modulus. Conventionally, σ is positive for tensile stresses and negative for compressive stresses.

When a shear stress is applied, we have

$$\tau = G\gamma \qquad (12.2)$$

analogous to Equation 12.1, where τ is the shear stress, γ the shear strain, and G is the shear modulus of the material. In practice, the stress on any engineering component is a combination of normal and shear components, and a detailed analysis of the relationship between the stresses is necessary to calculate elastic deformation. Stress analysis is beyond the scope of this text, but the reader is referred to the excellent account by Cottrell mentioned in the Bibliography. E and G are related by the expression

$$G = \frac{E}{2(1 + v)} \qquad (12.3)$$

where v is *Poisson's ratio*, i.e. the ratio of lateral contraction to longitudinal extension during elastic deformation. Equation 12.3 applies only to elastically isotropic materials. It is not applicable to single crystals of metals, which are strongly anisotropic, but may be used for ordinary polycrystalline metals provided that no preferred orientation of crystals has been caused by cold working (Section 12.2).

The elastic and shear moduli are fundamental constants with a unique value for each metal but, since they depend on the binding forces between atoms, they may be altered by any factors that may modify these forces. An increase in temperature, for example, will increase the mean spacing of the crystal lattice from r_0 to r_1 in Fig. 2.26, so that the elastic modulus now becomes the gradient of the curve at $r = r_1$; it can be seen that the result is a lowering of the value of E, by about 0.03% for every 1 °C rise in temperature for a typical metal. The presence of alloying elements may change the elastic modulus either way; if there is no strong affinity between the different species of atoms, then the strains set up by the size discrepancy of the solute atoms will serve to shift the solvent atoms from their equilibrium positions and make further elastic deformation easier. In this case, alloying lowers the elastic moduli. If, on the other hand, the A—B bonds in the alloy are appreciably stronger than A—A or B—B bonds, i.e. if the tendency is for the alloy to form a superlattice or an intermetallic compound, then the Condon—Morse potential trough (Fig. 2.26) will be deepened and sharpened by the increased negativity of the ϕ_E curve (it may be assumed that the ϕ_R curve is unaffected by the change in ϕ_E), and therefore the gradient of the force curve in Fig. 2.26 will be increased at $r = r_0$ with a corresponding increase in the elastic moduli.

Crystal imperfections such as dislocations also affect the elastic moduli by permitting small anelastic increments of slip to occur within the elastic range. Metals prepared by ordinary methods contain large numbers of dislocations, and it is doubtful if any can show a true elastic range of deformation in the sense that plastic flow is totally excluded. The occurrence of quite extensive slip considerably below the nominal yield point of alloys has been amply demonstrated by work on metal fatigue (Section 15.3). It is only recently, since it has become possible to grow perfectly crystalline metal 'whiskers', that true values of elastic moduli have been experimentally determined. Alternatively, if the dislocations can be locked in the structure so that their movement is restricted, an increase in the elastic moduli may be achieved; this may be accomplished by *strain ageing* (Section 12.4)

343

or age hardening (Section 16.2). Some relationships between composition and Young's Modulus are shown in Fig. 12.1.

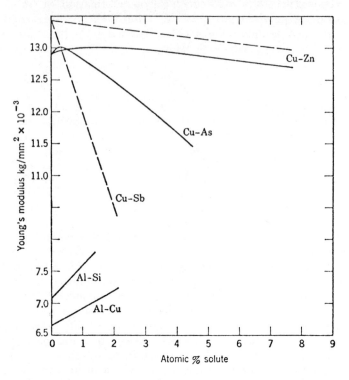

Figure 12.1   The effect of solute atoms on the elastic moduli of copper and
aluminium. (Reproduced with permission, from *Mechanical Properties of
Metals*, by D. McLean. Copyright © 1962, John Wiley & Sons Inc.)

12.2   PLASTIC DEFORMATION

When a metal specimen is stressed beyond its elastic limit, plastic
deformation occurs in addition to elastic deformation and the specimen acquires a permanent set. For the ordinary polycrystalline component made up of fine randomly orientated grains, the mode of
deformation is not obvious, but if a single crystal of a hexagonal
metal is plastically deformed then the superficial mechanism can be
seen with the naked eye. It will be seen that the metal has deformed
by a process of *slip* along a definite set of crystallographic planes,
blocks of metal sliding over one another to produce the extension
(Fig. 12.2). If the metal is sufficiently ductile, the process can be
continued until a specimen of circular cross section has changed to
a thin flat ribbon by continued slip of these flat sided 'blocks' as
shown, schematically in Fig. 12.3. Crystallographic examination has
established that the planes on which slip occurs (known as 'slip

344

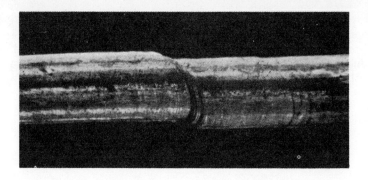

Figure 12.2   Localised slip in a single crystal of cadmium. (From A. H. Cottrell (1956) *Dislocations and Plastic Flow in Crystals,* by permission of Oxford University Press.)

Slip bands

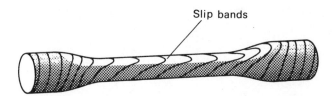

Figure 12.3   Plastic deformation on a single set of slip planes (schematic). The central portion has pulled out and flattened to form a ribbon section.

planes') are substantially the same for all metals and alloys of any one crystal structure, and that they tend to be the planes exhibiting the greatest degree of close packing in the structure. Thus, for h.c.p. metals the most usual slip plane is the (0001) plane, and for f.c.c. metals it is the (111) family of planes (see Section 2.42). Within these planes slip occurs in close-packed directions, and therefore the total number of ways in which slip can occur in a metal (each way is known as a 'slip system') is the product of the number of slip planes and the number of slip directions in each plane. For h.c.p. metals the combination is (0001) [$11\bar{2}0$] $\equiv$ 1 x 3 slip systems = 3; for f.c.c. metals, the combination is {111} [110] $\equiv$ 4 x 3 slip systems = 12. For the b.c.c. structure, which is *not* close-packed, the six {110} planes have the highest density and each of these slip planes has two [111] close-packed directions, making up 12 slip systems. However, slip can also occur on the [111] direction on the {112} and {123} planes, giving a total of 48 slip systems for b.c.c. metals. Despite the greater number of slip systems, b.c.c. metals are generally harder than f.c.c. metals because the lack of close packing gives a greater resistance to slip.

The stress required to initiate slip in any one slip system is known as the *critical resolved shear stress*, which will be the same for all slip systems in any given family of slip planes. However, the resolved shear stress acting on the slip systems in a crystal under stress will

345

differ widely from system to system because all will have differing orientations to the applied stress, and slip will be initiated on those systems in which the resolved shear stress first reaches the critical value, τ_c. Let us consider a cylindrical specimen of cross section A with an axial load P acting on it, as in Fig. 12.4. If the normal to

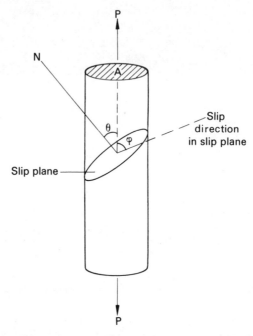

Figure 12.4 Calculation of the critical resolved shear stress. (N is the normal to the slip plane.)

the slip plane on which the stress is to act is inclined at an angle θ to the axis of the specimen, the area of slip plane on which the load acts is $A/\cos \theta$; hence the stress acting on the slip plane is given by

$$\frac{P}{\text{area of slip plane}} = \frac{P \cos \theta}{A}$$

However, for slip to occur, the significant value of stress is that resolved in a slip direction within the slip plane. Hence, if OS is the slip direction concerned, the resolved shear stress τ is given by

$$\tau = \frac{P}{A} \cos \theta \cos \phi$$

$$= \sigma \cos \theta \cos \phi \qquad (12.4)$$

where ϕ is the angle between OP and OS and σ is the axial stress acting on the specimen. Equation 12.4 is known as Schmidt's Law, and the function $\cos \theta \cos \phi$ as the Schmidt Factor.

From Equation 12.4 we can see that if either θ or $\phi = 90\,°$ the resolved shear stress becomes zero and no slip will occur — that is to say, if the applied stress axis lies either at right angles to or within the slip plane. The maximum value of τ is achieved when $\theta = \phi = 45\,^{\circ}$, from which

$$\tau = \tfrac{1}{2}\sigma \qquad\qquad (12.5)$$

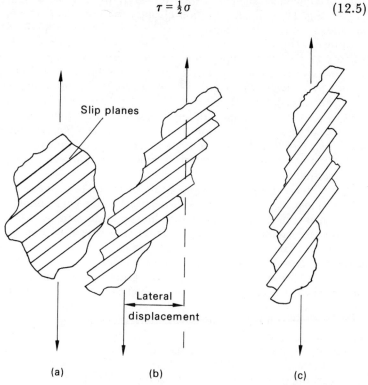

Slip planes

Lateral displacement

(a)   (b)   (c)

Figure 12.5 (a) Crystal under stress. (b) Free unconstrained yielding results in considerable lateral displacement. (c) Rotation of slip planes during deformation permits elongation with very little overall lateral movement.

Thus, as the stress on a specimen is increased, τ will reach the critical value to initiate slip first on those slip planes inclined at 45° to the stress axis in which a slip direction also lies at 45° to the stress axis, and plastic yield will first occur in grains satisfying this condition. The angular limitation on slip mentioned above can only occur in practice in single crystals of hexagonal metals, since these have only one slip plane; for f.c.c. and b.c.c crystals it is impossible for an applied stress to be at right angles to more than one set of slip planes, so that slip may occur on any of the remainder. The random grain orientations in normal polycrystalline specimens means, furthermore, that even with hexagonal metals only a very small proportion of grains are orientated so that slip is impossible.

When yielding occurs within some of the grains of a polycrystalline metal, it is never 'free' yielding because of the restriction superimposed by surrounding grains, which may be unfavourably orientated for

347

yielding. Locally 'free' yielding is shown in Fig. 12.5(a) and involves a lateral as well as a lengthwise deformation but in practice the lateral deformation is restricted, and as slip occurs the planes rotate so that they become more aligned to the stress axis; this is shown schematically in Fig. 12.5(c). The rotation changes the value of τ on the slip systems so that as deformation proceeds the value of τ is perhaps decreasing on the initially active set of planes and increasing on several other sets. Eventually slip commences on another set of planes whilst the first set becomes temporarily inactive; the slip system with highest value of τ is always the active one but, where two or more systems are nearly equal, slip will occur on them simultaneously. Those intersecting patterns of slip play an important part in the physical consequences of plastic deformation. In fact, in f.c.c. and b.c.c specimens with their large number of slip systems, it is almost impossible to avoid getting simultaneous slip on several systems, however carefully the stressing is carried out.

If a metal undergoes severe plastic deformation (as, for example, when an ingot is rolled down to form sheet) all the grains undergo rotation as well as strain and tend to line up in the same way in relation to the applied stress, and are said to have acquired a *preferred orientation*. Likewise, impurities and second phases become aligned along the stress axis and the material then gains a *texture* or *fibre structure*. Alignment of impurities is beneficial if it is along the axis on which the working load is to be applied, since their cross-sectional area (i.e. weakening effect) is thereby reduced; this occurs naturally in the rolling or forging processes. Preferred orientation in sheet metal has been turned to good account in the manufacture of transformer cores, since it provides a technique for aligning the easy directions of magnetisation with the magnetic field of the winding; if, however, the sheeting is to be used as the starting point for further manufacturing processes (e.g. pressing or deep drawing) then the effect may be extremely troublesome since it causes marked anisotropy of deformation. Some examples of preferred orientation are shown in Table 12.1.

Thus far we have considered only deformation resulting from uniaxial stress; however, engineering components are more often than not subjected to triaxial stressing. Under these conditions, the criteria for yielding are related to the three *principal stresses* σ_1, σ_2 and σ_3 acting on the component. Any three-dimensional state of stress — including both tensile and shear stress — can be reduced to three mutually perpendicular principal stresses acting on the mutually perpendicular planes, known as the *principal planes*. If the magnitudes of σ_1, σ_2 and σ_3 are known, the shear stress may be calculated using either the *Tresca criterion*:

$$\tau_{max} = \frac{\sigma_1 - \sigma_3}{2} \tag{12.6}$$

or the *von Mises criterion*:

$$(\sigma_1 - \sigma_2)^2 + (\sigma_2 - \sigma_3)^2 + (\sigma_3 - \sigma_1)^2 = 2\sigma_0^2 \tag{12.7}$$

348

Table 12.1　Development of Preferential Orientations in Metals as a Result of Cold Working Procedures

Note: these textures are idealized. In practice, considerable scatter occurs, and very frequently other textures are present.

Crystal structure	Mode of working		Texture
f.c.c.	Wire drawing and extrusion	$\langle 111 \rangle$ $\langle 100 \rangle$	Parallel to wire axis
b.c.c.	Wire drawing and extrusion	$\langle 110 \rangle$	Parallel to wire axis
c.p.h.	Wire drawing and extrusion	$\langle 10\bar{1}0 \rangle$	Parallel to wire axis
f.c.c.	Rolling	$\{110\}$	Parallel to rolling plane
		$\langle 112 \rangle$	Parallel to rolling direction
b.c.c.	Rolling	$\{001\}$	Parallel to rolling plane
		$\langle 110 \rangle$	Parallel to rolling direction
c.p.h.	Rolling	$\{0001\}$	Parallel to rolling plane
		$\langle 11\bar{2}0 \rangle$	Parallel to rolling direction

where σ_0 is the tensile yield stress. The von Mises criterion gives the most satisfactory correlation with practice, but for our purposes the Tresca criterion is simpler and quite adequate. It will be noted that for conditions of pure tension, where $\sigma_2 = \sigma_3 = 0$, the Tresca criterion reduces to Equation 12.5. The most important general conclusion from these yield criteria for our purposes is that where σ_2 or σ_3 becomes finite and of the same sign as σ_1, i.e. both positive (tensile) or both negative (compressive), the acting shear stress is *reduced*. In the extreme case where $\sigma_1 = \sigma_2 = \sigma_3$, the component is under pure *hydrostatic stress* and no plastic deformation is possible at all.

12.3　DISLOCATIONS IN METALS

An introduction to the concept of plastic deformation by dislocation movements has already been given in Section 2.6.3; we must now consider the nature of these imperfections in rather more detail. The basic concept of dislocation theory is that slip, when it takes place, does *not* occur simultaneously over the entire area of the slip plane, but starts at one point and then spreads over the plane at finite speed so that during the process of slip there will be at any instant a slipped region (increasing) and an unslipped region (decreasing) as shown in Fig. 12.6. The Burgers vector, b, of the dislocation, it will

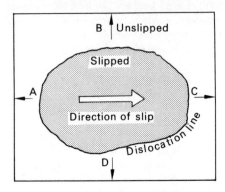

Figure 12.6   The different components of a dislocation loop in a slip plane: at A and C the loop is the edge-type dislocation, and at B and D it is the screw type.

be remembered, gives the distance and direction of slip generated by the dislocation loop, and if the direction of slip is that shown by the arrow in Fig. 12.6, then the nature of the loop at points A, B, C and D will be *negative* edge (symbol T), *right-handed* screw, *positive* edge (symbol ⊥), and *left-handed* screw respectively. The atomistic configurations associated with the four types are shown in Fig. 12.7 and it should be noted that if dislocations of similar type but opposite sign come together they will mutually annihilate each other. The reason for the designation 'screw dislocation' can be appreciated if we make a circuit through the atoms 123456789 in Fig. 12.7(d). In a perfect crystal this sequence would form a closed loop, atom 1 being coincident with atom 9; in the figure, however, the sequence is the first circuit of a right-handed spiral or screw, which, if continued, would follow the dislocation line right through the crystal.

Inspection of Fig. 12.7 should reveal another important fact: the slip plane of dislocation is the plane that contains *both the dislocation line and the Burgers vector*; from this it follows that an edge dislocation can move on only one slip plane, since b is at right angles to the dislocation, and therefore only one plane can contain both. A screw dislocation has greater freedom because b is parallel to the dislocation line, and therefore slip can occur in theory on *any* plane which contains the dislocation.

A dislocation line, being by definition a region of misfit in a crystal, possesses a higher energy than its surroundings, and is thus thermodynamically unstable. The strain energy of a screw dislocation is slightly lower than that of an edge dislocation; approximate values are as follows:

$$E_{screw} \simeq Gb^2 l \tag{12.8}$$

$$E_{edge} \simeq \frac{Gb^2 l}{1 - \nu} \tag{12.9}$$

350

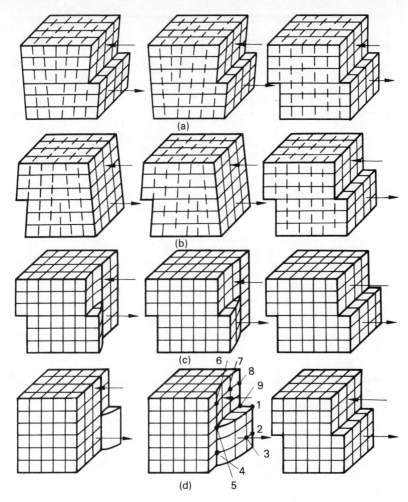

Figure 12.7 The ways that the four basic orientations of a dislocation move under the same applied stress: (A) positive edge, (B) negative edge, (C) left-hand screw, and (D) right-hand screw.

where G is the shear modulus, b the Burgers vector, l the length of dislocation line, and ν Poisson's ratio. It is important to note that $E \propto b^2$; thus a dislocation will always tend to move in the direction of minimum unit slip; that is, in close-packed directions in the slip plane. It also explans why dislocations of like sign do not easily join up. Fig. 12.8 shows the coalescence of two edge dislocations to form one dislocation of Burgers vector 2b; the energy of the combination is therefore double that of the two dislocations taken separately and hence there is a strong tendency for the dislocations to lower their total energy by dissociating. This mutual repulsion makes an important contribution to crack resistance, since the first stage of fracture is normally the running together of dislocations at some obstacle to slip (Section 15.1).

351

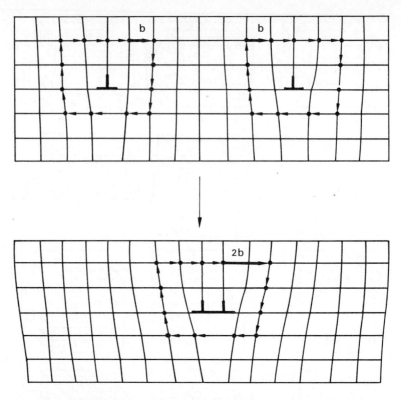

Figure 12.8 Combination of two positive edge dislocations of Burgers vector b. (From R. E. Reed-Hill (1973) *Physical Metallurgy Principles, 2nd edition,* by permission of D. Van Nostrand Co.)

A second consequence of the energy associated with dislocations is that they possess a *line tension* as an interface between the slipped and the unslipped regions of crystals, exactly analogous to the surface tension that operates on liquid surfaces. A dislocation line, therefore, will always try to shorten itself; curved regions tend to straighten themselves and loops shrink. Conversely, work must be done to lengthen a dislocation loop, and in particular to increase its curvature; both these effects contribute appreciably to the work required for plastic deformation. Any simple dislocation loop lying in a slip plane must therefore have a force F acting on it in the slip plane as in Fig. 12.9(a), if the slipped area is to be maintained or increased. Let us say that the loop lies in a slip plane of total area A, has a Burgers vector b, and is acted on by a shear stress τ in the plane. If a small element dl of the loop (Fig. 12.9(b)) moves a distance dx under the action of the shear stress, an additional area of slip d$l \cdot$ dx is generated, and the additional *mean* displacement on the entire slip plane within the crystal is given by

$$\frac{\text{b} \cdot \text{d}l \cdot \text{d}x}{A}$$

352

The work done, W, equals force × displacement, and hence

$$W = \frac{b \cdot dl \cdot dx}{A} \cdot A\tau$$

$$= \tau b \cdot dl \cdot dx$$

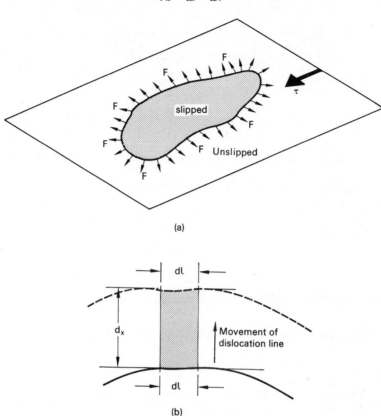

(a)

(b)

Figure 12.9 (a) A dislocation loop lying in a slip plane on which a resolved shear stress is acting. A force F, denoted by the small arrows, is acting outwards on the loop at right angles to it at all points. (b) The area of increment (shown shaded) swept out by the movement of a small element dl of a dislocation line.

But the force F acting on the dislocation does work equal to $F \cdot dx$ in producing the slip increment. Thus

$$F \cdot dx = \tau b \cdot dl \cdot dx$$

and hence

$$F = \tau b \cdot dl \qquad (12.10)$$

If the line is now caused to bow out (between, say, two obstacles to movement), we have (Fig. 12.10) the force F over a length l of the line balanced by the line tension T. Hence

$$\tau b l = 2T \sin\frac{\theta}{2}$$

353

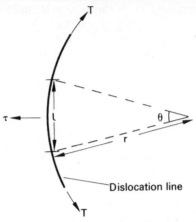

Figure 12.10 Forces acting on a small bowed element of a dislocation line.

For small values of θ, we can say $\sin \theta/2 \to \theta/2$, and therefore

$$\tau bl \simeq T \cdot \theta$$

But $\theta = 1/r$, and so

$$\tau \simeq \frac{T}{br} \tag{12.11}$$

The line tension T of the dislocation may be given approximately by

$$T \simeq Gb^2 \tag{12.12}$$

and so, substituting for T in Equation 12.11, we have

$$\tau \simeq \frac{Gb}{r} \tag{12.13}$$

Equation 12.13 is important, since it shows that the shear stress to move a dislocation line is inversely proportional to the radius of curvature. We shall have occasion to refer to this result when discussing strengthening techniques in Chapter 16.

It might be wondered why, if dislocations are inherently unstable, they should occur at all in cast metals. There are three reasons for this. Crystals will not grow readily unless there are steps on the surface to which the atoms may attach themselves. An emergent screw dislocation (Fig. 12.11) is a way of ensuring the permanent presence of a stepped surface, and the observation of surface spirals on crystals has given visible confirmation of this growth mechanism. The second reason is that the side-arms of *the dendrites* that form in the liquid (see Section 5.8) are exceedingly fragile as first formed, being thin and composed of metal that is only just below its melting point and hence very soft. They are thus easily damaged, for example by convection currents and gas evolution in the surrounding liquid, and although in theory the arms should knit together exactly on completion of solidification (Fig. 12.12(a)), in practice the situation is more akin to that shown in Fig. 12.12(b), with the junctions between bent

354

dendrite arms becoming regions of high dislocation density. Finally, cavities may occur by the diffusion together of vacancies, and these can then collapse with the formation of edge dislocations, as outlined in Fig. 12.13.

(a)

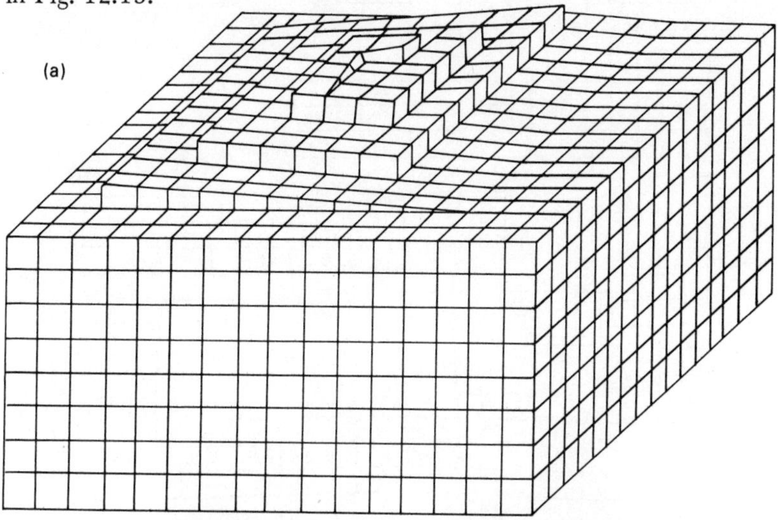

(b)

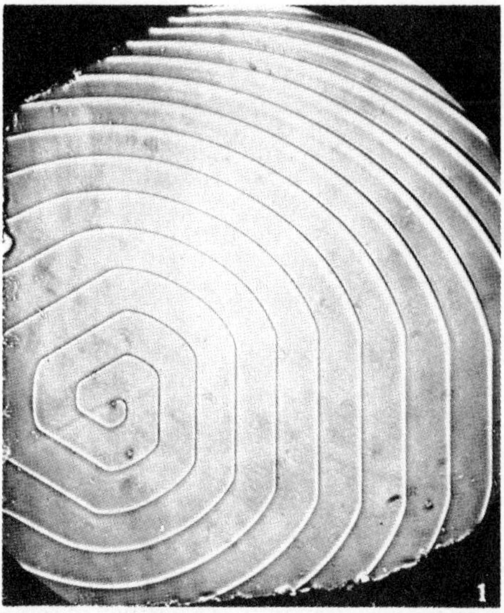

Figure 12.11 (a) Schematic diagram of crystal growth by an emergent screw dislocation. (b) Spiral growth on a silicon carbide crystal ($\times$ 60) (from Dr. A. R. Verma).

So far, we have considered only the movement of dislocations by simple slip (frequently referred to as *glide*). There are, however, several elaborations of the initial concept of mobility which must now be discussed. Although an edge dislocation can only move in its slip plane by simple slip, there is a way in which it can move normal

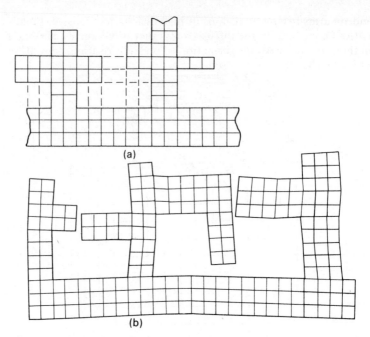

Figure 12.12 (a) Dendrite growth in ideal conditions. The arms 'fit' onto each other to give a perfect lattice. (b) Mismatch between distorted dendrite arms. Growth of this sort culminates in the mosaic structure shown in Fig. 5.29.

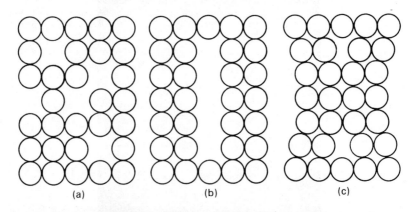

Figure 12.13 The formation of edge dislocations from vacancies. (a) Randomly scattered vacancies in crystal lattice. (b) Agglomeration of vacancies to give a small void. (c) Collapse of void resulting in two edge dislocations. (From K. J. Pascoe (1978) *An Introduction to the Properties of Engineering Materials, 3rd edition,* by permission of Van Nostrand Reinhold.)

to the slip plane — the process known as *climb*. Fig. 12.14 illustrates this movement, by the addition or removal of an atom from the edge of the extra plane which makes up the dislocation. Removal of atoms from the extra plane is known as *positive climb* and may take place by the atoms moving into nearby vacancies in the lattice or by

356

diffusing away to an interstitial position in the lattice. Conversely, an atom diffusing to and attaching itself to the extra plane results in *negative climb*. Climb by this method is therefore dependent on the diffusion of individual atoms or vacancies; at room temperatures, when diffusion rates are slow, it is relatively unimportant, but it can play a significant part in deformation by *creep* at high temperatures, when an edge dislocation which meets an obstacle in its path can avoid it by climb to the next parallel slip plane and thus continue its motion. This is one of the reasons for the softening of metals at high temperatures. Screw dislocations are able to move in a variety of slip planes and do not exhibit climb; they can change their slip plane by the process of *cross slip* (Fig. 12.15).

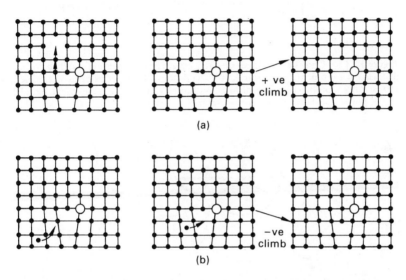

(a)

(b)

Figure 12.14 Climb of edge dislocation. (a) *Positive* climb, by the movement of an atom away from edge into a nearby vacancy. (b) *Negative* climb, by the migration of an interstitial atom onto the edge. Either mechanism enables the dislocation to get past the large blocking atom.

The second important point is that slip in the metal lattice does not necessarily involve the simple movement from position x to position y (Fig. 12.16). If two models of close-packed planes (e.g. from table-tennis balls) are slid over one another, it will be found that movement xy can be more easily accomplished by two separate steps xz and zy. These two steps may take place at separated intervals in the lattice, so that a sizeable block of atoms is situated in the 'half way' position z (Fig. 12.17). The two stages of slip xz and zy are each known as a *partial dislocation*, and the two taken together are referred to as an *extended dislocation*. It will be noted that the region between the two partials is out of phase vertically with the lattice below, breaking up the vertical sequence of atom positions, and is known in consequence as a *stacking fault*. Thus in an f.c.c. lattice

357

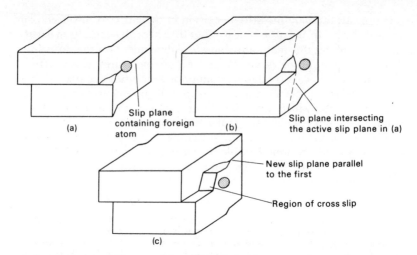

Figure 12.15 Cross slip avoid an obstacle by a screw dislocation. (a) The dislocation approaches an obstacle. (b) The dislocation transfers onto an inclined intersecting slip plane. (c) The dislocation resumes motion on a slip plane parallel to the initial plane and avoids the obstacle.

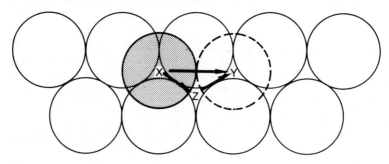

Figure 12.16 Movement of individual atoms during slip tends to occur in two stages XZ, ZY rather than the simple movement XY, XZ and ZY each constitute a *partial dislocation* movement.

the vertical sequence changes from abcabcabc to (say) abcabc* bcabcab, and in a hexagonal lattice the stacking would change from ababab to aba* cbcbc. (* indicates the position where the dislocation lies.) The extended dislocations can then move as ribbons through the lattice, intersecting or interacting with other dislocations in the same general way as a complete dislocation.

12.4 MECHANISM OF PLASTIC DEFORMATION

We must now examine the behaviour of dislocations when a metal is deformed plastically. Such behaviour must account for the various observed facts of deformation:

(a)   As deformation proceeds, the metal gets progressively harder. (This is shown by the fact that $d\sigma/d\epsilon$ is positive throughout the

358

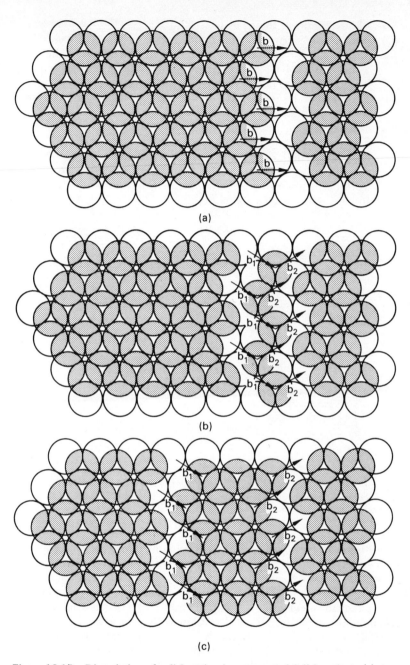

Figure 12.17 Dissociation of a dislocation into two partial dislocations. (a) An edge dislocation having close-packed slip plane. The extra half plane of atoms ends on the lower plane of atoms in the figure. (b) Slip can occur in two steps. The row of atoms shown has slipped by b_1, creating two partial dislocations. (c) The creation of a stacking fault between the two partial dislocations. (From M. M. Eisenstadt (1971) Introduction to Mechanical Properties of Materials, by permission of Macmillan Co.)

plastic region of the true stress—strain curve, Fig. 12.31).

(b) Microscopy shows that the slipping process occurs in bands, and is therefore confined to a very small proportion of the available slip planes. Schematically, it would be represented by Fig. 12.18(a) rather than by Fig. 12.18(b).

(c) X-ray and etch pit work has shown that the dislocation content of the metal rises steeply from a concentration of 10^4–10^6 dislocations/cm^2 for an annealed metal to about 10^{12} dislocations/cm^2 in the several deformed state. Since we have shown that the strength of a perfect crystal is very much higher than one containing dislocations (Section 2.6.1), at first sight the work hardening effect should indicate a *decrease* rather than an increase in dislocation content.

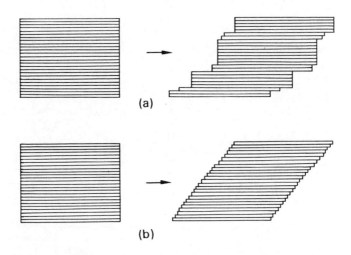

(a)

(b)

Figure 12.18 (a) Slip concentrated into bands of a few highly active planes. (b) Theoretical homogeneous slip. Deformation of this nature would (if it occurred) be more akin to twinning (Fig. 5.38) than to normal slip, although it would not be restricted by angular relationships with the undeformed lattice.

All these facts can be satisfactorily explained in terms of dislocation behaviour. The dislocations present in an annealed metal prior to plastic deformation are distributed over all the possible slip planes, forming a three-dimensional network, looking remarkably like chicken wire under the electron microscope (Fig. 12.19). Distribution may not be uniform because the dislocations will occur in high concentrations at low angle boundaries (e.g. dendrite joints) within the crystal, and will also attach themselves to any foreign atoms present. Whatever the precise distribution may be, those dislocations that move under a shear stress and cause slip must thread their way through stationary dislocations lying across their path in intersecting inactive planes. The most important result of these intersections is the formation of steps or *jogs* in the dislocation lines. Intersections between two edge dislocations are shown in Fig. 12.20. It should be

360

Figure 12.19 Dislocation network is a lightly deformed copper—aluminium alloy. Many of the loops approximate to an hexagonal habit (x 32 000). (From J. Kastenbach and E. J. Jenkins.)

noted that the intersection in Fig. 12.20(a) results in a jog lying in the plane of the dislocation, whereas in Fig. 12.20(b) the jog is formed *normal* to the slip plane. In both these cases, the jog is formed in the stationary dislocation and is of length b_1, where b_1 is the Burgers vector of the mobile dislocation, whilst the moving dislocation is unaltered in form. Consequently the results of the intersection are only apparent when the stationary dislocation begins to move through the crystal. At later stages of plastic deformation, when more than one set of slip planes become active, both intersecting dislocations become active, and in consequence jogs are produced in *both* dislocations. The jog shown in Fig. 12.20(a) is unstable, since movement of the portion AB to A'B', which merely involves the natural slip of the dislocation, will eliminate it, thereby shortening the overall length of the dislocation loop and lowering the strain energy of the crystal. The jog of Fig. 12.20(b) is not unstable (although it could be eliminated by a reversal of slip in the vertical plane), but since the jog has an edge orientation with its Burgers vector lying in the slip plane, it can move readily along with the rest of the dislocation. Hence the whole stepped line ABCD can move through the crystal, and the jog does not impede the mobility of the dislocation. We may say, therefore, that intersection of edge dislocations has little effect on plastic deformation, apart from the work required to form the jog, i.e. to lengthen the dislocation line.

361

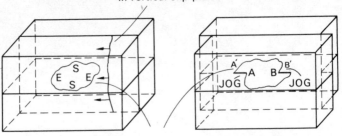

Moving edge dislocation
in vertical slip plane

Inactive dislocation loop in horizontal slip plane

(a)

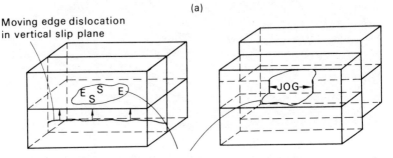

Moving edge dislocation
in vertical slip plane

Inactive dislocation loop in horizontal slip plane

(b)

Figure 12.20 Intersection of a stationary dislocation loop by a mobile edge dislocation, (a) to produce a jog in the plane of the dislocation loop, (b) to produce a jog at right angles to the dislocation loop.

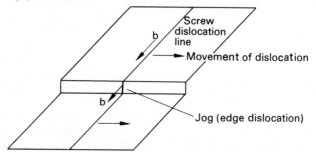

Figure 12.21 Edge-type job in a screw dislocation. The Burgers vector of the jog is wrongly oriented for movement of the jog with the dislocation line, and can therefore only move by the creation either of a row of interstitial atoms or of a row of vacancies.

If, however, two screw dislocations intersect (Fig. 12.21), the result is that one (or both) acquire a jog which has an edge orientation. Under these conditions the Burgers vector of the jog does not lie in the plane of the step (as it did in Fig. 12.20(a)) but at right angles to it. This means that the only way in which the jog can move in the direction of the rest of the dislocation is by the

362

laborious process of climb, which is dependent on the diffusion of either atoms or vacancies. There are two important consequences of this interaction. Firstly, movement of the dislocation is retarded very considerably by the drag of the jog, thereby making slip more difficult, and secondly, if the shear stress is sufficient to force the jog to move, the result will be the creation of a row of either vacancies or interstitial atoms. Movements of this nature are probably largely responsible for the heavy increase in vacancy concentration during plastic deformation. It has been estimated that 10% plastic strain can create 10^{18} vacancies/cm^3.

Dislocations may also react with one another to produce a new dislocation. When two dislocations meet on a slip plane, the Burgers vector of their product is equal to the vector sum of their original Burgers vectors. They will only tend to combine (i.e. will attract each other) if the total strain energy is reduced thereby; if the energy of their product is greater than the sum of their separate energies, they will repel each other. Thus, dealing with the simplest cases, two parallel dislocations of opposite sign on the same slip plane may meet with mutual annihilation, where dislocations of like sign on the same plane exert a mutual repulsion, making it difficult to pack them close together. An obstacle in a slip plane to the passage of a dislocation therefore serves to slow up following dislocations as well, until a complete jam occurs or until the forward pressure of the following dislocations is sufficient to force the leader past the obstacle. These jams, depicted in Fig. 12.22, are usually referred to as

Figure 12.22   Dislocation pile-up at an obstacle in a slip plane.

dislocation *pile-up*. If the load is removed after pile-up has occurred, there will be little recovery of the metal because we can assume that there are obstacles to the passage of the dislocation in either direction. However, when reverse loading takes place (i.e. causing slip to the left in Fig. 12.22) any obstacles to movement must be less than that which caused the pile up (since the dislocations have already passed them, and the back pressure from the pile-up acts to favour slip, with the result that yielding and flow take place at lower stress in the reverse than the forward direction). This is known as the *Bauschinger effect*.

Partial dislocations moving through the lattice may also interact if they meet on intersecting planes, and it has been shown that under certain conditions the resultant partial dislocation has a Burgers vector which does not lie in its slip plane and is therefore unable to move. Slip is thus blocked on the two slip planes, since the remaining partials cannot move forward across the intersection of the plane. This form of block is known as a Cottrell–Lomer barrier, and makes an appreciable contribution towards work hardening. A further form of sessile dislocation may occur by the migration of vacancies into a group. If sufficient collect on a plane (Fig. 12.23(a)), the result is a cavity on the atomic scale, and collapse may occur as in Fig. 12.23(b) so that the cavity now becomes a stacking fault. However, the resultant dislocation loop has its Burgers vector normal to the plane of the fault, and is therefore unable to glide. This form of block becomes increasingly probable as the number of vacancies builds up from the forced movement of jogs during deformation.

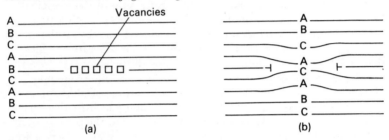

(a) (b)

Figure 12.23   (a) Vacancy cluster in a crystal lattice. (b) Collapse of the cluster to form a dislocation loop enclosing an area which is a stacking fault (reproduced by permission, from *The Structure and Properties of Materials*, by H. W. Hayden, W. G. Moffatt and J. Wulff (Vol. III). Copyright © 1965 by John Wiley & Sons Inc.)

From the above brief discussion, it will be seen that interaction of dislocations tends to block their movement, as an increasing number of sessile interactions occur which 'de-activate' slip planes. However, these effects by themselves can only partially account for the observed work hardening of metals; a large contribution is the result of the great increase in dislocation concentration which takes place. The result is that moving dislocations have an ever-increasing number of intersecting dislocations to thread through, requiring additional work to form the jogs, and greatly increasing the likelihood of sessile intersections. Dislocations can breed in a variety of ways, the most important of which is known as the 'Frank Read' source, which has actually been observed in crystals under the microscope. The mechanism, depicted in Fig. 12.24, is as follows. The dislocation line AB lies (say) between two anchoring jogs, CA and BD; although the portion AB can move readily under the action of the stress, CA and BD cannot, so the line AB is anchored at its ends. Action of a shear stress now causes the dislocation line to bulge, and the sequence of events is shown in Fig. 12.24(b–f), the shaded area representing the slipped region. The important feature of this arrangement is that after completion of the cycle which creates a complete free dislocation loop the dislocation line AB is still present and the process can be repeated an indefinite number of times. This affords an explanation of the restriction of slip to certain preferred planes (i.e. those

364

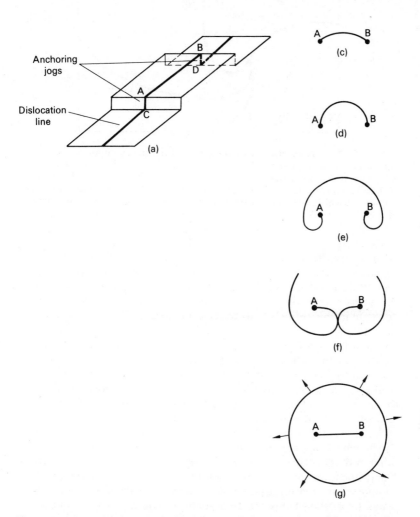

Figure 12.24   Operation of a Frank–Read dislocation source. (a) The conditions necessary to set up a source. (b)–(g) Dislocation movement under the action of a shear stress, leading to the formation of a complete loop, with the retention of the original line AB. When stage (f) is reached, the line can shorten itself by changing to the configuration in (g); hence this step occurs spontaneously.

which contain suitable dislocation anchors), of the fact that large amounts of slip occur on these planes, and also of the rapid increase of dislocation density. A source of this nature will operate until obstacles on the slip plane cause a sufficient pile-up of dislocations for the back pressure to prevent any further generation, or until the dislocation line between the jogs is threaded by other dislocations.

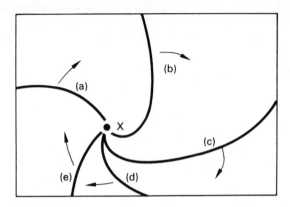

Figure 12.25 Slip generated by a dislocation with one end pinned by an anchoring jog (seen end-on at X in the diagram) and the other end free to move at the surface of the crystal. (a)—(e) represent successive positions of the line; each full revolution results in one increment of slip. By this mechanism a single dislocation whirling round repeatedly on its jog axis can generate a large amount of slip.

Resistance to the operation of the Frank—Read source is also afforded by the necessary bending of the dislocation line (see Equation 12.13), and reaches a maximum at the stage depicted in Fig. 12.25(d); thereafter the increase in the radius of curvatuve of the major part of the loop more than offsets energetically the further decrease in the radii close to the pinning jogs. It follows that the further apart the two jogs, the more readily will the source operate; on the other hand the further apart they are, the greater is the probability that the dislocation line between them will be disrupted by further threading as deformation proceeds. Thus Frank—Read sources tend to operate on smaller and smaller lengths of dislocation line in the later stages of plastic deformation, and this also contributes in some degree to the overall work-hardening effect.

An alternative source of slip is illustrated in Fig. 12.26, in which a jog anchors one end only of the dislocation in each slip plane, the other end being free to move and reaching the surface of the crystal. The dislocation can then rotate about its fixed and as shown, each complete rotation resulting in an increment of slip over the entire plane. Cross slip of this nature probably accounts for the fact that the active slip planes in metals tend to occur in relatively thick 'packets', which are observed under the microscope as slip bands.

Thus far we have considered interactions between dislocations, and obstructions to movement generated by the dislocations themselves. However, all metals will contain some amount of foreign inclusions (such as oxide or sulphide particles) and impurity atoms. These place additional obstacles to the free movement of dislocations, and may also act as sources of dislocations, especially in the presence of high localised stresses, which may arise after heat treatment from differential contractions. At high temperatures the dislocations may avoid these by the processes of climb or cross-slip, but at ordinary

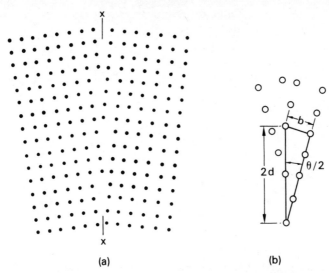

Figure 12.26 (a) Dislocation model of a small-angle grain boundary. (b) The geometrical relationship between θ, the angle of tilt, and d, the spacing between the dislocations. (From R. E. Reed-Hill (1973) *Physical Metallurgy Principles*, *2nd edition*, by permission of D. Van Nostrand Co.)

temperatures this is not possible, and such obstacles can therefore act as powerful strengthening agents. These are considered in more detail in Chapter 16.

Finally, we must consider the role played by grain boundaries in dislocation movements and plastic deformation. Grain boundaries vary in nature depending on the relative orientation of the grains they separate; low angle grain boundaries are extremely thin and may be considered as arrays of dislocations (Fig. 12.26), whilst high-angle boundaries are thicker and contain a narrow but definite disordered region of atoms (Fig. 5.28). Most grains contain networks of small-angle boundaries and these do not appreciably impede slip; on the other hand high-angle boundaries (which normally constitute the visible grain boundaries in metals) are a barrier to slip processes, and at ordinary temperatures are the strongest part of a polycrystalline structure. They may act as dislocation 'sinks' into which dislocations moving through a grain can vanish, and may also generate dislocations if the 'pressure' from a deforming grain is sufficient. This last point is important: large scale plastic deformation must involve cooperative deformation of the grains on either side of the grain boundary, and therefore general yielding (as opposed to the 'early' yielding of individual grains of favourable orientation) will only occur when deformation of one grain can be propagated across a grain boundary to generate appropriate deformation in a neighbouring grain.

Let us suppose that the grain in Fig. 12.27 has yielded on the slip plane shown but that the neighbouring grains are unfavourably orientated to accommodate the deformation. In these circumstances there will be a pile-up of dislocations at the grain boundaries, and the slip

367

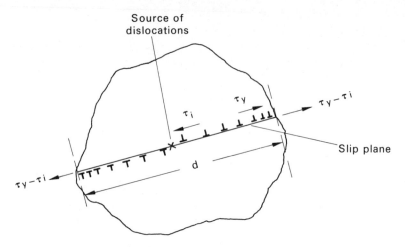

Figure 12.27 Yielding within one crystal in a polycrystalline metal. Yielding will become general when the stress across the grain boundary, $\tau_y - \tau_i$, is sufficient to trigger off dislocation sources in the adjacent grains.

plane may be thought of as acting like a very narrow crack, giving rise to a stress concentration across the boundaries into the neighbouring grains. If the stress to cause further yielding within the grain is τ_y, the actual stress at the head of the pile-up will be $\tau_y - \tau_i$, where τ_i is the sum of all the obstacles to further slip and may be regarded as a 'frictional' shear stress. The acting stress $\tau_y - \tau_i$ will then generate slip across the boundary when

$$\tau_y - \tau_i = K \cdot d^{-n} \qquad (12.14)$$

giving the *Hall-Petch relationship:*

$$\tau_y = \tau_i + K \cdot d^{-n}$$

where K is a constant depending on the *way* in which yielding occurs in the neighbouring grains (i.e. by generation at the grain boundary or by releasing dislocations from locked positions within the grain, whichever requires the less energy), and d is the grain diameter. This relationship between grain size and the yield strength of metals has been verified experimentally; $n = \frac{1}{2}$ for b.c.c. metals such iron but varies somewhat from this value for metals with other crystal structures.

From the above discussion it will be appreciated that work-hardening or *strain-hardening* is not a simple phenomenon but a highly complex and cumulative result of several different processes. As the dislocation density increases through the action of Frank–Read sources in the slip planes, so the number of threading operations increases that a mobile dislocation must perform in order to continue its motion. All these intersections require additional work and some of them will result in locks or the formation of more Frank–Read sources; thus further obstruction to movement builds up. Finally the stress to cause additional deformation reaches a level that compels dislocations of like sign to coalesce in pile-ups with the formation of

368

micro-cracks, and we have reached the first stage of *fracture*. This will be considered as a separate process in Chapter 15.

Many attempts have been made to develop a mathematical relationship between degree of strain and the hardness of metals, which really amounts to fitting an equation to the shape of the true stress—strain curve (see Section 12.6), but so far without any quantitative success. In view of the complexity of deformation processes, this is not very surprising. On the other hand, a reasonably satisfactory relationship has been derived between the resolved shear stress τ and the dislocation density ρ:

$$\tau = \tau_0 + G \cdot b \cdot C \cdot \rho^{\frac{1}{2}} \qquad (12.16)$$

where G is the shear modulus, b the Burgers vector, C a constant which varies in value from 0.3 to 0.6, and τ_0 the critical stress to move a single dislocation through a perfect lattice of the metal. From this expression it may be inferred that there is a relationship between ρ and the hardness of metals, but no fully satisfactory quantitative relationship has yet been established.

12.5   ANNEALING

There comes a point during the cold plastic working of metals when the beneficial effects of work-hardening are more than offset by the danger of embrittlement. Resistance to brittle fracture (Section 15.2) depends in part on the ability of slip processes to blunt a crack tip; if dislocations are sufficiently tangled and pinned, blunting cannot occur so readily and further deformation (or overloading during service) may result in failure. It may then be necessary to *anneal* the metal, either to increase its toughness or to permit further deformation to be carried out safely.

Annealing is a complex process made up of three components which may occur in sequence or simultaneously: *recovery*, in which the dislocations re-arrange themselves into patterns of minimum free energy; *recrystallisation*, in which near strain-free grains nucleate and absorb the original structure (see also Section 5.10); and *grain growth*, in which the larger grains grow at the expense of the smaller. All stages require some degree of thermal activation to proceed, which varies for each metal or alloy. The critical temperature is known as the recrystallisation temperature, and also varies somewhat with the degree of work hardening; large deformations give rise to greater internal disorder and a higher internal energy; hence annealing processes commence at a lower temperature.

Recovery occurs when the dislocations in the work-hardened metal, which are distributed through the structure in a randomly interlocking network of 'jams' and pile-ups, acquire sufficient activation energy to climb out of their slip planes (if necessary) and re-arrange themselves in a more ordered array. This process results in the formation of subgrains within the original grains (see Fig. 12.28) that are separated by low angle or *kink* boundaries, and is known as

369

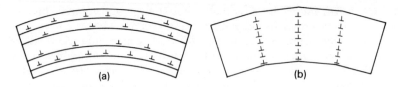

Figure 12.28   The realignment of edge dislocations during polygonisation.
(a) The excess dislocations that remain on active slip planes after a crystal is
bent. (b) The ordering of dislocations into vertical lines resulting in the formation
of subgrains within the original structure. (From R. E. Reed-Hill (1973)
Physical Metallurgy Principles, 2nd edition, by permission of D. Van Nostrand
Co.)

polygonisation. The overall dislocation density for metals that deform
on multiple slip-systems (e.g. f.c.c. and b.c.c. structures) does not
decrease appreciably at this stage, although a few dislocations of opposite sign may meet and annihilate each other, whilst others may
run into neighbouring grain boundaries and be obsorbed. Zinc and
cadmium crystals, which deform on single slip systems, can however
completely return to their undeformed hardness during recovery,
since there will be an absence of locking intersections and jogs between
dislocation lines. Recovery of the first sort does not involve any
appreciable loss in strength, but the dislocation movements do allow
some accommodation of residual elastic stresses; it is therefore used
industrially as a stress-relief anneal after the final stage of plastic
working.

The major changes during annealing occur in the recrystallisation
stage, which is a nucleation-and-growth process involving a change
from a strained to an unstrained lattice. The driving force for recrystallisation is the lower free energy of a strain-free lattice; the
difference amounts to roughly 5% of the energy used in cold working.
The grain size of the product depends on the degree of prior deformation; small amounts (circa 2%) produce relatively few nuclei for
grain growth and the result (usually undesirable) is a very large grain
size. This is one technique used for making metal single crystals. Larger deformations produce more nuclei, and so the recrystallised grain
size diminishes. Recrystallisation temperatures are raised by solute
atoms in the lattice; such atoms tend to stabilise grain boundaries
by collecting in them in rather the same way that they stabilise dislocations. Thus 1% silver is sometimes added to copper to prevent
it being softened by annealing during soft-soldering.

Recrystallisation also results in the growth of twins — the broad
prominent *annealing twins* which can be seen in copper and α-brass.
they are formed by a mechanism totally different from that of the
deformation twins discussed in Section 5.12. If we consider a strain-
free recrystallised grain growing out into a work-hardened structure,
it will form boundaries with the surrounding un-annealed grains.
These boundaries vary in energy depending on the angular relation
between the new and old grains, and there will always be some cases
where a high grain boundary energy can be reduced if the active
growing grain rotates to form a twin over that particular section of

370

boundary. Whether or not a twin will form there depends on whether the energy required to form a twin is more than offset by the lowered boundary energy, so that the energy of the system as a whole is lowered by twinning. This form of twinning is therefore only possible in metals with a low stacking fault energy (twins may be considered as stacking faults) and is related in consequence to the ease with which metals form partial dislocations. The number of twins will depend on the number of suitably orientated grains which the growing grain meets, and therefore increases as the grain size of the worked sample decreases. Metals with an f.c.c. structure and, to a lesser extent, the hexagonal structures exhibit this form of twinning; it is particularly common in copper, α-phase copper alloys and austenitic steels (see Fig. 5.39). It is not observed in b.c.c. metals such as iron.

Yet another result of recrystallisation is the formation of prominent *grain textures* in cold-rolled sheet metal, in which the grains have their crystallographic axes aligned in roughly the same direction. Behaviour of this sort is important in that it results in strongly anisotropic mechanical properties which can lead to problems of unhomogeneous flow when manufacturing components from sheet metals.

If a metal is held above its recrystallisation temperature for a sufficient time, a further structural change may take place known as *grain growth*, in which the driving force is a reduction in free energy by means of a decrease in the total grain boundary area. This is brought about by the action of the surface tension of the grain boundaries which, as shown in Section 2.6.4, possess a higher free energy than the rest of the metal. The effect is not peculiar to metals; it may be seen in an assembly of soap bubbles under low pressure, or perhaps more frequently in the bubble system left behind when a beer bottle is emptied. Grain boundary surface tension plays little part in controlling the initial shapes of grains, unless the cooling rate is particularly slow, since grain shape is determined by random obstruction from neighbouring crystals also growing from the melt (Fig. 5.27). The effect of surface tension on the grains is twofold: it always tends to straighten grain boundaries (to keep them to the minimum length), and it adjusts the angles at the corners of grains. Since the surface tension varies little from boundary to boundary (in a pure metal, at least), there is a tendency for the corner angles where three grains join to adjust themselves to 120° (by a consideration of the triangle of forces, see Fig. 12.29); this confers stability on six-sided grains — the only form that can have straight sides *and* 120° corner angles. For grains with more than six sides, the average internal angle must be greater than 120°, and therefore the average angles of neighbouring grains at the point of contact must be less. Hence, in order to equilibrate the three tensions and achieve stability, the grain boundaries move as shown in Fig. 12.30. Conversely it will be appreciated that grains with less than five sides will have average internal angles less than 120°, and will therefore shrink, frequently to the point of extinction.

Grain growth in general is undesirable since it results in inferior mechanical strength (see Equation 12.15) and may constitute a considerable problem in the heat treatment or hot plastic working

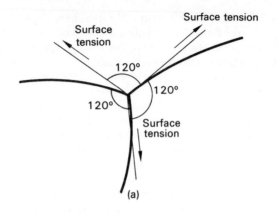

Figure 12.29 (a) Triangle of forces at the junction of three grain boundaries. (b) Grain boundary junction in equilibrium state in β-brass ($\times$ 500). The vertex is not quite normal to the page and thus angles are not quite equal to 120°. (From A. R. Bailey (1973) *The Role of Microstructure in Metals, 2nd edition*, by permission of Metallurgical Services Laboratories, Ltd.)

of thick sections in which the centre regions may remain at high temperatures far longer than the surface.

12.6 BEHAVIOUR OF METALS UNDER LOAD

Having considered the various mechanisms by which metals may deform plastically, it is now convenient to discuss briefly what happens when a specimen is loaded to failure in tension, and the information that may be obtained from such a test. Mechanical tests may be performed for a variety of loadings (e.g. tension, compression, torsion) but each test is usually restricted to a simple form of stressing and therefore cannot simulate service conditions exactly (which often

372

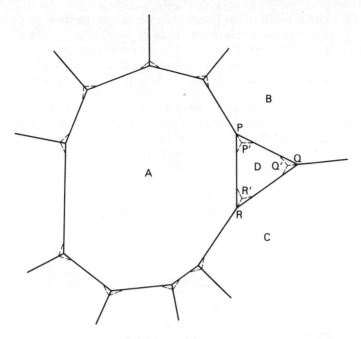

Figure 12.30 Illustrating the principle of grain growth and grain shrinkage in metals. Grains A, B, and C, having internal angles $> 120°$ will encroach on the small grain D in order to achieve equilibrium where three grain boundaries. This causes P, Q, R to move P', Q' and R' (new grain boundaries shown dotted). Further encroachment on D then occurs as $P'Q'$, $Q'R'$, and $R'P'$ straighten themselves under the action of their surface tension. But this movement then throws the junctions out of equilibrium, so that further movement of P', Q', R' takes place, and so on. Similar movements will occur at the other corners of grain A.

involve extremely complex combinations and fluctuations of stresses). If we consider the tensile test, which is probably the best known and most useful of the mechanical tests, the results may be plotted in three different ways: as a *load-extension curve*; as a *conventional stress—strain curve*; or as a *true stress—strain curve*.

The first of these is convenient for testing a finished component selected at random from a batch; it has the advantages of presenting a picture of the way in which that component will behave in practice, and of being quick and easy to plot, since no conversion of the direct measurements is required. On the other hand, such a test tells us little about the metal as such; it can only predict the behaviour of the metal when it is in the particular shape that has been tested. For design considerations we need to be able to predict behaviour for any shape of component, and therefore a more fundamental plot of the results is required, which takes into account the dimensions of the test piece. Such a plot is the *conventional stress—strain curve*, in which load and extension are converted into stress (load/cross-sectional area) and strain (increase in length per unit length) respectively. However, this test, although providing valuable design information, is not scientifically accurate since stress is measured throughout the test in terms of the *initial cross section* of the specimen, and strain is measured in terms of the *initial* length. The curve (see Fig. 12.31) is thus the same shape as the load-extension curve, since both load and extension are divided by constant denominators. The principal advantage of this is ease of plotting results; moreover, the errors introduced into the elastic and initial plastic deformation ranges are small enough to be negligible. In the later stages of the curve, however, the conventional stress becomes wildly inaccurate as the difference between the current and the initial cross section increases. In particular, the curve suggests that the stress reaches a maximum (the Ultimate Tensile Strength; P in Fig. 12.31) and

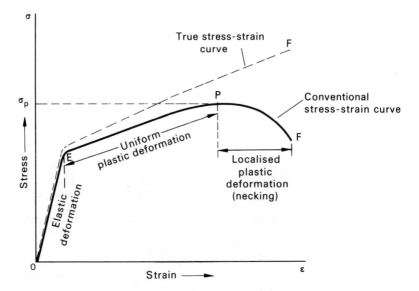

Figure 12.31   True and conventional stress—strain curves for a metal that has been worked: E is the elastic limit; F is the point of fracture.

then decreases until fracture occurs. Scientifically speaking, this is ridiculous; the stress at the narrowest point on the specimen goes on increasing throughout — it is only the *load* on the specimen which decreases beyond the point P. Oddly enough, it is this very inaccuracy which constitutes another advantage of the conventional curve, since the maximum on the curve represents the maximum conventional stress to which the metal can be subjected without fracture. Up to this point, work hardening will strengthen the metal adequately to support the load; under a greater stress, work hardening is no longer adequate and the specimen will continue to deform until fracture occurs.

The third method of plotting results is the *true stress—strain curve*, also shown in Fig. 12.31. Here both stress and strain are measured in terms of minimum cross section and length *at the time of measurement*. Such a curve is scientifically accurate (within the limits of the apparatus used) but is relatively lengthy to construct, and the slight

374

increase in accuracy is not sufficient to make it worth plotting except for fundamental research on the strength of metals. If the initial and instantaneous cross-sections of the specimen are A_0 and A, the relationship between the conventional stress, σ_c, and the true stress, σ_t, is given by

$$\sigma_t = \sigma_c \frac{A_0}{A} \qquad (12.17)$$

Since $A_0 > A$ for all states of tension, σ_t is always greater than σ_c and therefore the inaccuracy in σ_c always tends to safety in design.

For strain, the conventional value ϵ_c is given by $(l - l_0)/l_0$, where l_0 is the initial and l the final length of the specimen. The true value of strain is the sum of all the increments which go to make up the total strain, expressed as a fraction of the length at the time of their measurement. Thus

$$\epsilon_t = \frac{\delta l}{l_0} + \frac{\delta l}{l_1} + \frac{\delta l}{l_2} + \frac{\delta l}{l_3} + \ldots + \frac{\delta l}{l}$$

$$= \int_{l_0}^{l} \frac{\delta l}{l}$$

$$= \log_e \frac{l}{l_0} \qquad (12.18)$$

To obtain the relation between ϵ_c and ϵ_t, write $\epsilon = (l/l_0) - 1$, from which we obtain

$$\epsilon_t = \log_e (\epsilon_c + 1) \qquad (12.19)$$

It is necessary for the engineer to know the difference between the true and conventional values of stress and strain, but beyond that we need not consider further the true stress—strain curve.

Let us now consider the conventional tensile test in rather more detail, so that we can see what happens when a metal is tested to fracture in simple tension. The test is normally carried out on a specimen of the form shown in Fig. 12.32. All measurements are

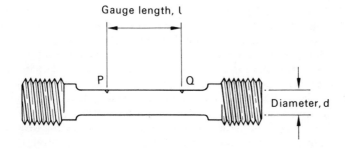

Figure 12.32 Tensile test specimen. In order to obtain comparable results on specimens of different size, l and d are related by the formula $l = 5d$.

made on the centre portion of the specimen between the points P and Q; the distance between these points (which are lightly punched on the specimen) is known as the *gauge length*. Two stress–strain curves for polycrystalline specimens are shown in Figs. 12.31 and 12.33. (Henceforth in this chapter, unless otherwise stated, the terms 'stress' and 'strain' refer to conventional rather than true values). Fig. 12.31 depicts the curve for worked metals and alloys, whilst Fig. 12.33 is the type of curve resulting from annealed non-ferrous metals.

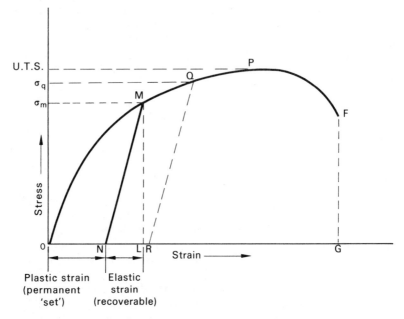

Figure 12.33   Conventional stress–strain curve for an annealed metal.

In Fig. 12.31, a pure elastic region (OE) is apparent, in which the strain is totally made up by the stretching of interatomic bonds, and OE is taken to be linear. However, it is very doubtful if a polycrystalline metal ever behaves in a purely elastic fashion; the more sensitive the measuring device, the lower on the curve is the apparent elastic limit, E. For many alloys in the annealed state (except steel) there is no elastic range at all and therefore not even an approximate yield point; the initial portion of the curve is of the form shown in Fig. 12.33. Over the early portion of the curve the deformation mechanism is as follows. As the stress rises it produces a proportional elastic strain (Equation 12.1) which is recoverable. In addition, dislocations which are already present within the metal are progressively pulled free of any anchoring impurity atoms or intersecting dislocations and move through the metal, permitting plastic deformation (slip) to commence. In high-purity annealed metals, anchoring impurity atoms and intersecting dislocations are scarce, and so plastic deformation commences readily, giving the curve of Fig. 12.33. The

376

presence of the alloying element affords copious anchors, so that appreciable plastic deformation does not occur until the stress has risen sufficiently to drag some of the dislocations free; the curve in this case is akin to Fig. 12.31. Since the dislocations are randomly distributed through the metal initially, they do not all start to give simultaneously. At first only those which are on suitably orientated slip planes and which have the least anchoring on those planes become operative. As these tangle and jam, the stress rises, and other dislocations start to move, and so the commencement of plastic deformation is normally a gradual process with no sharply defined yield point.

This state of affairs is structurally inconvenient since it means that, even at quite low loads, many alloys acquire a permanent deformation. However, if the stress is raised to σ_m (Fig. 12.33) and the specimen unloaded, the loading curve is *not* followed back to the origin since only the elastic component NL of the total strain OL is recoverable. At the point M the dislocations are tangled to a certain degree (i.e. the metal has work hardened), and on removal of the stress they will remain in that state, and will not move significantly until some further application of stress exceeds σ_m. Thus, if metal is stressed to M, released, and then reloaded, the second stress—strain curve will follow NM rather than OM. In respect of Fig. 12.33 this is particularly important, since we now have a metal with a definite elastic range and yield point, and therefore suitable for structural engineering. By analogy the metal could be further stressed to Q and released; it will be seen from this that *work hardening increases the elastic range*. This is also valuable, but obviously P represents an absolute limit to this process, and in fact it would be unsafe for the normal working load to approach as close to P as Q as in Fig. 12.33; should a chance stress exceeding Q be applied, the energy necessary for the metal to deform to the fracture point F would be relatively slight; i.e. the safety factor would be insufficient. Another way of looking at this is to consider the *area under the curve*, which represents the work required to cause fracture; it is thus a measure of the *toughness* of the metal. For a metal stressed to the point M and then released, the work to cause fracture is represented by the area MFGN. By further deforming it to the point Q we have admittedly raised the elastic range by a small amount, but the work to fracture has been reduced considerably to QFGR, and for most purposes this would be deleterious. It should be noted that stressing of this nature (to create or increase the elastic range) need not be in tension; it is normally an integral part of the fabrication process for a component, usually under compressive loading, and is carefully adjusted to give the optimum compromise between elastic range and toughness.

The Yield Strength. Despite the fact that we have just emphasised the absence of a precise yield point, in many engineering alloys (particularly work-hardened ones) the change of slope at about this point is sufficiently abrupt for a yield strength to be quoted. Where there is only a gradual change of slope (e.g. Fig. 12.33), an alternative value, the *proof stress*, is quoted. This is the stress required to produce a known small permanent strain — usually 0.1% or 0.2%. Thus in Fig. 12.33, if ON represents a strain of 0.1%, σ_m is the *0.1% proof stress.*

The yield strength is the most important criterion for engineering design, since it marks the normal design limit of useful engineering strength. Despite this fact, the *ultimate tensile strength* is still the most commonly referred to strength property of a metal; when we speak of a '100 ton steel' it is the ultimate tensile strength, σ_p in Fig. 12.31, to which reference is made. Gradual yielding of the sort already discussed is the commonest form of yielding, and occurs when the stress required to *start* the dislocations moving is less than that required to keep them moving through the dislocation network. However, the situation can arise in which the initial yield stress is greater than that required to keep them moving, and the metal then exhibits the phenomenon of the *double yield point.*

This state of affairs arises in mild steels, and its appearance on the stress–strain curve is shown in Fig. 12.34. Although it appears a

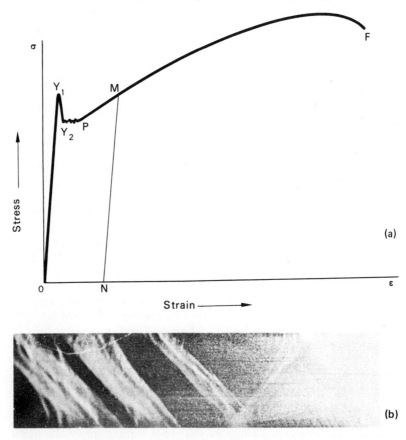

Figure 12.34 (a) Stress–strain curve for an annealed mild steel. Luders bands or 'stretcher strains' form over the strain range $Y_1 P$. (b) Luders bands on a tensile specimen of mild steel (from C. W. Richards (1961) *Engineering Materials Science,* by permission of Chapman and Hall Ltd.)

378

trivial modification to the normal curve, it has serious engineering consequences. Its cause has already been mentioned briefly in Section 12.4. The interstitial atoms carbon and nitrogen in steels exercise a strong mutual attraction with the dislocations present, and thereby act as anchoring points. At ordinary temperatures, when diffusion of these atoms is slow, the dislocations can only move by breaking away from these atoms, which requires a considerably higher stress than that required to keep them moving (initially, at least) once they have broken free. There are two immediate consequences of this mechanism. Firstly steels, even in the annealed state, have a much better approximation to a true elastic range than most alloys (Fig. 12.34, OY_1). Secondly yielding, always a gradual process, becomes spread over a very considerable strain increment. Yielding first occurs, as always, on those planes most suitably orientated for slip, i.e. at 45° to the stress axis. On polished specimens this may be visible as a band or bands of plastic deformation running at 45° across it; such bonds are known as *Luders Bands* (Fig. 12.34(b)). In this instance, however, the process of yielding has lowered yield stress to Y_2, and therefore a considerable degree of plastic deformation may occur in the first formed bands before work hardening raises the stress sufficiently to wrench dislocations free from their carbon atom anchors on other slip systems. Yielding thus occurs progressively in a series of little 'jerks' as the dislocations in other regions are pulled free, giving a serrated appearance to the curve in this region if a sufficiently sensitive instrument is used (Y_2 to P). Once all the dislocations have been freed, plastic deformation proceeds in the normal way until fracture.

The most serious practical consequence of this occurs when mild steel sheets are pressed to form components, e.g. for car bodies. Pressing to shape involves plastic deformation but, since the radius of curvature varies over the finished component, the metal goes a different distance along the stress—strain curve at different regions of the metal. In regions where bending is slight, the stress—strain state of the metal may lie between Y_1 and P, in which case Luders bands will appear on the sheet, and all the deformation necessary to form the curvature will occur within the bands. When formed in this way, they are usually referred to as *stretcher strains*. They are not mechanically detrimental, but they are unsightly since the band region will not show smooth curvature, and as they represent surface irregularities, they can only be removed by costly repolishing of the sheet. Stretcher strains may be avoided by uniformly prestressing the sheet into the ordinary plastic region; this has the effect of wrenching all the dislocations away from their interstitial atoms, and, providing that pressing is carried out reasonably soon after prestressing, the double yield point is avoided. The stress—strain curve then follows a reloading pattern NMF analogous to the reloading curve discussed previously. However, if the prestressed sheet is stored for a long time or annealed before pressing, the interstitial atoms migrate to the dislocations and anchor them again, thereby causing a return of the double yield point and the likelihood of stretcher strains during fabrication. This effect is known as *strain ageing*, and is principally restricted to b.c.c. metals and alloys; nitrogen is another interstitial atom which in addition to carbon can be responsible for the anchoring effect.

The plastic region. The plastic region of the stress—strain curve falls naturally into two sections, EP and PF in Fig. 12.31. EP represents the range of uniform plastic deformation over the gauge length; PF represents the range of nonuniform or localised plastic deformation (necking), and leads directly to fracture at the point F. From E to P the strengthening effect of work hardening is sufficient to counteract weakening by decrease in cross-section of the specimen. Deformation is considered as stable in this region, since any element of the specimen which deforms more than the rest is also strengthened more than the rest against further deformation. That element will therefore not deform further until the rest of the specimen deforms by a similar amount, and then only if the stress is raised. Thus the gradient of EP gives a measure of the ability of the metal to work harden; the steeper the gradient, the greater the rate of work hardening. For an accurate measure of work hardening, the true stress is needed, and therefore the true stress—strain curve should be used; the rate of hardening is always greater than suggested by the conventional curve. In fact, the plastic region of the true stress—strain curve can be plotted to a reasonable approximately by the algebraic relation

$$\sigma_t = K\epsilon_t^n \qquad (12.20)$$

where n is the *strain-hardening coefficient* (roughly equal to $0.1-0.5$ for most metals and alloys), and K is the constant which is equal to the stress when $\epsilon_t = 1$.

The shape of this curve shows that, although work hardening continues throughout the test, the *rate* of work hardening steadily decreases. It follows that there is a limit to the extent that work hardening can sustain a specimen of steadily decreasing cross-section. This limit is reached when a small increase in strain (and therefore decrease in cross-section) no longer actually increases the load supportable at that point. This stage is the maximum in the conventional stress—strain curve, and the conventional stress at this point is known at the *ultimate tensile strength* of the metal. Loading beyond this stage results in unstable deformation, since it is localised at the point of narrowest cross-section. Work hardening continues in this region (see the true stress—strain curve) but the true stress rises faster than the strength of the metal. As a result, 'necking' occurs (Fig. 12.35(a)) and eventually fracture according to the ductile mechanism outlined in Section 15.1.

It should be noted that as soon as a neck forms in the specimen the condition of a simple uniaxial tension is no longer valid. The neck (Fig. 12.35(b)) acts as a very blunt notch or 'crack' in the specimen, and a triaxial system of stress is set up. Elongation of the metal in the centre of the neck is restrained by the ring of unstressed metal surrounding it. Thus the shear stress at the neck is raised and there is a greater likelihood of failure being initiated by cracking.

380

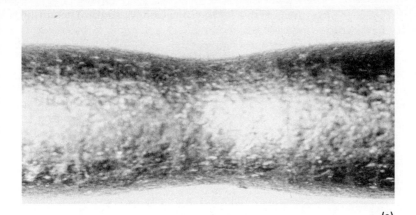

(a)

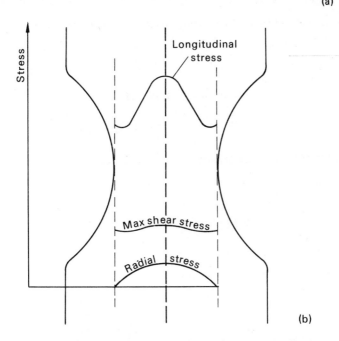

(b)

Figure 12.35 (a) Necking in a tensile specimen of mild steel. (b) Stress distribution in a necked tensile specimen. (Reproduced by permission, from *The Structure and Properties of Materials,* by H. W. Hayden, W. G. Moffatt and J. Wulff (Vol III). Copyright © 1965, by John Wiley & Sons, Inc.)

Although the ultimate tensile strength is not a very valid criterion for general design, it is useful to know the strain corresponding to it, i.e. the strain at which necking commences. Firstly, it denotes the absolute limit of safe loading in tension to which a strut can be subjected and, secondly, it marks the practicable limit of plastic strain during tensile fabrication processes, when uniform flow of the metal is important to maintain shape and homogeneity. This strain may be calculated as follows.

If W is the load supportable by the specimen, the critical condition for necking to commence is given by

$$\frac{dW}{d\epsilon} = 0$$

i.e. $dW = 0$, since $d\epsilon$ is finite. Now $W = \sigma_t A$, where A is the specimen cross section. Hence

$$dW = \sigma_t\, dA + A\, d\sigma_t$$

Therefore
$$\sigma_t\, dA + A\, d\sigma_t = 0$$

at the critical point. Therefore

$$\frac{d\sigma_t}{\sigma_t} = -\frac{dA}{A}$$

If we assume that deformation occurs at constant volume, we can write

$$-\frac{dA}{A} = \frac{dl}{l}$$

where l is the length of the specimen. Hence

$$\frac{d\sigma_t}{\sigma_t} = \frac{dl}{l} = d\epsilon_t \qquad \text{(from Equation 12.18)}$$

Therefore

$$\sigma_t = \frac{d\sigma_t}{d\epsilon_t}$$

Substituting for σ_t in Equation 12.20, we have then

$$K\epsilon_t^n = nK\epsilon_t^{n-1}$$

Therefore

$$\epsilon_t = n \qquad\qquad\qquad (12.21)$$

or

$$\log_e(\epsilon_c + 1) = n \qquad\qquad\qquad (12.22)$$

Hence the true strain at the point of necking is equal to the strain hardening coefficient.

Because of necking, some of the results of a tensile test are affected by the dimensions of the specimen. For example, the length of specimen taken up by neck will be the same for two identical specimens of equal cross section, irrespective of their initial length. Yet the length of neck *in terms of strain* will be less for the longer specimen, which will therefore *appear* less ductile over the necking range. A similarly misleading comparison will occur if identical specimens of equal length but differing diameter are tested to failure: the length of neck will be greater for the specimen with the larger diameter, and this will give an impression of greater ductility. A standard gauge length-to-cross sectional area ratio has therefore been agreed:

$$l_0 = 5.65\,A^{\frac{1}{2}}$$

giving $l_0 = 5d$ for a circular cross-section, where l_0, A and d are the gauge length, cross-sectional area and diameter respectively of the test piece. Until recently, different ratios were in common use: $l_0 = 4.51\,A^{\frac{1}{2}}$ in USA, $l_0 = 4\,A^{\frac{1}{2}}$ in Great Britain, and $l_0 = 11.3\,A^{\frac{1}{2}}$ in Germany, and so care must be taken in comparing test results from diferent nations or determined at different times.

Two further figures are frequently quoted from tensile tests, and, although they are of very dubious scientific value, they are often used as a basis for comparison of the quality of specimens. These are:

(a) *The percentage elongation at fracture.* This is just a synonym for the conventional strain at the point of fracture. It is usually obtained by fitting the two broken halves of the specimen together and then measuring the gauge length — a technique not conducive to high accuracy. Percentage elongation, together with reduction of area (see below) is usually taken as a broad indication of the ductility, or otherwise, of the specimen, but in fact elongation is valuable to the engineer only while it is uniform, i.e. up to the value corresponding to the ultimate tensile stress. At this point it gives the fabrication engineer some idea of the maximum tensile formability of the metal without annealing, but further strain increments between this point and fracture give values which are really meaningless.

(b) *The percentage reduction in area at fracture.* If A is the initial cross-section and a is the final cross-section at the fracture point, the percentage reduction in area is given by

$$\frac{A - a}{A} \times 100\%$$

It is also used as a rather loose measure of ductility, and is rather more valuable in this respect than elongation. Thus specimens that fail in the elastic or early part of the plastic ranges of the curve (i.e. tend to be brittle) show low values of reduction in area, whereas a specimen exhibiting ductile behaviour with necking will give a much higher value. However, the agreement is only qualitative. An unexpectedly low value is therefore a useful indication of a faulty specimen or poor-quality alloy.

12.7 CREEP AND STRESS RELAXATION

Hitherto we have assumed that when a metal is subjected to a stress less than its u.t.s. it will support that stress indefinitely. But this is not really true; slow plastic deformation, known as *creep*, can occur under constant load, even when the stress level lies below the elastic limit. Although creep can occur over a wide range of temperature (it has been observed down to 4 K), creep problems in engineering are normally associated with the dimensional stability of components under stress at high temperatures (e.g. in gas turbines) and are therefore regarded as the concern of the mechanical engineer and metallurgist. This is usually a problem of creep at *constant load*, involving a continuous (but not necessarily uniform) strain rate with the passage of time. For the civil engineer, however, it is creep *at constant strain* and at ordinary temperatures that is of importance, since this can have a considerable effect on the durability of structures. Two important examples of this type of creep are afforded by the behaviour of prestressing bars in concrete and of bolts or rivets in structural joints. In both cases, satisfactory performance depends on the ability of bars or bolts to maintain a high elastic tensile stress throughout the lifetime of the structure. Creep at constant strain results in the gradual replacement of elastic strain by plastic strain and a consequent relaxation of elastic stress. As a result, prestressed concrete loses some part of its residual compressive stress, so that its load-bearing capacity in tension is slowly weakened; bolts and rivets cease to clamp the structural members together rigidly, thus permitting vibration to develop at the joint, and the possible onset of fretting and fatigue failure.

First, let us consider briefly the general response of metals to stress over long periods of time. The form of a creep curve, in which strain is plotted against time, is shown in Fig. 12.36. It will be seen that the

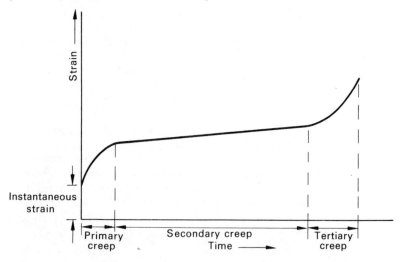

Figure 12.36   Generalised form of creep curve for metals.

curve comprises three parts: *primary creep*, in which the strain rate decreases with time; *secondary creep*, exhibiting a constant strain rate; and finally *tertiary creep*, where the strain rate increases until fracture occurs. It is the primary creep stage (sometimes referred to as *logarithmic creep*) that is of concern to civil engineers, since it is the stage that predominates at low temperatures.

Analysis of the first two stages of creep show that several deformation mechanisms operate. When stress is first applied, immediate plastic deformation occurs, as predicted by the stress–strain curve, until work hardening exactly balances the stress. The metal is then in a state of equilibrium where the stress tending to move the dislocations one way is opposed by the obstacles to that movement that the dislocations have themselves generated. But this is a very delicate balance and slight random thermal fluctuations are usually sufficient in the presence of the applied stress to trigger off further dislocation movements past the weaker obstructions. Thus plastic deformation will continue after the load has been applied, but the obstructions to dislocation movement will also increase in numbers and magnitude so that fewer and fewer dislocations can be activated in this way. The result is the logarithmic type of creep curve shown in Fig. 12.37 and which can be represented by one of the two relations

$$\epsilon = \epsilon_0 + \alpha \log_e t \tag{12.23}$$

or

$$\epsilon = \epsilon_0 + \beta t^{\frac{1}{3}} \tag{12.24}$$

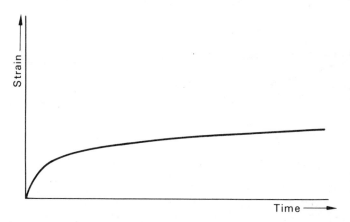

Figure 12.37　　Logarithmic or transient creep curve.

where ϵ_0 is the instantaneous strain, t is the time, and α and β are constants. Creep of this nature is often referred to as *transient creep*, and is the dominant feature of the primary section of the curve.

There is, however, a second pattern of deformation known as *steady-state creep* or *quasi-viscous creep*, which has a slight bearing on the primary stage and becomes dominant in the second stage (when transient creep has virtually ceased). Here the strain rate is constant because the processes of *recovery* and thermal softening can balance some part of the work hardening that accompanies flow so that the resistance to deformation remains the same. The strain rate is then given by the expression

$$\dot{\epsilon} = A\sigma^n \cdot e^{-Qd/RT} \qquad (12.25)$$

where A and n are constants and QD is the activation energy for self-diffusion, i.e. for atoms of metal M to move through the lattice of metal M. The constant n is roughly 4 and A depends on the micro-structure of the metal. Steady-state creep is thus principally dependent on temperature and becomes appreciable at temperatures above $T_m/2$ K, where T_m is the melting point of the metal. It occurs by the processes of dislocation *climb* (which enables them to evade obstacles by going round them), cross-slip of screw dislocations (see Section 12.3), and — at higher temperatures — vacancy diffusion and grain boundary slip. With the exception of cross slip, none of these processes occurs to any appreciable extent at ambient temperatures.

When considering bolts or prestressing bars, we are thus primarily concerned with stress relaxation as a result of logarithmic creep at constant strain, giving a curve of the form shown in Fig. 12.38.

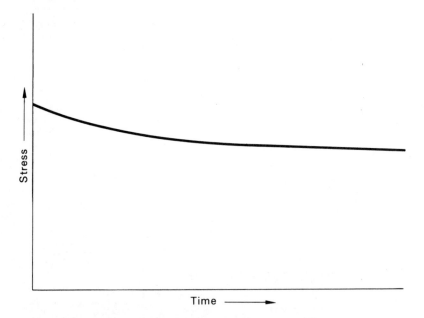

Figure 12.38   Stress relaxation due to creep at constant strain.

When the load is first applied, the strain is purely elastic; however, creep (i.e. *plastic* strain) will develop according to Fig. 12.37. If the total strain ϵ is constant, we may say that

$$\epsilon = \epsilon_e + \epsilon_p \qquad (12.26)$$

where ϵ_e and ϵ_p are the elastic and plastic components of strain respectively. Thus, as ϵ_p increases, ϵ_e must decrease in proportion. Since ϵ_e is related to the stress σ acting on the alloy by $\epsilon_e = \sigma/E$,

386

the curve for the stress relaxation is proportional to an inversion of that in Fig. 12.37.

We should note, however, that Equation 12.26 is not precisely true, as this simple treatment takes no account of the strain changes in the bolted components or concrete as the stress relaxes (in other words, the strain does not strictly speaking remain constant throughout), but provided that the elastic strain is large by comparison with the creep strain (a reasonable assumption in a structural material) the above treatment gives a valid approximation.

CHAPTER 13

Strength and Failure of Concrete

13.1 DIFFERENT KINDS OF STRENGTH

The first consideration in the design of concrete structures is that they should be strong enough to support the loads that they will be called upon to carry. That is to say the stresses in the materials must not be so great that their strength is exceeded and failure occurs.

As discussed in Chapter 14 (see Section 14.1), the terms strength and failure have different meanings in different circumstances. Thus strength can be taken to be the stress at which failure occurs, and failure coincides with an unacceptable loss of performance. In concrete this is usually synonymous with the degree of fracture at which the applied stress reaches its maximum value. In some cases, such as direct tension, failure is evident in the rupture of the testpiece, while in compression failure occurs when internal cracking is well developed but the testpiece is still whole and capable of carrying load. In a few cases, such as certain triaxial compressive states, failure is better defined in terms of unacceptable deformations.

Different types of loading give rise to different modes of failure, and it is vital that the relevant strength is chosen to match the application. For the flexural failure of reinforced concrete beams it is the uniaxial compressive strength of the concrete that matters, while for the cracking of a concrete ground slab it is the flexural tensile strength. Other situations may demand torsional strength or strength under multiaxial stress, fatigue strength or impact strength. The design data have in all cases been derived from empirical test programmes at the engineering level, performed on large concrete specimens, and it is a curious fact that the tests for this most important of properties are among the least satisfactory. For example, as described below, the test for the uniaxial compressive strength of concrete is, in fact, performed under conditions of poorly controlled triaxial stress, and there are no less than three tensile strength tests, each giving a different answer.

Interfacial interactions between the h.c.p. and the aggregate are of great significance in that cracking in concrete more often than not follows the aggregate-h.c.p. interface. Thus the strength of concrete

388

depends on the bond between aggregate and paste as well as on the strengths of the individual materials, and the interfacial effects, concerned with the formation and propagation of cracking, are given an extended treatment below. As might be expected, the strength of the h.c.p. increases with degree of hydration, and also varies with the moisture content, and these matters of structure and state are discussed later in the chapter. First of all, however, the test methods are described, and, while they refer specifically to concrete, they are equally applicable, on a reduced scale, to h.c.p.

13.2 STRENGTH TESTS

13.2.1 Compressive Strength

In Britain, the 'Test for Compressive Strength of Test Cubes', to quote BS 1881 (1970), is the predominant method of determining and checking concrete strength. The concrete cubes are usually cast in lubricated steel moulds, which are machined carefully to ensure that opposite faces are plane and parallel. The concrete is fed into the mould and fully compacted and the top face is trowelled smooth; the specified curing regime is usually storage under water at constant temperature.

The cube crusher consists essentially of two heavy flat platens, the lower one fixed in position and the upper one having a ball seating which allows the rotation of the platen to match the top face of the cube at the start of loading; the platen then remains fixed in that position while the load rises to failure. The cube must be placed centrally in the crusher, and the trowelled surface is at the side so that the load is applied through a pair of parallel faces. A very fast rate of loading gives overhigh strengths, and an appropriate standard rate is laid down. The crushing strength is given by

$$\frac{\text{maximum load recorded}}{\text{nominal area of cross-section of the cube}}$$

The test is open to the usual kinds of error arising from indifferent experimental techniques. Variability in the cube strength results from failure to mix, compact and cure the concrete in the cubes in the standard manner, and from a lack of precision in testing; for example, failing to centre the cubes accurately. In addition, the crusher itself may not perform in a reproducible way, the ball seating perhaps failing to lock in position, or the platens being domed or dished instead of truly flat. In addition, in extreme cases errors have been found of as much as 10% in indicated load because of incorrect calibration.

However, the real objection to the test concerns the imposed stress pattern. Ideally it should be a uniaxial compressive stress, and at first sight it appears that the test is quite satisfactory, since it is not too difficult to apply the load centrally and to ensure that the platens are sufficiently stiff to distribute that load over the whole loaded face. The trouble arises from another direction and is a matter of platen restraint. The Poisson's ratio of the steel of the platens is less than that of the concrete of the cube, and the friction between steel and concrete ensures that the concrete is restrained from expanding laterally under load. That is, it is subjected to inward acting forces on the top and bottom faces, thus setting up a complex three-dimensional system

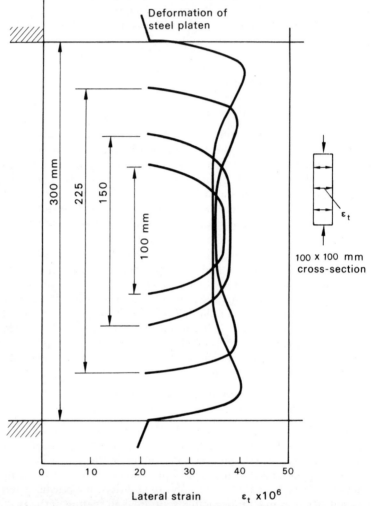

Figure 13.1 Variation of lateral deformation profile with height of prism specimen for an applied compressive stress of 9 N/mm^2. (From K. Newman (1966) Concrete control tests as measures of the properties in concrete, in *Symposium on Concrete Quality,* by permission of the Cement and Concrete Association.)

of stress in the cube. A typical distribution of lateral strains for various height/diameter ratios is shown in Fig. 13.1. It can be seen that for larger values of the ratio the effects of the end restraint die away towards the mid-depth of the cylinder, and that a purely uniaxial uniform stress exists over the centre portion of the specimen. Platen restraint improves the strength, and cubes have a strength some 1.25 times that of corresponding cylinders. The testing of cylinders has a clear advantage in that the stress state at failure is comparable to that in many real concrete structures. However, it has one disadvantage; the cylinders are cast in a vertical position so that the top surface has to be carefully capped before the load is applied through it. The failure load is very sensitive to the quality of the capping, and it is difficult to secure acceptable reproducibility under the conditions of everyday testing. Furthermore, the likelihood of segregation (see Section 13.7.2.3) is higher in cylinders than in cubes.

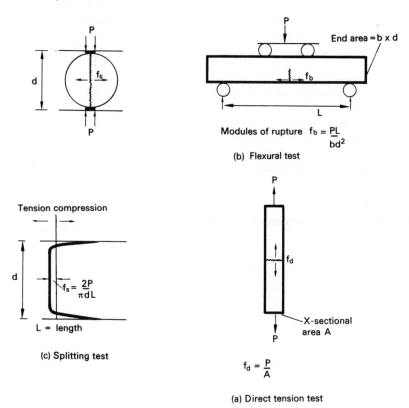

Figure 13.2 Tensile tests for concrete.

In spite of the defects of the compressive tests, the crushing strength of cubes or cylinders is the prime indicator of concrete quality, and it is found in practice that the tests provide a very tolerable way of comparing one concrete with another, and an acceptable measure of concrete strength for use in design.

13.2.2 Tensile Tests

The three tests for concrete are illustrated in Fig. 13.2.

391

13.2.2.1 *Tension Test*

For metals and timber the tensile strength is of first importance, but this is not so for concrete because it has a tensile strength that is much less than its compressive strength. In addition, the simple testing techniques that are used for timber and metals are not so satisfactory for concrete. There are three main difficulties.

(a) Test specimens which are representative of the concrete as a whole have a relatively large cross-sectional area, with the result that the axial alignment of the specimen in the testing machine becomes a real problem. Although special techniques have been used with success, eccentric loading can easily occur, causing both lower apparent strength and a greater variability in the test results.

(b) The grips of the testing machine cause stress concentrations where they react on the concrete, and local failure can occur, the specimen breaking near the ends rather than within the central zone.

(c) Once a crack forms it is likely to propagate rapidly through the specimen, so that the measured strength is that of the weakest section. This is not usually the case in concrete structures, and the hindrance to crack propagation offered in the other two tests is a closer approach to reality.

The direct tension test is not satisfactory as a standard, and its use is confined to research laboratories.

13.2.2.2 *Flexural Test*

As described in BS 1881, the test is performed on a prism of cross-section ($b \times d$) of 100 by 100 mm or 150 by 150 mm, and simply supported over a span (L) of 400 mm (or 600 mm). The load is applied at the third points, and failure occurs when a flexural tensile crack at the bottom of the beam, normally within the constant moment zone, propagates upwards through the beam. The strength is the calculated maximum tensile stress, f_b, under the total failure load, P. The calculation is based on simple beam theory, so that

$$f_b = \frac{My}{I}$$

where $M = PL/6$, $y = d/2$, $I = bd^3/12$.

f_b is known by the unlikely (and misleading) name of modulus of rupture, and is thus given by

$$f_b = \frac{PL}{bd^2} \tag{13.1}$$

This formula is based on the assumption that the distribution of tensile stress is linear from the neutral axis to the bottom of the beam, whereas, in fact, the stress-strain relation is curved; this means that the calculated stress is higher than that actually developed in the concrete. The stress gradient also has the effect of

392

inhibiting the upward growth of the flexural cracks, so that there is no question here of the immediate propagation of the first crack at the weakest section. A number of cracks may form and undergo limited growth before instability sets in, and one crack finally does propagate through to the top of the beam. For both the above reasons the modulus of rupture is greater than the direct tensile strength.

13.2.2.3 *Indirect Tensile (or Splitting) Test*

This test, like the flexural test, is included in BS 1881. The concrete specimen is a 150 mm diameter cylinder which is placed on its side and loaded across the vertical diameter through the platens of a compressive testing machine. Hardboard or plywood strips are inserted between the cylinder and the platens to ensure even loading over the full length of the specimen.

The theoretical stress distribution in the horizontal direction on the vertical diameter is shown in Fig. 13.2. As can be seen, there is a practically uniform tensile stress of $f_s = 2P/\pi dL$ over the majority of the diameter, and local high compressive stresses at top and bottom. The concrete fails by cracking in the plane of the vertical diameter, and it seems reasonable to accept that the tensile stress is responsible, and that the above expression gives the true tensile strength of concrete. Certainly, this view is supported by the fact that the splitting strength is, as might be expected, higher than the direct tensile strength and lower than the flexural strength. However, some doubt remains both because the stress in the cylinder is multiaxial rather than uniaxial, and because the local zones of compressive stress at the ends of the diameter may influence the failure load.

Although the test method is open to criticism, the test is simple to perform, it uses standard equipment that is available in most laboratories, and experience shows that the results are consistent and reproducible. There is thus a reasonable case for its use as a measure of the tensile strength of concrete.

13.2.3 Comparative Strengths

The relative magnitude of the strengths measured in the various tests described above are illustrated by the following typical figures:

Cube strength	40 N/mm^2	compressive
Cylinder strength	32 N/mm^2	
Flexural strength		
(modulus of rupture)	4 N/mm^2	tensile
Indirect tensile strength	3 N/mm^2	
Direct tensile strength	2 N/mm^2	

Many tests have been developed to measure other kinds of strength, such as impact strength or multiaxial compressive strength. The techniques are necessarily more complicated than those already described, and the experimental expertise required is of a high order. No further details of the apparatus will be given here, but some of the results will be introduced later. There can be little hope that these difficult techniques will become standard, and it must be the aim to relate the other kinds of strength to the compressive and tensile strengths as determined by the standard uniaxial methods.

13.3 BONDING FORCES

As stated in the first part of this chapter, failure in concrete is normally associated with some degree of fracture of the material. That is, there is a separation of surfaces to a greater or lesser extent within the material, which implies the breaking of the bonds that hold the material together. This behaviour is influenced by the manner in which the material deforms, and, in the limit, it is an integration of the interactions of the atoms and molecules of which the material is made. It is logical to begin the examination of the processes of fracture with the bonding forces at the molecular level.

13.3.1 Interparticle Forces

Bonding energies can be found in principle by considering the energy change when a pair of atoms, or molecules or ions, are brought together from an infinite distance apart. This matter has been discussed in some detail in Chapter 2 (see Section 2.5.1) and it can be summarised on the form of the Condon—Morse curve shown in Fig. 13.3, in which the interparticle force, F, is related to the interparticle distance, r.

The normal equilibrium position is that of minimum potential energy, that is of zero interparticle force, and corresponds to the distance r_0.

The application of an external force upsets the force equilibrium, and a change in interparticle spacing must occur to restore the balance. Thus the force—distance curve of Fig. 13.3 has the shape of a stress—strain curve of a material bonded together in the manner of the pair of particles. If the applied force is removed, the interparticle forces operate to return the spacing to r_0, thus exhibiting elastic behaviour. The curve is approximately linear on either side of the equilibrium position so that it is to be expected that materials behaving in the ideal manner of the pair of particles will be linearly elastic.

Further, if the applied force (in tension) is greater than the maximum value, F_{max}, the particles will separate. The cohesive strength of the atomic structure is exceeded and fracture occurs. Estimates can be made of these theoretical cohesive strengths for different pairs of particles, and the magnitude comes out at much the same order, between 10 and 100 KN/mm^2.

In practice the phenomenon of linear elasticity is far from unknown, particularly at low levels of applied stress but the strengths of real materials are much lower than the theoretical figures. It seems that the structure of real materials has some relation to the atomic structure, particularly at low levels of stress, but that this becomes less and less true as the applied tensile stress is increased. The Condon—Morse curve provides a useful start but it is essential to explain the

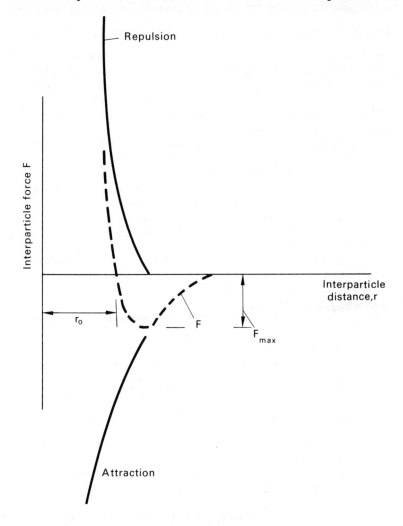

Figure 13.3   Interparticle force—distance relation.

difference (under tensile stress) between atomic behaviour and behaviour of the material in the mass.

Application of a compressive stress leads, as can be seen from the Condon—Morse curve, to a diminution of the interatomic spacing. The larger the applied compression the closer the particles move to each other, with no indication of fracture whatsoever. When the applied stress is hydrostatic this does indeed represent reality in that the strength of materials is virtually infinite under all-round compression, the increasing stress merely causing a decrease in volume. Under uniaxial (or biaxial or unequal triaxial) stress, in contrast to the evidence of the Condon—Morse curve, fracture does occur, and the reasons for this are examined in the next two sections.

13.3.2 Fracture in Hardened Cement Paste and Concrete

The fracture of h.c.p. under applied tension is brittle in nature (see Section 2.5.3), there being very little other than elastic deformation before the catastrophic propagation of a critical crack. The separation of the parts of the testpiece is a process of the integration of the pulling apart of pairs of particles in the manner discussed in the previous section. It is a typical cleavage failure.

As described in Chapter 3, h.c.p. has a large volume of pores so that there are large areas of potential failure surface over which the interparticle forces are minimal. Over other large areas there are van der Waals forces acting to bond together the parallel layers of the crumpled foils forming the gel, which are connected here and there by stronger interparticle bonds (see Fig. 3.5). Within the solid layers the molecules are held together by the strongest (covalent and ionic) bonds. There are no cleavage planes of the kind that are found in well ordered crystalline structures, but there are preferred failure paths, following the areas of greatest weakness. Thus the cleavage failure will pass through the maximum number of zones of pores, otherwise passing between the solid layers with the minimum of interparticle bonds, and seldom needing to tear apart any solid layer itself. On this basis, there is no hope of the tensile strength approaching the theoretical cohesive strength of the solid particles. Other considerations, discussed later in this chapter, make the situation even worse, so that the strengths quoted earlier in the chapter (see Section 13.2.3) are four to five orders of magnitude less than the theoretical.

The presence of aggregate offers two more possibilities of failure:

(a) *Within the aggregate.* This seldom occurs with normal good natural aggregates, but it does occur to some degree in many, perhaps most, lightweight aggregates. It implies that a lightweight aggregate concrete will have less strength than the h.c.p. but, because of the interface phenomenon in (b) below, it does not, of itself, mean that lightweight concretes are weaker than concretes made with normal aggregates.

(b) *At the aggregate — h.c.p. interface.* The cleavage strength of the interface is usually lower than that of the h.c.p. itself (and of the aggregate particles), and although the separation can occur in the surface layer outside the boundary of the aggregate, it is normal for the h.c.p. to come away cleanly from the aggregate. The path of the propagating crack follows the interfaces, usually of the largest pieces of aggregate, and crosses the smallest possible h.c.p. filled gaps between successive interfaces. The interface is usually the weakest part of concrete, and it is then inevitable that concrete is less strong than either the h.c.p. or the aggregate.

Interfacial strength can be very low without any stress at all being applied, either because voids (air- or water-filled) form on the interface, or because differential movements (shrinkage or thermal) occur between the h.c.p. and the aggregate. Apart from this special, though common, phenomenon, low bond strengths occur under tensile load when various influential factors are unfavourable. Thus a smooth

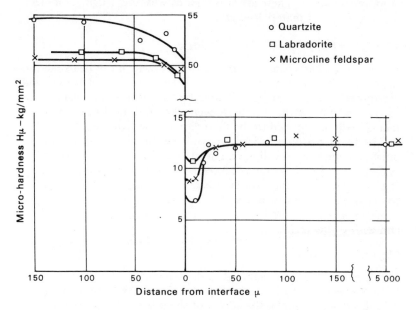

Figure 13.4   Variation in micro-hardness across the aggregate — h.c.p. interface. (From K. M. Alexander, J. Wardlaw and D. J. Gilbert (1968). Aggregate—cement paste bond and the strength of concrete, in *Conference on the Structure of Concrete,* by permission of the Cement and Concrete Association.)

aggregate is less effective than a rough one, surface contamination of the aggregate has, not surprisingly, deleterious results, and, as is generally the case for h.c.p. strength, young pastes made with a high water/cement ratio have less good interfacial strength than their older, denser counterparts.

Under the most favourable conditions the h.c.p. — aggregate cleavage strength reaches that of the h.c.p. itself, indicating that the interface attraction is more than just physical. There is some direct evidence that chemical interaction does take place between the h.c.p. and even so-called inert aggregates. For instance, the presence of an interfacial layer, extending into both the h.c.p. and the aggregate, is demonstrated by measurements of micro-hardness as shown in Fig. 13.4. It is tempting to believe that herein lies the path to a real improvement in concrete strengths. If only reactions between paste

397

and aggregate could be encouraged . . .! Unfortunately, it is not as easy as it might seem, if only because reactions of this kind can be the reason for the *deterioration* of concrete (see Chapter 17).

So far, it is clear that concrete will not perform well in tension, but no reason has been given for other than an admirable performance in compression. In fact, as demonstrated in the typical figures in Section 13.2.3, the uniaxial compressive strength is at least ten times the uniaxial tensile strength. This is useful but it is disappointing in that the molecular logic points to an infinite reserve of strength under compressive stress. There are two reasons why the delightful molecular prospects do not materialise.

13.3.2.1 *The Effect of Heterogenetity*

H.c.p. has a most irregular porous structure, which can be likened to a random space frame. Very little knowledge of structural engineering is required to appreciate that it is perfectly possible for tensile forces to arise in some of the members, even when nothing but inward facing loads are applied to the boundary of the structure. Similarly, it is perfectly possible, indeed unavoidable, that h.c.p. subjected to uniaxial compression will have a most complicated internal stress system, including not only compressive and shear stresses, but also tensile stresses. In other words, it is possible for local cleavage failure to occur in h.c.p. under uniaxial compression.

When aggregate is added the degree of heterogeneity increases, and with it the likelihood of local cleavage fracture.

13.3.2.2 *The Effect of Shear Stresses*

The curves of Fig. 13.3 referred to the separation of particles one from another in the direction of the bonding force between them. It is also possible, under the action of shear force, for particles to separate by moving at right angles to the direction of the bond. Although the pair of particles are separated, there may be no separation of the shear surface as a whole, because, after the bond with one particle is broken, a new bond may form with an adjacent particle. That is, slip occurs, but not immediate fracture. This is the basis of plastic deformation preceding ductile fracture.

In concrete under compression, slip does occur, but not without significant local failure. Molecular bonds are broken, but are not usually reformed after slip, although, if no further slips occurs for some time (measured in months rather than hours), there may be some remaking of bonds, and this may be accentuated by renewed hydration causing autogenous healing. In general, however, the cohesion between shear surfaces remains broken, though separation of the surfaces may not happen immediately. In fact, frictional sliding resistance is still provided, but the surface is susceptible to cleavage fracture which may then occur as the stress system adjusts itself to increasing applied load, or to changes induced by progressive time-dependent failure. Slip is most likely ot happen at the h.c.p. — aggregate interface, and this is called shear-bond fracture.

The shear-bond resistance is provided initially by the cohesive molecular strength, but once the bonds are broken it has the characteristic frictional resistance. Thus shear-bond resistance is approximately proportional to the normal stress across the interface, and the roughness of the aggregate becomes of importance; as shown in Fig. 13.5,

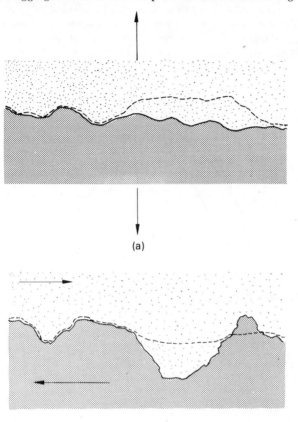

(a)

(b)

Figure 13.5 Diagrammatic representation of h.c.p.-aggragate interface, showing typical fracture path for (a) cleavage failure, (b) shear-bond failure. (From K. M. Alexander, J. Wardlaw and D. J. Gilbert (1968) Aggregate-cement paste bond and the strength of concrete, in *Conference on the Structure of Concrete*, by permission of the Cement and Concrete Association.)

rough aggregates have an enhanced coefficient of friction through their ability to develop mechanical interlock. The failure surface may then, as shown, pass both through the h.c.p. and through spurs of aggregate.

H.c.p. and concrete both exhibit some degree of slip in the course of loading to failure in compression. The effect is more marked in concrete, with its greater degree of heterogeneity. This can be seen in Fig. 13.6, which demonstrates the more usual result that the aggregate and h.c.p. are both stronger than the concrete made from them; this may not be true if a weak aggregate is used, or if the

399

paste has a high water/cement ratio, or if there is a particularly strong reaction between the aggregate and the h.c.p. The stress—strain curve for h.c.p. is curved throughout, but it can be seen from the figure that concrete exhibits greater inelasticity leading to greater curvature, and the capacity to carry a reduced stress at higher strain levels. This gives rise to the descending branch of the stress—strain curve. In passing, it should be noted that the arguments given above with reference to compression can be equally well applied to tensile stress. Thus, compressive stresses exist in concrete subjected to a uni-axial tensile load, and some slip does occur, but the effects are small,

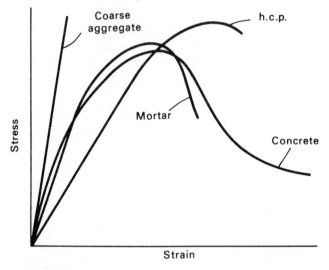

Figure 13.6 Typical stress—strain characteristics of aggregate, hardened cement paste, mortar and concrete. (From R. N. Swamy and C. B. S. Kameswara Rao (1973) *Cement and Concrete Research*, **3**, 413, by permission of Pergamon Press.)

and there is very little evidence of other than straightforward cleavage fracture.

The failure process of concrete has some superficial similarity to that of metals, but the differences are considerable. Slip and cleavage occur in both, but the molecular structures are very different. Metals (see Section 12.2) exhibit plasticity in addition to their elastic response, and are tough, requiring high levels of work to fracture. They are thus ductile materials, whereas concrete is essentially brittle in tension. In compression, it is less brittle in that considerable energy is required for the formation and extension of the system of multiple cracks that form before fracture occurs, and under some triaxial states of stress it may even behave in a ductile manner.

13.4   FEATURES OF FAILURE IN CONCRETE

The previous two sections dealt with fracture at molecular level, and with separation and sliding in h.c.p. and concrete. In this section the discussion is taken further to include the findings of experimental work on the failure of concrete specimens, and the validity of theoretical ideas in relation to these observations. Attention is confined to uniaxial states of stress, two- and three-dimensional systems coming later, in Section 13.5.2.

13.4.1 Tensile Stress

As seen earlier, the tensile strength of concrete is lower than the molecular cohesive strength. The behaviour is brittle giving a cleavage fracture and, in a testpiece, failure occurs by the rapid propagation of a crack. This observed process of failure fits well the concepts of crack propagation and brittle fracture put forward by Griffith as long ago as 1920.

13.4.1.1 *Crack Propagation*

The basis of fracture mechanics for static loadings is the premise that the two necessary and sufficient conditions for a crack to grow are:

(a) The stress must be sufficient to initiate fracture.
(b) The energy balance must be favourable for crack growth.

That is, the energy released by crack growth must be at least as much as that required to form the new fracture surfaces.

The Griffith approach to this problem has been described in some detail in Chapter 2 (see Section 2.5). It is based on the consideration of a single elliptical crack, for which the stress at the tip, σ_{max}, is given by

$$\sigma_{max} = 2\sigma \left(\frac{c}{\rho}\right)^{\frac{1}{2}} \tag{13.2}$$

where σ is the uniform tensile stress (on a thin plate) containing the crack of length $2c$, with a radius of curvature of ρ at the end of the major axis.

Hence, as shown in Section 2.5.3, the stress for crack propagation, σ, is a function of the crack length, c,

$$\sigma = \left(\frac{2\gamma_w E}{\pi c}\right)^{\frac{1}{2}} \tag{13.3}$$

where γ_w is the energy required for the formation of a fracture surface, and E is the elastic modulus.

In most cases, γ_w is not just the surface free energy (γ). It may for instance include plastic work, which in metals is very much greater than the free energy, or chemical energy if fracture is accompanied by a chemical reaction.

The Griffith formulae are not too easy to apply to real materials. The first (Equation 13.2) requires knowledge of the radius of curvature at the tip of the crack which can hardly be known, and the second (Equation 13.3) needs the value of γ_w, the energy required to form the new fracture surfaces, which must be determined experimentally.

Nevertheless, the form of the equations gives useful clues to practical considerations, such as the prevention of the propagation of cracks.

In Equation 13.2, it is apparent that the stress at the crack tip will be reduced, and hence the applied stress for extension of the crack raised, if the radius of curvature is increased. The likelihood of failure can therefore be reduced if some kind of crack blunting device can be introduced, such as inducing the crack to run into a larger void or pore. Similarly, from Equation 13.3 it is apparent that the energy balance can be tilted in favour of nonpropagation if γ_w is increased by, say, the crack encountering a zone of ductile material requiring the input of considerable plastic deformation energy. These methods of crack stopping can provide real advantages and it should be noticed that they will occur only if the material is not homogeneous, either by its own nature or by deliberate manufacture of a composite structure.

The ideas discussed above give a qualitative view of what happens when cracks propagate, but they do not satisfy the essence of any scientific approach, which demands the quantitative expression of the phenomenon and the possibility of setting up experiments to determine numerical values of the relevant material properties.

A more profitable approach is through the stress intensity factor, K, which characterises the spatial stress distribution around the tip of the crack. It is proportional to the applied stress and has a critical value for the onset of crack propagation K_c, which is equivalent to the satisfying of both the conditions given above. K has units of $N/m^{1.5}$ and, for elastic conditions, it can be simply related to the critical strain energy release rate, G_c, at which crack propagation begins.

Then

$$G_c = \frac{dU}{dA} \tag{13.4}$$

(with units of N/m) where A is the area of the crack, and U is the strain energy.

The expressions for strain energy and surface energy (given in Section 2.5.3) lead to

$$G_c = 2\gamma_w \tag{13.5}$$

where $2\gamma_w$ is the (elastic) work to fracture/unit volume.

The precise form of the relationship between G_c and K_c depends on the system of stress. For example,

$$K_c^2 = EG_c \qquad \text{for plane stress} \tag{13.6}$$

$$K_c^2 = \frac{EG_c}{(1 - \nu^2)} \qquad \text{for plane strain} \tag{13.7}$$

Thus G_c and K_c provide alternative measures of the resistance of a material to cracking, and experimental procedures for their determination have been developed, one of which is described below. Their values are useful as properties of the material in the presence

402

of a crack, and thus provide a means for predicting the fracture behaviour of real structures. The results of the previous paragraphs are strictly applicable only to the elastic state, whereas G_c often includes terms for plastic work, etc. In this situation the relation between G_c and K_c becomes tenuous, and experimental results require careful interpretation.

K_c (and G_c) can be found by various experimental techniques, allied to appropriate analytical expressions. One of the most popular

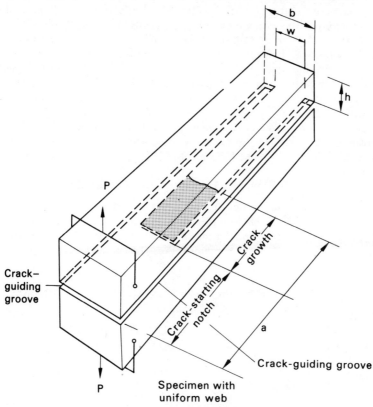

Figure 13.7   General arrangement of the double-cantilever beam (DCB) specimens. (From J. H. Brown (1972) *Magazine of Concrete Research*, 24, 191).

is the double cantilever beam (DCB) method in which a specimen, as shown in Fig. 13.7, has two arms which are prised apart by the load, P. This results in the formation and propagation of a crack, constrained to occur within the preformed slot. P is related to K_c through

$$K_c^2 = 12Pa\ \zeta\ /bwh^3 \qquad (13.8)$$

in which all other quantities are known from the geometry, h and b being the cantilever height and width, w the web width, a the crack length and $\zeta = 1 + 1.32\ (h/a) + 0.532\ (h/a)^2$. Hence K_c may be determined.

13.4.1.2 *Crack Propagation in Concrete*

The derivations and formulae given above assume a certain regularity and order about the material. Unfortunately, this is not true of h.c.p. and concrete. Both are materials in which there are cracks and voids of more than molecular dimensions, even before any load is applied. The tensile load itself gives rise to a wide range of stresses of both senses, so that the remarks of C. R. Calladine ring true. He said (Calladine, 1968) 'Cement paste, we know, is a very complex non-homogeneous material, and therefore, I submit, it is quite irrational to apply Griffith's crack ideas to it! Certainly, the pure form of Griffith's theory must fail, for there is little doubt that even under uniaxial tension the failure of concrete (and h.c.p.) consists of the eventual linking of a number of cracks, both pre-existing and formed under load, rather than the rapid propagation of a single crack. This behaviour stems from both the complicated stress field and the crack-stopping efficiency of the nonhomogeneous structure'.

Nevertheless, in the end, a single crack does form and cause the failure of the testpiece, and it is not altogether foolish to attempt to measure values of G_c and K_c. In Table 13.1, results are given for

Table 13.1   Fracture Parameters for Various Materials

γ = surface free energy per unit area, γ_w = work to fracture. (From F. Radjy and T. C. Hansen (1973) Fracture of hardened cement and concrete, *Cement and Concrete Research*, 3, 343, by permission of Pergamon Press Ltd.)

	γ $\times 10^{-9}$ MN/m	γ_w $\times 10^{-9}$ MN/m	γ_w/γ	K_c MN/m$^{1.5}$
Mild steel ($-150°$C)	2 000	4×10^6	2×10^3	
Silica glass	1 200	3 000	2.5	
h.c.p.	400	2 300	6	0.2
Concrete		10 000		0.6

h.c.p. and concrete with comparative magnitudes for other materials. The figures are approximate only, but they can be used to draw conclusions about one material relative to another. Thus the value of the ratio γ_w/γ is a measure of brittleness. A completely brittle material will have a value of unity while a highly ductile metal such as copper has a value of 10^5. Thus the h.c.p. is seen to be not so brittle as glass, but it certainly can not be called ductile.

The relative magnitudes of γ_w and K_c for h.c.p. and concrete show that concrete has the higher fracture toughness, attributable to the greater crack-stopping ability introduced by the presence of the aggregate. These are but the crudest conclusions, and it may be that experimental investigations will eventually reveal a more detailed set of relationships between fracture toughness and the parameters of concrete structure and state.

13.4.2 Compressive Stress

Engineers do their best to take advantage of the higher strength of concrete in compression, and to avoid the disadvantages of the lower strength in tension. They incline towards geometries and support conditions that give predominantly compressive stresses in their concrete members so that it is the compressive strength of concrete that is of greatest importance in the design of structures to carry the imposed loads.

Crack propagation occurs in compression as well as in tension, and it is possible to extend the ideas of the Griffith theory from the simple case of uniaxial tension to the more complicated situation of uniaxial compression, and even to biaxial states of stress. However, as has been seen, the theory is less than adequate when applied to concrete in tension, and it becomes even more irrational to try to apply it to concrete in compression. Hence the treatment in this section is confined to the engineering level of information, based on the scrupulous observations of careful experiments.

13.4.2.1 *Progress of Breakdown*

Both kinds of fracture occur in concrete and h.c.p. under compressive stress — cleavage under the tensile stresses induced by heterogeneity of the material and slip, especially in form of shear-bond failure at the aggregate-h.c.p. interface, caused by shear stresses. This is demonstrated in the lattice model of Fig. 13.8. The concrete is represented by the regular two-dimensional arrangement of four pieces of aggregate set in a mortar matrix. Under the action of forces

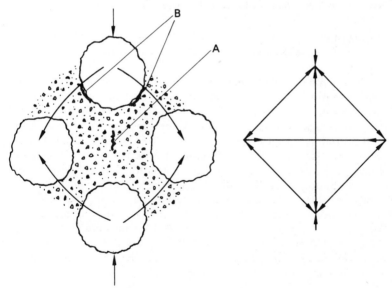

Figure 13.8   Lattice model for concrete. (From A. L. L. Baker (1959) *Magazine of Concrete Research*, 11, 119.)

applied vertically to the top and bottom pieces, the internal stress system can be likened to that of an equivalent pin-jointed lattice. Then the tensile force developed in the horizontal member represents a tensile stress in the mortar at the centre of the mode, tending to cause a vertical crack (A). Similarly, the aggregate pieces tend to move relative to the matrix (and this may be interpreted as a movement between the ends of the lattice members meeting at a joint),

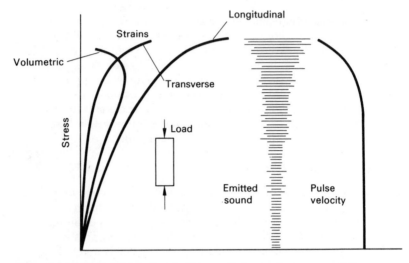

Figure 13.9   Measurements in a load-controlled test of concrete to failure. (From K. Newman (1966) Concrete systems, in L. Holliday (Ed.) *Composite Materials,* by permission of Elsevier Publishing Co.)

through the occurrence of shear-bond cracks at the interface (B). These cracks, unlike the tensile cracks, are at an angle to the direction of loading.

The breakdown of the material consists of the formation of cracks, their extension with arrest (stable propagation), and extension without arrest in which fracture of the test specimen must occur eventually (unstable crack propagation). This is a continuous process which may be followed in a displacement controlled test (see Fig. 13.11) way beyond the strain at which maximum stress is recorded. The breakdown of the material is gradual with a steadily increasing degree of damage (Spooner and Dougill, 1973). However, in the normal load controlled test which terminates at maximum stress, particular characteristics of the behaviour predominate at different stress levels. Appropriate limits can then be set which have meaning in terms of the likelihood of failure of the test specimen (or structural member). Measurements taken during such a (normal) test are shown in Fig. 13.9. They include:

(a)   strains — longitudinal, transverse and volumetric;
(b)   acoustic emission, recording the noise given out by the concrete in breaking down;

406

(c)   ultrasonic pulse velocity across the specimen.

The progress towards failure is as follows:

(1)   *Initial condition*. Before any load is applied, cracks are present
as previously described. They could cause premature failure if
they happen to present a plane of weakness to the local stress
system, but this is unlikely because they will probably form
no more than one part of a chain of cracks making up a
fracture path, the remainder of the chain forming in the normal
way under load.

(2)   *At low stress levels*. Under low load cracks are initiated at
molecular level at isolated sites throughout the material. The
evidence is conflicting as to whether they are primarily cleavage
or slip fractures, but it is certain that there is very little crack
propagation at this stage, so that the process might be described
as incipient cracking. There is very little effect on the stress—
strain relation, no significant irrecoverable strain is developed,
and, under short-term load, the concrete behaves in an approximately elastic manner. Little noise is emitted and there is no
change in the ultrasonic pulse velocity.

(3)   *At intermediate stress levels*. The crack system multiplies, and
the crack propagation that occurs is stable in that, if the load
is maintained constant, propagation ceases. In concrete, numerous shear-bond cracks develop giving a multidirectional pattern
of cracking. Such cracks do not run out into the surrounding
matrix. In h.c.p., while slip-type fractures may be present at
the microscopic level, the cracks show a general orientation in
the direction of the load.

(4)   *Approaching maximum stress*. Once this level is reached the
crack propagation becomes unstable, and failure will occur if
the load is sustained, the time to failure decreasing as the level
of the sustained load is increased. The shear-bond (or slip)
cracks either join up with mortar cracks or branch out into the
mortar, so that numerous lengthy cracks exist in the testpiece
in the general direction of the load. As failure draws near, adjacent cracks coalesce and open, individual cracks link to form
a chain, and finally a general failure surface develops, often at
$20-30°$ to the direction of the load, as shown in Fig. 13.10.
The strain rates both in contraction (longitudinally in the
direction of the load) and in extension (transverse to the load)
increase, and the change in volume of the testpiece switches
sign, so that the volume starts to increase instead of diminishing,
and it may exceed its preloading volume before failure of the
specimen occurs. As shown in Fig. 13.9, the noise levels increase
dramatically and the ultrasonic pulse velocity, measured across
the direction of loading, drops significantly as failure approaches,
thus giving another indication of the rapid extension of cracking.

The boundary between stress levels 3 and 4 is important in that it
represents the level of stress below which the test specimen (or con-

Figure 13.10 Concrete specimen on failure in compression. (From K. Newman (1966) Concrete systems, in L. Holliday (Ed.) *Composite Materials*, by permission of Elsevier Publishing Co.)

crete member) will not fracture. It has been named the point of Final Breakdown by Newman (1973) and is usually 80—90% of the maximum stress in a short-term test such as that of Fig. 13.9. It corresponds to the point of minimum volume.

Newman also defined the limit between levels 2 and 3 as the Initial Cracking level at about 50% of the short-term maximum stress. This is less easily identified in the characteristics of strain development, and is, in any case, a less satisfactory concept in that the predominating features of levels 2 and 3 overlap and should not be thought to be sharply separable. Nevertheless, it proves convenient to refer to the Initial Cracking level later (see Section 13.5.2).

If a stiff loading machine is used, the test can be displacement controlled and the specimen does not fail instantaneously in an unstable manner when the maximum stress value is attained. Load can be removed gradually and the descending branch of the stress—strain curve traced, as shown in Fig. 13.11. The total work to fracture depends on the ability of the aggregate to act as a crack-stopper or retarder. The softer aggregate is not very effective in this way, and the stress—strain curve is similar to that of h.c.p. with a long nearly linear portion as the stress is raised, and a very steep descending branch. In contrast, the harder aggregate gives a more rounded curve, and a long tail to the descending branch. In this latter case, the strain is, of course, increasing as the stress is removed, and this means that

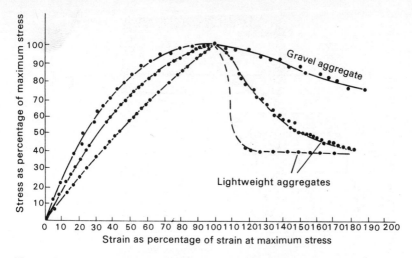

Figure 13.11 Comparative relations between stress and strain for prisms of concrete. (Reproduced with permission, from F. J. Grimer and R. E. Hewitt, in *Conference on Civil Engineering Materials*, edited by M. Te'eni. Copyright © 1971 by John Wiley & Sons, Inc.)

cracks continue to coalesce and widen during the descent from the summit. As stated earlier, the degree of damage increases in a continuous manner throughout the test.

13.5 OTHER FORMS OF STRESS

The previous sections have covered the failure of concrete under short-term uniaxial stress. The basic modes of fracture detailed therein are also apparent when the conditions of loading are less simple; two out of a number of possibilities are dealt with in this section.

13.5.1 Sustained Load

If the applied uniaxial compressive stress is below the level of Final Breakdown, as defined in the previous section, failure will not occur, even if the load is sustained indefinitely. Creep strain does occur (see Chapter 10) and, if the stress is near to the Final Breakdown limit, the time-dependent strain can be large. If the stress is above the limit, unstable crack propagation continues with time, the cracks coalesce and widen, and failure is inevitable. The time to failure is shorter the nearer the stress to the short-term ultimate value.

The situation is summed up admirably in Fig. 13.12, based on experimental results of Rusch (1960). The figure shows the coordinated results of tests in which the strain response was recorded in a number of specimens each at a different level of constant sustained (uniaxial compressive) stress. All the results lie within three boundaries:

(a) the instantaneous modulus, E_c, which gives the least possible strain corresponding to the effect of instantaneous loading;

409

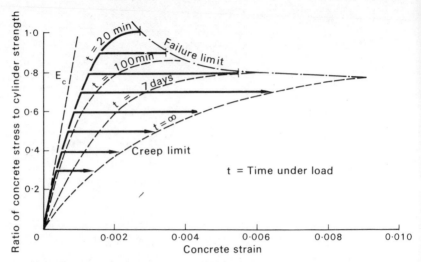

Figure 13.12 Stress–strain curves for concrete subjected to different levels of sustained compressive stress. (From H. Rusch (1960) *Proceedings of the American Concrete Institute,* **57**, 1, by permission of the American Concrete Institute.)

(b) the creep limit, giving the greatest possible strain for specimens below the First Breakdown limit, under load for an infinite time;

(c) the failure limit, which gives the strain (and time) at failure for levels of stress above the First Breakdown limit. For these tests the limit is at about 80% of the short-term ultimate strength. The failure limit is found directly from the tests, whereas the other two boundaries are defined by a time dimension that cannot be realised experimentally, and they have to be determined by extrapolation. This phenomenon of time-dependent fracture is of even greater significance in Timber (see Section 14.7.12).

13.5.2 Multiaxial Stress

Various theoretical criteria such as the Coulomb or the Von Mises (see Jaeger, 1972) endeavour to express the occurrence of failure (or yield) in terms of the three-dimensional stress system. They cannot expect to be universally successful where there are alternative modes of fracture, shear or tension, cleavage or shear-bond. In addition, for concrete, it is found that within the limited zone in which they might apply with some consistency, it is difficult to match the theoretical criteria to the practical results. For the engineer empiricism is the best hope, and the remainder of this section is devoted to the presentation of test results.

Tests under uniaxial compression are difficult enough, as described in Section 13.4, and under multiaxial systems of stress the difficulties multiply. It requires excellent technique to obtain consistent and reliable results, and it is inevitable that there is considerable scatter when the results of different investigations are gathered together. There is then no justification for precise mathematical or numerical representation of the failure surface.

The simplest form of multiaxial stress is the biaxial system in which the third principal stress is zero. The complete envelope for the failure of concrete under biaxial stress is given in Fig. 13.13. These are a par-

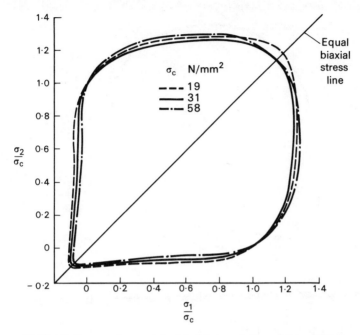

Figure 13.13 Failure envelopes under biaxial stresses σ_1 and σ_2, relative to uniaxial strength σ_c. (From H. Kupfer, H. K. Hilsdorf and H. Rusch (1969) *Proceedings of the American Concrete Institute,* **66**, 660, by permission of the American Concrete Institute.)

ticular set of results which fit well the general tenor of other investigations. The applied stresses σ_1 and σ_2 are plotted nondimensionally as proportions of the uniaxial compressive strength, σ_c.

The state of equal biaxial compression is of interest in that theoretical criteria indicate that failure occurs when the biaxial stresses equal the uniaxial strength, that is $\sigma_1 = \sigma_2 = \sigma_c$. The experimental results for concrete show that this is not the case, the biaxial stresses being rather larger than the uniaxial. This may be related to the mode of failure, as discussed below and indicated in Fig. 13.14.

Under the stress systems that include tension, within three of the four quadrants (Zone 1), the failure is in the cleavage mode, except when the state approaches the uniaxial compressive. Over a short length of the curve on either side of the uniaxial compression the failure is as described above, with shear-bond fracture contributing to the breakdown, and cracks forming all round the specimen in the general direction of the load (Zone 2). In the compression-compression quadrant (Zone 3) the crack pattern becomes more regular in that

411

the cracks form in the plane of the applied loads, splitting the specimen into slabs rather than into the columns of uniaxial compression.

The salient points of the failure surface in the three-dimensional compression zone are demonstrated well by the plot of results on the so-called Rendulic plane, which is defined as the plane containing the two lines, the σ_1 axis ($\sigma_2 = \sigma_3 = 0$) and the equal biaxial stress

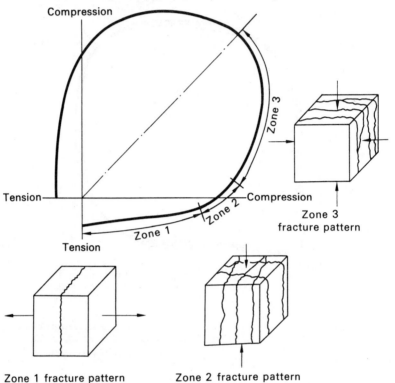

Zone 1 fracture pattern   Zone 2 fracture pattern

Figure 13.14 Modes of failure and fracture patterns for concrete under biaxial stress. (From G. W. D. Vile (1968) The strength of concrete under short term biaxial stress, in *Conference on Structure of Concrete*, by permission of the Cement and Concrete Association.)

line ($\sigma_2 = \sigma_3$, $\sigma_1 = 0$) (see Fig. 13.13). A typical plot on the Rendulic plane is shown in Fig. 13.15. The scatter of experimental results from different investigations is again considerable and the shape of the failure envelope depends on, for instance, the type of aggregate, the water/cement ratio and the volume concentration of aggregate.

In the Rendulic plane the line of equal triaxial stress, $\sigma_1 = \sigma_2 = \sigma_3$ (see Fig. 13.15) occurs at an angle of arctan $1/\sqrt{2}$ to the horizontal, and it is apparent that the ultimate strength state is unattainable under equal all-round stress. On the other hand, Initial Cracking *can* occur, and this is attributable to the compression of the pore structure and a reduction of porosity.

The upper and lower failure lines correspond to the two different

412

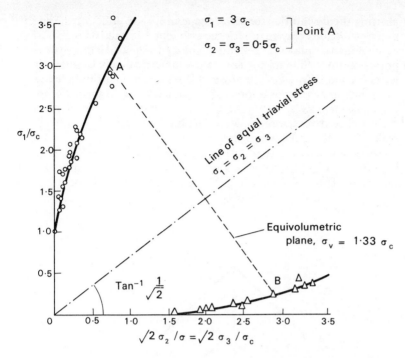

Figure 13.15 Typical failure envelopes on the Rendulic plane. (From Y. Niwa and S. Kobayashi (1967) *Memoirs of the Faculty of Engineering, Kyoto University*, XXIX, 1–15, by permission, as quoted by K. Newman and J. B. Newman (1971) in *Conference on Civil Engineering Materials*, John Wiley.)

modes of failure described for the biaxial stress states. They can be defined in relation to the usual form of triaxial test in which the applied loads impose an axial stress together with a lateral confining pressure. The lower failure line comes from

$$\sigma_1 < \sigma_2 = \sigma_3$$

and the upper from

$$\sigma_1 > \sigma_2 = \sigma_3$$

Any stress system can be taken as the sum of volumetric and deviatoric components. The volumetric stress is given by

$$\sigma_v = (\sigma_1 + \sigma_2 + \sigma_3)/3$$

and the deviatoric stresses in each of the three principal directions by

$$\sigma_{1d} = \sigma_1 - \sigma_v \qquad \sigma_{2d} = \sigma_2 - \sigma_v \qquad \sigma_{3d} = \sigma_3 - \sigma_v$$

Thus the hydrostatic stress is purely volumetric, and planes at right angles to the line of equal stress, $\sigma_1 = \sigma_2 = \sigma_3$, are planes of equal volumetric stress. Any other state of stress can be identified firstly by its equivolumetric plane, and secondly by the distance from the equal stress point on that plane. The distance is found by

413

plotting the deviatoric stresses in the appropriate directions. The final position can be compared with the envelope for failure for that equivolumetric plane. The Rendulic plot of Fig. 13.15 gives two points on the boundary for any equivolumetric plane, as shown by letters A and B (shown here for $\sigma_v = 1.33\sigma_c$). In fact this is really six points since there is a threefold symmetry with respect to the three principal axes. Note that in locating an equivolumetric plane the distance from the origin along the line of equal stress is, to the scale of the Rendulic plot, $\sqrt{3}\sigma_v$.

The whole closed boundary for a given equivolumetric plane can be found from a plan view looking down the equal stress line towards the origin. This is shown in Fig. 13.16 for short-term strength only.

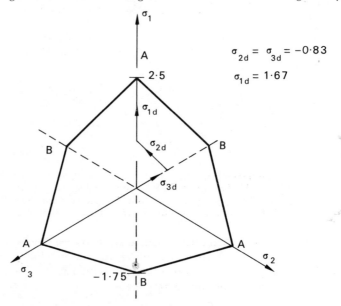

Figure 13.16 Failure envelope on the equivolumetric plane. Values are given in terms of σ_c. (From Fig. 13.15).

The three pairs of points A and B are indicated. Similar envelopes can be drawn for all other equivolumetric planes, including tensile volumetric stresses as well as compressive.

As shown in Fig. 13.16, the view of the principal stress axes is foreshortened in the equivolumetric plane and the scale within the plane is $\sqrt{2}/\sqrt{3}$ times that along the principal stress axis. Hence, to locate the triaxial stress position on the plane, the deviatoric stresses must be plotted to the foreshortened scale rather than to the scale of the stress axis as it appears in the Rendulic plot. This is demonstrated for three deviatoric stresses, σ_{1d}, σ_{2d} and σ_{3d} corresponding to the stress system for point A, namely $\sigma_1 = 3\sigma_c$, $\sigma_2 = \sigma_3 = 0.5\,\sigma_c$.

13.5.3 Design

The failure envelopes described in the last section can be adapted to

414

provide suitable safe limits for stress systems, and can thus be presented as design charts. However, any minor inaccuracies in treatment (theoretical or empirical) are covered by the universal factor of safety (or more correctly load factor) applied when calculating the loads that the structural member must bear and this justifies a simpler approach.

It was pointed out earlier in connection with failure under biaxial stress systems that the intermediate principal stress does affect the

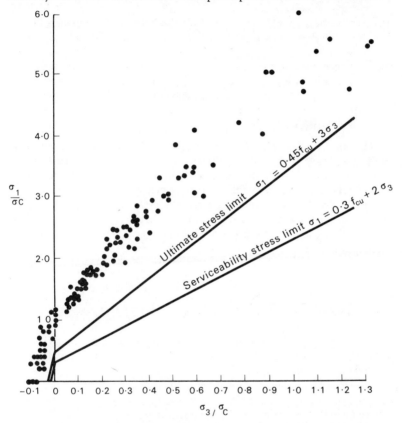

Figure 13.17 Design chart for the strength of concrete under multi-axial stress. (From D. W. Hobbs, C. D. Pomeroy and J. B. Newman (1977) *Design Stresses for Concrete Structures Subject to Multi-axial Stresses*, by permission of the Institution of Structural Engineers.)

strength and thus contributes to the inapplicability of certain theories of fracture or yield. With references to Fig. 13.13 this implies that with the minimum principal stress, $\sigma_3 = 0$, the magnitude of σ_1 changes as the intermediate principal stress, σ_2, varies within the range 0 to σ_1. As can be seen, the change in σ_1 is comparatively small so that it becomes possible, for the purposes of design, to ignore the effect of σ_2.

Experimental values are plotted in Fig. 13.17, giving the relation

415

between σ_1 and σ_3 for a variety of values of σ_2. The strength σ_c is a general measure of uniaxial strength covering both cubes and cylinders. Results are given for both the compression-compression and the compression-tension quadrants, and it is clear that the scatter resulting from the neglect of σ_2 is not too large.

If the uniaxial strength is taken to be the characteristic cube strength, f_{cu}, (see Section 13.7.2.4) then the strength of the concrete in the structural member will be less because:

(a) the cube strength is greater than the cylinder strength and greater than the strength of the same concrete in the structure (see Section 13.2.1);

(b) the concrete in the structure is *not* the same concrete in that its conditioning (including compaction and weathering) is different from the concrete in the cubes. It is less strong and this is allowed for by means of the material load factor, γ_m (see Section 13.7.2.4.).

The sum of these effects is, according to the suggested design procedure of Fig. 13.17, embodied in a factor of 0.45. The resulting relation for design (ultimate) strength is shown on the figure, together with a relation for serviceability. Here serviceability corresponds to the Final Breakdown level discussed earlier (see Section 13.4.2.1) and may be regarded as the upper bound to continuous or fluctuating stresses that can be maintained without the concrete displaying any signs of distress.

13.6   INFLUENCE OF STRUCTURE AND STATE

It was shown earlier that the skeletal structure of h.c.p. develops from the cement grains (see Section 3.2.3), and that the development is a gradual process.

The density of the solid, and hence its strength, depends on the original spacing of the cement grains and on the degree of hydration. These two effects are both reflected in the porosity, so that the connection between strength and porosity is likely to be very significant. Further, this is likely to point to engineering conclusions, for instance on the importance of entrapped or entrained air or the effect of high-temperature curing on the rate of development of strength.

Environmental changes in the ambient atmosphere cause heat and moisture transfer within the h.c.p., resulting in changes in the state and quantity of water in the gel. This modification of the structure has the inevitable consequence of a change of strength. The presence of aggregate introduces the inherent weakness of the particle-paste interface, and a general lowering of strength in concrete in comparison with the h.c.p. contained in it. Nevertheless the trends established for h.c.p. in relation to porosity and moisture state are still valid for concrete.

13.6.1 Porosity

In a material system consisting of a solid phase and pores, it is a matter of common observation that the strength drops off rapidly as the porosity increases; a porosity of 0.1 is enough for the strength to be halved, and all strength has gone when the porosity reaches about 0.5. H.c.p. is no exception, and Powers introduced the concept of the gel-space ratio, α_g, which is related to capillary porosity (see Chapter 3). It is defined as

$$\alpha_g = \frac{\text{volume of cement gel}}{\text{sum of volumes of gel and capillary space}}$$

In terms of the symbols used earlier (see Section 3.2.5) this is:

$$\alpha_g = \frac{m v_g (1 + w_n/c)}{m v_c + w_0/c} \tag{13.9}$$

Empirically, the compressive strength of h.c.p. is then given by

$$\sigma = \sigma_0 \alpha_g^n \tag{13.10}$$

where σ_0 is the strength of the gel and included unhydrated cement (and sand).

The index n is approximately 3, and the value of σ_0 depends on the type of cement; Powers and Brownyard's results show that a higher C_3A content gives a lower strength. Their values of σ_0 for mortars varied between 90 and 130 N/mm^2.

The gel-space concept is effective for normal h.c.p.s where the remaining unhydrated cement forms only a small proportion of the whole. It predicts that the maximum strength, for $\alpha_g = 1$, will be attained whenever the gel has grown to fill all the capillary space. However, when special high-pressure techniques of compaction are used on pastes of very low water/cement ratio (down to 0.08) much higher strengths are reached, even though not all the capillary space has been filled. Since, for complete hydration, the initial water/cement w_0/c needs to be greater than 0.38 (see Fig. 3.7) there must be a great deal of unhydrated cement in these very high strength pastes.

The concept of the gel-space ratio ignores any possible effect on strength of the gel pores, and a more relevant measure is the total porosity (see Section 3.2.5.3). A plot of compressive strength of h.c.p.s against total porosity is given in Fig. 13.18 and the results include various methods of preparation used by several different experimenters. The scatter is agreeably small. Powers' results for h.c.p.s with practical water/cement ratios and prepared by normal mixing methods are on the extreme right of the figure. At the other extreme are specimens prepared by a special hot pressing technique giving porosities of less than a tenth of the normal. The strengths realised by this technique are then over five times the h.c.p. strengths (up to say 100 N/mm^2) encountered in concretes used in everyday practice.

Extrapolation to the right of the figure indicates that there is a critical porosity, p_{crit}, at which the strength drops to zero. The equation to the mean line drawn in the figure gives a strength, f,

$$f = A \log_{10} \frac{p_t/p_{crit}}{} \qquad (13.11)$$

where p_t is the total porosity, the constant A is $310\ N/mm^2$, and p_{crit} is 0.55.

Clearly this relation cannot be extrapolated to the left to give a strength of the solid material but it is interesting to note that, if anything, the strength at the left extreme is underestimated by the equation.

Very high strengths can also be achieved by impregnating the h.c.p. with polymers or by forming compacts from ground up particles of h.c.p. Here again the strengths relate to the total porosity,

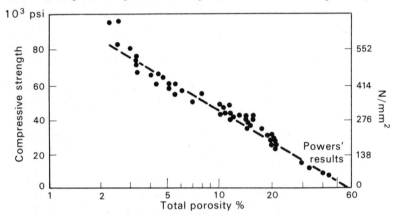

Figure 13.18 The dependence of strength of h.c.p. on porosity. (From D. M. Roy and G. R. Gouda (1975) *Cement and Concrete Research*, 5, 153, by permission of Pergamon Press.)

and it can be concluded that the essence of achieving high strength is 'to fill up the holes'.

Most practical concretes include stiff aggregates which absorb little water, and the range of aggregate contents lies within a comparatively narrow range. Thus, although the type of aggregate and its volume concentration do affect the strength of the concrete, their influence in practice is so much less than that of the porosity of the mix that they may be considered as secondary variables. Porosity, unfortunately, is not a practical quantity, and neither is the degree of hydration; these more fundamental quantities are replaced by free water/cement ratio (see Section 3.5.2) and age of concrete and the strength/porosity relation appears in the everyday form of Fig. 13.19. The concrete is assumed to be wet-cured at normal temperatures, and to contain no entrapped or entrained air. The porosity increases with water/cement ratio and decreases with concrete age, and the strength varies inversely.

13.6.2 Moisture Content and Maturity

Firstly, in the long term, if the concrete is kept wet and the capillary

418

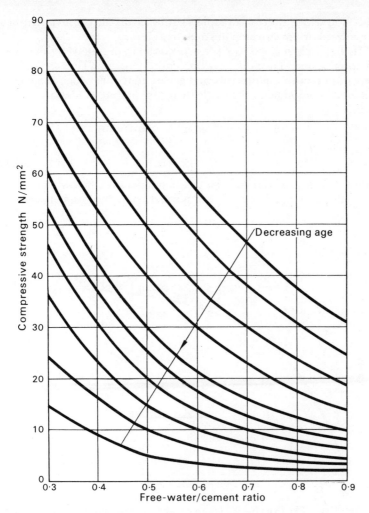

Figure 13.19 Practical chart for strength to water/cement ratio for concrete, also showing the effect of age (other variables excluded). (Reproduced from D.C. Teychenne, R. E. Franklin and H. C. Erntroy (1975) *Design of Normal Concrete Mixes*, by permission of the Director, Building Research Establishment. Crown copyright, Controller HMSO.)

pores remain full of water, hydration continues, the density of the solid structure of the h.c.p. increases, and the strength continues to develop. On the other hand, if drying occurs and the capillaries empty, hydration ceases and the strength development is inhibited. These effects are shown in Fig. 13.20, in which drying occurs in the course of air-curing; it is also shown that if moist-curing is resumed hydration and strength development recommence.

Secondly, if water is removed from the h.c.p. structure, the solid layers draw together and are more difficult to force apart under load, giving enhanced strength. Then, for a given state of hydration the drier the concrete, the higher the strength. This too is shown in Fig.

419

13.20, in which short-term rapid wetting or drying causes a significant downward or upward step in strength.

Higher temperatures of curing speed the hydration reactions (see Chapter 3) and hence the development of strength. Thus strength depends on both temperature and age of concrete, which can be conveniently combined in the single quantity, maturity. Maturity is defined as:

$$maturity\ (M) = age\ (t) \times temperature\ (T)$$

or in general

$$M = \int T \cdot dt \tag{13.12}$$

where T is measured from $-10°C$, the temperature below which strength development ceases. For normal temperatures, between

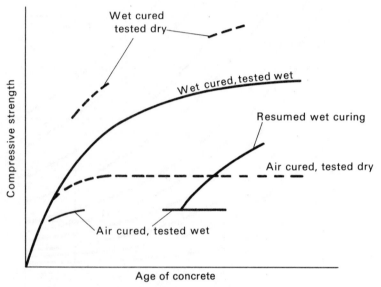

Figure 13.20 Relative effects of curing and testing conditions on the compressive strength of concrete. (From H. J. Gilkey (1937) *Engineering News Record*, 119, 630. Copyright McGraw-Hill. Used with permission of McGraw-Hill Book Company.)

5°C and 40°C, and assuming that the concrete does not dry out, the strength is related directly to the maturity. Then approximately equal strengths are found for concrete with the same maturity. For example, a concrete cured for 20 days at 10°C has a maturity of

$$M = 20 \cdot (10 - [-10]) = 400 \ °C \ days$$

and should have much the same strength as the same concrete cured for 10 days at 30°C.

The strength—maturity relation is logical but nevertheless empirical, and is of limited application. In particular, if high temperatures are applied immediately after casting, the rates of setting and hardening are raised, but the effect on long-term strength is likely to be detrimental.

13.7   VARIABILITY OF CONCRETE STRENGTH

13.7.1   Weibull Weakest Link Theory (Weibull, 1939)

The Griffith concept of brittle fracture, although by no means
directly applicable to concrete, contains the essential grain of truth
about the crack propagation that leads to its failure. Likewise, the
Weibull weakest link theory, although too simple for the complexity
of concrete, does embody the concept that explains certain important
experimental facts.

The Griffith and Weibull ideas are linked in that the Weibull
theory is a statistical approach to the failure of a 'Griffith' material,

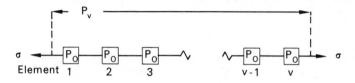

(a) The weakest link model

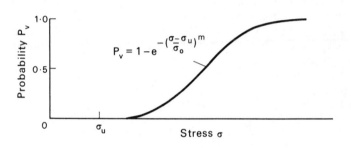

$$P_v = 1 - e^{-\left(\frac{\sigma - \sigma_u}{\sigma_o}\right)^m}$$

(b) Weibull cumulative distribution

Figure 13.21   The Weibull weakest link theory.

with the appropriate brittle habit of fracture. In their simplest forms
both consider the effects of a uniform direct tensile stress. The statistical approach views the testpiece as a chain of elements, each at
the same stress level, with the same probability of failure, P_0 (see
Fig. 13.21(a)). This is equivalent to the probability of the presence
of a Griffith crack of critical length, and if the crack propagates the
element fails instantaneously. Further, since the testpiece is made
up of a chain of elements, the failure of one element signals the failure
of the whole testpiece. Thus, the strength of the testpiece is the
strength of its weakest link.

421

The probability of survival of an element is $1 - P_0$, and the probability of testpiece survival is, for v elements,

$$1 - P_v = (1 - P_0)^v$$

so the probability of testpiece failure is

$$P_v = 1 - (1 - P_0)^v = 1 - \exp \left[v \log (1 - P_0) \right]$$

The bracketed term, referred to as the risk of rupture, is proportional to the total volume, V. Further, P_0 is a function only of the stress, σ. So, for the case of uniform tensile stress,

$$P_v = 1 - \exp \left[-f(\sigma) \cdot V \right]$$

Weibull found that a successful form of the function is

$$f(\sigma) = \left[\frac{\sigma - \sigma_u}{\sigma_0} \right]^m$$

where σ_u is a threshold stress below which there is zero probability of failure, and σ_0 and m are material constants. The final form of the probability of failure is

$$P_v = 1 - \exp \left\{ -\left[(\sigma - \sigma_u)/\sigma_0 \right]^m \cdot V \right\} \tag{13.12}$$

This is represented by the cumulative distribution plot shown in Fig. 13.21(b) which expresses, for a given stress value, the probability of failure before the stress reaches that value. The constant σ_0 determines the spread of the curve, and the constant m affects its slope. If $m = 1$, the distribution is exponential, and represents an underlying normal distribution of strengths.

For two testpieces, alike in every respect except that they have different volumes, V_1 and V_2, the probability of failure will be the same if

$$\left[\frac{(\sigma_1 - \sigma_u)}{\sigma_0} \right]^m \cdot V_1 = \left[\frac{(\sigma_2 - \sigma_u)}{\sigma_0} \right]^m \cdot V_2$$

This implies that, for $V_2 > V_1$, $\sigma_1 > \sigma_2$, so that a larger testpiece will fail at a lower stress than a smaller one. This appeals to common sense in that a critical crack is more likely to be present in a larger volume of material.

As seen earlier, the failure process in concrete is progressive even under direct tension, and is thus more complicated than that assumed in the simple Weibull model. It is possible to extend the model so that it contains elements in parallel as well as in series, but even this cannot be relied on to yield realistic material properties or accurate numerical predictions of material behaviour. Nevertheless, important features anticipated by the theory do appear in concrete. Even when concrete cubes are manufactured and tested with the greatest regard for uniformity, they do not have the same strength, but vary in manner not dissimilar from that postulated by the Weibull theory. Concrete strength does diminish with the size of the testpiece; for instance the cube strength of a 150 mm cube is about 95% of that of a 100 mm cube.

Any truly successful statistical theory would have to be very sophisticated, and for practical purposes it is necessary to approach the problem from the other end; that is, to make measurements of variations in strength, and to draw on the experience of practising engineers.

13.7.2 Variability in Practice

The last section dealt with variations in strength of concrete testpieces which differed only in the random arrangement of their internal structure. None of the variations could be ascribed to different mix proportions, or to different treatment in casting, or to changes in testing technique between one cube and another. In practice the engineer must take account of the possible variations in strength within the structural members, and between structural members which may be made over a period of months by a labour force which may change and which may be inexperienced. There is little chance that the uniformity of material assumed in the last section can be achieved, and the variability in strength of cubes from a site can be attributed much more to changes in the proportions of the constituents of the material than to the random effects of the internal structure. The causes of these larger variations in strength can be listed under four headings.

13.7.2.1 *Materials*

Over a prolonged period the concreting materials will not retain complete uniformity of quality. This is not usually serious as far as the aggregate itself is concerned since its grading and the quality of the particles themselves are unlikely greatly to affect the strength of the concrete. On the other hand, if the aggregate becomes dirty, or if its moisture content increases (so that it feeds water to the cement paste) the strength of the concrete may suffer significantly. More important is the variation in the quality of the cement. This derives from the changes in the raw materials together with day-to-day adjustments in the manufacturing process. To control such alterations is very difficult, and it is occasionally possible for the strength of cement to vary by as much as 10% in the course of a week or two, even when coming from the same works.

13.7.2.2 *Batching and Mixing*

Concrete strength is especially suceptible to changes in the water/cement ratio, which means that inaccuracy in weighing either the water or the cement can cause serious loss of strength. This can result from either carelessness in weighing or deficiencies in the weighing machines, and it is remarkably difficult either to train human beings to work in a consistently error-free manner or to design weighing apparatus that can stand up to the exigencies of site conditions and practice. The ingredients of the concrete must be thoroughly mixed to ensure uniformity, so that an effective mixing action is required of the mixer, and a certain minimum time of mixing must be adhered to. The former can be (and in general is) achieved by good design, but the latter may prove difficult as it tends to run counter to the axiom of site performance that time is money, and that each operation should be completed as quickly as possible.

13.7.2.3 *Transportation, Placing and Curing*

The danger during transportation of the concrete is that jolting will cause the materials to segregate, so that the coarse aggregate falls to the bottom and the water rises to the top. This is avoided by normal good practice such as transfer in a crane skip, in the rotating drum of a mixer truck, or by pumping, Dropping the concrete from a great height can also cause segregation through the different momenta of the various sizes of particle when the concrete stops abruptly on hitting the ground; this too can be easily avoided in deep walls or columns by controlling the rate of descent through a pipe or zig zag chute. More important, incomplete compaction leaves an unduly high air content in the concrete, with a corresponding loss of strength. This can be avoided if the right means of vibrating the air out of the concrete is available and properly used. The normal instrument for in situ concrete is the poker vibrator, but under appropriate conditions vibrating tables, shutter vibrators or compressed air hammers are effective.

Curing is probably the most neglected part of the whole concreting operation. The intention is to prevent the drying out of the concrete, with the consequent cessation of hydration and strength development, or to inhibit the development of temperature gradients through the structure, with the consequent danger of cracking. The measures that are adopted include sealing the surface with a waterproof membrane, often sprayed on, or lining with insulating material, or covering with an absorbent material, such as hessian, which is then supposedly kept wet by regular hosing down.

13.7.2.4 *Testing*

The strength of the concrete is measured from cubes, the concrete for which is sampled in situ between the mixing and placing operations. The cube strengths are thus open to the same sources of variation as the concrete in the structure, but the transportation, compaction and curing hazards are likely to be less severe. Testing introduces a further set of possible variations. The calibration of the cube crusher may be in error or the platens of the machine may carry particles of the previous cube crushed, and this may cause local high stresses and premature failure. The cube may not be correctly centred, or the ball seating may not function successfully.

This is a formidable list of malfunctions or malpractices, and there is little chance of concrete variability being reduced to negligible proportions. It is like road accidents in that the technology is generally adequate; the disaster, although often blamed on bad design, is more often than not caused by human failing.

424

The strengths of a large number of cubes can be plotted in the form of the histogram of Fig. 13.22, and it is found that the distribution approximates to the normal or Gaussian distribution, as shown

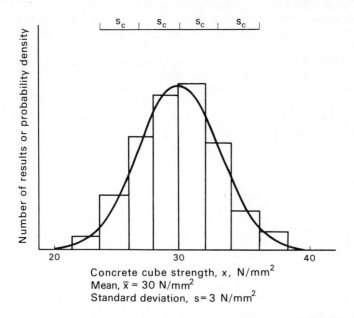

Concrete cube strength, x, N/mm²
Mean, x̄ = 30 N/mm²
Standard deviation, s = 3 N/mm²

Figure 13.22   Histogram of distribution of cube strengths, with equivalent normal distribution. Mean $\bar{x}$ = 30 N/mm². Standard deviation, s_c = 3 N/mm².

by the bell-shaped mean curve. The equation of the normal distribution is

$$y = \frac{1}{s_c \sqrt{(2\pi)}} \exp\left[-\frac{(x - \bar{x})^2}{2 s_c^2} \right] \tag{13.14}$$

where y is the probability density and x is the variate, in this case cube strength. Concrete strength cannot be satisfactorily expressed in terms of one number. It requires two, both of which appear in Equation 13.14. They are:

(a)   the mean strength, $\bar{x}$, usually represented for concrete by f_{cm} where for n cubes, $\bar{x} = \Sigma x / n$;

(b)   the range of strength, given by the standard deviation, s_c, for cube strength, defined by

$$s_c^2 = \frac{\Sigma (x - \bar{x})^2}{n - 1} \tag{13.15}$$

The standard deviation has the same units as strength; for different concretes with the same variability it will often be greater for the concrete with the higher strength. For the purposes of comparison it is then better to use the nondimensional coefficient of variation,

425

S_v, expressed as a percentage,

$$S_v = \frac{s_c}{\bar{x}} \times 100\% \qquad (13.16)$$

For standard conditions (wet curing at normal temperature, testing at 28 days, complete compaction, etc.) the mean strength depends on the mix proportions, and particularly on the water/cement ratio.

The coefficient of variation depends on the random variability of the standard mix, and, to a greater extent, on the degree to which the standard conditions are met. It is thus a function of the degree of control over the concreting operation. If the concrete work on a site is performed by skilled men using efficient equipment under conscientious supervision, the degree of control can be described as excellent, and the coefficient of variation could be as low as 10%. At the other extreme poor control can lead to a coefficient of variation as high as 25%.

The design for strength of concrete structural members is based on the strength of the concrete (and steel). The design cube strength, f_c^1, represents the reasonable minimum cube strength of the concrete in the structure at any time during its life. This differs from the cube strength found by sampling concrete at the site, firstly because the transportation, compaction and curing are likely to be less well controlled in the structural member; and secondly because the member in the course of time suffers weathering or possibly more severe forms of chemical attack. The potentially worse performance of the concrete in the structural member is recognised by the inclusion of a material load factor, γ_m. Then

$$f_c^1 = \frac{f_{cu}}{\gamma_m} \qquad (13.17)$$

where f_{cu} is the characteristic cube strength, which is the reasonable minimum cube strength of the concrete sampled on site. It is defined more rigorously as a strength which has an agreed probability of not being attained, so that it can be expressed in terms of the statistical properties of the normal distribution. The word reasonable is included above to circumvent the difficulty that there is a theoretical statistical probability of zero, or even negative strength. The characteristic strength is related to the mean strength through the standard deviation, s_c.

$$f_{cu} = f_{cm} - K \cdot s_c \qquad (13.18)$$

where the value K expresses the agreed probability and $K \cdot s_c$ is referred to as the margin. From the properties of the normal distribution, if $K = 1.64$ there is a 5% probability of cubes having a strength below f_{cu}. That is 1 cube in 20 can be expected to have a strength less than f_{cu}. Similarly, for $K = 1.96$, 1 cube in 40 is likely to fail below f_{cu}.

The design strength is thus related to the mean strength by

$$f_c^1 = \frac{f_{cm} - K \cdot s_c}{\gamma_m} = \frac{f_{cm}\left(1 - \frac{K \cdot S_v}{100}\right)}{\gamma_m} \qquad (13.19)$$

Normal practical values for γ_m and K are 1.5 and 1.64.

426

The statistical approach means that, for a chosen design strength, the mean cube strength of the concrete (and hence the mix proportions) depends on the coefficient of variation (and hence the degree of control that can be maintained on site). As shown in Fig. 13.23,

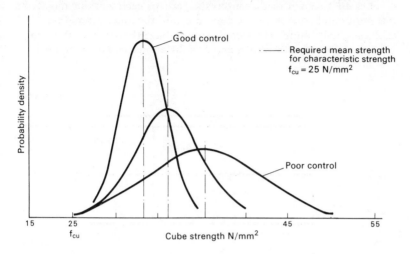

Figure 13.23   Distributions of cube strengths for different degrees of control.

if the degree of control is poor, the coefficient of variation is high, and the required strength of the concrete becomes appreciably greater than for good control. Both good control and high strength are expensive, so the engineer, in determining the mix and the organisation of the concreting operation, is faced with another of those tricky optimum decisions that are so common in civil engineering.

In the last paragraph the problem of the designer was considered, and the mean strength of the concrete was deduced from the design strength and the degree of control. The complementary problem of the site engineer is to ensure that the concrete as produced does indeed reach the standard expected. The most obvious acceptance test consists, as previously suggested, of making cubes from concrete sampled on the site, and measuring the strength after 28 days. In practice the delay of 28 days from casting to testing can be crucial in that site construction continues in the intervening period. If the cubes, after 28 days, are found to be too weak, removal of defective material from the structure also involves removing later (and probably sound) material added on top in the meantime.

The penalty is too great and the temptation is very strong to seek unsatisfactory alternative solutions. This dilemma is hard to resolve. Attempts have been made to reduce the time to testing by, for instance, accelerating the curing of the cubes in hot water, but such methods may not be acceptable, and the 28-day strength is the normal acceptance criterion.

427

Statistical techniques offer a way of identifying a systematic deterioration of concrete quality before it becomes critical, and corrections to the concreting operation can then hopefully be made before any substandard concrete is actually placed in the structure. Samples of n cubes are taken at regular intervals, perhaps each working day, from different batches of the same concrete, and the mean strength of each sample is found. The sample should have the same mean strength, $\bar{x}$, as the concrete as a whole, and any difference can be compared

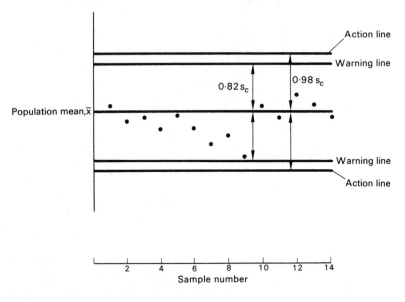

Figure 13.24 Control chart for monitoring concrete strength.

with the standard deviation of the means of the samples, s_{cm}. This is less than that of the cube strengths taken one at a time, i.e. s_c, and the two are simply related:

$$s_{cm} = \frac{s_c}{\sqrt{n}} \qquad (13.20)$$

The number of cubes in the sample does not have to be great and values of n between 4 and 10 are suitable. The sampling technique has the dual advantage of softening the gross effect of the inevitable occasional rogue cube, and of improving the distribution of results — if the distribution of single cube results departs somewhat from the Gaussian, that of the means of samples will be distinctly closer to the normal.

Each mean m (of n cube strengths) can be plotted on a control chart of the kind demonstrated in Fig. 13.24, which has time as its horizontal axis, and thus shows the trends in mean strength m as concreting proceeds. The mean strength expected of the concrete is also shown so that it is immediately obvious whether the concrete

428

is strong enough to satisfy the criterion of structural safety, or so strong as to violate the criterion of economy of material cost. To aid judgement on the seriousness of any departure from the mean, warning and action lines are drawn, the former to draw attention to a likely error in the system, and the latter to show unequivocally that the concrete is unsatisfactory and that the appropriate action must be taken. The distances of these lines from the mean $\bar{x}$ are related to the standard deviation of the means, in much the same way as described earlier using the properties of the normal distribution. Thus suitable values are

$$\text{warning line } \bar{x} \pm 1.64\, s_{cm}$$

$$\text{action line } \bar{x} \pm 1.96\, s_{cm}$$

which can be expressed in terms of the overall standard deviation, s_c, for $n = 4$

$$\text{warning line } \bar{x} \pm 1.64\frac{s_c}{\sqrt{n}} = \bar{x} \pm 0.82\, s_c$$

$$\text{action line} \qquad\qquad \bar{x} \pm 0.98\, s_c$$

In the figure, a series of results are plotted with a distinct downward trend, and a wideawake site engineer would be looking for the cause long before even the warning line is reached. As can be seen, an optimistic view has been taken of his efforts so that there is a sudden return to good mean quality.

13.8   REFERENCES

Alexander, K. M., Wardlaw, J. and Gilbert, D. J. (1968) Aggregate-cement paste bond and the strength of concrete, *Proceedings of an International Conference on The Structure of Concrete* (Ed. A. E. Brooks and K. Newman) Cement and Concrete Association, London, p. 59.

Baker, A. L. L. (1959) An analysis of the deformation and failure characteristics of concrete, *Magazine of Concrete Research*, 11, 119.

British Standard 1881 (1970) Part 4, *Methods of Testing Concrete for Strength*, British Standards Institute, London.

Brown, J. H. (1972) Measuring fracture toughness of cement paste and mortar, *Magazine of Concrete Research*, 24, 185.

Calladine, C. R. (1968) Discussion of crack propagation and failure, *Proceedings of an International Conference on the Structure of Concrete* (Ed. A. E. Brooks and K. Newman) Cement and Concrete Association, London, p. 211.

Gilkey, H. J. (1937) The moist curing of concrete, *Engineering News Record*. 119, 630.

Grimer, F. J. and Hewitt, R. E. (1971) The form of the stress—strain curve of concrete interpreted with a diphase concept of material behaviour, *Proceedings of Conference on Civil Engineering Materials* (Ed. M. Te'eni) Wiley-Interscience, p. 681.

Hobbs, D. W., Pomeroy, C. D. and Newman, J. B. (1976) *Design Stresses for Concrete Structures Subject to Multi-axial Stresses*, the Institution of Structural Engineers. Jaeger, J. C. (1962) *Elasticity, Fracture and Flow*, Methuen, London, 208 pp.

Kupfer, H., Hilsdorf, H. K. and Rusch, H. (1969) Behaviour of concrete under biaxial stresses, *Proceedings of the American Concrete Institute*, 66, 656.

Newman, K. (1966a) Concrete control tests as measures of the properties of concrete, *Proceedings of Symposium on Concrete Quality*, Cement and Concrete Association, London, p. 120.

Newman, K. (1966b) *Concrete Systems*, Chapter VIII, Composite Materials, Elsevier Publishing Co.

Newman, K. and Newman, J. B. (1971) Failure theories and design criteria for plain concrete, *Proceedings of Conference on Civil Engineering Materials*, (Ed. M. Te'eni) Wiley-Interscience, p. 963.

Niwa, Y. and Kobayashi, S. (1967) Failure Criterion of Cement Mortar under Triaxial Compression', *Memoirs of Faculty of Engineering*, Kyoto University, Japan XXIX.

Radjy, F. and Hansen, T. C. (1973) Fracture of hardened cement paste and concrete, *Cement and Concrete Research*, 3, 343.

Roy, D. M. and Gouda, G. R. (1975) Optimization of strength in cement pastes, *Cement and Concrete Research*, 5, 153.

Rusch, H. (1960) Researches towards a general flexural theory for structural concrete, *Proceedings of the American Concrete Institute*, 57, 1.

Spooner, D. C. and Dougill, J. W. (1975) A quantitative assessment of damage sustained in concrete during compressive loading, *Magazine of Concrete Research*, 27, 151.

Swamy, R. N. and Kameswara Rao, C. B. S. (1973) Fracture mechanism in concrete systems under uniaxial loading, *Cement and Concrete Research*, 3, 413.

Teychenné, D. C., Franklin, R. E. and Erntroy, H. C. (1975) *Design of Normal Concrete Mixes*, HMSO, London.

Vile, G. W. D. (1968) The strength of concrete under short-term static biaxial stress, *Conference on The Structure of Concrete* (Ed. A. E. Brooks and K. Newman) Cement and Concrete Association, London, p. 275.

Weibull, W. (1939) A statistical theory of the strength of materials *Proceedings of Royal Swedish Institute for Engineering Research*, Stockholm, 151, 5.

CHAPTER 14

Strength and Failure in Timber

14.1 INTRODUCTION

While it is easy to appreciate the concept of deformation primarily because it is something that can be observed, it is much more difficult to define in simple terms what is meant by the *strength* of a material. Perhaps one of the simpler definitions of strength is that it is a measure of the resistance to failure, providing of course that we are clear in our minds what is meant by failure.

Let us start therefore by defining failure. In those modes of stressing where a distinct break occurs with the formation of two fracture surfaces failure is synonomous with rupture of the specimen. However, in certain modes of stressing, fracture does not occur and failure must be defined in some arbitrary way such as the maximum stress that the sample will endure or, in exceptional circumstances such as compression strength perpendicular to the grain, the stress at the limit of proportionality.

Having defined our end point it is now easier to appreciate our definition of strength as the natural resistance of a material to failure. But how do we quantify this resistance? This may be done by calculating either the stress necessary to produce failure or the amount of energy consumed in producing failure. Under certain modes of testing it is more convenient to use the former method of quantification while the latter tends to be more limited in application. Before describing some of the more frequently applied modes of loading timber, mention must first be made of the use of standard sizes of sample and the adoption of sampling techniques for the characterisation of different timbers.

14.2 SAMPLE SIZE AND SELECTION

Although in theory it should be possible to determine the strength properties of timber independent of size, in practice this is found not to be the case. A definite though small size effect has been established

431

and in order to compare the strength of a timber sample with recorded data it is advisable to adopt the standard sizes set out in the literature.

Two standard procedures have been used internationally; the original was introduced in the USA as early as 1891 using a test sample 2 × 2 inches in cross-section; the second, European in origin, employs a test specimen 20 × 20 mm in cross-section. Prior to 1949 the former size was adopted in the UK, but after this date this larger sample was superseded by the smaller, thereby making it possible to obtain an adequate number of test specimens from smaller trees. Because of the differences in size the results obtained from the two standard procedures are not strictly comparable and a series of conversion values has been determined. With the exception of stiffness which has a ratio of 1.07 for the mean of the 2-inch results divided by the mean of the 20-mm results, most of the strength properties have ratios of from 0.79 to 0.96 (Lavers, 1969).

The early work in the UK on species characterisation employed a sampling procedure in which the test samples were removed from the log in accordance with a cruciform pattern. For 2 × 2 inch samples this method is described fully in ASTM Standard D143-52 which is still used at the current time in the USA and in some of the South American countries. After 1949, the practice in the UK was at first to retain the cruciform sampling scheme but to modify it slightly. However, this was subsequently abandoned and a method devised applicable to the centre plank removed from a log; 20 × 20 mm sticks, from which the individual test pieces are obtained, are selected at random in such a manner that the probability of obtaining a stick at any distance from the centre of a cross-section of a log is proportional to the area of timber at that distance; the cruciform method is retained for use on very small diameter logs. Test samples are cut from each stick eliminating knots, defects and sloping grain; this technique is described fully by Lavers (1969).

14.3   STRENGTH TESTS

The methods of test in the UK are set out in BS 373:1957 'Methods of Testing Small Clear Specimens of Timber'. A number of improvements in methods have been introduced over the years though the standard has not been fully revised. The important points in the standard for each test, together with the improvements, are listed below for the more important properties, though it should be appreciated that this is not the complete list of tests.

14.3.1   Tension Parallel to the Grain

In this test the sample is 300 × 20 × 6 mm and waisted in the central region to 6 mm × 3 mm; a sample under test is illustrated in Fig. 11.4 and the calculated stress at rupture is recorded as the ultimate tensile strength. This test is performed only infrequently since the amount

432

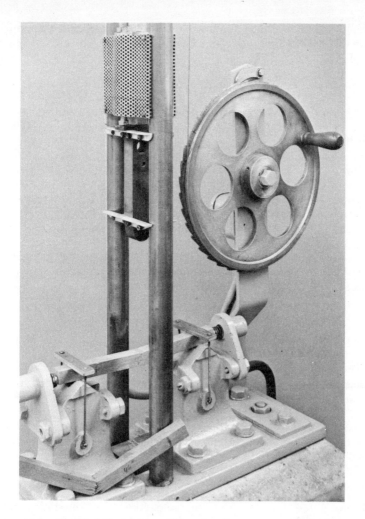

Figure 14.1 Modified Hatt–Turner impact bending test on a 300 x 20 x 20 mm specimen. (From the Princes Risborough Laboratory, Building Research Establishment. © Crown copyright.)

of timber loaded in tension under service conditions is quite small. A further reason for the lack of tensile data is the difficulty experienced in performing the tensile test; first, due to the very high tensile strength of timber, it is difficult to grip the material without crushing the grain, and secondly, failure is frequently in shear at the end of the waisted region rather than in tension within the waisted region. It is almost impossible to conduct the standard tensile test in 'green' timber.

14.3.2 Compression Parallel to the Grain

The sample size is 60 x 20 x 20 mm and the load is applied at a rate of 0.01 mm/s through a ball contact and plunger. The maximum

strength parallel to the grain is obtained by dividing the maximum
load recorded during the test by the cross-sectional area of the
specimen.

14.3.3   Static Bending

A specimen 300 x 20 x 20 mm is supported over a span of 280 mm
in trunnions carried on roller bearings and the load is applied to the
centre of the beam at a constant strain rate of 0.11 mm/s. The orientation of the growth rings is parallel to the direction of loading and an
extensometer is usually attached to provide a load-deflection diagram
from which is calculated the modulus of elasticity.

Three strength properties are usually determined from this test.
The first and most important is the *modulus of rupture*, which is a
measure of the ultimate bending strength of timber for that size of
sample and that rate of loading. This modulus is actually the
equivalent stress in the extreme fibres of the specimen at the point
of failure, calculated on the assumption that the simple theory of
bending applies: in three-point bending the modulus is given by:

$$\text{MOR (in N/mm}^2) = \frac{3PL}{2bd^2} \qquad (14.1)$$

where P is the load in newtons, L is the span length in mm, b is the
width of the beam in mm, and d is the thickness of the beam in mm.

The second strength parameter is *work to maximum load*, which is
a measure of the energy expended in failure and is determined from
the area under the load-deflection curve up to the point of maximum
load. This parameter is consequently a measure of the toughness of
timber, as is also the third parameter, namely *total work*, where the
area under the load-deflection curve is taken to complete failure.
Both values have the units mm N/mm^3.

14.3.4   Impact Bending

We have noted above how two energy parameters of the static bending test provide a measure of the toughness of the material. A number
of more direct methods have been employed in timber, many of
these borrowed from metallurgy. Thus both Charpy and Izod
methods have been used, but, owing to the variability of the structure
of timber, the small samples used in these tests can be unrepresentative of the material. Consequently larger samples have been adopted
and tested in a modified Hatt-Turner type of machine (Fig. 14.1). In
this test a beam 300 x 20 x 20 mm is supported over a span of
240 mm on chair supports radiused to 15 mm. A weight of 1.5 kg,
also radiused to 15 mm, is dropped onto the beam from increasing
heights until failure occurs, or the deflection equals 60 mm.

Criticism has been levelled at this technique principally on the
grounds that the sample is subjected to repeated blows rather than
one single blow. Nevertheless the test, though so empirical in nature,
appears to give a good indication of the toughness of a timber in
practice.

14.3.5 Shear Parallel to the Grain

This test is made on a 20 mm cube using a pivoted-arm shear test rig as illustrated in Fig. 14.2. The cube is loaded through a ball seating at

Figure 14.2 Shear parallel to the grain using a 20 mm cube. (From the Princes Risborough Laboratory, Building Research Establishment. © Crown copyright.)

a rate of 0.008 5 mm/s. Tests are made on matched pairs of specimens in the radial and tangential planes and averaged to give an *ultimate shear strength*.

14.3.6 Cleavage

The sample is 45 × 20 × 20 mm with a transverse groove at one end into which are slotted the tensile grips. The rate of loading is 0.042 mm/s and tests are made on matched pairs of specimens to give failures in both radial and tangential planes. The failing load is usually expressed as force per unit width of test sample. A similar test in which the sample is grooved at both ends is used to measure the tensile strength of timber perpendicular to the grain.

14.3.7 Hardness

This is an important property in the use of timber for both domestic and industrial flooring. A specially hardened steel tool rounded to a diameter of 11.3 mm is embedded in the timber to a depth corresponding to half its diameter at a rate of 0.11 mm/s.

435

Table 14.1   Average and Standard Deviation of Various Mechanical Properties
of Selected Timbers at 12% Moisture Content

	Density	Static bending*				Impact	Compression	Hardness	Shear	Cleavage	
	Dry (kg/m³)	Modulus of Rupture (N/mm²)	Modulus of elasticity (N/mm²)	Energy to max. load (mmN/mm²)	Energy to fracture (mmN/mm²)	Drop of hammer (m)	Parallel to grain (N/mm²)	On side grain (N)	Parallel to grain (N/mm²)	Radial plane (N/mm width)	Tangential plane (N/mm width)
HARDWOODS											
Balsa	176	23	3200	0.018	0.035		15.5		2.4		
		7.3	1060	0.007	0.017		4.43		0.62		
Obeche	368	54	5500	0.058	0.095	0.48	28.2	1910	7.7	9.3	8.4
		6.5	620	0.010	0.015	0.072	3.00	268	0.67	1.82	1.58
Mahogany (*Khaya ivorensis*)	497	78	9000	0.070	0.128	0.58	46.4	3690	11.8	10.0	14.0
		15.0	1520	0.026	0.044	0.149	8.45	816	2.56	2.08	2.90
Sycamore	561	99	9400	0.121	0.163	0.84	48.2	4850	17.1	16.8	27.3
		11.0	1160	0.028	0.049	0.136	4.83	639	2.32	2.95	3.91
Ash	689	116	11900	0.182	0.281	1.07	53.3	6140	16.6		
		16.6	2170	0.045	0.097	0.216	7.73	1158	2.52		
Oak	689	97	10100	0.093	0.167	0.84	51.6	5470	13.7	14.5	20.1
		16.8	1960	0.026	0.051	0.209	7.98	911	2.38	2.86	2.08
Afzelia	817	125	13100	0.100	0.203	0.79	79.2	7870	16.6	10.5	13.3
		26.6	1760	0.043	0.087	0.215	12.02	914	2.28	2.00	2.49
Greenheart	977	181	21000	0.213	0.395	1.35	89.9	10450	20.5	17.5	22.2
		20.9	1990	0.047	0.088	0.207	8.49	1531	3.06	4.79	4.97

SOFTWOODS

Norway spruce (European spruce)	417	72	10200	0.086	0.116	0.58	36.5	2140	9.8	8.4	9.1
		10.2	2010	0.022	0.040	0.116	5.26	353	1.44	1.07	1.20
Yellow pine (Canada)	433	80	8300	0.089	0.097	0.56	42.1	2050	9.3	8.2	11.6
		10.9	1440	0.015	0.019	0.100	6.14	473	1.61	1.57	1.77
Douglas fir (UK)	497	91	10500	0.097	0.172	0.69	48.3	3420	11.6	9.5	11.4
		16.9	2160	0.038	0.081	0.200	8.03	865	2.29	1.90	2.17
Scots pine (UK)	513	89	10000	0.103	0.134	0.71	47.4	2980	12.7	10.3	13.0
		16.9	2130	0.032	0.053	0.167	9.25	697	2.45	1.82	2.47
Caribbean Pitch pine	769	107	12600	0.126	0.253	0.91	56.1	4980	14.3	12.1	13.3
		14.5	1800	0.042	0.060	0.196	7.76	1324	2.81	1.23	1.58

* In three-point loading.

14.4 STRENGTH VALUES

For those strength properties described above, with the exception of tensile parallel to the grain, the mean values and standard deviations (see below) are presented in Table 14.1 for a selection of timbers covering the range in densities to be found in the hardwoods and softwoods. Many of the timbers whose elastic constants were presented in Table 11.1 are included. All values relate to a moisture content which is in equilibrium with a relative humidity of 65% at 20 °C; these are of the order of 12% and the timber is referred to as 'dry'. Modulus of elasticity has also been included in the table.

Table 14.1 is compiled from data presented in Bulletin 50 of the Forest Products Research Laboratory (Lavers, 1969) which lists data for both the dry and green states for 200 species of timber. The upper line for each species provides the estimated average value while the lower line contains the standard deviation.

In Table 14.2 tensile strength parallel to the grain is listed for certain timbers and it is in this mode that timber is at its strongest.

Comparison of these values with those for compression parallel to the grain in Table 14.1 will indicate that, unlike many other materials, the compression strength is only about one-third that of tensile strength along the grain.

14.5 VARIABILITY IN STRENGTH VALUES

In Chapter 4 attention was drawn to the fact that timber is a very variable material and that for many of its parameters, e.g. density, cell length and microfibrillar angle of the S_2 layer, distinct patterns of variation could be established within a growth ring, outwards from the pith towards the bark, upwards in the tree, and from tree to tree. The effects of this variation in structure are all too apparent when mechanical tests are performed.

An efficient estimator of the variability which occurs in any one property is the *sample standard deviation*, denoted by *s*. It is the square root of the variance and is derived from the formula:

$$s = \sqrt{\frac{\Sigma x^2 - \frac{(\Sigma x)^2}{n}}{n-1}} \tag{14.2}$$

where *x* stands for every item successively and *n* is the number of items in the sample.

The frequency diagram is typical of the manner in which many strength properties are distributed, and in general it approximates to the theoretical normal distribution curve. Results for compression strength of western hemlock are presented on a frequency basis in Fig. 14.3; superimposed on this is the plot of a normal distribution and the degree of fit will be seen to be very good. Now in a normal distribution approximately 68% of the results should lie in theory

Table 14.2 Tensile Strength Parallel to the Grain of Certain Timbers

Timber		Moisture content %	Tensile strength (N/mm²)
Hardwoods			
Ash	(Home grown)	13	136
Beech	(Home grown)	12.6	180
Yellow poplar	(Imported)	15	114
Softwoods			
Scots pine	(Home grown)	16	92
Scots pine	(Imported)	15	110
Sitka spruce	(Imported)	15	139
Western hemlock	(Imported)	15	137

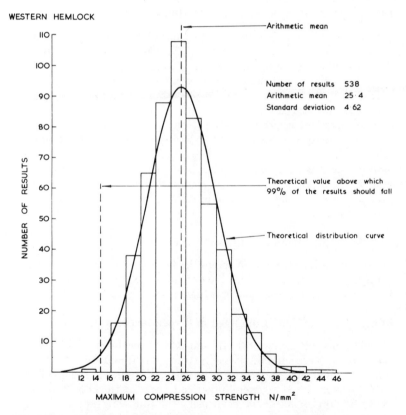

Figure 14.3 Frequency distribution of maximum compression strength. (From the Princes Risborough Laboratory, Building Research Establishment. © Crown copyright.)

within $+1s$ and $-1s$ of the mean and 99.8% should fall within $\pm 3s$ of the mean.

The standard deviation provides us with a measure of the variability, but in itself gives little impression of the magnitude unless related to the mean. This ratio is known as the *coefficient of variation*, i.e.

$$CV = \frac{s}{\text{mean}} \%$$ (14.3)

The coefficient of variation varies considerably, but is frequently under 15% for many biological applications. However, reference to Table 14.1 will indicate that this value is frequently exceeded. For design purposes the two most important properties are the moduli of elasticity and rupture which have coefficients typically in the range of 10–30%.

14.6 INTER-RELATIONSHIPS AMONG THE STRENGTH PROPERTIES

14.6.1 Modulus of Rupture and Modulus of Elasticity

A high correlation exists between the moduli of rupture and elasticity for a particular species: Fig. 14.4 shows an example of this, but in this particular case redwood and whitewood are considered as a single population because of their similarity in strength properties.

It is doubtful whether this correlation between MOR and MOE represents any causal relationship; rather it is more probable that the correlation arises as a result of the strong correlation that exist between density and each modulus. Whether it is a causal relationship or not, it is nevertheless put to good advantage for it forms the basis of the stress grading of timber by machine. The deflection of timber as it is fed between rollers is used to grade the timber on a strength basis. Further reference to stress grading will be made later in this chapter.

14.6.2. Impact Bending and Total Work

Good correlations have been established between the height of drop in impact bending test and both *work to maximum load* and *total work*: generally the correlation is higher with the latter property.

14.6.3 Hardness and Compression Perpendicular to the Grain

Correlation coefficients of 0.902 and 0.907 have been established between hardness and compression strength perpendicular to the grain of timber of 12% moisture content and timber in the green state respectively. It is general practice to predict the compression strength from the hardness result using the following equations:

$$Y_{12} = 0.00147x_{12} + 1.103$$ (14.4)

$$Y_g = 0.00137x_g - 0.207$$ (14.5)

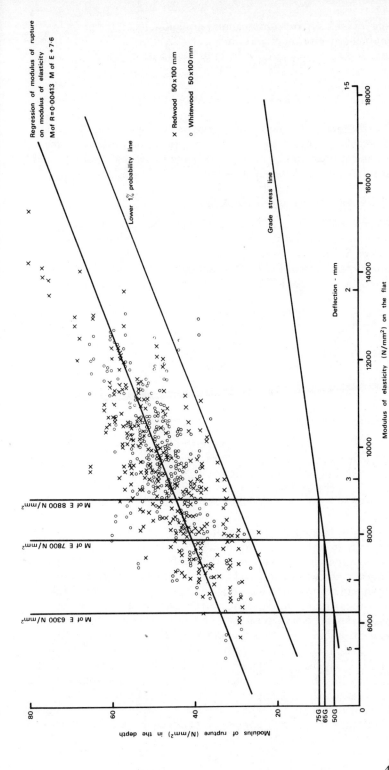

Figure 14.4 The relationship between modulus of rupture and modulus of elasticity for both redwood (*Pinus sylvestris*) and whitewood (principally *Picea abies*). (From the Princes Risborough Laboratory, Building Research Establishment. © Crown copyright.)

where Y_g and Y_{12} are compression perpendicular to the grain in N/mm^2 for green timber and timber at 12% moisture content respectively, and x_g and x_{12} are hardness in Newtons.

14.7 FACTORS AFFECTING STRENGTH

Many of the variables noted in the previous chapter as influencing stiffness also influence the various strength properties of timber. Once again, these can be regarded as being either material dependent or manifestations of the environment.

14.7.1 Anisotropy and Grain Angle

The marked difference between the longitudinal and transverse planes in both shrinkage and stiffness has been discussed in previous chapters. Strength likewise is directionally dependent and the degree of anisotropy present in both tension and compression is presented for Douglas fir in Table 14.3. Irrespective of moisture content the

Table 14.3 Anisotropy in Strength

Timber	Moisture content (%)	Tension			Compression		
		$//$ (N/mm^2)	$\perp$ (N/mm^2)	$// : \perp$	$//$ (N/mm^2)	$\perp$ (N/mm^2)	$// : \perp$
Douglas fir	>25	131	2.69	48.7	24.1	4.14	5.82
	12	138	2.90	47.6	49.6	6.90	7.19

highest degree of anisotropy is in tension (48:1): this reflects the fact that the highest strength of timber is in tension along the grain while the lowest is in tension perpendicular. A similar degree of anisotropy is present in the tensile stressing of both glass reinforced plastics and carbon-fibre-reinforced plastics when the fibre is laid up in parallel strands.

Table 14.3 also demonstrates that the degree of anisotropy in compression is an order of magnitude less than in tension. Whilst the compression strengths are markedly affected by moisture content, tensile strength appears to be relatively insensitive. The comparison of tension and compression strengths along the grain in Table 14.3 reveals that timber, unlike most other materials, has a tensile strength considerably greater than the compression strength.

Anisotropy in strength is due in part to the cellular nature of timber and in part to the structure and orientation of the microfibrils in the wall layers. Bonding along the direction of the microfibrils is covalent whilst bonding between microfibrils is by hydrogen bonds. Consequently, since the majority of the microfibrils are aligned at only a small angle to the longitudinal axis, it will be easier to rupture the cell wall if the load is applied perpendicular than if applied parallel to the fibre axis.

442

Since timber is an anisotropic material it follows that the angle at which stress is applied relative to the longitudinal axis of the cells will determine the ultimate strength of the timber. Fig. 14.5 illustrates that tensile strength is much more sensitive to grain angle than is compression strength. However, at angles as high as 60° to the

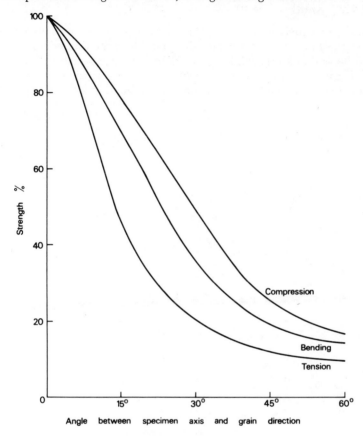

Figure 14.5 Effect of grain angle on the tensile, bending and compression strength of timber. (After R. Baumann (1922).)

longitudinal both tension and compression strengths have fallen to only about 10% of their value in straight grained timber. The sensitivity of strength to grain angle in timber is identical with that for fibre orientation in both glass and carbon-fibre-reinforced plastics.

It is possible to obtain an approximate value of strength at any angle to the grain from knowledge of the corresponding values both parallel and perpendicular to the grain using the following formula, which, in its original form, was credited to Hankinson:

$$\sigma_\theta = \frac{\sigma_p}{\sigma_p \sin^n \theta + \sigma_q \cos^n \theta} \tag{14.6}$$

where σ_θ is the strength property at angle θ from the fibre direction,

σ_p is the strength parallel to the grain, σ_q is the strength perpendicular to grain, and n is an empirically determined constant. In tension $n = 1.5-2$; in compression $n = 2-2.5$. The equation has also been used for stiffness where a value of 2 for n has been adopted.

14.7.2 Knots

Knots are associated with distortion of the grain and since even slight deviations in grain angle reduce the strength of the timber appreciably it follows that knots will have a marked influence on strength. The significance of knots, however, will depend on their size and distribution both along the length of a piece of timber and across its section.

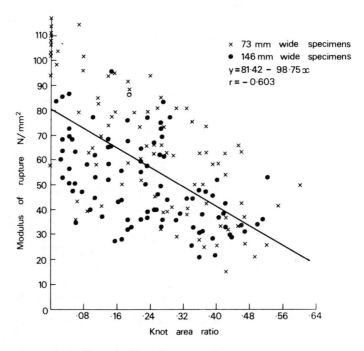

Figure 14.6 Effect of knot area ratio and on the strength of timber. (From the Princes Risborough Laboratory, Building Research Establishment. © Crown copyright.)

Thus knots in clusters are more important than knots of a similar size which are evenly distributed, while knots on the top or bottom edge of a beam are more significant than those in the centre; large knots are much more critical than small knots.

It is very difficult to quantify the influence of knots; one of the parameters that has been successfully used is the *knot area ratio*, which relates the sum of the cross-sectional area of the knots at a cross-section to the cross-sectional area of the piece. The loss in bending strength that occurred with increasing knot area ratio in 200 home-grown Douglas fir boards is illustrated in Fig. 14.6.

444

14.7.3 Density

In Chapter 4 density was shown to be a function of cell wall thickness and therefore dependent on the relative proportions of the various cell components and also of the level of cell wall development of any one component. However, variation in density is not restricted to different species, but can occur to a considerable extent within any one species and even within a single tree. Some measure of the interspecific variation that occurs can be obtained from the limited amount of data in Table 14.1. It will be observed that as density increases so the various strength properties increase. Density continues to be the best prediction of timber strength since high correlations between strength and density are a common feature in timber studies.

Most of the relations that have been established throughout the world between the various strength properties and timber density take the form of

$$\sigma = kg^n \qquad (14.7)$$

where σ is any strength property, g is the specific gravity, k is a proportionality constant differing for each strength property, and n is an exponent that defines the shape of the curve. An example of the use of this expression on the results of over 200 species tested in compression parallel to the grain is presented in Fig. 14.7: the correlation coefficient between compression strength and density of the timber at 12% moisture content was 0.902.

Similar relationships have been found to hold for other strength properties though in some the degree of correlation is considerably lower. This is the case in tension parallel to the grain where the ultrastructure probably plays a more significant role.

Over the range of density of most of the timbers used commercially the relationship between density and strength can safely be assumed to be linear with the possible exception of shear and cleavage; similarly, within a single species the range is low and the relationship can again be treated as linear.

14.7.4 Ring Width

Since density is influenced by the rate of growth of the tree it follows that variations in ring width will change the density of the timber and hence the strength. However, the relationship is considerably more complex than it appears at first. In the ring-porous timbers such as oak and ash (see Chapter 4) increasing rate of growth (ring width) results in an increase in the percentage of the latewood which contains most of the thick-walled fibres; consequently, density will increase and so will strength. However, there is an upper limit to ring width beyond which density begins to fall owing to the inability of the tree to produce the requisite thickness of wall in every cell.

In the diffuse-porous timbers such as beech, birch and khaya, where there is uniformity in structure across the growth ring, increasing rate of growth (ring width) has no effect on density unless, as before, the rate of growth is excessive. In the softwoods, however,

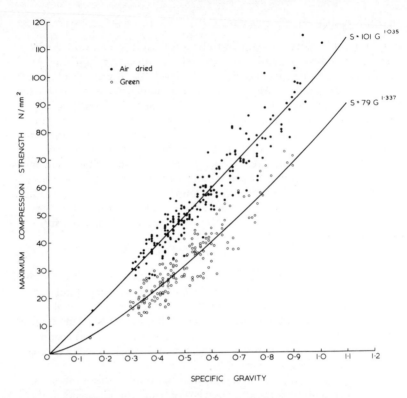

Figure 14.7 The relation of maximum compression strength to specific gravity for 200 species tested in the green and dry states. (From the Princes Risborough Laboratory, Building Research Establishment. © Crown copyright.)

increasing rate of growth results in an increased percentage of the low-density earlywood and consequently both density and strength decrease as ring width increases. Exceptionally, it is found that very narrow rings can also have very low density: this is characteristic of softwoods from the very northern latitudes where latewood development is restricted by the short summer period. Hence ring width of itself does not affect the strength of the timber: nevertheless, it has a most important indirect effect working through density.

14.7.5 Ratio of Latewood to Earlywood

Since the latewood comprises cells with thicker walls it follows that increasing the percentage of latewood will increase the density and therefore the strength of the timber. Differences in strength of 150–300% between the late and earlywood are generally explained in terms of the thicker cell walls of the former: however, some workers maintain that when the strengths are expressed in terms of the cross-sectional area of the cell wall the latewood cell is still stronger than the earlywood. Various theories have been advanced to

446

account for the higher strength of the latewood wall material; the more acceptable are couched in terms of the differences in micro-fibrillar angle in the middle layer of the secondary wall, differences in degree of crystallinity and, lastly, differences in the proportion of the chemical constituents.

14.7.6 Cell Length

Since the cells overlap one another it follows that there must be a minimum cell length below which there is insufficient overlap to permit the transfer of stress without failure in shear occurring. Some investigators have gone further and have argued that there must be a high degree of correlation between the length of the cell and the strength of cell wall material, since a fibre with high strength per unit of cross-sectional area would require a larger area of overlap in order to keep constant the overall efficiency of the system.

14.7.7 Microfibrillar Angle

The angle of the microfibrils in the S_2 layer has a most significant effect in determining the strength of wood. Fig. 14.8 illustrates the

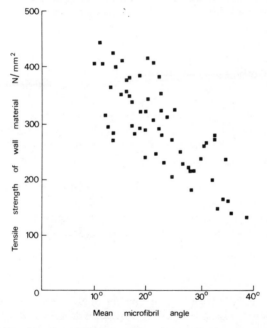

Figure 14.8 Effect of microfibrillar angle on the tensile strength of *Pinus radiata* blocks. (From I. D. Cave (1969) *Wood Science and Technology*, 3, 40–48, by permission of Springer-Verlag.)

marked reduction in tensile strength that occurs with increasing angle of the microfibrils: the effect on strength closely parallels that with changing grain angle.

447

14.7.8 Chemical Composition

In Chapter 4 the structure of the cellulose molecule was described
and emphasis was placed on the existence in the longitudinal plane of
covalent bonds both within the glucose units and also linking them
together to form filaments containing from 5 000 to 10 000 units.
There is little doubt that the high tensile strength of timber owes
much to the existence of this covalent bonding. Certainly experiments
in which many of the interunit bonds have been ruptured by gamma
irradiation resulting in a decrease in the number of units in the
molecule from over 5 000 to about 200 have resulted in a most
marked reduction in tensile strength; it has also been shown that
timber with inherently low molecular lengths, e.g. compression wood,
has a lower than normal tensile strength.

Until recently it has been assumed that the hemicelluloses which
constitute about half of the matrix material play little or no part in
determining the strength of timber. However, it has recently been
demonstrated that some of the hemicelluloses are orientated within
the cell wall and it is now thought that these will be load-bearing.

It is known that lignin is less hydrophillic than either the cellulose
or hemicelluloses and, as indicated earlier, at least part of its function
is to protect the more hydrophillic substances from the ingress of
water and consequent reduction in strength. Apart from this indirect
effect on strength, lignin is thought to make a not too insignificant
direct contribution. Much of the lignin in the cell wall is located in
the primary wall and in the middle lamella. Since the tensile strength
of a composite with fibres of a definite length will depend on the
efficiency of the transfer of stress by shear from one fibre to the
next, it will be appreciated that in timber the lignin is playing a most
important role: compression strength along the grain has been shown
to be affected by the degree of lignification not between the cells
but rather within the cell wall, when all the other variables have been
held constant.

It would appear, therefore, that both the fibre and the matrix components of the timber composite are contributing to its strength as in
fact they do in most composites: the relative significance of the fibre
and matrix roles will vary with the mode of stressing.

14.7.9 Reaction Wood

14.7.9.1 *Compression Wood*

The chemical and anatomical properties of this abnormal wood,
which is found only in the softwoods, were described in Chapter 4.
When stressed, it is found that the tensile strength and toughness is
lower and the compressive strength higher than that of normal timber.
Such differences can be explained in terms of the changes in fine
structure and chemical composition.

14.7.9.2 *Tension Wood*

This second form of abnormal wood, which is found only in the

448

hardwoods, has tensile strengths higher and compression strengths lower than normal wood: again this can be related to changes in fine structure and chemical composition.

14.7.10 Moisture Content

The marked increase in strength on drying from the fibre saturation point to oven-dry conditions was described in detail in Chapter 7 and illustrated in Fig. 7.2; recent experimentation has indicated the probability that at moisture contents of less than 2% the strength of timber may show a slight decrease rather than the previously accepted continuation of the upward trend.

Confirmatory evidence of the significance of moisture content on strength is forthcoming from Fig. 14.7 in which the regression line for over 200 species of compression strength of green timber against density is lower than that for timber at 12% moisture content; strength data for timber are generally presented for these two levels of moisture content (Lavers, 1969).

Within certain limits the regression of strength, expressed on a logarithmic basis, and moisture content can be plotted as a straight line. The relationship can be expressed mathematically as:

$$\log_{10} \sigma = \log_{10} \sigma_f + k(\mu_f - \mu) \tag{14.8}$$

where σ is the strength at moisture content μ, σ_f is the strength at fibre saturation point, μ_f is the moisture content at fibre saturation point, and k is a constant. It is possible therefore to calculate the strength at any moisture content below the fibre-saturation point, assuming σ_f to be the strength of the green timber and μ_f to be 25%. This formula can also be used to determine the strength changes that occur for a 1% increase in moisture content over certain ranges.

This relation between moisture content and strength may not always apply when the timber contains defects. Thus it has been shown that the effect of moisture content on strength diminishes as the size of knots increase. The relation, even for knot-free timber, does not always hold for the impact resistance of timber. In some timbers, though certainly not all, impact resistance or toughness of green timber is considerably higher than it is in the dry state; the impact resistances of green ash, cricket bat willow and teak are approximately 10%, 30% and 50% higher respectively than the values at 12% moisture content. This increase could be related to the increased flexibility (decreased stiffness) in the wet state, but since the latter is common to all timbers it is difficult to understand why toughness of green timber is not higher than dry timber for all species. Certainly it is not possible to account for the disparity in behaviour in terms of differences in microscopic structure.

14.7.11 Temperature

At temperatures within the range +200 °C to −200 °C and at constant moisture content strength properties are linearly (or almost linearly)

449

related to temperature, decreasing with increasing temperature. However, a distinction must be made between short- and long-term effects.

When timber is exposed for short periods of time to temperatures below 95 °C the changes in strength with temperature are reversible. These reversible effects can be explained in terms of the increased molecular motions and greater lattice spacing at higher temperatures. Fig. 14.9 illustrates the increase in strength that occurred in three

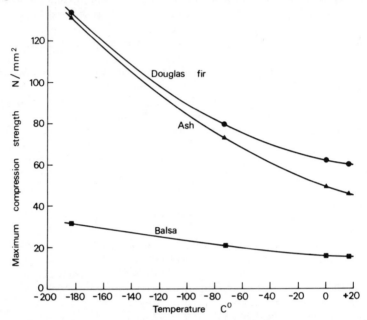

Figure 14.9   The effect of temperature on the maximum compression strength parallel to the grain in two hardwood and one softwood timbers. (From the Princes Risborough Laboratory, Building Research Establishment. © Crown copyright.)

timbers loaded in compression along the grain as temperature was progressively reduced.

At temperatures above 95 °C or at temperatures above 65 °C for very long periods of time there is an irreversible effect of temperature due to thermal degradation of the wood substance, generally taking the form of a marked shortening of the length of the cellulose molecules and chemical changes within the hemicelluloses. All strength properties show a marked reduction with temperature, but toughness is particularly sensitive to thermal degrade. Repeated exposure to elevated temperature has a cumulative effect and usually the reduction is greater in the hardwoods than in the softwoods.

The effect of temperature is very dependent on moisture content, sensitivity of strength to temperature increasing appreciably as moisture content increases (Fig. 14.10), as occurs also with stiffness (Fig. 11.13). The relationship between strength, moisture content and temperature appear to be linear over the range 6–20% and −20°

450

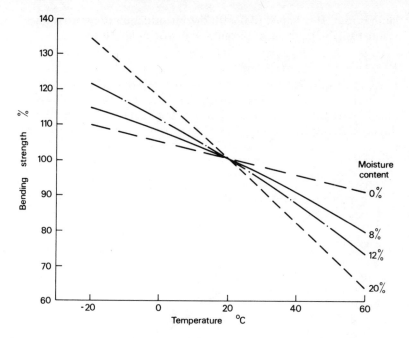

Figure 14.10 The effect of temperature on the bending strength of *Pinus radiata* timber at different moisture contents.

to 60 °C, thereby allowing transformation of results. However, in the case of toughness, while at low moisture content it is found that toughness decreases with increasing temperature, at high moisture contents toughness actually increases with increasing temperature.

14.7.12 Time

In Chapter 11 timber was described as a viscoelastic material and as such its mechanical behaviour will be time dependent. Such dependence will be apparent in terms of its sensitivity to both rate of loading and duration of loading.

14.7.12.1 *Rate of Loading*

Increase in the rate of load application results in increased strength values, the increase in 'green' timber being some 50% greater than that of timber at 12% moisture content; strain to failure, however, actually decreases. A variety of explanations have been presented to account for this phenomenon, most of which are based on the theory that timber fails when a critical strain has been reached and consequently at lower rates of loading viscous flow or creep is able to occur resulting in failure at lower loads.

The various standard testing procedures adopted throughout the world set tight limits on the speed of loading in the various tests: unfortunately, such recommended speeds vary throughout the world, thereby introducing errors in the comparison of results from different

451

laboratories. It is hoped that with the introduction of international standards (ISO) greater uniformity will arise in the future.

14.7.12.2 Duration of Load

In terms of the practical use of timber the duration of time over which the load is applied is perhaps the single most important variable. Many investigators have worked in this field and each has recorded a direct relationship between the length of time over which a load can

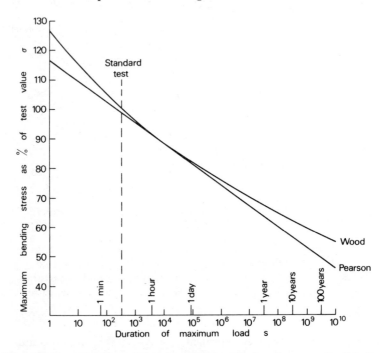

Figure 14.11 The effect of duration of load on the bending strength of timber. (After L. W. Wood (1951) *Forest Products Laboratory Report* No. 1916 and R. G. Pearson (1972) *Holzforschung*, **26** (4).)

be supported at constant temperature and moisture content and the magnitude of the load. This relation appears to hold true for all forms of stress but is especially important for bending strength.

The modulus of rupture (maximum bending strength) will decrease in proportion, or nearly in proportion, to the logarithm of the time over which the load is applied; failure in this particular time-dependent mode is termed *creep rupture* or *static fatigue*. Wood (1951) indicated that the relation is slightly curvilinear and that there is a distinct level-ling off at loads approaching 20% of the ultimate short-term strength such that a critical load or stress level occurs below which failure is unlikely to occur; the hyperbolic curve that fitted Wood's data best for both rapid and sustained loading is indicated in Fig. 14.11.

Other workers have reported a linear relation, though a tendency to nonlinear behaviour at very high stress levels has been recorded by some of them. Pearson (1972), in reviewing previous work in the field of duration of load and bending strength, plotted on a single graph the results obtained over the last 30 years and found that despite differences in species, specimen size, moisture content, or whether the timber was solid or laminated, the results showed remarkably little scatter from a straight line described by the regression:

$$\sigma = 91.5 - 7 \log_{10} t \qquad (14.9)$$

where σ is the stress level (%), and t is the effective duration of maximum load (h). This regression, which is only slightly different from those of Clouser (1959) to which reference was made in Chapter 11, is also plotted in Fig. 14.11 for comparison with the curvilinear line. Pearson's findings certainly throw doubt on the existence of a critical stress level below which creep rupture does not occur.

These regressions indicate that timber beams which have to withstand a dead load for 50 years can be stressed to only 50% of their ultimate short-term strength. However, the reduction in strength with time can be accelerated if instead of steady-state conditions cyclic changes in moisture content occur. This effect has already been described in Chapter 11 and illustrated in Fig. 11.20: a beam loaded to only $\frac{3}{8}$ of its short-term ultimate strength, but subjected to repeating wetting and complete drying, failed in a very short period of time. This work has been confirmed on a number of species and it appears that the significant factor is the *range* in moisture contents through which the samples were cycled rather than the actual values of moisture content obtained.

Variations in moisture content do occur in beams under service conditions, but, since the cross-section of the beams is usually considerable, the effect of daily changes in moisture content appear to be restricted to only the outer layers. Certain seasonal changes, however, will have an effect and will contribute to an increase in deflection and decrease in time to creep rupture; it should be appreciated that these contributions will be small since the range in moisture content change will be very much lower than the range used in the experiments described above.

In Chapter 11 attention was drawn to the fact that viscoelastic behaviour could manifest itself not only as creep (deformation increasing with time at constant stress) but also as stress relaxation (stress decreasing with time at constant strain). Very few experiments on stress relaxation have been conducted on timber, but in one of these samples of Douglas fir at 8% moisture content and 22 °C were strained to 90% of the estimated ultimate strain before allowing the samples to stress relax; over 50% failed within a period of 7 days (Bach, 1967).

Failure Theories for Time-dependent Fracture.   Timber has been shown to be a viscoelastic material and in the formulation of a failure theory to account for creep rupture (static fatigue) the general theory of viscoelasticity has been adopted using some particular boundary criterion which is defined in appropriate rheological terms.

453

Generally a critical strain criteria has been adopted; thus Ylinen (1957) combined a viscoelastic model, namely Maxwell's differential equation for relaxation, with St Venant's fracture criteria. The relationship between stress and time is given by the equation

$$\sigma_f = E_\infty \epsilon_f + kT(E - E_\infty)(1 - e^{-\epsilon_f/kt}) \qquad (14.10)$$

where σ_f is the failure stress, ϵ_f is the strain at failure, t is time, T is the relaxation time, E is the instantaneous elastic modulus, and E_∞ is the elastic modulus for infinitely long loading $<E$. This theory was shown to agree well with data in the literature.

A similar approach has utilised the combination in series of a spring and a Maxwell element and applied the Coulomb hypothesis of a critical elastic strain: this provided the time of onset of plastic flow which in turn was used to calculate the long-term load of the timber.

James (1962) in questioning the general applicability of Ylinen's results suggested replacement of the linear model by a nonlinear one. Even so, this critical strain approach is not sufficiently general since the total strain to rupture is dependent on the time history of loading.

An entirely different approach has been the use of some energy criterion of fracture since it has been demonstrated that the total mechanical work to fracture may be constant when uniform rates of loading are adopted in the various excitation modes. These observations have led to the adoption of the Reiner–Weisenberg failure criterion in conjunction with linear viscoelastic theory. This particular criterion states that the total sum of stored elastic energy at unstable fracture is a constant which is independent of the time to failure; this criterion then is an extended form of the von Mises failure criterion for elastic materials and is applicable to viscoelastic materials that do not change in volume under long term stressing. Although this latter requirement is not entirely met in the case of timber, this failure criterion has been examined by Bach (1973) who found that it was acceptable for timber in a slightly modified form:

$$w_c^v(t_f) = -w_c^e(t_f) + \int_0^{t_f} (\dot{w} - \dot{D}) \, dt = R_v \qquad (14.11)$$

where $w_c^v(t_f)$, the viscoelastic conserved elastic potential at failure, equals the materials constant, R_v; $w_c^e(t_f)$ is the instantaneously recoverable elastic potential at failure; t_f is the time to failure; $\dot{w}$ is the rate of work done on the system; and $\dot{D}$ is the rate of dispersion of nonelastic energy. Bach has used this in conjunction with viscoelastic theory to calculate time to failure under different methods of loading.

14.8 STRENGTH, FAILURE AND FRACTURE MORPHOLOGY

There are two fundamentally different approaches to the concept of strength and failure. The first is the classical strength of materials approach, attempting to understand strength and failure of timber in terms of the strength and arrangement of the molecules, the fibrils, and the cells by thinking in terms of a theoretical strength and attempting to identify the reasons why the theory is never satisfied.

The second approach is much more practical in concept since it considers timber in its present state, ignoring its theoretical strength and its microstructure and stating that its performance will be determined solely by the presence of some defect, however small, which will initiate on stressing a small crack; the ultimate strength of the material will depend on the propagation of this crack.

Both approaches have been applied to timber though the second to a lesser degree than the first primarily because it is fairly recent in concept and the theories require considerable modification for the different fracture modes in an anisotropic material; both approaches are discussed below for the more important modes of stressing.

14.8.1 Classical Approach

14.8.1.1 *Tensile Strength Parallel to the Grain*

Over the years a number of models have been employed in an attempt to quantify the theoretical tensile strength of timber. In these models it is assumed that the lignin and hemicelluloses make no contribution to the strength of the timber; in the light of recent investigations, however, this may be no longer valid for some of the hemicelluloses. One of the earliest attempts modelled timber as comprising a series of endless chain molecules, and strengths of the order of $8\,000$ N/mm^2 were obtained. More recent modelling has taken into account the finite length of the cellulose molecules and the presence of amorphous regions. Calculations have shown that the stress to cause chain slippage is generally considerably greater than that to cause chain scission irrespective of whether the latter is calculated on the basis of potential energy function or intrachain linkbond energies; preferential breakage of the cellulose chain is thought to occur at the C–O–C linkage. These important findings have led to the derivation of minimum tensile stresses of the order of $1\,000$–$7\,000$ N/mm^2 (Mark, 1967).

The ultimate tensile strength of timber is of the order of 100 N/mm^2, though this varies considerably between species. This figure corresponds to a value between $\frac{1}{10}$ and $\frac{1}{70}$ of the theoretical strength of the cellulose fraction. Since this accounts for only half the weight of the timber (Table 4.1) and since it is assumed, perhaps incorrectly, that the matrix does not contribute to the strength, it can be said that the actual strength of timber lies between $\frac{1}{5}$ and $\frac{1}{35}$ of its theoretical strength.

In attempting to integrate these views of molecular strength with the overall concept of failure it is necessary to examine strength at the next order of magnitude, namely the individual cells. It is possible to separate these by dissolution of the lignin-pectin complex cementing them together (Chapter 4 and Figs. 4.5 and 4.7). Using specially developed techniques of mounting and stressing, it is possible to determine their tensile strengths: much of this work has been done on softwood tracheids, and mean strengths of the order of 500 N/mm^2

have been recorded by a number of investigators. The strengths of the latewood cells can be up to three times that of the corresponding earlywood cells.

Individual tracheid strength is therefore approximately five times greater than that for solid timber. Softwood timber also contains parenchyma cells which are found principally in the rays, and lining the resin canals, and which are inherently weak; many of the tracheids tend to be imperfectly aligned and there are numerous discontinuities along the cell; consequently it is to be expected that the strength of timber is lower than that of the individual tracheids. Nevertheless, the difference is certainly substantial and it seems doubtful if the features listed above can account for the total loss in strength especially when it is realised that the cells rupture on stressing and do not slip past one another.

When timber is stressed in tension along the grain, failure occurs catastrophically with little or no plastic deformation (Fig. 11.2) at strains about 1%. Visual examination of the sample usually reveals an interlocking type of fracture which can be confirmed by optical microscopy. However, as illustrated in Fig. 14.12, the degree of interlocking is considerably greater in the latewood than in the earlywood; whereas in the former the fracture plane is essentially vertical, in the latter the fracture plane follows a series of shallow zig-zags in a general transverse plane; it is now thought that these thin-walled cells contribute very little to the tensile strength of the timber. Thus failure in the stronger latewood region is by shear, while in the earlywood, though there is some evidence of shear failure, most of the rupture appears to be in straight tension.

Examination of the fracture surfaces by electron microscopy reveals that the plane of fracture occurs either within the S_1 layer or, as is more common, between the S_1 and S_2 layers. Since shear strengths are lower than tensile strengths these observations are in accord with comments made previously on the relative superiority of the tensile strengths of individual fibres compared with the tensile strength of timber. By failing in shear this implies that the shear strength of the wall layers is lower than the shear strength of the lignin-pectin material cementing together the individual cells.

Confirmation of these views is forthcoming from the work of Mark (1967) who has calculated the theoretical strengths of the various cell wall layers and has shown that the direction and level of shear stress in the various wall layers was such as to initiate failure between the S_1 and S_2 layers. Mark's treatise has received a certain amount of criticism on the grounds that he has treated one cell in isolation, opening it up longitudinally in his model to treat it as a two-dimensional structure; nevertheless, the work marked the beginning of a new phase of investigation into elasticity and fracture and the approach has been modified and subsequently developed. The extension of the work has explained the initiation of failure at the S_1–S_2 boundary, or within the S_1 layer, in terms of either buckling instability of the microfibrils, or the formation of ruptures in the matrix or framework giving rise to a redistribution of stress.

456

Figure 14.12   Tensile failure in spruce (*Picea abies*) showing mainly transverse cross wall failure of the earlywood (left) and longitudinal intrawall shear failure of the latewood cells (right) x 110, polarized light). (From the Princes Risborough Laboratory, Building Research Establishment. © Crown copyright.)

Thus, both the microscopic observations and the developed theories appear to agree that failure of timber under longitudinal tensile stressing is basically by shear. However, under certain conditions the pattern of tensile failure may be abnormal. At temperatures in excess of 100 °C the lignin component is softened and its shear strength is reduced. Consequently, on stressing, failure will occur within the cementing material rather than within the cell wall.

Secondly, in timber that has been stressed in compression before being pulled in tension, it will be found that tensile rupture will occur along the line of compression damage which, as will be explained below, runs transversely. Consequently, the tensile fracture will be horizontal giving rise to a brittle type fracture.

457

In the literature is recorded a wide range of tensile failure criteria, the most commonly applied being some critical strain parameter, an approach which is supported by a considerable volume of evidence, though its lack of universal application has been pointed out by several workers. The situation is therefore very similar to that described for the criteria used to explain failure under sustained load (Section 14.7.12.2).

14.8.1.2 *Compression Strength Parallel to the Grain*

Compression failure is a slow yielding process in which there is a progressive development of structural change. The initial stage of this sequence appears to occur at a stress less than 25% of the ultimate failing stress (Dinwoodie, 1968) though Keith (1971) considers that these early stages do not develop until about 60% of the ultimate. There is certainly a very marked increase in the amount of structural change above 60% which is reflected by the marked departure from linearity of the stress—strain diagram illustrated in Fig. 11.2. The former author maintains that linearity here is an artefact resulting from insensitive testing equipment and that some plastic flow has occurred at levels well below 60% of the ultimate stress. Compression deformation assumes the form of a small kink in the microfibrillar structure, and because of the presence of crystalline regions in the cell wall it is possible to observe this feature using polarisation microscopy (Fig. 14.13). The sequence of irreversible anatomical changes leading to failure originates in the tracheid or fibre wall at that point where the longitudinal cell is displaced vertically to accommodate the horizontally running ray. As stress and strain increase these kinks become more prominent and increase numerically, generally in a preferred lateral direction, horizontally on the radial plane (Fig. 14.14) and at an angle to the vertical axis of from 45° to 60° on the tangential plane. These lines of deformation, generally called a crease and comprising numerous kinks, continue to develop in width and length; at failure, defined in terms of maximum stress, these creases can be observed by eye on the face of the block of timber. At this stage there is considerable buckling of the cell wall and delamination within it, usually between the S_1 and S_2 layers. Models have been produced to simulate buckling behaviour and calculated crease angles for instability agree well with observed angles (Grossman and Wold, 1971).

At a lower order of magnification, Dinwoodie (1974) has shown that the angle at which the kink traverses the cell wall varies systematically between earlywood and latewood, between different species, and with temperature. Almost 72% of the variation in the kink angle could be accounted for by a combination of the angle of the microfibrils in the S_2 layer and the ratio of cell wall stiffness in longitudinal and horizontal planes.

Attempts have been made to relate the size and number of kinks to the amount of elastic strain or the degree of viscous deformation. Under conditions of prolonged loading, total strain and the ratio of

458

Figure 14.13 Formation of kinks, in the cell walls of spruce timber (*Picea abies*) during longitudinal compression stressing. The angle, θ, lying between the plane of shear and the middle lamella varies systematically between timbers and is influenced by temperature (x 1600, polarized light). (From the Princes Risborough Laboratory, Building Research Establishment. © Crown copyright.)

creep strain to elastic strain (the creep function), appear to provide the most sensitive guide to the formation of cell wall deformation; the gross creases appear to be associated with strains of 0.33% (Keith, 1972).

The number and distribution of kinks is dependent on temperature and moisture content: increasing moisture content, though resulting in a lower strain to failure, results in the production of more kinks, although each smaller in size than its 'dry' counterpart; these are to be found in a more even distribution than they are in dry timber. Increasing temperature results in a similar wider distribution of the kinks.

14.8.1.3 *Static Bending*

In the bending mode timber is subjected to compression stresses on the upper part of the beam and tensile on the lower part. Since the strength of clear timber in compression is only about one third that in tension failure will occur on the compression side of the beam long before it will do so on the tension side. In knotty timber, however, the compressive strength is often equal to and can actually exceed the tensile strength. As recorded in the previous section, failure in

459

compression is progressive and starts at low levels of stressing: consequently the first stages of failure in bending will frequently be associated with compression failure and as both the bending stress and consequently the degree of compression failure increase so the

Figure 14.14 Failure under longitudinal compression at the macroscopic level. On the longitudinal radial plane the crease (shear line) runs horizontally, while on the longitudinal tangential plane the crease is inclined at 65° to the vertical axis. (From the Princes Risborough Laboratory, Building Research Establishment. © Crown copyright.)

neutral axis will move progressively downwards from its original central position in the beam (assuming uniform cross-section) thereby allowing the increased compression load to be carried over a greater cross-section. Fracture occurs when the stress on the tensile surface reaches the ultimate strength in bending.

460

14.8.1.4 *Toughness*

Timber is a tough material, and in possessing moderate to high stiffness and strength in addition to its toughness it is favoured with a unique combination of mechanical properties emulated only by bone, which, like timber, is a natural composite.

Toughness is generally defined as the resistance of a material to the propagation of cracks; various methods are available to measure toughness or some index of toughness and these have been described in Section 14.3.4. In the comparison of materials it is usual to express toughness in terms of *work of fracture*, which is a measure of the energy necessary to propagate a crack thereby producing new surfaces.

In timber the work of fracture involved in the production of cracks at right angles to the grain is about $10^4 \, \text{J/m}^2$; this value is an order of magnitude less than that for ductile metals, but is comparable with that for the man-made composites. Now the energy required to break all the chemical bonds in a plane cross section is of the order of $1-2 \, \text{J/m}^2$; that is, four orders of magnitude lower than the experimental values. Since pullout of the microfibrils does not appear to happen to any great extent it is not possible to account for the high work of fracture in this way (Gordon and Jeronimidis, 1974).

One of the earlier theories to account for the high toughness in timber was based on the work of Cook and Gordon (1964) who demonstrated that toughness in fibre-reinforced materials is associated with the arrest of cracks made possible by the presence of numerous interfaces. As these interfaces open, so secondary cracks are initiated at right angles to the primary thereby dissipating its energy. This theory is applicable to timber as Fig. 14.15 illustrates, but it is doubtful whether the total discrepancy in energy between experiment and theory can be explained in this way.

Some recent and interesting results have contributed to a better understanding of the problem (Gordon and Jeronimidis, 1974). Prior to fracture it would appear that the cells separate in the fracture area; on further stressing these individual and unrestrained cells buckle inwards generally assuming a triangular shape. In this form they are capable of extending up to 20% before final rupture thereby absorbing a large quantity of energy. Inward buckling of helically wound cells under tensile stresses is possible only because the microfibrils of the S_2 layer are wound in a single direction. Observations and calculations on timber have been supported by glass-fibre models and it is considered that the high work of fracture can be accounted for by this unusual mode of failure. It appears that increased toughness is possibly achieved at the expense of some stiffness, since increased stiffness would have resulted from contrawinding of the microfibrils in the S_2.

So far, we have discussed toughness in terms of only clear timber. Should knots or defects be present timber will no longer be tough and the comments made earlier as to viewpoint are particularly relevant here. The material scientist sees timber as a tough material: the structural engineer will view it as a brittle material because of its inherent defects and this theme will be developed in the following section.

Figure 14.15 Crack-stopping in a fractured rotor blade. The orientation of the secondary cracks corresponds to the microfibrillar orientation of the middle layer of the cell wall (× 990, polarized light.) (From the Princes Risborough Laboratory, Building Research Establishment. © Crown copyright.)

Loss in toughness, however, can arise not only on account of the presence of defects and knots, but also through the effects of acid, prolonged elevated temperatures or fungal attack on wood, or the presence of compression damage resulting from overstressing within the living tree or in the handling or utilisation of timber after conversion. Under these abnormal conditions the timber is said to be *brash*.

14.8.2 Engineering Approach to Strength and Fracture

Two basic aspects will be discussed: the first is *fracture mechanics*, which is concerned with the strength of actual materials as limited by

462

the propagation of cracks; while the second is assessing variability in materials relates strength to the weakest link in the structure of the material. The two concepts are obviously related though each may be applied independently to different aspects of strength.

14.8.2.1 *Fracture Mechanics and its Application to Timber*

The concept of fracture mechanics has already been discussed in detail in Section 2.5, and those readers unfamiliar with this concept would be well advised to read that previous section. Suffice it here to remind readers of the most salient points of the concept and to discuss its application to timber.

Griffith postulated firstly that in a material there are flaws, and it is the worst flaw which causes failure; and, secondly, that a balance exists between the energy due to elastic strain and the energy necessary to produce new surfaces. The largest flaw will become self-propagating when the rate of release of strain energy exceeds the rate of increase of surface energy of the propagating crack. The theory was later modified to take account of both plastic work and kinetic energy.

Now assuming a continuum structure and that all deformation around the crack tip obeys linear elasticity theory, it follows from Griffith's concept that the crack will propagate when

$$\sigma^2 \pi c = EG_c \qquad (14.12)$$

where $2c$ is the crack length, σ is the stress applied, E is Young's modulus, and G_c is the critical strain energy release rate ($G_c = 2\gamma$, where γ is the energy required to form new fracture surfaces). Now the left-hand side of Equation 14.12 contains engineering parameters (stress + distance) while the right-hand side contains material properties. Now since

$$EG_c = K_c \qquad (14.13)$$

for plane stress (or $EG_c/(1 - \nu^2)$ for plane strain) it follows that K_c, the *stress intensity factor* at the critical state, is also a material property: most important, K_c is independent of crack length and this means for each material it is possible to obtain the relationship between the critical stress and the length of crack.

Experimentally, by measuring stress to failure for a series of crack geometries, it is possible by plotting $\sigma \times 1/c$) to get K_c. Knowing K_c and setting limits on c, the limiting stress σ can be calculated from Equations 14.12 and 14.13.

Three modes of failure are defined: the opening mode is usually denoted as 1 and the appropriate critical intensity factor as K_{1c}; the forward shear mode is 2 and the transverse shear mode is 3.

Now it has been shown that the critical stress intensity factors for orthotropic materials in tension, bending and shear are similar to those for isotropic materials, but in Equation 14.13, relating the critical stress intensity factor with the critical strain energy release rate, E must be replaced by the appropriate elastic constant. Furthermore, there does not appear to be any interaction between the three basic modes in an orthotropic body and hence the superposition of these three modes is sufficient to describe the most general case.

A major difference in the fracture of isotropic and orthotropic materials in tension is that a crack in the latter does not necessarily propagate in a direction perpendicular to the direction of maximum stress. Timber, because of its very low cleavage strength, will tend to fracture with splitting along the grain irrespective of the direction of the initial crack.

Early investigations on western white pine indicated that the concept of fracture mechanics in the opening mode form could be applied to fracture in the LR and LT planes, i.e. cleavage along the grain. G_{1c} was shown to be independent of both crack length and specimen geometry, but dependent on moisture content and temperature. Fracture energy in the LR plane is frequently higher than in the LT plane which can probably be related to the microfibrillar reinforcement around the pits.

A value of K_{1c} of 60 MN/m$^{3/2}$ has been recorded for a mode 1 fracture in the LT direction in utile (Williams and Birch, 1975); such a value is appreciably lower than that for aluminium alloy (7075—T6) with a K_{1c} of 940 MN/m$^{3/2}$ and for medium carbon steel with a K_{1c} of 1700 MN/m$^{3/2}$ though very much higher than the values quoted for ceramics in Table 13.1.

In the assessment of the tensile strength of Douglas fir perpendicular to the grain it was found that the presence of checks or splits, frequently as a result of shrinkage on drying out, had a very marked affect in reducing strength. Comparison of actual strengths with those predicted using fracture mechanics and treating the splits as cracks which would operate in mode 1 was exceedingly good (Schniewind and Lyon, 1973).

A further practical application of fracture mechanics has been the evaluation of the effect of knots on the tensile strength of timber *along* the grain. Correlation coefficients of between 0.87 and 0.96 were obtained between the actual test values and calculated values for timber containing isolated knots: the choice of a crack length equal to $\frac{3}{4}$ of the knot diameter for a centrally placed knot, or $\frac{7}{8}$ for an edge knot gave estimates of the stress at fracture which were on average within 1% and 3% respectively of the average test values (Pearson, 1974).

It appears, therefore, that within certain boundaries defined in terms of moisture content, temperature and probably time, fracture mechanics when applied to the mode 1 type of fracture has definite advantages over the critical stress or strain theories previously proposed for timber.

Recent work has been directed at extending the theory: thus it is interesting to note that fracture toughness in timber has now been shown to be related to the crack velocity, the resistance to crack propagation decreasing as crack velocity increases. Interest in mode 2 fracture has revealed that the energy release rate is very dependent on the different elastic properties in the principal directions.

Interest is now centred on the possible application of fracture mechanics to the determination of strength under long-term stressing. Initial results on slow crack growth are promising and predicted results have shown good agreement with the curves presented in Fig. 14.11 and discussed in Section 14.7.12.2 (Mindness *et al.*, 1975).

So far the discussion has been restricted to tensile rupture, either along the grain or transverse to it as cleavage. Until fairly recently it was thought that fracture mechanics could be employed only where a distinct crack is involved. However, this tool has now been applied to compression failure of carbon-fibre-reinforced plastics where the compression crease is treated as a crack for the purpose of the analysis. The treatment seems to have been successful and no doubt will be applied to timber in the near future.

14.8.2.2 *Weibull Distribution*

As discussed in Section 13.7.1, the Griffith and Weibull ideas are connected in that the Weibull theory is a statistical approach to the failure of a brittle material. Using the weak-link concept as described in Section 13.7.1, Weibull presented the first theories capable of quantifying effects of stress distribution and volume on strength of materials. The distribution which arises from a set of nonlinear hazard functions is called the Weibull distribution. This is an extreme value distribution for minimum values and consequently is strongly influenced by the lowest value unlike most other distributions. It is to be preferred for predicting the lower end of a distribution which quite frequently is the critical area in practical terms.

Two and sometimes three parameters are required covering the shape, scale and location of the distribution; although methods are established for estimating these parameters no one method appears ideal under all circumstances. Computer programs for calculating these parameters have recently been published (Pierce, 1976).

The complexity of the concept has done little to prevent its fairly extensive usage in ceramics but so far it has been used only infrequently for timber. However, this concept has been used in the derivation of working stresses for timber, about which more will be said in a later section.

One of the first more specialised applications of the Weibull distribution was in the evaluation of the bending strengths of beams when it was found that the changes in strength resulting from changes in span and beam depth could be explained satisfactorily by a weakest-link model. More recently it has been successfully applied to predict the relation between specimen volume and strength in timber stressed in tension perpendicular to the grain (Barrett, 1974). Timber appears to exhibit size effects to a greater extent than that in most other materials.

14.9 PRACTICAL SIGNIFICANCE OF STRENGTH DATA

The engineer contemplating the use of timber will automatically ask himself:

465

(a) How do the mechanical properties of timber compare with other materials? and

(b) Are there working stresses for timber and, if so, how are they derived and how can they be used?

These questions are discussed below.

14.9.1 Comparison of Strength Data of Timber and Other Materials

Timber has been used as a construction material for centuries and continues to be used extensively in many forms of construction despite the increase in number of competing materials. Its ability to withstand competition is a reflection not just of the level of strength it possesses, but more important the level of strength in terms of both cost and weight.

In Table 14.4 the tensile and compression strengths of whitewood, one of the principal timbers of construction, are set out in comparison with many other traditional materials together with some of the

Table 14.4   Strength and Stiffness of Timber
(Constructional 'Whitewood') in Comparison
with Other Materials

Material	SG	Tensile strength (MN/m^2)	Specific tensile (MN/m^2)	E (GN/m^2)	$\dfrac{E}{SG}$ (GN/m^2)	Compression (MN/m^2)
Wood (spruce)	0.46	104	226	10	22	37
Concrete	2.5	4	2	48	19	69
Glass	2.5	50	20	69	28	50
Aluminium (Duralumin)	2.8	247	88	69	25	—
Cast iron	7.8	138	18	207	26	120
Steel (Mild) (0.06% carbon)	7.9	459	58	203	26	800
ABS	1.1	50	45	3	3	50
PVC (rigid)	1.5	59	39	2.4	1.7	55
Polyester resin/G cloth	1.8	276	153	18	10	270
Epoxy uni-directional roving	1.8	1 100	611	45	25	400
CFRP	1.5	1 040	693	180	120	1 040

new constructional materials: stiffness (in bending) is included for completeness. It will be observed that the tensile strength of whitewood is in the middle of the range, but when strength is quoted in terms of specific gravity so that the strength of an equal mass of material is compared, timber is far superior to the traditional materials, though not as good as the man-made composites. In both stiffness, which at best is only moderate in comparison to other materials, and also compression parallel to the grain, which is poor in comparison, the position of timber is enhanced when the property is expressed in terms of weight. The use of timber for the Mosquito aircraft during the war and its continued, though declining, use for glider production bears testament to its high strength to weight ratio.

When cost is taken into the comparison process, the position of timber relative to other materials is certainly favourable. In the comparison of timber and steel beams to carry the same load with similar deflections the cost of the former will be slightly lower than its steel counterpart. However, where strength and stiffness are not so important and where timber has to be machined to a particular profile and subsequently painted, the competition from extruded rigid plastic foams is very acute and, as labour costs escalate, timber will tend to be replaced in the area of mouldings and skirtings.

14.9.2 Working Stresses for Timber

Timber, like many other materials, is graded according to its anticipated performance in service; because of its inherent variability, distinct grades of material must be recognised and the derived stresses for these are usually referred to as *grade stresses*.

The derivation of these stresses underwent considerable modification in 1973 when new stress values were produced and material could be assessed either visually or by machine. Over the last two years a major revision of the stress values and their subsequent use in design has commenced and so it is proposed below to present not only that currently adopted but also the previous and possible post forms of this progression.

14.9.2.1 *Basic Stresses and Grade Stresses Prior to 1973*

Visual Grading. Basic stress can be defined as the stress which can be permanently sustained with safety by an ideal structural component containing no strength-reducing characteristics. In the derivation of basic stresses from the tests on the small clear samples described in Section 14.3 consideration was given to both the variability in the strength figures for clear timber and the need to ensure that the imposed load was a safe one for that particular set of conditions.

The effect of moisture on strength has been discussed previously in this and other chapters and in the derivation of the basic stresses cognisance of this effect was evident by the derivation of stresses for two levels of moisture content, namely 'green' and 'dry': the latter has a value of 18%, which is taken as a conservative estimate of the maximum value likely to occur within a building.

Timber has been described as a variable material and in Section 14.5 a measure of this variability was described and values of this parameter, the standard deviation(s) are included in Table 14.1. It was shown how the frequency distribution of a set of test results approximates closely to a normal distribution curve which can be used to calculate the value below which a certain percentage of the results will not fall. For most strength properties it was reasonable to assume that a chance of getting a lower value than the estimated

467

minimum once in a hundred times was not taking too high a risk. Now in the normal distribution curve 98% of the values will be within the range: mean $-2.33s$ to mean $+2.33s$, i.e. only 1% will lie below this range.

The apparent strength of timber is influenced by rate of loading, specimen size and shape, and duration of loading. Rather than apply a series of factors, a single factor derived mainly from experience was used: generally a value of 2.25 is used for most properties, excepting compression parallel to the grain where it is 1.4; these factors are in effect a safety factor for pieces of minimum strength, and also cover the possibility of slight overloading.

The basic stress, therefore, for each property of each species was obtained primarily from the results of the standard tests on small clear specimens by dividing the statistical minimum (mean $-2.33s$) by the appropriate safety factor (2.25 or 1.4). However, the measured values of modulus of elasticity were usually taken as the basic values and are not reduced.

Most structural timbers are not without defects and it was necessary to apply a factor known as the *strength ratio* to the basic stress in order to obtain a safe stress for these members.

Earlier in this chapter the various factors that affect strength were described and sets of grading rules have been devised which are used to assess the strength of timber visually in terms of the size and distribution of knots, slope of grain, rate of growth, and presence of fissures and wane. The timber was assigned a numerical grade, e.g. 40, 50, 65 and 75, which was really the strength ratio: thus the *grade stresses* for green timber were derived from the basic stresses using these ratios, i.e. green timber of grade 50 was assessed as being only half as strong as that of clear, straight-grained timber. Dry stresses were then derived from these green stresses using the strength/water relations referred to in Section 14.7.10. Both sets of stresses were published in the British Standard Code of Practice CP 112:Part 2: 1971 'The Structural Use of Timber'. The Code also demonstrates how the *permissible stresses* used in structural design in timber are calculated from the product of the grade stresses and a series of modification factors which cover aspects such as load sharing between members, duration of load, slenderness ratio and bearing area.

Mechanical Grading. Visual grading is a laborious process since all four faces of the timber should be examined. Furthermore, it does not separate naturally weak from naturally strong timber and hence it has to be assumed that pieces of the same size and species containing identical defects have the same strength: such an assumption is invalid and leads to a most conservative estimate of strength.

Many of the disadvantages of visual grading can be removed by mechanical grading, a process which was introduced commercially in the early 70's. The principal underlying the process is the high correlation that has been found to exist between the moduli of elasticity and rupture; this was described in Section 14.6.1 and illustrated in Fig. 14.4. It has been proven that this relation provides a more

468

accurate measure of strength than can be obtained by visual inspection.

In Fig. 14.4 it will be noted that the relation for two species can be represented by a single regression line below which is drawn the lower confidence limit at a vertical distance of 2.33 x Standard Error of the Estimate of the modulus of rupture: approximately 99% of the results lie above this line. When mechanical grading was still in its infancy the magnitude of the safety factor was set as 3: however, as more experience was gained in the method, the factor was reduced to near its original level. When applied to the lower confidence limit, as indicated in Fig. 14.4, a grade stress line is produced and when deflection limits are imposed, different grade stresses can be read off from this grade stress line.

Having established the relation for each species, a stress-grading machine can then be set up to automatically grade the timber. As each length passes through it is loaded laterally and the deflection measured automatically: a small computer ascribes grade stresses which are indicated along the length of the plank by splashes of dye. A final colour strip indicates the lowest grade stress that was present in the plank. Two grades of timber, namely M75 and M50, are specified in CP 112.

14.9.2.2   *Grade Stresses 1973 to 1981 at Least*

The year 1973 saw significant changes in timber grading and in the derivation of stresses. BS 4978 'Timber Grades for Structural Use' was published, which specified both visual and machine grading of structural timbers and detailed the characteristics of a reduced number of stress grades. Three visual grades for laminating timber were also specified in this new standard.

In terms of visual grading this standard meant that the same basic features that were used previously to determine the strength ratio were retained in concept though not in detail; for example, the influence of knots is now reflected in the *knot area ratio*, which is the sum of the projected cross-sectional areas of all the knots at a cross-section in terms of the cross-sectional area of the timber. Three visual grades have been defined, namely SS (Select Structural), GS (General Structural) and REJECT: to qualify for SS grade, in addition to satisfying requirements for slope of grain, distortion, rate of growth and wane, the total knot area ratio at any cross-section in the timber sample must not exceed one third, or, where more than half of the cross-section of either margin (the outer quarters of the depth of the cross-section) is occupied by knots, the total knot area ratio must not exceed one fifth; the corresponding total knot area ratios for GS grade are one half and one third. Although a direct comparison between the new and old grades is not strictly acceptable owing to differences in mode of derivation of the stress values, it is generally recognised that the GS corresponds to btween the old 30 and 35 stress ratio and the SS to between the 50 and 60 stress ratios. The corresponding grades for machine graded timber are MGS and MSS. However, since other grades had been derived prior to this date, it was decided to retain the old M75 and M50 grades. Timber purporting to comply with these grades must be marked to identify the grader or company responsible for grading together with the grade of the piece.

The grades for horizontally laminated members specified in BS 4978 are known as LA, LB and LC grades; they have limiting knot area ratios of 10, 25 and 50% respectively and are essentially the same grades as specified in CP 112.

As the amount of experimental data on mechanical stress grading accumulated so it became obvious that grade stresses could be derived from structural size pieces rather than small clear specimens. In the wake of BS 4978 specifying the new grades, an amendment to CP 112 (Amendment Slip No 1, 28 September 1973) was published which gave the stresses for the new MSS, SS, MGS and GS grades and for the established M75 and M50 grades. These tables of stresses cover the 'green' and 'dry' conditions of exposure for the more common softwood species under different loading conditions. In the derivation of these stresses the 1% exclusion limits were obtained using the Weibull distribution. An extract from the tables is presented in Table 14.5, from which it will be noted that with the exception of modulus

Table 14.5   Dry Stresses and Moduli of Elasticity

	Grade	Redwood, Whitewood N/mm^2	Douglas fir (home-grown) N/mm^2	Sitka spruce (home-grown) N/mm^2
Bending	M75	10.0	12.4	6.6
	SS, MSS	7.3	9.0	5.2
	M50	6.6	8.3	4.5
	GS, MGS	5.1	6.3	3.6
Tension parallel	M75	7.0	8.7	4.6
	SS, MSS	5.1	6.3	3.6
	M50	4.6	5.8	3.2
	GS, MGS	3.5	4.4	2.5
Compression parallel	M75	10.8	12.4	6.4
	SS, MSS	8.0	9.0	5.0
	M50	7.1	8.3	4.3
	GS, MGS	5.6	6.3	3.5
Mean modulus of elasticit	M75	10 700	12 500	9 000
	MSS	10 200	10 800	8 200
	M50	9 000	10 500	7 700
	MGS	8 800	9 500	7 200
	SS	10 000	10 700	7 800
	GS	8 600	9 800	7 000

of elasticity, the strength values for the comparable machine and visual grades are identical. An important point which is not revealed by the tabulated stress values is that the yield of any one grade is considerably higher when the timber is machine rather than visually graded.

470

CP 112 deals with many aspects of the structural use of timber and is the principal text for the architect and structural engineer. In addition to the basic and grade stresses presented for different timber, a whole range of modification factors is included, thereby allowing the derivation of permissible stresses, i.e. the stress which can safely be sustained by a structural component under the particular condition of service and loading. Thus these modification factors cover a wide range of variables among which may be listed duration of load, length and position of bearing surfaces of beams, load sharing in beams, slenderness ratios in compression members, and degree of curvature in laminated beams.

A considerable volume of data on the basic loads in shear and tension is presented for the various components used to fasten timber members — nails, bolts, and a wide range of connectors. Modification factors covering duration of load, moisture content level, and stress concentrations are also included.

Certain types of plywood are now well established as structural sheet materials and this code presents dry grade stresses and moduli for both Douglas fir and Finnish birch plywood of different thicknesses and methods of board construction. Recent amendments to the code cover the use of laminated timber and the adoption of finger jointing for the construction of the laminae.

14.9.2.3   *Future Trends*

With increasing emphasis on the mechanical grading of timber it is likely that sooner or later the visual grades will disappear from CP 112. A certain degree of rationalisation among the grade stresses for machine-graded timber appears to be necessary and it is highly probable that the number of grade stresses will be reduced, possibly by the removal of the lowest grade, MGS. It is likely that the future will see a greater degree of harmonisation of the various international standards, especially those pertaining to the mechanical stress-grading of timber.

More immediate than any international agreement will be the implications arising from the current major revision of CP 112 in which limit state and probabilistic methods of design are being considered (Sunley, 1974). Collapse and excessive deformation are the two most important limit states to be considered in timber design, though other states such as excessive vibration, fatigue, biodeterioration, and corrosion in fasteners could well be considered.

Characteristic loads and strengths are derived by taking into account statistical variability, and these are then modified by applying partial factors to obtain the design values. Partial factors reflect the degree of uncertainty involved in loading and construction, and from the materials aspect they allow the designer flexibility to obtain the design strength appropriate to a particular design situation by modifying for the effects of load design, size and exposure conditions. Partial factors for the different materials will be different; concrete and steel have recommended values of 1.5 and 1.15 respectively and

471

these values apply to all designs in those materials. In timber, however, it is likely that there will not be a universal partial factor for material strength because of the anisotropic nature of timber and because different types of test are used for different properties.

14.10 REFERENCES

Bach, L. (1967) Static fatigue of wood under constant strain, *FPL* (Vancouver) *Info Rpt* VP-X-24.

Bach, L. (1973) Reiner—Weisenburg's theory applied to time-dependent fracture of wood subjected to various modes of mechanical loading, *Wood Science*, 5(3), 161—171.

Barrett, J. D. (1974) Effect of size on tension perpendicular to grain strength of Douglas fir, *Wood and Fiber*, 6(2), 126—143.

Baumann, R. (1922) Die bisherigen Ergebnisse der Holzprüfungen in der Materialprüfungsanstalt an der Tech Hochschule Stuttgart, *Forsch. Gebiete. Ingenieurw.*, H231, Berlin.

Cave, I. D. (1969) The longitudinal Young's modulus of *Pinus radiata*, *Wood Science and Technology*, 3, 40—48.

Cook, J. and Gordon, J. E. (1964) A mechanism for the control of crack propagation in all brittle systems, *Proceedings of the Royal Society*, A 282, 508.

Dinwoodie, J. M. (1968) Failure in timber, Part 1: Microscopic changes in cell wall structure associated with compression failure, *Journal of the Institute of Wood Science*, 21, 37—53.

Dinwoodie, J. M. (1974) Failure in wood, Part 2: The angle of shear through the cell wall during longitudinal compression stressing, *Wood Science and Technology*, 8, 56—67.

Gerhards, C. C. (1977) Effect of duration and rate of loading on strength of wood and wood-based materials, *FPL Madison*, Report 283, 24 p.

Gordon, J. E. and Jeronimidis, G. (1974) Work of fracture of natural cellulose, *Nature* (London), 252, 116.

Grossman, P. U. A. and Wold, M. B. (1971) Compression fracture of wood parallel to the grain, *Wood Science and Technology*, 5, 147—156.

James, W. L. (1962) Dynamic strength of elastic properties of wood, *Forest Products Journal*, 11(9), 383—390.

Keith, C. T. (1971) The anatomy of compression failure in relation to creep-inducing stress, *Wood Science*, 4(2), 71—82.

Keith, C. T. (1972) The mechanical behaviour of wood in longitudinal compression, *Wood Science*, 4(4), 234—244.

Lavers, G. M. (1969) The strength properties of timber, *Bulletin 50, Forest Products Res Lab (2nd Edn.)* HMSO.

Mark, R. E. (1967) *Cell Wall Mechanics of Tracheids*, Yale Univ. Press, New Haven.

Mindness, S., Nadeau, J. S. and Barrett, J. D. (1975) Slow crack growth in Douglas fir, *Wood Science*, 8(1), 389—396.

Pearson, R. G. (1972) The effect of duration of load on the bending strength of wood, *Holzforschung*, 26(4), 153—158.

Pearson, R. G. (1974) Application of fracture mechanics to the study of the tensile strength of structural lumber, *Holzforschung*, 28 (1), 11–19.

Pierce, C. B. (1976) The Weibull distribution and the determination of its parameters for application to timber strength data, *Building Research Establishment Current Paper* 26/76.

Schniewind, A. P. and Lyon, D. E. (1973) A fracture mechanics approach to the tensile strength perpendicular to grain of dimension lumber, *Wood Science and Technology*, 7, 45–59.

Sunley, J. G. (1974) Changes in basic assumptions in UK timber design codes, *Building Research Establishment Current Paper 49/74*.

Williams, J. G. and Birch, M. W. (1975) Mixed mode fracture in anisotropic media, Preconference volume, *The Properties of Wood in Relation to its Structure*, NATO Science Committee.

Wood, L. W. (1951) Relation of strength of wood to duration of load, *Forest Products Laboratory (Madison) Report No 1916*.

Ylinen, A. (1957) Zur Theorie der Danerstandfestigkeit des Holzes, *Holz als Roh- und Werkstoff*, 15 (5), 213–215.

CHAPTER 15

Metallurgical Aspects of Fracture

We must now examine the ways in which a metal may rupture under load. The majority of metals exhibit very considerable fracture toughness, i.e. a considerable amount of work (represented by the area under the stress strain curve) is expended before final rupture occurs. Such behaviour is of infinitely greater structural worth than the higher UTS values but lower toughness exhibited by brittle materials such as glass. Fractures of this nature are known as *ductile fractures*. Unfortunately, however, metals do not invariably show fracture toughness; under certain environmental and loading conditions they may fail in a *brittle* mode, with almost negligible expenditure of energy. The possibility of such failure represents a potential disaster to the structural engineer, and it is therefore of the utmost importance that he should understand thoroughly the factors that may determine the failure mode of metals.

15.1 DUCTILE FAILURE

It is possible for failure to occur purely as a result of shear processes within a metal. If a particular slip system has favourable orientation, fracture may occur by the method shown in Fig. 15.1; once initiated, the process is favoured by the stress concentrations that develop at the steps where the slip bands cut the surface. Such failure is virtually confined to single crystals since only these can have the slip planes extending continuously from surface to surface. More commonly, slip takes place on intersecting sets of slip planes, and leads to necking, as discussed in Section 12.6. If the metal is sufficiently ductile, deformation can then proceed as in Fig. 15.2 until 'point' or 'chisel-edge' failure occurs.

Practical experience and investigation, however, has shown that failure in normal polycrystalline metals and alloys of average purity has features derived both from the shear mechanism just outlined and from the Griffith theory set out in Section 2.5. It will be remembered that the Griffith expression for fracture strength is based on

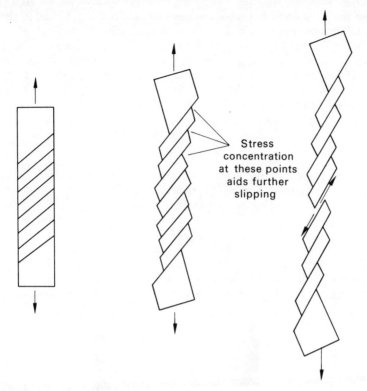

Stress
concentration
at these points
aids further
slipping

Figure 15.1 Fracture following yielding in pure shear. Fractures of this nature are rare and confined to single crystals with no intersecting slip planes.

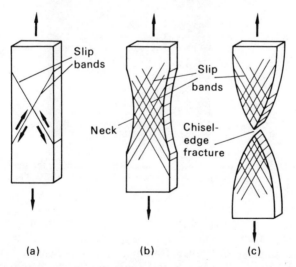

Slip
bands

Slip
bands

Neck

Chisel-
edge
fracture

(a)   (b)   (c)

Figure 15.2 Deformation by slip on several planes. (a) Crystal oriented for double slip. (b) Development of a neck. (c) Chisel edge fracture. (From R. E. Reed-Hill (1973) *Physical Metallurgy Principles, 2nd edition*, by permission of D. Van Nostrand Co.)

the stress required to propagate a microcrack through a solid in the total absence of any plastic deformation, so that all the work of fracture goes into creating fracture surface. If a crack is introduced into a metal, however, the growth of the crack is always accompanied by plastic deformation at the vicinity of the crack-tip, and a much better correlation with reality is obtained if the Griffith expression is modified to:

$$\sigma_f \simeq E \left(\frac{\gamma + p}{c}\right)^{\frac{1}{2}} \tag{15.1}$$

where p represents the work done in the plastic deformation associated with crack growth. For most metals and alloys, $p \gg 100\,\gamma$, and we may therefore further approximate to

$$\sigma_f \simeq \left(\frac{Ep}{c}\right)^{\frac{1}{2}} \tag{15.2}$$

The correlation of the expression with practice indicates that internal cracks must develop in metals as part of the failure mechanism, and

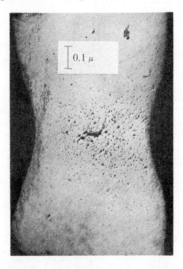

Figure 15.3 Incipient ductile fracture. Section through the neck of a tensile specimen of tough-pitch copper. (From K. E. Puttick (1959) *Philosophical Magazine*, 4, 964.)

this has been confirmed by microscopic examination (Fig. 15.3). We must therefore consider how these cracks may originate.

The simplest mechanism leading to crack nucleation is the result of dislocation pile-up at an obstruction in the slip plane. If the shear stress acting on the plane is sufficient to overcome the strong mutual repulsion of the dislocations, they will coalesce to form a crack as shown in Fig. 15.4. The critical shear stress τ_c for this to occur has

476

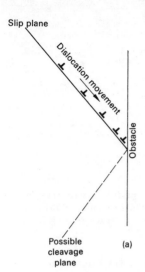
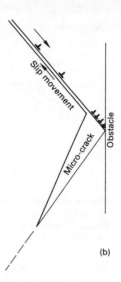

Figure 15.4 Pile-up of dislocations leading to crack nucleation. (a) Dislocations piling up at an obstacle on the slip plane. (b) Coalescence of dislocations leads to the opening up of a microcrack.

been shown to be given by

$$\tau_c = \tau_i + \frac{12\gamma}{n\mathrm{b}} \qquad (15.3)$$

where τ_i is a 'frictional' stress opposing dislocation movement, n is the number of dislocations of Burgers vector b in the pile-up, and γ is the surface energy of the resultant crack. Further calculation leads to the approximate relationship that $n \cdot \tau > \sim 0.7G$ for a crack to form; if numerical values are assigned to τ and G, it appears that between 10^2 and 10^3 dislocations are involved in these critical pile-ups. Obstacles may take the form of impurity inclusions, hard-phase constituents or grain boundaries; if we assume that the average length of slip plane occupied by the pile-up is half the grain diameter d, i.e. that on average the source of the dislocations lies in the centre of the grain, it can be shown that the critical shear stress for crack formation is related to the grain size by the expression

$$\tau_c = \tau_i - kd^{\frac{1}{2}} \qquad (15.4)$$

which bears a strong resemblance to Equation 12.14 relating grain size to yield point, and justifies the well known fact that a fine grain size raises the fracture toughness of metals. A similar expression could be derived to relate τ_c to the mean distance between impurities or precipitate phases on the·slip planes; the cavities shown in Fig. 15.3 have in fact all originated at oxide particles.

The interaction of dislocations may also lead to crack nucleation; Cottrell has shown how slip on $\{110\}\langle111\rangle$ slip systems can lead to cleavage on $\{100\}$ planes in iron (Fig. 15.5). Mention has already

477

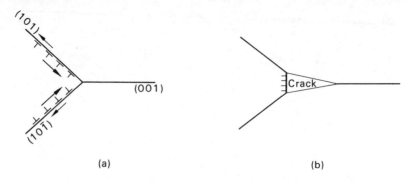

(a)   (b)

Figure 15.5   Crack nucleation by the combination of dislocations travelling on two intersecting {101} planes. The result is a crack nucleated on the 001 basal plane. This mechanism is applicable to alloys with b.c.c. structures (e.g. irons and steels) which slip on the {110} family of planes.

been made in Section 5.12 of cracks formed at the intersection of deformation twins.

We must now consider what determines whether the metal will fail in the brittle or ductile manner, and this in turn depends on the behaviour of the microcracks. Reference has already been made to the inherent instability of cracks; this arises from the stress concentration at their tip, $\sigma_{max} \simeq 2\sigma \, (c/\rho)^{\frac{1}{2}}$, and also to the fact that triaxial stress conditions are set up at the tip, even where the overall loading

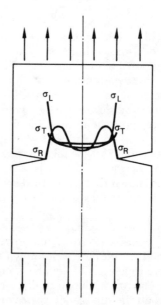

Figure 15.6   Stress distribution produced in notched cylinder under uniaxial loading. σ_L = longitudinal stress; σ_r = transverse stress; σ_R = radial stress. (From G. E. Dieter, *Mechanical Metallurgy*. Copyright 1961, McGraw-Hill Ltd. Used with permission of McGraw-Hill Book Company.)

of the metal is uniaxial. The three stresses (longitudinal, radial and transverse) are shown in Fig. 15.6; since they are all tensile (i.e. positive in sign) they must result in a *lowering* of the shear stress τ acting on the slip planes in the vicinity of the crack tip, according to the Tresca yield criterion as set out in Section 12.2. Thus conditions are less favourable for shear and hence more favourable comparatively for crack propagation.

However, shear occurs readily in the majority of metals, and its operation will serve to blunt the crack tip, increasing ρ rapidly for

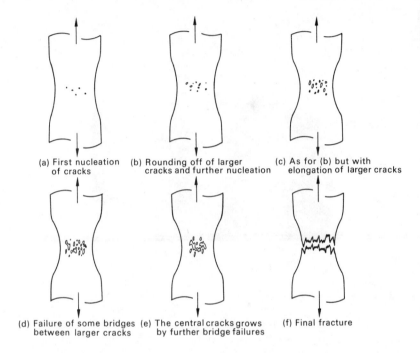

(a) First nucleation of cracks   (b) Rounding off of larger cracks and further nucleation   (c) As for (b) but with elongation of larger cracks

(d) Failure of some bridges between larger cracks   (e) The central cracks grows by further bridge failures   (f) Final fracture

Figure 15.7   Schematic sequence of events leading to ductile fracture.

very little change in c and thereby lowering σ_{max} below the critical value for crack propagation. The crack is thus arrested and the whole process of dislocation pile-up, crack nucleation, and arrest will be repeated elsewhere. These cracks appear to nucleate only in the latter stages of necking, and their location in the neck is obvious, since this is the region of highest stress. That they should form preferentially in the centre of the neck rather than at the surface (see again Fig. 15.3) is less immediately obvious, but a logical result of the stress distribution in the neck (Fig. 12.35(b)). This shows that the shear stress (necessary to force the dislocations together) is as high at the centre as anywhere, whilst the longitudinal stress (necessary to open up the cracks when nucleated) is very much higher than elsewhere.

The whole process of ductile fracture may thus be seen as a gradual honeycombing of the neck with internal cracks which are drawn out

479

until they become voids elongated in the direction of the stress. As more and more form, so the strands of metal between the voids get thinner and thinner and also undergo tensile necking ('internal necking'). Finally these strands fail in shear, so that the pores coalesce by a series of fractures at 45° to the stress axis, and when the central cavity reaches a sufficient size, the remaining metal at the periphery finally fractures in pure shear by the crack running at 45° to the

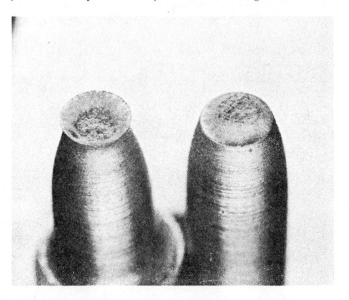

Figure 15.8 Cup and cone fracture. (From R. E. Reed-Hill (1973) *Physical Metallurgy Principles, 2nd edition*, by permission of D. Van Nostrand Co.)

surface, giving the well known cup-and-cone fracture. This sequence of events is depicted schematically in Fig. 15.7, and a typical cup-and-cone fracture is shown in Fig. 15.8.

15.2 BRITTLE FRACTURE AND THE DUCTILE-BRITTLE TRANSITION

Ductile fracture is a rarity in engineering structures, since its mechanism requires a steady loading far in excess of any reasonable design requirements; unfortunately brittle fracture is not so rare, and it is doubly unfortunate that the high strength metals with a b.c.c. structure such as iron, chromium, molybdenum, tungsten and vanadium are the most prone to sudden brittle failure. The initial stage of fracture — that of crack nucleation — is basically the same as for ductile failure; however, if the crack blunting mechanism fails to operate, the crack may propagate rapidly through the metal in the manner of a Griffith crack (although there is always a small amount of plastic deformation associated with brittle failure in metals.)

480

A variety of factors may contribute to this very undesirable result, the commonest being a high strain rate, low temperature and the presence of a stress raising feature such as a notch. However, the internal metallurgical structure of a component may play a major part in its response, and thus fabrication techniques must be taken into account. This is particularly true of welding, with its profound effect on microstructure and possible introduction of brittle phases and residual stresses (see also Section 16.6). Finally, a size factor operates: thick sections are more prone to brittle failure than thin sections of similar materials, which is troublesome in that it makes it difficult to interpret the results of laboratory tests in terms of large-scale structures.

Concise discussion of these factors presents problems, since all are closely interrelated; however, certain salient features should be noted. High strain rates (e.g. impact loading) not only raise the yield stress, but may result in the presence of very high elastic stresses in a metal before the dislocation movements have *time* to accommodate the stress by plastic deformation. In consequence, avalanches of dislocations may be released along favourably orientated slip planes, and if there is an obstacle they may pile up with such rapidity that the resultant crack is very much larger than the minimum size postulated by Equation 15.3, and may in fact be large enough to satisfy the Griffith conditions for propagation. In these circumstances, failure will occur as soon as the shear necessary to nucleate the crack has taken place; that is to say, the fracture stress drops to the level of the shear stress, but not below it. This mechanism agrees with the known brittle vulnerability of steel, in which the dislocation networks are locked initially by their atmospheres of carbon atoms (Section 12.6) to give a double yield point. This effect can only exaggerate the 'avalanche' effect referred to above. By contrast, if there is little or no dislocation locking, the dislocations move in smaller numbers and pile up to generate small cracks that have time to blunt themselves before satisfying the Griffith criterion.

Large crack nucleation is not necessarily sufficient in itself to generate total fracture; cleavage cracks have been observed in steels that merely split individual grains and have been checked at the grain boundary. However, other factors may raise the shear stress in the vicinity of the crack and thus promote brittle failure.

Of these, *temperature* is the most important, since the critical shear stress is highly sensitive to variations in temperature, whilst the critical fracture stress is not. The general effect is plotted in Fig. 15.9 and shows that below a temperature T_c the metal will respond in brittle fashion if the applied stress is high enough. For most metals, T_c is well below the normal range of working temperatures, or (in the case of f.c.c. metals) they are immune from the effect. However, where there is a *notch*, the effect of the plastic constraint at its tip may raise the yield stress up to three times its normal value: this has the effect of raising the ductile-brittle transition temperature to T_N (Fig. 15.9), and it is very unfortunate that for plain carbon steels $T_N \sim 0$ °C. High strain rates, by raising the

481

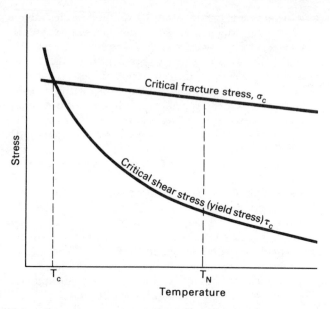

Figure 15.9 Relation of fracture stress and yield stress to temperature. T_c is the critical temperature for the ductile-brittle transition. Under notched conditions (T_N) it is raised considerably because τ_c is raised at the tip of the notch.

value of Y, may still further raise the critical transition temperature.

Grain size is also significant, the toughness of the metal decreasing as the grain size increases; the relationship is plotted in Fig. 15.10, and results from Equation 15.4. Grain size on occasion is partly responsible for the *size effect*: the fact that large structures are more

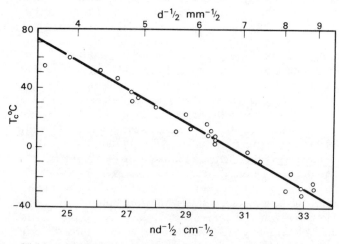

Figure 15.10 The dependence of the transition temperature of mild steel on the grain size. (From R. W. K. Honeycombe (1971) *The Plastic Deformation of Metals*, by permission of Edward Arnold.)

482

prone to brittle failure than small ones. Size plays a part in two ways:
the microstructure of a thick section is more difficult to control,
and is more likely to contain undesirable impurity segregations or
residual stress concentrations that may aid the nucleation or propagation of brittle cracks. Under impact, a thick section will also contain
a greater stored elastic energy than a thin section of similar composition; this energy is available for crack propagation.

Numerous tests have been devised to reveal a potentially brittle
response from metals; of their nature they are essentially qualitative

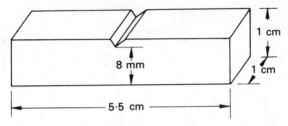

(a) V– notch Charpy impact test specimen

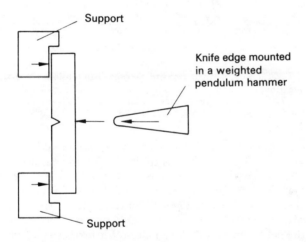

(b) Method of applying the impact load to a Charpy specimen

Figure 15.11　The Charpy impact test. (From R. E. Reed-Hill (1973) *Physical
Metallurgy Principles, 2nd edition*, by permission of D. Van Nostrand Co.)

and, particularly in view of the size factor, the results must be interpreted with great care. The best known methods are the *Charpy* and
Izod tests, in which a notched specimen is subject to impact loading
at varying temperatures (Fig. 15.11). Thus two of the three most
embrittling factors are present and the very important ductile-brittle
transition temperature may be determined under extremely hostile
conditions. The notch is normally of standard depth and tip radius,
but keyhole notches may be used where it is necessary to investigate

483

the effect of the tip radius on the transition temperature. Machining the notch should be done with especial care since there is always the possibility that secondary cracks may be started at the notch tip, thereby invalidating the exact conditions of the test; this is particularly the case with hard alloys. For this reason, and because of possible variations in microstructure, these tests should always be considered on a statistical basis, similar to fatigue tests and in contrast with the simple tensile test. Some typical results are shown in Fig. 15.12. Other

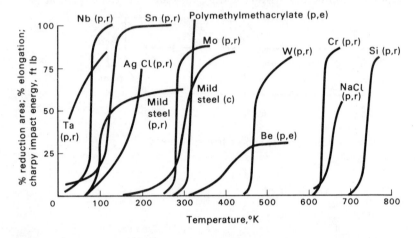

Figure 15.12   Typical ductile-brittle transitions. (p = plain tension; r = reduction of area; e = elongation; c = Charpy V-notch impact energy.) (Reproduced by permission, from *The Mechanical Properties of Matter* by A. H. Cottrell. Copyright © 1964 by John Wiley & Sons, Inc.)

tests include the *Robertson* test, in which a crack is started by impact from a fine saw cut in a steel which is subject to a temperature gradient (Fig. 15.13). The temperature of the steel at the point where the crack stops is then noted as the *crack-arrest temperature*; at this point the steel is sufficiently plastic for the crack blunting mechanism to operate. In other words, the elastic energy of the impact is no longer sufficient to generate the necessary plastic deformation for further crack growth. Full size plates may be tested by the *Pellini test*; a brittle weld-bead of metal is deposited in the centre of the plate, notched, and then a heavy weight is dropped on it. The bead cracks on impact and the ability of the underlying plate to resist crack propagation is noted for varying temperatures; this test, in which deflection of the plate is limited to a maximum of 5°, was developed to simulate possible service conditions of ships' hulls. Ordinary tensile tests at low temperatures and slow bend tests are also sometimes employed where service conditions are likely to be less arduous; the results usually lie somewhere between those from the Charpy test and the ordinary tensile test.

General techniques for strengthening metals are considered in Chapter 16; these are based primarily on ways of restricting disloca-

484

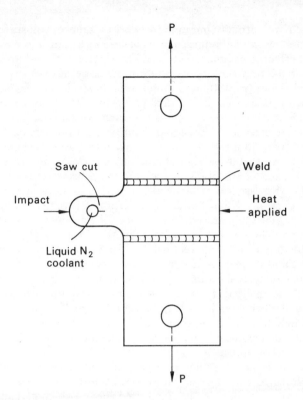

Figure 15.13   The specimen used for a Robertson test. (From G. E. Dieter, *Mechanical Metallurgy*. Copyright 1961, McGraw-Hill. Used with permission of McGraw-Hill Book Company.)

tion movements and thereby increasing resistance to plastic flow. These methods, however, are not necessarily suitable to improve the notch toughness of metals; too great a resistance to plastic flow may merely increase the likelihood of brittle cleavage cracks under adverse conditions, whilse a *decrease* in the yield strength may even improve resistance to fracture. For example, duralumin and 70/30 brass, although considerably stronger than aluminium and copper respectively in simple tension, show very much poorer notch impact resistance at all temperatures.

Vulnerability may arise from defects in each of the three basics of an engineering creation: design, materials, and construction. At the design level, great care is necessary to eliminate or at least minimise any stress raising features; any re-entrant angles or notches that occur as design features should have the maximum possible radius. Major failures have resulted from relatively trivial features such as poorly designed hatches in ships. The materials of construction must be selected with great care, since toughness is very sensitive to impurities and to grain size. Impurities are at their most damaging when they segregate to grain boundaries; if they lower γ appreciably, even the most ductile of metals (such as those with the f.c.c. struc-

485

ture which are normally immune from brittle failure) will suffer brittle *intercrystalline* failure. Traces of lead or bismuth in copper have this effect. Oxygen, nitrogen and to some extent carbon in steels raise the transition temperature in steels because of their strain ageing behaviour (p. 380), and therefore mild steels that have been deoxidised with aluminium ('*killed*' *steels*) and steels containing manganese or nitrogen-fixing elements (such as titanium, niobium or boron) are essential if low-temperature impact is a service risk. Killed steels are additionally tough in that they have a fine grain size (due to the internal dispersion of alumina particles acting as solidification nuclei) and are resistant to grain growth. This latter feature means that they are less prone to the size effect than ordinary steels; in the absence of a grain-stabilising dispersion it is virtually impossible to prevent some grain growth in thick sections during recrystallisation after rolling. Where possible, a large reduction in the final rolling pass is desirable to ensure a fine recrystallised grain size. In addition, where size and shape permit it and expense warrants it, quenched and tempered steels show greatly enhanced notch toughness compared with normalised steels, because of their very fine structure.

Spectacular brittle failures — the most noteworthy being the breaking in half of nineteen ships during the 1939—45 war — have been very much associated with welded assemblies and this has led to somewhat unjustified suspicions of welding as a constructional technique. Welding is a skilled process that is certainly prone to flaws if not performed with due care; however, it must be emphasised that the root cause of failure in the ships cited above (which had all-welded hulls) was not faulty welding workmanship but the fact that a welded hull is metallurgically one piece of metal, and therefore that a brittle crack could travel right round the entire hull without hindrance. By contrast, a crack nucleated in a plate of a riveted hull (which would be just as susceptible in itself to brittle failure) is automatically arrested at the edges of the plate, with — probably — no serious consequences. This is not to say that welding is blameless as a technique; it carries its own risks, several of which are particularly relevant to the likelihood of brittle failure. Imperfect penetration of weld metal into the joint gap leaves a cavity which can act as a dangerous internal notch; residual stresses, particularly if associated with the formation of a brittle phase such as martensite, can result in the formation of brittle cracks; even the small notch at the point where the welder strikes his arc has been known to initiate a crack. The problems of welding in this context have been intensified in recent years by the increasing use of ultrahigh-strength steels which have high hardenabilities (and therefore increasing risk of martensite formation) and are also more sensitive to embrittlement by impurities picked up during welding.

15.3 FATIGUE FAILURE

If a metal is subjected to cyclical loading, there is a strong probability

486

that it will ultimately fail, even though the peak load in the cycle is well within the elastic range. Such failures occur suddenly, in the manner of brittle failures, perhaps after many years of service and several million stress cycles, and are thus anathema to the engineer. They are known as *fatigue failures*. As an extreme example, most of us have probably employed the fatigue principle quite unconsciously when bending a piece of wire repeatedly to break it in the absence of wire cutters. Failures of this nature are not uncommon — in fact it has been suggested that as much as 80% of service failures are due to

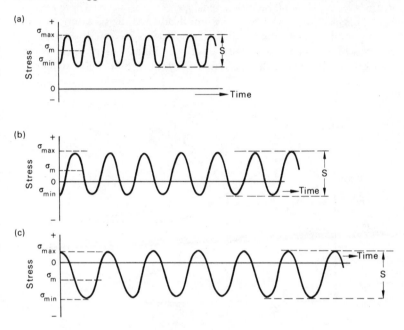

Figure 15.14 Possible alternating stress patterns in components subject to fatigue loading. (a) Entirely tensile. (b) Tensile-compressive, mean stress tensile. (c) Tensile-compressive, mean stress compressive.

fatigue — and, while they are commonest in mechanical devices where engine vibrations are an inescapable part of the working environment, many civil engineering structures are subject to vibrational loading and thereby carry the risk of fatigue failure.

Before considering the mechanism by which fatigue proceeds, we must consider the nature of the cyclical stressing imposed on components. Clearly there is an infinite number of permutations of loading that falls within the definition of fatigue, and clearly also, as with assessments of notch toughness, it is extremely difficult to reproduce service conditions accurately in laboratory tests. Happily, one important variable, the *frequency* of the stress cycle, has no apparent effect on the life of the specimen except at very high frequencies, e.g. more than 1 kcycle/s, when the life of the specimen is prolonged. It must be noted, however, that the life of the specimen

487

is expressed in terms of N, the number of cycles to failure, and not the *time* to failure, which may in fact be shorter at very high frequencies. Although irregular stress cycles are perfectly possible in tests, the stress is normally varied regularly and sinusoidally, either by a push—pull tensile loading (in which case the stress is uniform across the cross-section) or by cantilever loading of a rotating specimen (in which case the stress varies across the specimen at any instant and evaluation of the results is more complex). The stress varies in the manner shown in Fig. 15.14, where σ_{max}, σ_{min} and the mean stress level σ_m may all be either tensile or compressive. Where σ $\sigma_m \sim 0$, as is usually the case, the amplitude S is the most important stress parameter, and the results of fatigue tests are usually plotted in terms of S against $\log N$. Ferrous alloys exhibit curves of the form A in Fig. 15.15, where there is a definite *fatigue limit* below which

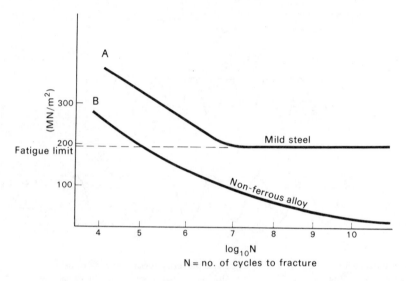

Figure 15.15   Typical fatigue curves for ferrous and nonferrous alloys.

value of S the specimen will support the load indefinitely. Nonferrous alloys, however, follow curve B and show no such fatigue limit, which means that a definite safe working lifetime must needs be assigned to nonferrous components. For practical purposes nonferrous materials are assigned an *endurance limit*, which is the stress required to cause failure for a specific value of N, usully taken as 10^8 cycles, but this of course may be varied according to the needs of the case. Where σ_m varies substantially from zero, an alternative plot may be employed, as in Fig. 15.16. The stress is then divided into two components: σ_a, the alternating component, and σ_m, the mean or static component; results are plotted to show the combinations of σ_a and σ_m that cause failure at some fixed value of N. A straight-line relationship was proposed by Goodman and is shown in the figure as the Goodman line; experimental results point to a curve lying somewhat above this line and beneath a proposed parabolic relationship developed by Gerber.

Perhaps the most important single feature of the fatigue response of metals and alloys is their sensitivity to surface condition. It seems to be an almost invariable rule that fatigue cracks nucleate on the surface (in the absence of any sizeable internal flaws), and very small imperfections such as fine scratches or tooling marks can cause sufficient stress concentration to start a crack. There are two important consequences from this sensitivity: firstly, a high-quality surface

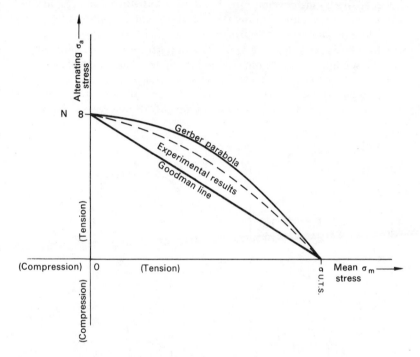

Figure 15.16   Goodman fatigue diagram, used in assessing fatigue behaviour where $\sigma_m \neq 0$. Experimental findings lie between the Goodman and Barber relationships, and approach the Goodman line for specimens of poorer surface quality. Hence the Goodman line affords a safer engineering design criterion.

finish is essential for fatigue resistance and, secondly, that fatigue testing has to be done on a statistical basis since, no matter how similar a number of specimens may be in terms of composition and microstructure, they can never have exactly the same qualities of surface, and an appreciable scatter of results may be expected. Fatigue strength can therefore only be assessed after a considerable number of tests have been carried out at each stress level and the curves shown in Fig. 15.15 plot mean values of N. In this way fatigue failure, although not strictly brittle failure, bears a strong resemblance to the failure of brittle nonmetallic materials, since we have to deal in terms

489

of the *probability* of failure, which stems from the probability of occurrence of a surface flaw of a particular depth.

Fatigue failure affords excellent evidence that some plastic deformation takes place in all metals at stresses far below the yield point. If the polished surface of a specimen is examined microscopically at intervals during a fatigue test, slip lines will be seen to develop on certain grains, presumably because of favourable orientation or weakening surface imperfections. In a comparatively short space of time (usually less than 10% of the fatigue life) a proportion of the deeper slip bands develop into a system of *intrusions* and *extrusions*. It is still not certain just how these curious features develop; two possible mechanisms are outlined in Fig. 15.17(a) and (b).

In the first example they result from patterns of cross slip, and in the second from random oscillating slip on a single set of planes. Once an intrusion has formed, it acts as a sharp notch and thereby encourages the development of the intrusion into an actual crack as a result of plastic constraint and concentration at its tip. This

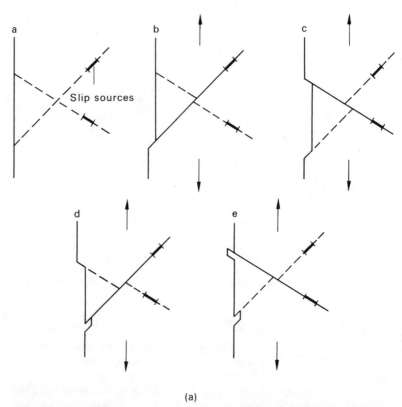

(a)

Figure 15.17 (a) Mechanism for formation of extrusions and intrusions during fatigue by cross-slip (from A. H. Cottrell and D. Hull, *Proceedings of the Royal Society*, A, 242). (b) Formation of extrusions and intrusions by slip on a single system of slip planes (from R. E. Reed-Hill (1973) *Physical Metallurgy Principles, 2nd edition*, by permission of D. Van Nostrand Co.)

crack then slowly grows into the specimen; material at its tip undergoes intensive plastic working and hardening, but may fail in the normal manner before the hardening process is sufficient to counterbalance the applied load. The crack then opens up a little further and the whole process is repeated. The ability of ferrous alloys to strain-age appears to build up sufficient strength for cracks to become nonpropagating and probably accounts for the fatigue limit that they exhibit. Fatigue failure may thus be envisaged as occurring in three stages: first, the formation of the intrusions and extrusions along certain slip planes; secondly the nucleation of a definite crack (which does not follow the slip planes) and its slow growth into the specimen; and finally the failure of the remaining cross-section when it becomes too narrow to support the peak loading in the fatigue cycle. The slowest part of the process involves the latter stage of intrusion growth and the early stages of true crack growth; this

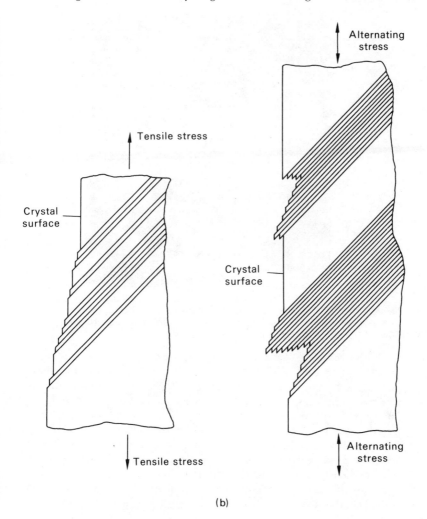

(b)

may occupy as much as 80% of the total fatigue life of the specimen. Once launched, however, it is reasonable that the crack should spread at a gradually increasing rate (see Equation 2.16) if we assume the crack tip radius to remain roughly constant. Experimental evidence suggests that the growth rate is given by

$$\frac{dc}{dN} = A\sigma^{2x} \cdot c^x \tag{15.5}$$

where σ is the peak stress based on the initial specimen cross section, c is half-depth of the crack, and A is a constant. The power x varies with the magnitude of σ; at low stresses $x \simeq 2$ but at high stresses it can rise to approximately 3.

For each stress cycle, the crack grows a further increment, and these give rise to the very characteristic appearance of a fatigue failure (Fig. 15.18) in which the final fracture surface shows rough and

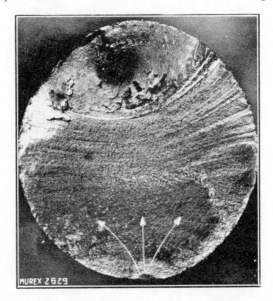

Figure 15.18 Fatigue fracture of an automobile axle shaft. (From E. C. Rollason (1956) *Metallurgy for Engineers, 2nd edition*, by permission of Edward Arnold (Publishers) Ltd.)

granular in texture, whilst the section through which the crack has grown is flat, and frequently shows growth lines which focus on the original surface imperfection that nucleated the failure. These lines may sometimes be obscured if the sides of the crack have rubbed together during its growth (in which case this part of the surface will appear polished) or if corrosion has played a part in the failure. However, the division of the fracture surface into two very different areas is characteristic and diagnostic of fatigue.

By far the most important single factor affecting fatigue strength is the surface condition of the metal; therefore the best preventative

492

measures are those that aim to preserve the surface in its pristine state. Components that may suffer from fatigue loading should be subject to very careful control and inspection during the final stages of machining, and then (if economically viable) given a high polish. Designs should avoid any built-in stress-raising features; in particular, if re-entrant angles and grooves are necessary, they should be given the maximum possible radii. Corrosion attack, by virtue of its pitting action on surfaces, is particularly damaging; in the presence of corrosion, no alloy exhibits a fatigue limit, and fatigue life is usually cut by a factor of at least 2 (see Section 19.7.1 and Fig. 19.15). Civil engineering structures are usually more at risk than mechanical ones in this respect, since the oily environment of many mechanical components serves to keep moisture at bay.

Since it is the intrusions that generate fatigue cracks, it is to be expected that any surface treatment designed to obstruct their development by blocking slip will be beneficial to fatigue life. Steels are *case hardened* by allowing carbon (*carburising*) or nitrogen (*nitriding*) to diffuse into their surface layers at high temperatures. This not only gives a very hard surface, but also puts it into a state of residual compressive stress which will serve to keep any incipient microcracks closed up and prevent them acting as stress raisers. Improvements in fatigue life averaging 60% and reaching a maximum of 230% have been recorded as a result of the treatment. A similar rather milder effect may be obtained by *shot-peening*, in which steel shot is blasted against the surface of the component; this serves to harden the surface by cold working, and also leaves it in a condition of mild compressive stress. By contrast, any steels that have suffered softening of their surface layers by decarburisation are particularly prone to fatigue failures, as is also the case with Alclad (duralumin sandwiched between thin layers of pure aluminium to improve corrosion resistance). It has already been noted that strain ageing is probably the reason for the existence of a fatigue limit with steels; alloy additions that favour strain ageing are therefore beneficial, and a fatigue limit has been induced in some aluminium alloys by this method. It must, however, always be born in mind that, whilst case hardening and strain ageing may be excellent treatments for fatigue resistance, they markedly increase susceptibility to impact failure, and the probability of each form of loading must be assessed before deciding on the appropriate preventative measures.

General, as opposed to surface, metallurgical treatments that improve fatigue resistance include the development of a fine grain size (grain boundaries will delay the growth of cracks, although it is rare for them to cause total arrest during fatigue), and the quenching and tempering of steels.

15.4 FRETTING

It is the mechanical rather than the civil engineer who is primarily concerned with the problems of friction and wear, since these are

493

normally associated with slides and bearings in power and transport engineering. However, there is one aspect — *fretting* — which can be a source of considerable trouble in civil engineering structures, the more so since it is insidious in its attack and by its nature tends to occur in unexpected situations. Fretting — or *fretting corrosion*, as it is frequently termed — involves the rubbing of two surfaces together, but the relative movements of the surfaces are small and *alternating*, rather than large and basically unidirectional as in bearings. It therefore occurs in supposedly rigid joints where the bolts or rivets have worked slightly loose, between the laminations of springs, and wherever metals not in rigid contact are subject to vibration, as for example in ball races or coiled metal sheet during transport. Fretting rarely, if ever, results in failure directly, but it is a very powerful instigator of fatigue cracks, and has probably been the root cause of far more fatigue failures than is realised.

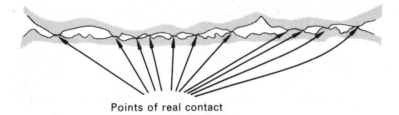

Points of real contact

Figure 15.19   Contact between two nominally flat metal surfaces.

When two metals are in contact, the true (as opposed to the nominal) contact area is confined to those high spots or asperities that make contact (Fig. 15.19). It has been estimated that this area may vary from 1/400 to $1/10^5$ of the nominal value, and therefore pressures at these points are considerable even under quite moderate loading. If the contacts are then subjected to relative movements of the two surfaces, a form of intense 'micromachining' takes place together with localised plastic working, gouging away any dirt or oxide layers and permitting clean metal-to-metal contact to occur. The combination of pressure and localised heating (due to the working effect) results in welds being formed between the surfaces, but if relative movement continues, these will fracture either directly or, in the case of small alternating movements, by fatigue failure. As fracture normally occurs not at the contact interface but at the limit of the work hardening induced (Fig. 15.20), fragments of metal become detached from both surfaces and remain in the joint as loose debris. Such fragments are extremely small (the size ranging from $1\,\mu$ to $10^{-2}\,\mu$) and therefore extremely active chemically; in the presence of air of normal humidity they oxidise rapidly to form ferric oxide, Fe_2O_3, in the case of steels, and alumina, Al_2O_3, with aluminium alloys. Unfortunately, the oxides are harder than the parent alloys — in the case of aluminium very considerably harder — and the alternating movement prevents, or at least delays,

494

their escape from the joint. They then act as abrasives, scouring and wearing down the surfaces so that (in the case of a bolted joint) the fit becomes looser and looser and the vibrations are able to increase in intensity. Thus does the damage build up; but the long-term engineering danger lies in the machining and gouging action of the individual particles. This creates a pattern of residual stresses and stress-raising scratches; where the engendered stress is tensile, conditions are ideal for the nucleation of fatigue cracks, and particularly so with bolted joints where the tensile stress in the bolt and the re-entrant angle of the threads make a ready-made site for the onset of fatigue.

Preventative measures may adopt one of two approaches. Either particular precautions may be taken to ensure rigid joints (by the

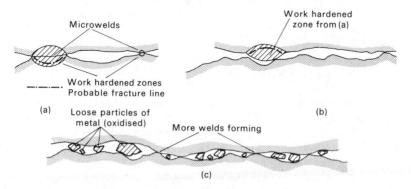

Figure 15.20   Schematic mechanism of fretting. (a) Welds and work-hardened zones at two points of contact between fretting surfaces. Fracture of the welded junction occurs at the limit of the work-hardened zone. (b) The work-hardened zone from (a), now attached to the upper surface, meets an asperity on the lower surface and breaks away. (c) Loose debris formed from (a) and (b) oxidises and acts as an abrasive between the joint surfaces.

use of high tensile or taper bolts, or welding wherever possible) or the metal surfaces may be partially or completely kept apart by interposing layers of lubricant or even polymer or paper sheeting between them. Lubricants cannot have quite the same effectiveness as in a well designed bearing (where circulation of lubricant helps to remove debris and minimise wear), but they have been shown to reduce fretting damage very considerably. It is important that the lubricant should be at least partially chemically bonded with the metal surface (chemisorption); extreme pressure lubricants that form sulphide or phosphide films are better in this respect than the fatty acid lubricants that form metallic soaps. If bonding is insufficient, the lubricant may be gradually lost from the joint interface, and the protective action, although good initially, will gradually cease and allow fretting to progress unimpeded.

495

CHAPTER 16

Metallurgical Strengthening Techniques

16.1 INTRODUCTION

For most engineering applications the working strength of an alloy is determined by its yield point, although plastic design, in which one or more members of a structure are loaded beyond their yield point but supported by the elastic reaction of neighbouring members, is becoming increasingly used in civil engineering. The ultimate tensile strength is also important in that it sets an upper limit to the load-bearing of a member; the relative magnitudes of yield point and u.t.s. are valuable in determining safety margins. The levels of both these features may be raised (not necessarily by equal amounts) by impeding the movement of dislocations; frequently such methods also increase the rate of work hardening during plastic strain.

The best known of these techniques — the *quenching and tempering* of steels — was developed to a fine art centuries before the existence of dislocations was established. Other strengthening processes also dependent on obstructing the movement of dislocations are *solution strengthening, precipitation hardening* (or *age hardening*), and *dispersion hardening.* Work hardening, in which the dislocations generate their own barriers to movement, and the effect of grain size on strength, have been considered in Section 12.4.

There is a limit, however, to techniques such as these. Strength gained by reducing dislocation mobility must automatically reduce the ability of dislocations to blunt a crack tip and the benefits are therefore inseparable from risks of embrittlement, if pushed too far. Thus the maximum potential of materials in pure strength can never be realised by engineers. *Fibre reinforcement* affords a method whereby maximum strength can be approached more nearly without embrittlement risk; strong fibres, which by themselves would be far too brittle to be considered as practicable materials, are encased in a plastic crack-arresting matrix so that to a considerable extent a compromise involving the best of both worlds may be achieved. These

496

various ways of improving mechanical strength will now be considered in more detail. It must always be born in mind that strengthening
techniques, other than work hardening, change the composition of
an alloy and are unlikely to leave other properties unaffected; in
some cases, ancillary properties may be very considerably worsened.
For example, the addition of 4% copper to aluminium to give the
age-hardening Duralumin alloys makes a very desirable increase to
the yield point and ultimate tensile strength, but at the same time
greatly lowers corrosion resistance and makes anodisation as a protective technique impossible. The whole question of strengthening is
therefore a complex one which must be considered beyond the immediate question of yield point before a fully satisfactory answer
can be derived.

16.2 STRENGTHENING BY DISLOCATION OBSTRUCTION

16.2.1 Principles

The impedance of dislocation movements is conveniently done by
introducing suitably dispersed irregularities into the crystal structure.
There are three normal ways of achieving this. On the smallest scale,
any foreign atom in a crystal disrupts the slip planes in its vicinity by
virtue of its differing atomic radius, and the result, which is known
as *solution strengthening*, may be brought about either by natural
impurities or by deliberate alloy additions. On a somewhat larger
scale, the precipitation of atoms in small clusters on the slip planes
of crystals may be achieved by suitable heat treatment of some
alloys; this is *age hardening* or *precipitation hardening*. Because of
the increased scale of the distorting structure, the obstruction effect
is very much greater than with solution hardening. On a still larger
scale, completely alien particles, usually consisting of oxides or
ceramics, may be introduced into crystals, either by pressing and
sintering a mixture of the finely powdered components, or by internally oxidising a particular alloying addition. This is known as *dispersion hardening*, and may be applied over a wide variation in oxide
content; where the oxide or ceramic content is greater than about
10%, the result is known as a *cermet*, but there is no rigid dividing
line between cermets and dispersion-hardened alloys. The behaviour
of dislocations in the face of these various obstacles is partly dependent on the spacing of the obstacles and partly on their resistance to
shear; it is not related to the method by which the obstacles are
introduced to the structure. We can accordingly consider the underlying principles of strengthening mechanisms in relation to these two
factors, and subsequently relate them to practical techniques.

16.2.1.1 *Solution Strengthening*

The stress field associated with any foreign atom in a lattice results
in a mutual attraction between the atom and any edge dislocation
line, since each to some extent relieves the stress field of the other.

In a substitutional solid solution, if the solute atom is larger than the solvent atom, it is attracted to the tensile portion of the dislocation field; conversely, relatively small solute atoms migrate to the compressive side (Fig. 16.1). Interstitial solutes, which always result in compressive stresses, gather on to the tensile side of the line.

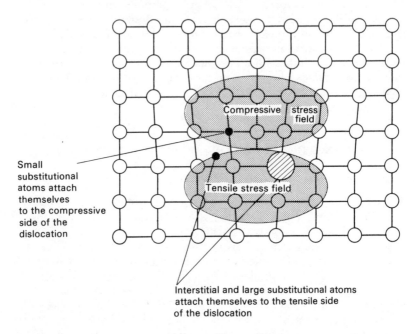

Figure 16.1   Interactions between solute atoms and the stress fields of an edge dislocation.

The result of the stress interaction is twofold: The impurity atoms exercise a 'pinning' effect on the dislocation which raises the yield point (already considered in Section 12.6 in connection with the double yield point and strain ageing), and also a 'frictional' resistance to their movement along the slip planes. The pinning effect arises because the dislocation cannot drag its attendant atmosphere of solute atoms with it as it moves through the crystal; it must break away from them if it is to become mobile. At ordinary temperatures the movement of the solute through the lattice, which must be diffusion dependent, is far too slow for normal deformation rates; hence breakaway is the only possibility, and the upper yield point observed in mild steels marks the critical stress for dislocations and solute carbon atoms to pull apart.

The degree of strengthening conferred by solute atoms varies considerably (see Fig. 16.2), depending partly on the degree of misfit of the solute and partly on the effect of the solute on the electron-atom ratio of the solvent. These effects are limited in extent, however, since excessive misfit precludes solubility, and excessive

498

valency differences lead to the formation of chemical compounds. So far, theory has been unable to account exactly for the pattern of the curves shown in Fig. 16.2; this is probably because several different strengthening mechanisms operate to different degrees in any one system. In addition to the pinning effect, the major resistance to dislocation movement arises from the extra stress necessary to propel the line through the stress fields created by the obstructions. If the dislocation line were inflexible, obstructions would have little direct effect on its movement, since statistically (for a reasonably dense dispersion of obstructions) as many would be repelling the line forwards as backwards. However, the line can bow out between the obstructions, and will do so to maintain its state of minimum free energy, which will be a balance between the energy associated with the curvature of the dislocation and the energy of its interaction with the stress fields. The 'frictional' resistance to movement is then determined by the average radius of curvature of the dislocation line, and this will depend on the size and spacing of the obstructions.

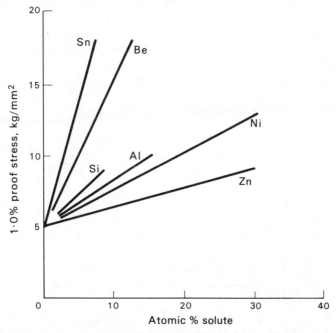

Figure 16.2 The effect of soluble alloying elements on the 1% proof stress of polycrystalline copper at room temperature. (From R. S. French and W. R. Hibbard (1950) *Transactions A.I.M.M.E.*, **188**, 53, by permission of A.I.M.E.)

The relationship between the shear stress τ and the radius of curvature r of the dislocation line is given by

$$\tau \simeq \frac{Gb}{r} \tag{12.13}$$

where G is the shear modulus of the metal and b the Burgers vector

499

of the dislocation. If we take the average spacing of the obstructions in a slip plane as x, then the radius of curvature of a dislocation

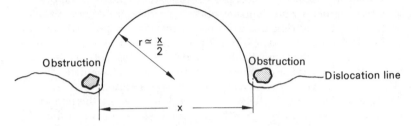

Figure 16.3   Dislocation line bowing out between two obstructions in a slip plane.

bowing out between the obstructions will be roughly $x/2$ (Fig. 16.3) and hence Equation 12.13 is modified to

$$\tau \simeq \frac{2Gb}{x} \qquad (16.1)$$

For a solid solution of atomic concentration 0.1%, the spacing of the solute atoms will be approximately 10 atomic spacings apart, and if we put $b = 1$ atomic spacing, then from Equation 16.1 $\tau \cong G/5$. Clearly this is a very much higher stress than is normally realised (see Section 2.6.1), and therefore the dislocation cannot curve around the individual obstructions. It probably moves forward in sections (Fig. 16.4), the exact movements conforming to the multiple re-

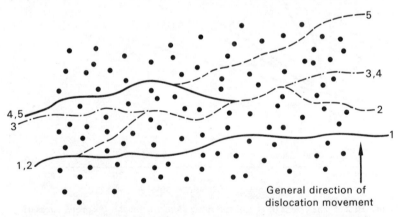

Figure 16.4   Dislocation line moving forward in sections by shear through an array of solute atoms in a metal crystal. The figures indicate the successive positions of the line.

quirements of minimum length, minimum curvature and avoidance of obstructions. For this mechanism of slip, the theoretical yield stress has been shown to be

$$\tau = 2.5G\epsilon^{4/3}c \qquad (16.2)$$

500

where c is the atomic concentration and ϵ the misfit factor of the solute atoms in the solvent lattice. The expression is in agreement with linear relationship between τ and c shown in Fig. 16.2, but predicts a yield strength that is somewhat too high — possibly because the solute atoms may not be very evenly spaced out through the lattice.

16.2.1.2 *Precipitation Hardening*

If we now turn to *precipitation hardening*, we have to consider clusters of atoms (GP Zones, see Section 16.2.2) rather than individual atoms. These exert a very much greater stress field, but at the same time are more widely spaced than individual solute atoms, so that the dislocation may be able to bend partially around individual clusters and then move forward in the manner of Fig. 16.4 but past one cluster at a time. The theoretical yield stress is then given by

$$\tau = 2Gec \tag{16.3}$$

It will be seen that this expression predicts a somewhat lower yield strength than Equation 16.2; this is at variance with the known fact that precipitation hardening is a much more powerful strengthening technique than solution hardening. However, there is in this instance an additional resistance to movement if the dislocation line is forced to shear through a cluster of atoms. The possible results of shear are shown schematically in Fig. 16.5, and additional work (known as the *cutting stress*) may be required for a variety of reasons, as follows.

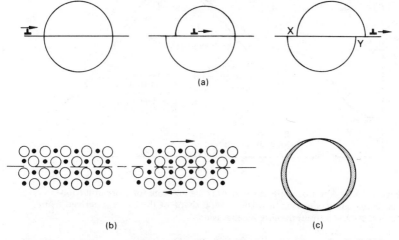

(a)

(b)   (c)

Figure 16.5   Shear by a dislocation of a precipitate particle or spherical cluster of atoms. (a) The process of shear. Unless the Burgers vector for the precipitate exactly corresponds to the Burgers vector of the matrix, there will be residual elastic stresses created at X and Y. (b) Where the precipitate has some internal ordering of different atomic species, the passage of a dislocation will disrupt the structure along the slip plane. (c) Additional interface (shown shaded) is created by shear of the precipitate; the particle is viewed looking down on the slip plane.

(a) The shear modulus of the cluster may be greater than that of the matrix, so that it has a greater natural resistance to dislocations. Obviously this effect could work the other way, i.e. $G_{cluster} < G_{matrix}$ but, with a pure metal matrix and a cluster that is an intermediate stage in the formation of an alloy phase or a compound, this is unlikely.

(b) If there is some ordering of atomic species in the cluster, the passage of a dislocation will result in disorder along the slip plane and a consequent increase in internal energy (Fig. 16.5(b)).

(c) Additional interface between the clusters and the matrix (Fig. 16.5(c)) is created as a result of shear.

(d) The Burgers vector of the dislocation in the matrix is unlikely to be the same as that when it is in the cluster; that is to say, the unit distance of slip may vary between matrix and cluster. The passage of dislocations will then create high elastic stresses at the interface.

The individual contribution of all these effects to the total increase in resistance to shear is not easy to estimate, but a qualitative relation between hardness and cluster growth is shown in Fig. 16.6. This

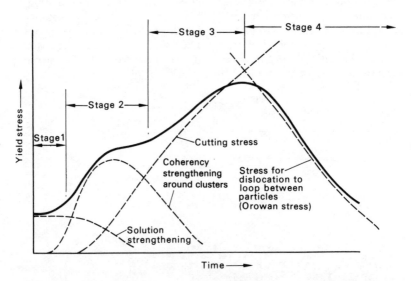

Figure 16.6 Plot of yield stress against ageing time for a precipitation-hardening alloy, showing how the general shape of the curve is derived from several separate strengthening mechanisms.

agrees with the general shape of hardness-time curves established in practical age hardening systems.

Stage 1 shows the alloy in the quenched state, when normal separation of a second phase has been suppressed and the hardness is due to the solid solution effect. As ageing commences and the solute atoms begin to form clusters that are coherent with the matrix, the stress fields surrounding each cluster increase in magnitude and resistance to deformation grows (stage 2); at the same time the solid solution contribution decreases as fewer and fewer solute atoms remain outside clusters. Beyond a certain size, however, the clusters become semi-coherent and ultimately incoherent, and the stress fields decrease, finally becoming negligible. However, with increase in size, the cutting stress increases and becomes the dominant strength-determining factor (stage 3). Growth of the clusters into an incoherent precipitate with a high cutting stress is, however, accompanied by a decrease in numbers of precipitate particles, and therefore an increase in their spacing. This makes it progressively easier for a dislocation loop to pass between the precipitate particles as the critical radius increases, and ultimately this requires a lower stress than the cutting of the particles. The critical shear stress for the alloy is then determined by Equation 16.1 and, as growth of particles and their spacing continue to increase, the yield stress will drop (stage 4). Alloys in this condition are said to be over-aged. Maximum strength is then achieved when it is equally possible for the dislocation to loop between or to cut through the particles.

Passage of a dislocation line through an array of precipitate particles at this stage of hardening then occurs by the mechanism put

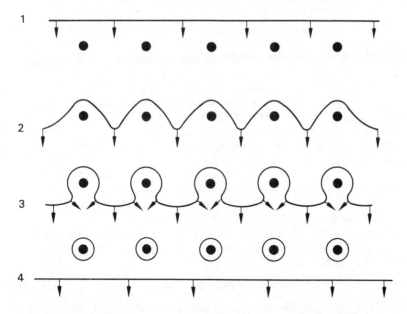

Figure 16.7   The passage of a dislocation line through an array of precipitate particles which cannot be sheared. Note the residual dislocation loops around the particles in stage 4 which increase the resistance to the passage of subsequent dislocations.

forward by Orowan, shown schematically in Fig. 16.7. The dislocation loops bow out once they have passed between two particles

503

until neighbouring loops meet; the line can then shorten itself by forming closed rings around each particle and at the same time free itself for further passage through the crystal. It will be seen from this mechanism that the passage of each dislocation reduces the effective interparticulate spacing x, so that work hardening occurs more rapidly in aged than in single phase alloys.

16.2.1.3 *Dispersion Hardening*

Dispersion-hardened alloys deform in the same pattern as overaged alloys; the individual particles are normally coarser and considerably harder than those produced by precipitation and the interparticulate spacing is wider, so that yield stress and deformation mechanisms depend entirely on dislocations looping between the particles. They have an advantage over age-hardened alloys in that the particle size is stable at high temperatures (i.e. there is no tendency for over-ageing and consequent reduction in yield stress), and for this reason most development work has been centred on their potential as creep-resistant alloys; at ambient temperatures their strength tends to be inferior to aged structures.

16.2.2 Techniques

We must now consider how best to obtain an optimally strengthening array of particles within an alloy matrix, i.e. so that it is equally possible for dislocation lines to loop between or to shear through the particles. This requires hard particles (to resist shear) and close spacing. Solid solutions can provide the close spacing but single atoms cannot provide much shear resistance; the introduction of a hard oxide dispersion may give adequate shear resistance but the spacing is too wide to prevent dislocation looping. Thus, neither solution hardening nor dispersion hardening is suited to achieve the peak strength in Fig. 16.6. There remains the technique of age hardening, in which the strengthening phase is precipitated out of solid solution by heat treatment; it has the great advantage, moreover, that the process can be terminated when the precipitate size and spacing have reached their optimal values.

In Chapter 5 the conditions for the precipitation of an alloy phase from a solid matrix were put forward; these were (a) the thermodynamic condition

$$E_a + E_b - \Delta F < 0 \qquad (5.2)$$

where E_a is the surface energy, E_b the strain energy, and ΔF is the free energy change associated with the formation of a nucleus of precipitate; and (b) the condition that the atoms must have sufficient mobility to diffuse and rearrange themselves into the pattern necessary for the precipitating phase. Let us suppose we have a binary alloy of metals M and N and that the M-rich end of the phase diagram is as shown in Fig. 16.8. When cooling an alloy of composition x from temperature T, where the structure is entirely α, both the above conditions will be satisfied after the alloy has cooled through BC to a suitable degree of supercooling. β phase will then be precipitated, and, if we assume that the β phase is harder than α (it may well have a more complex crystal structure and hence greater resistance

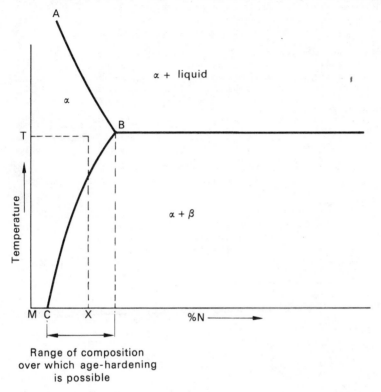

Range of composition
over which age-hardening
is possible

Figure 16.8 Suitable equilibrium diagram for an age-hardening alloy system.

to shear), the alloy will be strengthened as a result of this transformation. However, such precipitation is usually coarse and frequently strongly biased towards formation in the grain boundaries; thus the bulk of the grains is relatively unaffected, and where precipitation does occur within the grains the interparticulate spacing is far too large to provide any appreciable resistance to dislocation looping. Increasing the cooling rate of the alloy will help to some extent by refining the precipitate and causing more β formation within the grains, but the general scale of the structure is still too coarse to give more than a mild strengthening effect.

In order to achieve the requisite degree of fineness, it is necessary to quench the alloy from the α region and thereby suppress precipitation totally by not meeting condition (b), above. We then have a metastable supersaturated α solid solution retained at room temperature. The activation energy necessary for precipitation may then

505

be supplied by heating rather than cooling the alloy, and it can be supplied in small amounts — depending on the degree of heating. If the heat put into the system is insufficient to overcome the energy barrier for direct precipitation of the β phase (curve 1, Fig. 16.9), it

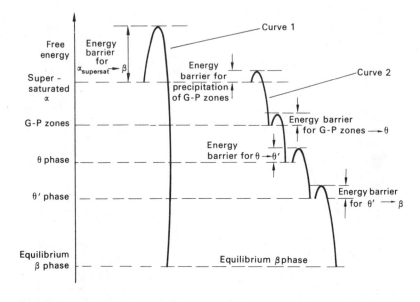

Figure 16.9   Schematic diagram of the energy changes accompanying precipitation: Curve 1, direct precipitation of equilibrium phase; Curve 2, the successive stages of age hardening.

may yet be sufficient to achieve an intermediate stage of precipitation requiring a lower activation energy (first peak of curve 2, Fig. 16.9). This stage is marked by the formation of clusters of atoms of metal N; usually they nucleate at discontinuities such as dislocations and lie in the slip planes of the matrix M. They are extremely small (30—50 atoms in diameter and 2—5 atoms thick) and since, even at this stage, total precipitation of N has been achieved, they are very closely spaced. These clusters are known as *Guinier—Preston* zones (usually abbreviated to GP zones), and after this initial stage of precipitation changes occur by growth of the larger zones (which are thermodynamically more stable) at the expense of the smaller. Thus, as ageing proceeds, both the size of individual particles and their spacing increase.

The low activation energy for their formation is due to the fact that they are *coherent* with the parent matrix (Fig. 16.10(a)); that is to say, the crystal structure remains that of the matrix and the crystal planes are therefore continuous through the zones, although strained out of shape by the different interatomic spacing within the zone. Thus the term E_a in Equation 5.2 is zero; the strain energy term E_b exists, but is balanced to a great extent by the relief of the supersaturation strain energy in the matrix.

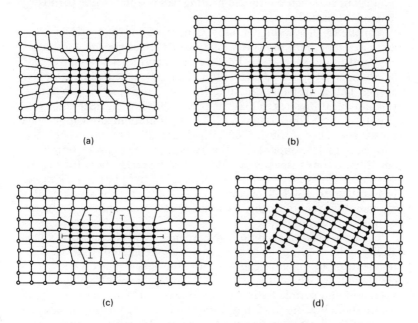

Figure 16.10   Schematic diagram of four stages in precipitation hardening. (a) Coherent precipitate: Guinier–Preston zone. (b) Part coherent, part semi-coherent precipitate. (c) Semicoherent precipitate. (d) Incoherent precipitate: the final (over-aged) stage of precipitation. Note that there would be considerably greater matrix strain than is implied in these sketches, especially in (b), (c) and (d).

Further heating permits modification of the precipitate to take place, by the surmounting of more secondary energy barriers (stages 2 and 3 of curve 2, Fig. 16.9). As growth of individual zones progresses, the strain fields surrounding them also increase until the coherency strains become too great, and the clusters then become *semi-coherent* with the matrix. Several stages of semi-coherency are possible, depending on the alloy system concerned. The precipitates may be coherent on one set of planes but mismatching on others (Fig. 16.10(b)) or it may take the form of a ring of dislocations surrounding the precipitate (Fig. 16.10(c)); the internal structure of the precipitate will also change and achieve the correct composition of the β phase. Both these stages are shown in the best known of all age-hardening systems — the Duralumin series of alloys based on 4% copper in aluminium — and the peak strengthening effect shown in Fig. 16.6 is reached in the semi-coherent state.

Further heating and growth of precipitate finally result in the formation of the equilibrium β phase and the formation of a true incoherent interface with the matrix. This relaxes much of the stress fields surrounding the particles (although there will be some residual

507

elastic energy as a result of volume changes) and consequently the hardness drops: the alloy is now *over-aged* and in its equilibrium condition, although continued heating will result in some further increase of particle size.

For the small amounts of alloying necessary (usually <5%), the gains in strength by the correct use of this technique are considerable, not to say spectacular in some cases, and further benefits can be obtained by plastic deformation of the quenched alloy prior to ageing. This is probably due to dislocation tangles setting up additional nucleation sites to that the precipitate is even finer and more closely spaced. Thus commercially pure aluminium and copper have tensile strengths of roughly 83 MN/m^2 and 220 MN/m^2; when alloyed with 4% copper and 2% beryllium, these figures rise to 240 and 450 MN/m^2 respectively. Age hardening raises them further to 414 and 1 034 MN/m^2, whilst age hardening and plastic working brings them to 450 MN/m^2 and 1.25 GN/m^2; thus the strength of aluminium is raised roughly by a factor of 5 and copper by a factor of 6 by this technique.

The principal limitation of age-hardened alloys lies in their vulnerability to heat. Should a heat-treated alloy be heated to its ageing temperature during fabrication or use, the ageing process will continue from where it ceased and the probable result will be an over-aged and permanently weakened material. This can only occur where there is some small but finite solubility of the precipitate phase in the matrix, so that the larger particles can continue to grow at the expense of the smaller, and the more insoluble the precipitate phase, the greater will be the temperature stability of the alloy. For this reason, dispersions of hard phases that are known to be totally insoluble in the matrix alloy have been developed under the title of *dispersion-hardened* alloys. They are prepared either by intimately mixing finely powdered matrix alloy and dispersion phase together, pressing in a mould, and then sintering so that alloy-phase particles unite by solid-phase welding; or by internal oxidation, in which solute atoms in the matrix are preferentially oxidised by the diffusion of oxygen into the alloy, to give an oxide dispersion. Thoria-dispersed nickel (TD nickel) and aluminium/aluminium oxide mixtures (SAP aluminium) have been the most successful products, but it has proved difficult to obtain an even dispersion of hard particles, and their mechanical properties at ordinary temperatures have not proved the equal of more conventional alloys.

16.3   STEELS AND THEIR HEAT TREATMENT

Steel is a major load-bearing constructional material for civil engineers. In terms of tonnage used it is second only to concrete — and is of course frequently employed to improve the tensile behaviour of concrete by means of reinforcement or prestressing. The great bulk of steels used fall into the category known as *structural* or *mild* steels with carbon content varying from 0.05% to 0.3%. They are usually employed in the *normalised* (i.e. unheat-treated) condition, in which state they are the cheapest structural alloys, and also satisfy the two prime engineering requirements: adequate strength and notch ductility. Of recent years, as riveting has tended to give way to welding in constructional techniques, a third critierion, *weldability*, has been added to the above requirements. Fortunately, the standard low-carbon structural steels are satisfactory in this respect, but over the past 20 years there has been an increasing demand for higher-strength steels which has brought the three main criteria into some mutual conflict. Obviously, in specific cases, other characteristics than those mentioned above may be of paramount importance, such as wear resistance or corrosion resistance, and strengths may on occasion be required which are far in excess of those attainable with mild steels. For that reason, after discussing the structure and properties of plain carbon steels, we shall consider briefly the principles underlying the development of the high-tensile alloy steels.

16.3.1 Plain Carbon Steels

The formal basis of steel structures obtainable by slow cooling is best understood by reference to Fig. 16.11, which outlines the rele-

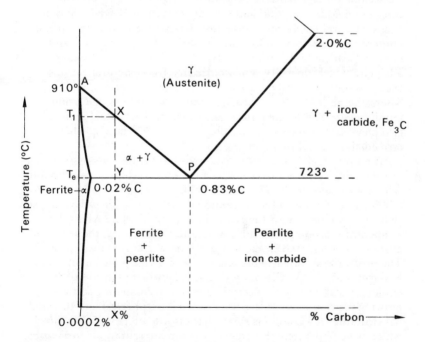

Figure 16.11 The relevant portion of the iron–carbon equilibrium diagram for slowly cooled (normalised) steels.

vant portion of the iron-carbon diagram. Iron exists in two crystalline forms, the high temperature variety being known as *austenite* or *γ-iron* and having an f.c.c. structure. Despite the fact that the size ratio of the carbon to the iron atom lies slightly above that predicted for interstitial alloying, austenite can dissolve up to about 2% carbon. *Ferrite* or *α-iron*, however, which is the stable allotrope at low temperatures, has a b.c.c. structure, and paradoxically, although it is a less close-packed structure than austenite, it can only dissolve a maximum of 0.02% carbon at 723 °C. This value drops to about 0.2 p.p.m. at 25 °C (the exact equilibrium solubility is difficult to establish), and so carbon may be regarded as virtually insoluble in ferrite. Consequently, carbon always appears in slowly cooled steels as iron carbide, Fe_3C, otherwise known as *cementite*. Cementite is an extremely hard phase, typical of the carbides of the transition elements, and possesses a complex crystal structure; it is the phase that is primarily responsible for the strength of steels, ferrite being far too ductile (yield stress 50 MN/m^2) for structural purposes.

Figure 16.11 shows that austenite decomposes by a eutectoid reaction; the triple point lies at 723 °C and 0.83% carbon, and the product, known as *pearlite*, consists of alternate platelets of ferrite and cementite, the ferrite platelets being approximately seven times the thickness of the cementite. The fineness of the eutectoid structure varies with the cooling rate of the steel but the relative thicknesses remain constant. Steels with less than 0.83% carbon are designated *hypoeutectoid* and contain a mixture of ferrite and pearlite; structural steels come into this category. Steels with carbon contents greater than 0.83% are classed as *hypereutectoid* and contain pearlite and cementite. Hypereutectoid steels have inadequate notch ductility and weldability for structural purposes; they are used primarily for tool steels, springs, cutters, and rails, where the ability to keep an edge and resist wear is the prime requirement. The temperature range over which a steel is in the two-phase $\alpha + \gamma$ region of the diagram (e.g. XY in Fig. 16.11) is termed the *critical range* of the steel.

When the steel is allowed to cool naturally from the austenitic range in air, ferrite grains will nucleate in the austenite grain boundaries at temperature T_1 (Fig. 16.11). These, being virtually pure iron, will reject carbon into the surrounding austenite as they develop, so that the composition of the austenite follows the line AP as the temperature drops. When the temperature reaches T_e, the remaining austenite contains 0.83% carbon, and it then transforms to pearlite. The process is shown schematically in Fig. 16.12. Once nucleation has occurred, the pearlite grows by an edgewise extension of the eutectoid platelets, with a certain amount of branching so that growing pearlite colonies are usually roughly spherical and the internal structure of a colony is somewhat disordered (Fig. 5.14). Steels which have undergone this transformation are said to be *normalised*.

The final structure and strength of a steel of any given carbon content is strongly dependent on two factors: the grain size of the parent austenite, and the cooling rate. A fine-grained austenite presents many sites for ferrite nucleation and therefore results in a fine-grained product. If the austenite grains are coarse, however, nucleation of ferrite may occur within the grains as well as at the

510

boundaries. Ferrite nucleated in this way grows in large disc-shaped crystals (see Section 5.7) that follow the slip planes of the austenite and give rise to a characteristic appearance known as Widmanstatten structure (Fig. 5.22). The occurrence of this structure in steels (particularly likely in thick sections, large castings, in the vicinity of heavy welded sections, and in welds) confers both low strength and low notch ductility and so is highly undesirable — although it may be difficult to avoid in large components. A faster cooling rate acts

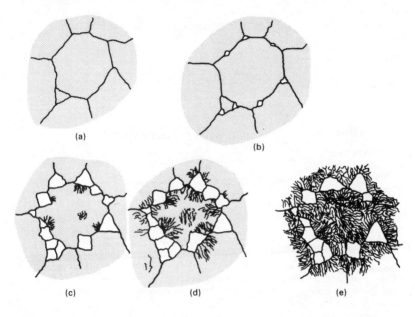

(a)

(b)

(c)

(d)

(e)

Figure 16.12 Schematic sequence of events as a hypoeutectoid steel cools through the critical range. (a) Structure 100% austenite. (b) Just below T_1^0, ferrite nucleates in the austenite grain boundaries. (c) Ferrite growth virtually complete; pearlite nucleation commences. (d) Continued growth of pearlite. (e) Transformation complete.

on the structure by increasing the degree of supercooling (thereby increasing the number of ferrite and pearlite nuclei) and thus giving a fine-grained product. Furthermore, because the transformation takes place at a lower temperature, the carbon atoms within the austenite are less mobile. Now the size and spacing of the pearlite platelets depends on the ability of the carbon to diffuse from a more-or-less homogeneous distribution in the austenite into separate regions of high concentration, and movement of carbon atoms occurs just in front of the advancing pearlite colony as shown in Fig. 16.13. In general, the pearlite structure will be as coarse as circumstances permit, since the fewer the individual lamellae, the lower will be the total interfacial energy of the eutectoid. However, if the diffusion is restricted by increased supercooling and/or less

511

time for the transformation, then the distance the carbon atoms can travel to the nearest cementite platelet is also curtailed and the result is a finer pearlite, giving a harder stronger steel.

The size of the austenite grains is a problem because austenite in plain carbon steels exists well above the recrystallisation temperature, and therefore grain growth — particularly for thick sections, which are rolled above 1 000 °C — is highly probable. The hot working of the steel does break up the austenite grains, however, and if the final

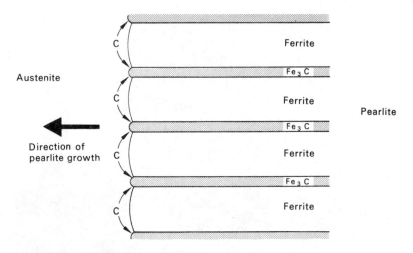

Figure 16.13 Diffusion of carbon atoms in austenite in front of an advancing colony of pearlite. Clearly, the finer the internal structure of the pearlite, the shorter is the maximum distance that the carbon atoms must diffuse to the nearest cementite lamella.

pass can be achieved at a sufficiently low temperature (850–900 °C) and with a sufficiently high reduction in thickness ($>20\%$), a fine ferrite grain size is obtained. Unfortunately, there are various practical problems in the way of achieving this, notably in the design of suitable mills for dealing with large reductions of thick sections; for structural girders also, the rolling sequence is a complex one and it is not possible to finish all parts with a high reduction at the same time: hence the resulting structure is likely to show considerable variations in grain size. When a Widmanstatten structure forms in spite of precautions, it may be removed by heating the steel to just above the critical range and then allowing it to cool in air.

The grain size of austenite may also be stabilised by 'killing' the steel, a deoxidistion process in which aluminium is added to the steel and forms a fine dispersion of alumina particles through the structure. The prime purpose is to remove oxygen from the steel by precipitation and thus prevent the formation of carbon monoxide bubbles in the ingot, but alumina particles have a valuable side effect in that they act as nuclei for the solidification of the steel (thus giving an initial fine-grained austenite) and they also obstruct the migration of grain boundaries during grain growth. The drawback to the process is that killed steel ingots have a larger contraction cavity (known as a *pipe*) than untreated steels; this means that a larger portion of the ingot has to be cropped off (since the pipe cannot be rolled) and results in a smaller yield and higher price to pay for the undoubted advantages of killed steels. Because of the economic factor, a whole range of 'semi-killed' steels have been developed as a compromise between cost and quality.

16.3.1.1 *Transformation Products of Austenite*

The importance of diffusion during the decomposition of austenite has already been commented on; it is now necessary to examine its effects rather more closely. Austenite need not always decompose immediately below the eutectoid temperature, 723 °C; providing a specimen is small enough it can be cooled rapidly and then held at any desired temperature below 723 °C by immersion in a salt bath or hot oil or water. It will then be found that both the rate of decomposition and the decomposition products will vary considerably according to the temperature, the basic reason for this being the progressive immobilising of the carbon in the austenite as the temperature drops. If systematic experiments are carried out for a particular steel, quenching specimens to a number of different temperatures, and then allowing the austenite to transform isothermally (i.e. at constant temperature), the rate of transformation may be plotted in a *time temperature transformation curve* (or *TTT curve* for short). The results of such an investigation for a eutectoid steel are shown in Fig. 16.14, and the diagram is best explained by considering the decomposition of the austenite at the five temperatures shown.

If the steel is held at T_1, the normal eutectoid reaction takes place and the end product is pearlite. Since the supercooling of the austenite is slight, the free-energy change is small and the decomposition rate is low; however, as the carbon is mobile a coarse pearlite begins to form when the T_1 isotherm cuts the P_s (pearlite start) line. Transformation is complete when the isotherm cuts the P_f (pearlite finish) line.

At T_2 the mobility of the carbon is lower and the free energy change is greater: the result is a shorter nucleation time and a faster reaction, whilst the product is finer because the carbon cannot move so readily in the austenite. Reaction rate and fineness of pearlite reach their maximum for an isotherm cutting the P_s line at N, known as *the nose* of the curve; the pearlite structure formed at this temperature is unresolvable under the optical microscope and is often referred to as *nodular pearlite* since it grows in the form of nodules in the austenite grain boundaries.

At lower temperatures, diffusion becomes the limiting factor and both nucleation and the rate of reaction become slower again. Now, however, the formation of pearlite is suppressed because the carbon cannot move the necessary distance to form the cementite plates.

This does not prevent transformation; instead, the pattern of transformation changes to permit the formation of cementite with smaller diffusion distances. The result at temperature T_3 is a structure known as *upper bainite* (from its position on the TTT diagram) or *feathery bainite*, and therefore the curves in the diagram become the B_s and B_f curves below the nose. In contrast to pearlite, ferrite is the nucleating phase for bainite, and forms very long thin crystals; the carbon is rejected to the edges of these crystals where it forms

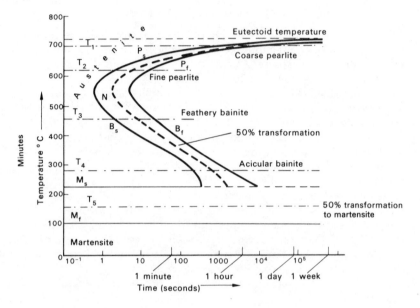

Figure 16.14   Time—temperature—transformation curve for a eutectoid steel.

chains of cementite particles. Unfortunately the alignment of cementite in this manner provides ready paths for cracks, and feathery bainite is consequently brittle and unreliable in service.

Transformation at a still lower temperature T_4, in the range $250°–300°$C, results again in bainite, but a different form of bainite: *lower* or *acicular bainite*. Nucleation is again by ferrite formation, but now the carbon does not even move to the edge of the crystals; it is precipitated within them as a very fine intermediate form of iron carbide of approximate composition $Fe_{2.4}C$, known as ε-carbide. The exact mechanism of this transformation is still not fully understood; although it is primarily a nucleation-and-growth process, it also appears to contain elements of a shear transformation, and is somewhat unique in this respect. With the precipitation of carbide taking place within the grains, there is now no tendency for dangerous alignment of the particles. The structure has really undergone a form of precipitation hardening, and possesses an attractive

514

combination of high strength and notch ductility. Lower bainite is in fact an excellent structural alloy, but unfortunately, as we shall see below, it is not easy to obtain in plain carbon steels in an adequate section thickness or without embrittling traces of upper bainite.

Thus far, the form of the transformation starting and finishing lines are what would be expected for a diffusion-dependent process. At about 230 °C, however, the starting line (M_s in Fig. 16.14) becomes horizontal, and the M_f line is separated from it by an interval of *temperature* rather than *time*. Clearly we are dealing with an entirely different type of transformation now, a transformation which is temperature rather than time dependent and which therefore must be a shear transformation (see Section 5.10). The product of the transformation is *martensite*, and the form of the diagram indicates that if the austenite is quenched to T_5, where T_5 lies approximately midway between the M_s and M_f temperatures, transformation will take place to about 50% martensite and the remaining austenite will not transform until the temperature is lowered further, or until sufficient time has elapsed to nucleate lower bainite. Only if quenching is taken to below the M_f line will the austenite transform entirely to martensite. For steels with 0.5% carbon and more, the martensite formed is the conventional acicular variety (Fig. 16.15) with a high degree of internal twinning to minimise distortion

Figure 16.15 Typical martensitic structure of a water-quenched plain carbon steel (x 400). (From A. R. Bailey (1972) *The Role of Microstructure in Metals, 2nd edition*, by permission of Metallurgical Services Laboratories Ltd.)

on the habit plane. When the carbon content is less than 0.2%, however, martensite forms in larger crystals that accommodate the habit plane by internal slip rather than by twinning; it is sometimes referred to as *massive* or *slipped* martensite to distinguish it from the commoner variety. Between 0.2% and 0.5% carbon, a mixture of these two structures is formed very much in the manner of a two-phase region in a phase diagram.

16.3.1.2 *Tempered Structures*

Although martensite is ideally a very strong material, it is exceedingly hard and brittle in practice, and therefore totally unsuited for structural use. However, if strengths are required in excess of those obtainable in normalised structural steels, they may be obtained by

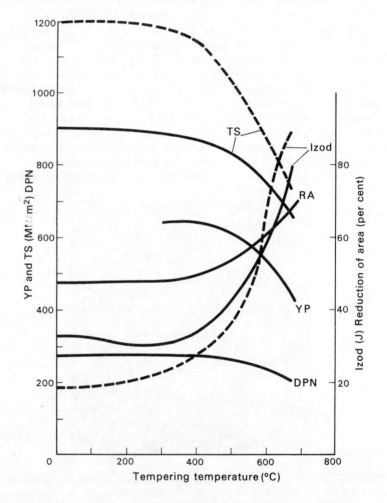

Figure 16.16 Variation in properties of 1% Ni steel with varying tempering temperatures. (From O. H. Wyatt and D. Dew-Hughes (1974) *Metals, Ceramics and Polymers*, by permission of Cambridge University Press.)

quenching a rather higher carbon steel (0.4–0.5%) and then heat treating (*tempering*) the martensite formed for an appropriate time

516

and temperature. This stage is analogous to the ageing process discussed in Section 16.2, and the metallurgical precipitation is broadly
similar in principle. Physically, however, the result is the antithesis
of age hardening; we are now concerned with *softening* an excessively
hard and brittle material. The ideal strength of the quenched structure is somewhat reduced (Fig. 16.16), but the precipitation serves
to relieve some of the internal strains and confers an acceptable
degree of notch ductility, so that the product is superior both in
strength and toughness to a normalised steel.

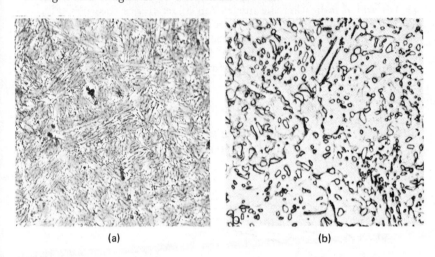

(a) (b)

Figure 16.17   (a) General appearance of sorbite in a tempered steel (x 200).
(From A. R. Bailey (1967) *The Structure and Strength of Metals*, by permission
of Metallurgical Services Laboratories Ltd.) (b) Spheroidised cementite in a
1.1% carbon steel (x 900). (From R. E. Reed-Hill (1973) *Physical Metallurgy
Principles, 2nd edition*, by permission of D. Van Nostrand Co.)

During the initial stage of tempering, at $100°-150°C$, ϵ-carbide
is precipitated along the twin boundaries within the martensite, and
the structure therefore bears a close resemblence to lower bainite.
The tempered structure, however, has a finer precipitate size and is
usually somewhat stronger. At $250°C$ any austenite that has been
retained in the structure after quenching (see below) decomposes to
lower bainite. Unfortunately this is accompanied by a volume expansion that sets up internal stresses, and as a result the notch ductility
actually decreases during this stage of tempering; this is one form of
temper-brittleness. At $300°C$ and above, the ϵ-carbide transforms to
cementite, and any carbon remaining trapped in the lattice is also
precipitated as cementite, thus allowing the martensite lattice to
relax back gradually towards the b.c.c. ferrite structure. As tempering
continues, the larger cementite particles then grow at the expense of
the smaller, and eventually the matrix relaxes completely back to
ferrite. The optimum combination of strength and notch ductility
for most applications occurs when the carbide particles are still too
small to be resolved under the optical microscope; this structure is

517

known as *sorbite* (Fig. 16.17(a)). Eventually the cementite particles become clearly visible, and the structure is then called *spherodite* (Fig. 16.17(b)). It is less used than sorbite since the gain in strength over a normalised steel is now hardly sufficient to warrant the expense of the heat treatment, except perhaps where very high notch ductility is required.

Figure 16.14 refers to a eutectoid steel, so there is no formation of primary ferrite or cementite, as there would be in hypo- or hypereutectoid steels. For a hypoeutectoid steel an additional line (F_s) is added indicating the nucleation of ferrite, as shown in Fig. 16.18. This line rises above the eutectoid temperature and is asymptotic to the upper limit of the critical range of the steel in question (for example, temperature X in Fig. 16.11). A similar line indicating cementite nucleation would appear in the diagram for a hypereutectoid steel.

16.3.1.3 *Use of Time–Temperature Transformation Curves*

TTT curves are of great importance to both designer and manufacturer since they indicate (with some adjustment, see below) the

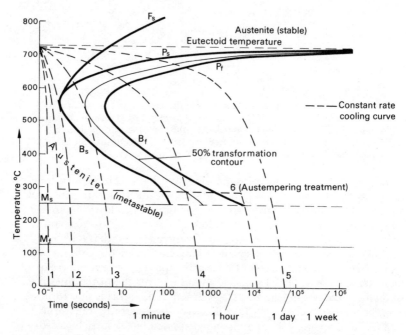

Figure 16.18 Time–temperature–transformation curve for a hypoeutectoid steel.

structures that will be formed at various cooling rates. Hitherto, we have tacitly assumed that a virtually instantaneous quench followed by isothermal transformation is possible and therefore that any of the structures discussed are readily attainable. In practice, of course,

518

this is a totally unrealistic approach. Only the very thinnest sections can be quenched at a rate remotely approaching instantaneous, and steels will normally be cooled at roughly constant rates, as indicated by curves 1 to 5 in Fig. 16.18. Thus if a steel is cooled at rates 1 and 2 it will be transformed totally to martensite; at rate 4 fine pearlite will be formed, and at rate 5 coarse pearlite. Steel cooled at rate 3 would transform 50% to very fine pearlite with perhaps a trace of bainite, since it just touches the 50% transformation contour at the nose of the TTT curve. Thereafter the cooling curve moves towards the start rather than the finish line and no further transformation will occur until the cooling curve cuts the martensite lines, when the remaining austenite will transform to martensite. This is an important point of interpretation: all changes in the diagram below the eutectoid temperature are irreversible unless the steel is reheated into the austenite region. The fact that curve 3 re-enters the metastable austenite region after passing through part of the upper bainite section does *not* mean that first-formed pearlite transforms to bainite and then back to austenite; it merely means that the time-dependent transformation will cease at the point where the cooling curve reaches its maximum transformation contour and will not recommence until the temperature has dropped sufficiently for martensite formation to commence.

The important cooling curve in Fig. 16.18 is curve 2, which just misses the nose of the TTT curve: this indicates the minimum rate of cooling necessary to form martensite and is known as the *critical cooling rate* of the steel. This is the cooling rate that must be achieved in the centre of a component (where cooling will be slowest) if we are to obtain a satisfactory quench to 100% martensite. This immediately introduces the highly important concept of a *size factor* in quenching problems, since the larger the section of the steel, the slower will be the cooling rate in its centre. Consider the TTT curve in Fig. 16.19. If curve 1 represents the maximum rate cooling attainable by a quench medium *on the surface* of the steel, the cooling rate in the centre may be no faster than curve 2. Thus we cannot consider a real steel component as cooling along a single line as in Fig. 16.18; it will cool over a band of rates, and the larger the component the wider will be the band. In the example just cited in Fig. 16.19, the quench is satisfactory; in fact it is a good quench on all counts since it is important to quench at the *minimum* satisfactory rate in order to minimise thermal shock and the risks of quench cracking. But if the steel has a greater cross-section, the centre cooling rate may well be no faster than curve 3, in which case the quench is not satisfactory. Three responses are possible in this situation. We may accept that the steel is not fully hardenable in that section thickness, and make due allowance in the design; or we may increase the quench intensity if possible (but this will increase thermal shock and may be undesirable on that count); or we may move the whole TTT curve to the right by the addition of suitable alloying elements to the steel, so that the nose just misses curve 3. The position of the TTT curve is also sensitive to grain size. A fine-grained steel provides many more sites for pearlite nucleation than a coarse-grained one, and thus decreases the section thickness that can be satisfactorily quenched. Conversely a coarse grain size increases it, but the drop in notch ductility and increased risk of quench cracking make this an unacceptable expedient for obtaining a full quench.

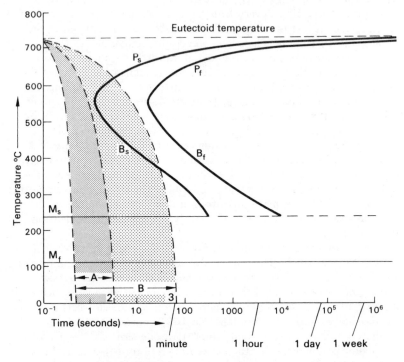

Figure 16.19   Time—temperature—transformation curve for a plain carbon steel showing the effect of section thickness on structure. Two components, A and B, are quenched under identical conditions: the thinner component A has a narrow band of cooling rates and the quench is satisfactory, whereas for the thicker component B with a wider band, it is not.

16.3.2   Alloy Steels

We are now in a position to consider the effect that alloying elements have on steels. One point should be made straight away: with the exception of silicon and manganese, alloy elements raise the cost of steels, sometimes very considerably, and therefore wherever possible the use of plain low-carbon structural steels is preferable on economic grounds. However, there are a number of instances where this type of steel is not adequate, either on grounds of mechanical strength (e.g. steel rods for prestressed concrete) or for some other necessary property such as corrosion resistance, and it is important then to understand how the structure and heat treatment potential of the steel may be affected by alloying additions.

520

16.3.2.1 *General Effects of Alloying Elements in Steels*

The principal effect of alloying elements in steel (including carbon) is to slow down the rate at which carbon can diffuse in the austenite phase. This has the effect of moving the TTT curves to the right, so that the critical cooling rate for the steel is decreased. This means either that we can get a thicker section of steel fully quenched at a given quench rate (i.e. we have increased the *hardenability* of the steel) or we can employ a milder quench for a given section thickness, thus reducing the risk of quench cracking. The maximum diameter that can be fully quenched for a given steel is termed the *limiting ruling section*, and this therefore increases as the hardenability increases. A further effect that follows logically is that the pearlite structure will be refined for a given normalising cooling rate, thereby improving the strength of thick sections; this is shown in Fig. 16.20. Conversely, mild steels with no alloy additions have the

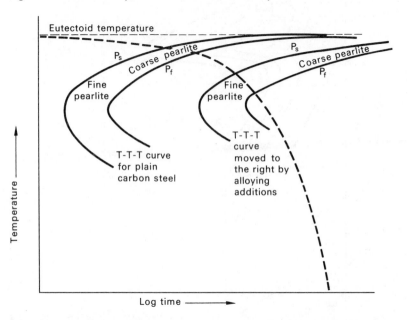

Figure 16.20 Schematic diagram to show the effect of alloying on pearlite structure. At a similar cooling rate (dotted curve) an alloy steel will give a finer pearlite than a plain carbon steel because of the timewise shift in the position of the T–T–T curve.

TTT curve well over to the left so that the nose is very close to the vertical axis. This makes full quenching impossible in all but the very thinnest sections, and is the reason why higher carbon or low alloy steels have to be employed if a heat-treated structure is desirable.

As the alloy content of a steel is increased, the TTT curve in addition to moving to the right develops a double nose and then splits into two separate curves, one for pearlite and one for bainite (Fig. 16.21). In general, the bainite curve is shifted less than the pearlite –

this is to be expected as bainite formation is less dependent on diffusion than pearlite — and so it is possible to obtain bainitic structures in alloy steels by continuous cooling. The shapes of the TTT curves (Fig. 16.13) show that this is impossible for plain

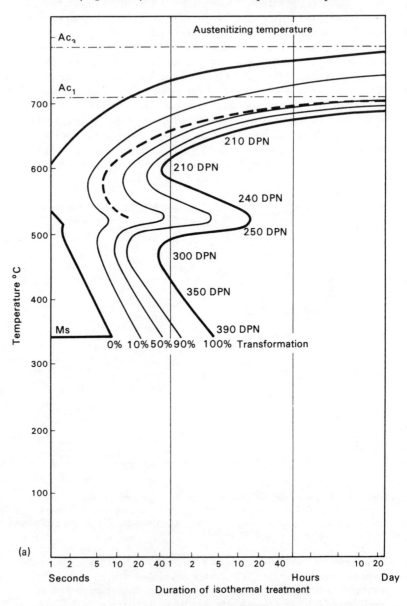

Figure 16.21 (a) T—T—T curve for a 1.25 Ni/0.5 Cr medium carbon low alloy steel showing the first signs of a double nose on the curve. (b) T—T—T curve for 2.5 Ni/0.7 Cr/0.5 Mo/0.3C steel, showing the complete separation of the pearlite and bainite curves. Note the very much greater retarding effect on the pearlite curve. (By permission of Inco Europe Ltd.)

carbon steels because of the 'undercut' shape of the curve in the bainite region. Where bainite is required in a plain carbon steel, it can only be produced by *austempering*; the steel is quenched to just above the M_s temperature and then held at constant temperature to transform as shown on curve 6 in Fig. 16.18. However, because the

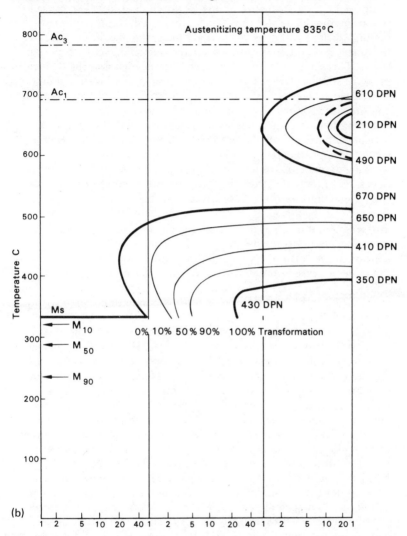

critical cooling rate for these steels is fairly high and because a quench in a salt bath at about 300 °C cannot be as intense as a quench in water at about 25 °C, austempering except in alloy steels is only practicable in thin sections.

The separation of the pearlite and bainite curves in Fig. 16.21(b) opens up an 'austenite bay' in which the austenite, although metastable, will remain untransformed for several hours. This has led to the development of the *ausforming* technique for the production of ultrahigh-strength steels. The steel is held in the bay at the lowest possible temperature (to discourage recrystallisation) and plastically deformed, preferably to about 75% deformation. It is then quenched and tempered in the normal way and may develop a tensile strength of up to 3 GN/m^2 whilst retaining its notch ductility. The cause of this strengthening is still to be fully elucidated, but it is thought that the high dislocation density created during the plastic working of the austenite is retained throughout the quenching and tempering stages. This serves to refine the martensite needles and the dislocation network stabilised by the fine carbides precipitated during tempering, so that we achieve a very satisfactory combination of work hardening and precipitation hardening in the final structure.

Strengthening techniques such as ausforming that combine deformation hardening with heat treatment have been called *thermo-mechanical treatments* (TMT) and have been extended to the hot working of austenite in its stable range prior to quenching. This produces only moderate strength increases, but is reported to improve the fatigue limit by up to 200 MN/m^2, probably because of grain refinement of the austenite. A third technique of this type is known as *isoforming* and involves deforming the steel during the transformation of austenite to ferrite and pearlite. The pearlitic cementite appears in the form of a very fine spherodite, and the result is a considerable increase in notch ductility whilst the strength remains little affected. Techniques of this type are likely to increase in importance in the structural field as demands for high strength coupled with good notch ductility increase.

So far we have only considered the consequences of the shift in the pearlite and bainite lines in a TTT curve. However, the martensite lines are also shifted downwards by alloying additions, and by increased carbon content, as shown in Fig. 16.22. Once the M_f temperature has dropped below the temperature of the quench, it is impossible to get 100% martensite, and the residual austenite is known as *retained austenite*. Since the effect of alloying elements on the M_s and M_f temperatures is additive, retained austenite becomes an increasing problem in high strength steels. The M_s temperature for a low-alloy steel can be estimated from the formula $M_s(°C)$ = $561 - 474C - 33Mn - 17(Ni + Cr) - 21Mo$. Even where the total alloy content is not sufficient to cause appreciable retention, segregation of alloying elements can lead to local pockets of austenite after quenching. This results in a poor quality steel since neither quenching nor tempering can go fully to completion, and the retained austenite may transform at a later stage, embrittling the steel and setting up internal stresses because of the expansion that accompanies the transformation. It is also reported to lower the fatigue strength.

Having considered the general effects of alloying elements in steels, we may now examine the individual behaviour of those that are more commonly present. These can be classified as (a) impurities, (b) carbide-forming elements, and (c) elements that remain in solid solution in the iron.

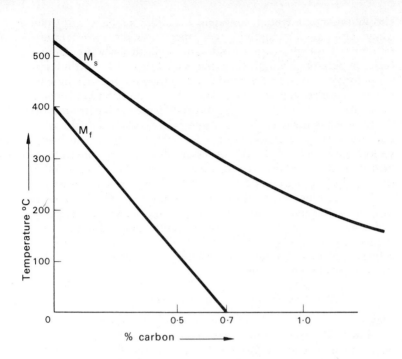

Figure 16.22 Effect of carbon on the M_s and M_f temperatures. Other alloying elements also affect the M_s and M_f values, but individually their effect is less than 1/10 that of carbon (see expression in text).

16.3.2.2 *Impurity Elements*

Impurity elements are the nonmetals oxygen, sulphur, phosphorus and nitrogen. Sulphur and phosphorus, which form low melting point impurities (see Section 5.13) are the most harmful as they embrittle the steel and make it hot-short, and are normally reduced to <0.05% during the steel-making process. Nitrogen is troublesome because it contributes to strain ageing (Section 12.6) and may therefore raise the brittle-ductile transition temperature after a structural member has gone into service. This problem is avoided by fixing the nitrogen as either aluminium or vanadium nitrides, both of which then contribute to grain refinement and stabilisation. Oxygen is also precipitated by aluminium, but if a fully killed steel is not required (see p. 486) silicon and manganese may be added as deoxidisers; manganese will also fix any residual sulphide as harmless manganese sulphide.

16.3.2.3 *Carbide-Forming Elements*

These will dissolve in the ferrite phase of the steel and form mixed carbides of the general form $(Fe,X)_3C$, where X is the alloying element; in order of increasing affinity for carbon, they are manganese, chromium, molybdenum, tungsten, vanadium, niobium and titanium. They confer great strength by the dispersion-hardening effect of extremely hard and stable carbides; more particularly they improve high-temperature strength since the insolubility of the carbides prevents (or slows down very considerably) the onset of softening by over-ageing mechanisms. The carbide dispersion also has a secondary strengthening effect by refining and stabilising the ferrite grain size. All these elements have the usual general effects on the TTT curves, but only if they are in solution; hence prolonged high-temperature soaking (at about 1 300 °C) is necessary if these steels are to be fully heat treated. Tempering effects are also retarded by the general slowing down of carbon diffusion, so that tempering is normally carried out in the range 550—600 °C. Of the elements listed above, manganese has less affinity for carbon than iron, and is therefore added more for its general solution strengthening and to increase the hardenability of low-alloy high-tensile steels. Molybdenum, though expensive, is important because it lowers the ductile-brittle transformation temperature.

16.3.2.4 Solution-Strengthening Elements

This group comprises silicon, manganese, nickel, copper and aluminium. All show the general effects listed earlier, but to a lesser degree than the carbide formers. Manganese is best considered to have a foot in either camp; it strengthens steel principally by solid solution, but manganese carbide also forms in association with pearlite and increases the percentage of pearlite for a given carbon content. It is also beneficial in lowering the brittle-ductile transition temperature, but only in the case of high-strength steels if the carbon content is kept low; a carbon/manganese ratio of less than 1 : 5 is recommended. The upper limit of manganese in structural steels is about 1.5%; above this amount the steel loses ductility and problems are likely to be encountered in welding (see below) because of increased hardenability. Silicon improves the strength of steels, but ductility decreases when it is in excess of 0.5% and in excess of 1% its action as a graphitising agent (made use of deliberately in cast irons) risks embrittling the steel in thick slowly cooled sections.

Although more expensive, nickel has greater potential as an alloying element that either silicon or manganese; it is the principal addition (together with smaller amounts of chromium and molybdenum) in most low-alloy high-tensile steels that are used in the heat-treated condition and develop strengths of the order of 1 000 MN/m^2. Nickel, like manganese and cobalt, acts as an austenite stabiliser.(see Fig. 16.23(a)), and high nickel-chromium steels such as the 18Cr 8Ni stainless steel retain austenite down to room temperature. The austenite is still metastable at room temperature (Fig. 16.23(b)), but the transformation occurs so sluggishly that it is suppressed at all but exceptionally slow cooling rates. Because of the natural ductility of the austenite, its retention has considerable advantages for ease of fabrication. Austenitic steels cannot be heat treated, but in some

526

stainless steels, and in Hadfields manganese steel (1% C, 12% Mn), the austenite may be induced to transform to martensite by *strain*; such steels have exceptionally high work hardening and wear resistant properties whilst retaining the natural toughness of austenite. They also exhibit good notch ductility down to very low temperatures.

Nickel is also the principal alloying element in the recently developed ultrahigh-strength *marageing* steels. With more than 15% Ni,

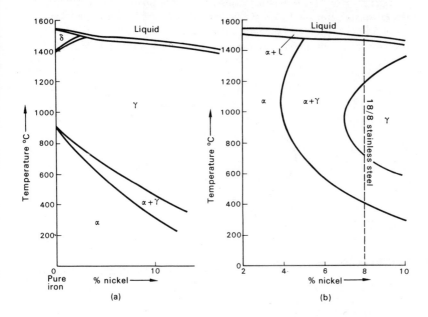

Figure 16.23 (a) Iron-nickel equilibrium diagram. (b) Effect of nickel on Fe/18% Cr alloy. This diagram indicates that ferrite is still the *equilibrium* phase for 18/8 stainless steel at room temperature. Diffusion rates are so slow, however, that austenite is retained to room temperature.

and in the absence of carbon (which has an embrittling effect), iron-nickel alloys will transform to martensite even under conditions of air cooling. Because of the absence of carbon the martensite is relatively soft, and on ageing at 500 °C it is hardened by precipitation of intermetallic compounds formed with the other alloying elements. A typical marageing steel might contain 18% Ni, 9% Co, 5% Mo, 0.5% Ti and 0.1% Al, in which the precipitate consists of Ni–Al and Ni–Ti compounds. Such steels can reach strengths of 2GN/m^2 with high notch ductility and, although their expense at the moment precludes their use in structural applications, they could well be of use for small components that have to withstand very high stress, the more so as they have good corrosion resistance. Copper and aluminium both have a mild strengthening action in solid solution but are principally added for other effects: aluminium for killing steels and grain refining, and copper because it improves

527

the corrosion resistance under normal atmospheric conditions. Steels containing copper for this reason are sometimes known as 'slow-rusting steels', and there is a good case for their more widespread adoption in structural engineering.

Boron has recently been introduced to steels in very small amounts (0.002%) in order to facilitate the formation of bainite by straight cooling and so take advantage of its excellent mechanical properties. Boron inhibits the formation of pearlite but has little effect on bainite, so that if sufficient alloying elements are added (0.5% Mo is typical) to give a double nose to the TTT curve, the addition of boron produces a curve of the form shown in Fig. 16.24.

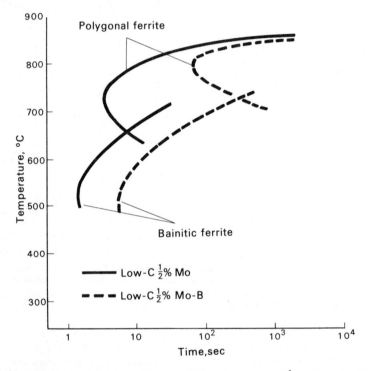

Figure 16.24   Isothermal transformation diagrams for low-C/$\frac{1}{2}$% Mo steels with and without boron (showing P_s and B_s lines only). (From K. J. Irvine, J. B. Pickering, W. C. Heselwood and M. Atkins (1957) *Journal of the Iron and Steel Institute*, **186**, 54–67.)

This greatly simplifies manufacturing problems, and also eliminates the stringent size factor that operates against bainite in plain carbon steels. In fact, a range of low-carbon bainitic steels with manganese and chromium additions has been developed to give strength up to 1.2 GN/m².

16.3.2.5   Stainless Steels

Brief mention should also be made of the corrosion-resistant *stainless*

528

steels. These depend primarily on the presence of chromium in the steel; providing that there is at least 10% Cr *in solution*, the chromium will oxidise preferentially on the surface of the steel to give a tough adherent impermeable invisible and inert skin of chromium oxide (see also Section 19.2), which will quickly repair itself if it is scratched or ruptured, *providing oxygen is available.* The protection conferred by the chromium increases as its percentage in the steel is raised, and there are three types available for general use: austenitic, based on 18% Cr, 9—12% Ni, <0.1% C and up to 3.5% Mo; ferritic, based on 17—20% Cr and <0.1% C; and martensitic, based on 13% Cr, and up to 0.2% C. Only the martensitic type (which undergoes a $\gamma \rightarrow \alpha$ transition on cooling) is heat treatable; the austenitic type is austenitic at all temperatures and the ferritic type is ferritic at all temperatures. They may be strengthened by precipitation hardening or plastic working. For specialised structural work and increasingly for cladding and architectural purposes the austenitic variety is preferred because it is exceedingly tough and easily shaped; the other types are restricted largely to cutting tools and high-temperature oxidation-resistant steels.

16.3.3 Weldability

At the commencement of Section 16.4 it was pointed out that good weldability is now one of the main criteria demanded of structural steels. It is a difficult term to define, but can be regarded broadly as implying that the presence of welds in the steel does not impair any of its engineering properties. Now the creation of a weld is an extremely complex process, involving a miniature casting between the components being joined, and a region on either side of the casting that has been intensively heat treated, and is known as the *heat-affected zone* (HAZ) as seen in Fig. 16.25. Faults in welds are especially dangerous in that they frequently take the form of very fine hair-line cracks and fissures which give the maximum degree of stress raising and thus render welded structures particularly liable to fatigue or brittle failure. They are attributable to three basic causes, individually or more frequently in combination: casting faults, residual stresses and unsuitable steel (i.e. steel of low weldability). The first of these categories includes contraction cavities, entrapment of metal oxide, and the formation of large low-strength columnar crystals or Widmanstatten structures in the solidified weld metal. (Fig. 16.25 shows these effects schematically.) They may arise from faulty welding technique (which places them outside the scope of this text) or — in the case of large columnar grains — simply because of the difficulty of getting sufficiently rapid cooling in thick sections. Grain-refining additions to the filler metal help to eliminate this problem in the weld pool, but cannot prevent grain growth in the HAZ. It cannot be too strongly emphasised that any modifications in design to make the task of the welder easier will contribute very positively to the strength and reliability of the final structure.

Residual stresses are likewise partly the result of design, partly of

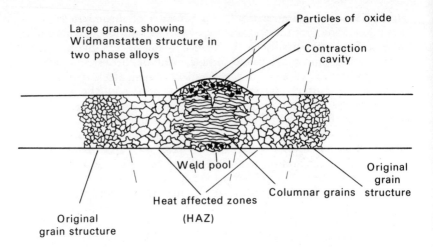

Figure 16.25   Schematic diagram of a fusion weld, showing some possible sources of weakness.

technique. They arise very simply from two causes: long-range *macrostresses* due to thermal contraction under constraint after completion of the weld, and short-range *microstresses* as a result of local phase transformations in the weld pool or the HAZ. Figure 16.26 shows some instances of the ways in which they may arise.

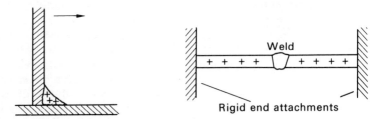

Tensile contraction stresses in the weld metal will tend to pull the vertical member over to the right. If both members are rigidly constrained, hairline cracks may develope.

Figure 16.26   Residual stresses in welded structures. (+ indicates tensile stress.)

They may lead to unacceptable distortion, but the principal risk is their imposition in addition to the working load on any flawed metallurgical structure in the vicinity of the weld.

16.3.3.1   *Hydrogen Embrittlement in Welds*

The final cause of trouble may lie in the selection of a steel which, although excellent in its general engineering properties, is unsuited

530

for welding. This is where the concept of weldability becomes relevant and where the economic factor becomes important, since steels of good weldability need to be manufactured within more stringent composition limits than would otherwise be the case. *Hydrogen cracking* (Fig. 16.27) in the HAZ is a major problem in

Figure 16.27   HAZ hydrogen crack from the root of a fillet weld in a high yield strength carbon manganese structural steel (x 8). (From F. Watkinson (1968) The practical implications of good weldability, *Metals and Materials*, 2, No. 7.)

welding and typifies the sort of conjoint action that is likely, since it depends on take up of hydrogen, residual stresses, and the microstructure of the HAZ. Unfortunately, hydrogen is not an easy element to eliminate from the atmosphere at a weld. With arc welding, it may be present from the decomposition of water vapour on the surface of the hot metal, whence atomic hydrogen diffuses into the steel. The source of the water vapour is usually adsorbed moisture in the flux material on the filler rod or, in the case of argonarc welding, imperfectly dried argon. For this reason, filler rods should always be carefully dried out in an oven shortly before making the weld. In the case of gas welding, hydrogen is present both from the decomposition of unburnt hydrocarbons and from the decomposition of the water vapour that is a major combustion product.

Once in the steel, hydrogen atoms act in two ways to embrittle it. They will diffuse into any internal cavities and there recombine to form molecular hydrogen. This may build up sufficient pressure to start a crack moving out from the cavity. Once started, the crack is likely to continue as more hydrogen diffuses in at its tip until eventually the crack reaches the critical size for brittle failure. The ability of hydrogen thus to cause sudden brittle failure after a period

531

of time under *steady* (rather than impact) loading conditions and at ordinary rather than low temperatures is one of its most dangerous features. This was amply demonstrated in the failure of the King's Bridge at Melbourne in 1961. Hydrogen also acts in solution as a mild austenite stabiliser. This means that the volume changes associated with the decomposition of the austenite take place at lower temperatures when the steel is harder and less able to accommodate them. In combination with residual stresses, this may well be sufficient to cause crack growth.

16.3.3.2 *Effect of Alloying Elements on Weldability*

Clearly the risk of cracking is increased if the microstructure of the steel is in any way hard and brittle, and this is one of the areas in which the increasing use of high-alloy steels in engineering has caused very much increased problems for the welder. The problem arises less in the weld pool itself than in the HAZ, which in large structures may suffer very rapid cooling after the weld as heat flows

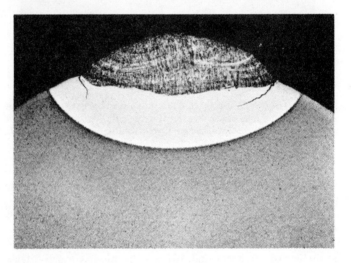

Figure 16.28   Martensite formation (white area) under an arc weld bead laid on a cold alloy steel bar. Note cracking (x 3). (From J. G. Tweeddale (1969) *Welding Fabrication*, Vol. 1.)

from it into the surrounding metal. Cooling rates can reach quench rates, and then martensite will form to give the microstructure least able to cope with the twin problems of hydrogen take up and residual stresses. As the alloy content of the steel is increased so the hardenability of the steel increases, and with it the likelihood of finding unwanted martensitic zones on either side of the weld (see Fig. 16.28 for a rather exaggerated example).

Since the brittleness of martensite also increases with carbon content (massive martensite has been found to be much less vulnerable than twinned martensite), a rough rule-of-thumb estimate of this aspect of weldability may be made by determining the *carbon equivalent* (CE) of a steel according to the formula

$$CE = \% \, C + \frac{\% \, Mn}{6} + \frac{\%(Cr + Mo + V)}{5} + \frac{\%(Ni + Cu)}{15} \quad (16.4)$$

The value of the CE should not exceed ~ 0.25 for heavy structural welds; if $CE > 0.25$, controlled cooling of the weld is necessary to avoid risks of embrittlement. The above expression is only applicable to low-alloy high-tensile steels; for high-alloy steels, inspection of the appropriate TTT diagram and a knowledge of the cooling rate that can be expected in the vicinity of the weld give the best estimate of weldability.

Equation 16.4 shows that carbon is the most potent factor in determining weldability, and it is somewhat ironic to note that, because of this and its effect in lowering the notch ductility of steels, carbon has become virtually a dirty word in advanced steel technology! The present-day tendency in high-strength steels is to cut the carbon content to an absolute minimum and rely increasingly for strength on other alloying elements; this trend has reached its extreme in the maraging steels which stipulate a maximum 0.03% carbon content.

The dangers of hydrogen embrittlement are reduced if time and/or heat treatment are permitted to accelerate the diffusion of hydrogen out of the steel. Treatment at 500 °C or allowing a time lapse of three days before putting the weld under service loading have been found to render the HAZ immune from cracking.

Sulphur, after hydrogen, is the element most likely to have an adverse effect on weldability. At high temperatures (near the fusion boundary), ferrous sulphide melts and spreads along the grain boundaries as a thin film which drastically lowers the toughness of the metal, and renders the HAZ vulnerable to *hot tearing* under the action of cooling stresses. To avoid this trouble it is advisable to use steels with a manganese/sulphur ratio of at least 20 and a sulphur content not exceeding 0.03%. Paradoxically, however, steels with very low sulphur content have proved more prone to hydrogen cracking because MnS inhibits hydrogen diffusion, and a *minimum* sulphur content (0.015% has been suggested) appears to be advisable.

Lamellar tearing is a further aspect of low weldability, and occurs as a series of cracks on the edge of the HAZ running parallel to the rolling direction of the steel. It is thus a particular problem where structural girders (which have a very pronounced directional microstructure) are welded. Joints should therefore be designed as far as possible so that the edge of the HAZ is not aligned with the lamellar structure of the girders, although this is not always easy to achieve. Lamellar tearing appears to be associated with nonmetallic inclusions that are elongated during rolling; manganese silicate has been found to be frequently associated with the failures, which suggests that silicon-killed steels should not be used for girders that are to be welded.

533

16.3.3.3 *Weld Decay*

Austenitic stainless steels have a particular problem in that they are subject to *weld decay*. In that part of the HAZ that is held in the region 500–800 °C during the welding process, chromium reacts with carbon in the steel to precipitate chromium carbide. This reaction occurs most readily at the grain boundaries because of the greater ease of diffusion there, so that, whilst the grain boundaries may be strengthened mechanically by the precipitate, they are denuded of the chromium on which their corrosion resistance depends. If the steel is then exposed to a corrosive environment, the grain boundaries are attacked preferentially and the steel may be rapidly penetrated (Fig. 16.29). The solution to this trouble is to *stabilise* the steel by additions of niobium or titanium, which because of their greater affinity for carbon prevent the chromium being precipitated out of solution.

Figure 16.29   Grain boundary precipitation in an austenitic stainless steel exhibiting weld decay. This region occurs a little distance away from, and on either side of, the weld pool (x 200).

16.4   FIBRE REINFORCEMENT

Although fibre-reinforced materials (in the form of reinforced concrete) have been known and widely used for many years in civil engineering, the concept employed is the old-established one of imparting tensile strength and toughness to a brittle material by means of tough fibres. In this section we shall consider the inverse technique: that of strengthening a ductile matrix by the use of high-strength fibres that would be too brittle to use by themselves. Briefly,

the principle underlying this form of strengthening is that the matrix
should transfer the load to the fibres and at the same time protect
the fibres from damage and act as a crack-arresting medium should
any break. When these roles are fulfilled satisfactorily, quite remarkable combinations of strength, toughness and strength/weight ratios
are achieved.

Two cautionary notes should be sounded, however. Firstly, the
highest strength fibres have to be grown in very carefully controlled
conditions, 'cropped' and then graded; this is very expensive, and so
are the techniques that have to be used for accurately aligning the
fibres and then impregnating them with the softer matrix material.
Together these techniques add up to a cost for the strongest composite materials that puts them out of the question economically for
general structural work. Whether they will become viable in the
future must depend largely on the development of cheaper mass
production techniques. Secondly, one of the principal attractions of
these materials to the engineer is that high strength can result in a
great saving of weight. Thinner sections and fewer load-bearing
struts may be employed to support the same payload. However, the
full strength potential of these composites can only be realised when
large elastic strains are present as well; if weight saving has been
achieved by lighter structural members, the elastic deflections may
be unacceptably large and much of the weight saving may be lost from
the use of additional struts to maintain rigidity. In other words
ultrahigh strength, if it is to be used to its full potential, requires the
development of comparable increases in the elastic moduli — or else
new design concepts to take into account ultrahigh working strains.

16.4.1 Loading Response of Fibre-Reinforced Composites

Let us consider first the case of a composite with a metallic matrix
containing uniaxial continuous fibres (Fig. 16.30). If we assume

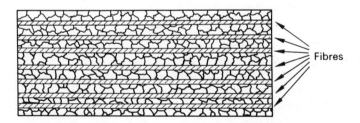

Figure 16.30   Schematic view of a metallic matrix reinforced with strong
continuous fibres.

that $\epsilon_m = \epsilon_f = \epsilon_c$ under all conditions of loading (the subscripts refer
to matrix, fibres and composite, respectively), the principle of load
transference is seen from Fig. 16.31; clearly, the greater the value of
E_f, the more efficient will be the load transference.

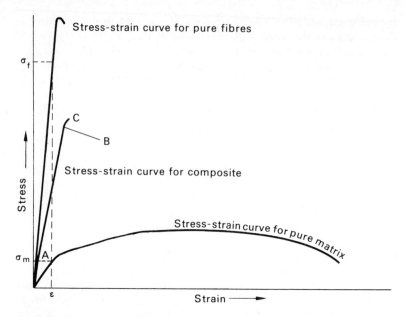

Figure 16.31   The principle of load transference in a fibre-reinforced composite. At an overall strain ϵ, the stress in the fibres is given by σ_f and in the matrix by σ_m. The magnitude of the *load* born by the fibres depends also on the respective volume fractions of fibre and matrix.

Over the initial stages of the stress-strain curve (Fig. 16.35), both fibres and matrix deform elastically, and the composite value for Young's Modulus is given by

$$E_c = E_f V_f + E_m V_m \qquad (16.5)$$

where V is the appropriate volume fraction of fibres or matrix, i.e. $V_f + V_m = 1$. At the point A the mode of deformation changes because the matrix has now reached its yield point, and becomes quasi-elastic. From now on the matrix deforms plastically, and the composite modulus is given by

$$E_c = E_f V_f + \frac{d\sigma_m}{d\epsilon} V_m \qquad (16.6)$$

where $d\sigma_m/d\epsilon$ is the gradient of the stress strain curve for the matrix at the particular stress level to be measured. Since $d\sigma_m/d\epsilon$ is not a constant in the plastic range, the composite stress-strain curve deviates slightly from a straight line, but as $d\sigma_m/d\epsilon$ is usually of the order of $E_m/100$, we can approximate to

$$E_m \simeq E_f V_f \qquad (16.7)$$

The third mode of deformation, when both fibres and matrix deform plastically, is always short and frequently nonexistent since many fibre materials deform elastically up to the fracture point. When it occurs (BC in Fig. 16.31), we have a mild strengthening effect by the matrix (depending on the magnitude of V_m) since the gradient of the

536

matrix curve will continue to be positive at strains where the fibres have already passed their ultimate tensile stress, and will therefore balance the negative gradient of the fibre curve. However, this effect is usually small enough to be neglected.

The mode of fracture in a composite is dependent on the volume fractions of constituents. If we consider how this varies with composite composition, we may start with the extreme case $V_f = 0$. The UTS of the composition is then given by

$$\sigma_{uc} = \sigma_{um}$$

where σ_{um} is the UTS of the pure matrix. If a very small amount of fibre is now incorporated in the matrix and the composite tested, it will be found that fracture of the fibres will occur at a level of load that the matrix is still able to support. In that case fibre fracture does not spell total composite fracture, and putting $V_m = 1 - V_f$ the strength of the composite is given by

$$\sigma_{uc} = \sigma_{um}(1 - V_f) \tag{16.8}$$

which shows that the introduction of the fibres has actually *weakened* the matrix! This mode of failure changes when V_f is sufficiently large for the fibres to support a load greater than the breaking load of the matrix. Fracture of the fibres (we are assuming the ideal case where all the fibres break at an identical stress) will then result in immediate total fracture of the composite, and expression 16.8 now becomes

$$\sigma_{uc} = \sigma_{uf} V_f + \sigma_m^*(1 - V_f) \tag{16.9}$$

where σ_m^* is the stress in the matrix at the ultimate tensile strain of the fibres.

There will be a minimum strength of the composite when the volume fraction of fibres is such that the two fracture modes are equally likely. Putting this volume fraction equal to V_{min} we can then say

$$\sigma_{uf} V_{min} + \sigma_m^*(1 - V_{min}) = \sigma_{um}(1 - V_{min})$$

from which, solving for V_{min}, we obtain

$$V_{min} = \frac{\sigma_{um} - \sigma_m^*}{\sigma_{um} + \sigma_{uf} - \sigma_m^*} \tag{16.10}$$

If the fibre content is to be sufficient to have an overall strengthening effect on the matrix, then $\sigma_{uc} > \sigma_{um}$, and the critical volume fraction, V_{crit} for this to be so will be given by

$$\sigma_{uc} = \sigma_{uf} V_{crit} + \sigma_m^*(1 - V_{crit}) = \sigma_{um}$$

Solving for V_{crit}, we then have

$$V_{crit} = \frac{\sigma_{um} - \sigma_m^*}{\sigma_{uf} - \sigma_m^*} \tag{16.11}$$

Inspection of Equations 16.10 and 16.11 shows that V_{crit} must always be greater than V_{min}; V_{crit} is therefore the important fibre content that must be exceeded if useful reinforcement is to be achieved. Thereafter the strength rises linearly as V_f increases (as shown in Fig. 16.32), the theoretical maximum fibre content for

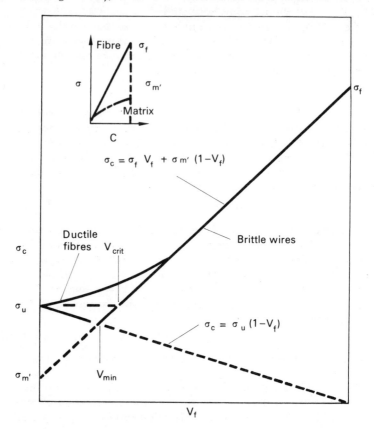

Figure 16.32 Theoretical variation of composite strength, σ_c, with volume fraction V_f for reinforcement with continuous ductile and brittle wires. (From A. Kelly and G. J. Davies (1965) *Metallurgical Reviews*, **10**, No. 37.)

fibres of circular cross section being $V_f = 0.906$. In practice the difficulties in fabrication of obtaining high V_f values *without* the fibres touching (this is important in order to avoid surface damage and brittle crack propagation where fibres touch) limits fibre contents to $V_f \simeq 0.75-0.8$, whilst expense may well limit V_f further.

16.4.2 Reinforcement by Discontinuous Fibres

Although long fibres are available in several materials now (e.g. high-strength steel or tungsten wires, carbon, boron and glass fibres), it is not always possible to have fibres continuous throughout the length of a component and we must therefore consider the benefits that may be obtained from discontinuous fibres. We will confine discussion to the most important stress range, when deformation in the

538

fibres is elastic and in the matrix is plastic. The fundamental difference between this situation and that analysed above is that we can
no longer say that $\epsilon_c = \epsilon_f = \epsilon_m$ at all points; there will be regions in
the composite where it still holds true, but there are considerable
deviations at the ends of the fibres, where the matrix flows round the
tip, as in Fig. 16.33.

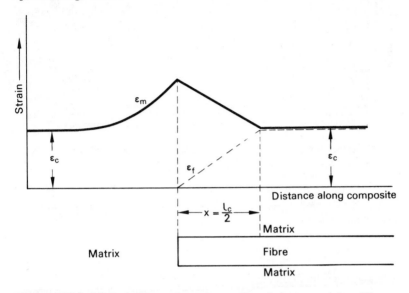

Figure 16.33   Strains in fibre and matrix at a fibre tip in a composite with
discontinuous fibres.

The stress at the fibre tip is now zero, but it builds up by stress
transference along the length of the fibre through the shear stress at
the fibre-matrix interface. We may put this shear stress equal to τ,
the flow stress of the matrix, since the shear stress at the interface
can never exceed that value, and we are considering the conditions
when the matrix is in a state of general plastic yield. The load on the
fibre at a distance x from its tip is then given by

$$P = 2\pi r \tau x \qquad (16.12)$$

when r is the fibre radius, and neglecting any work-hardening effects
in the matrix. Hence under steady load the stress on the fibre increases
linearly and at distance x is

$$\sigma_x = \frac{2\tau x}{r} \qquad (16.13)$$

This reaches a maximum when $\epsilon_f = \epsilon_c$ and hence when the stress on
the fibre has reached the level $\epsilon_c E_f$; it will be appreciated that this
value is the same as that for continuous fibres and therefore that
stress transference is being fully utilised. The situation is shown
graphically in Fig. 16.34. The value of x when stress transference is

539

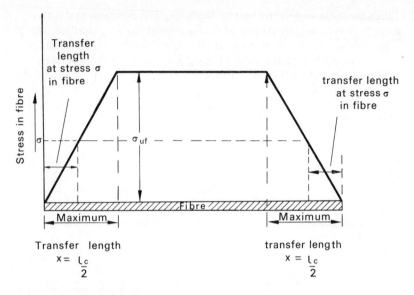

Figure 16.34 Stress variation along a fibre in a composite reinforced with discontinuous fibres.

complete is known as the *transfer length*, and it can be seen that as the stress is raised on a composite, so the transfer length increases. If the fibre is to give its maximum strengthening effect, then stress transference should act up to the ultimate tensile stress of the fibre, i.e. $\sigma_x = \sigma_{uf}$. The transfer length at this stress level is given by

$$x = \frac{\sigma_{uf} r}{2\tau}$$

From Fig. 16.34 we can see that the overall length of the fibre must be $2x$ if the stress is to reach σ_{uf} in the centre of the fibre. Hence the critical fibre length l_c is given by

$$l_c = 2x = \frac{\sigma_{uf} \cdot r}{\tau} \tag{16.14}$$

Usually this value is expressed in terms of the aspect ratio of the fibre, i.e.

$$\frac{l_c}{d} = \frac{\sigma_{uf}}{2\tau} \tag{16.15}$$

For composites with a brittle polymer matrix, such as an epoxy plastic or resin, failure occurs at the fibre-matrix interface and stress transference then depends on the coefficient of friction as the matrix slides over the fibre. This leads to a modification of Equation 16.15 and results in transfer lengths that are about an order of magnitude greater than for composites with a metal matrix.

540

For discontinuous fibres, the *average* stress in the fibres, $\bar{\sigma}$, is less than σ_{uf} at the point of failure (see Fig. 16.34), being given by

$$\bar{\sigma}_f = \frac{1}{l} \int_0^l \sigma_f \, dx$$

where l is the length of the fibre. This reduces to

$$\bar{\sigma}_f = \sigma_{uf} \left(1 - \frac{l_c}{2l}\right) \tag{16.16}$$

For the best use of the fibres we require that $\sigma_f \rightarrow \sigma_{uf}$, and therefore

$$l \gg l_c$$

i.e. it is important to use fibres with large aspect ratios.

The strength of a composite with discontinuous fibres is then (see also Equation 16.9)

$$\sigma_{ucd} = \sigma_{uf} V_f \left(1 - \frac{l_c}{2l}\right) + \sigma_m^* (1 - V_f) \tag{16.17}$$

If we ignore the matrix contribution to the ultimate strength of a composite and divide this by Equation 16.9, we obtain

$$\frac{\sigma_{ucd}}{\sigma_{uc}} = 1 - \frac{l_c}{2l} \tag{16.18}$$

which gives an approximate idea of the loss of strength on using discontinuous fibres. This expression assumes similar values of V_f in both composites, where $V_f > V_{min}$. Figure 16.35 shows the plot of

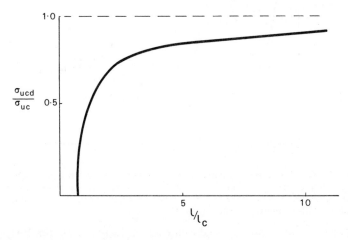

Figure 16.35   The variation in the strength ratio of composites reinforced with discontinuous and continuous fibres as a function of the length of the reinforcing fibres.

Equation 16.18; providing that $l \gg 5l_c$ there is no very serious weakening.

541

16.4.3 Design of Reinforced Composites

Fibre-reinforced components are strongly anisotropic, with their optimum strength in alignment with the long axis of the fibres; they are more anisotropic than any other structural material (although all metals in the worked state exhibit varying degrees of anisotropy that result from manufacturing techniques) and this is a factor that must be taken into account in structural design.

If the angle between the stress axis and the axis of fibre alignment is ϕ, the strength of the composite drops rapidly as ϕ increases, once ϕ has exceeded a small critical value ($\approx 4°$). This is shown in Fig. 16.36, from which it can be seen that the overall curve is the combination of three separate curves, each representing a different mode

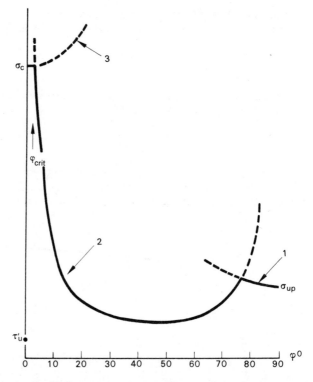

Figure 16.36 Variation in composite strength with fibre orientation. 1, Matrix fails by flow transverse to fibres. 2, Shear failure of matrix on a plane parallel to fibres. 3, Matrix flows parallel to fibres until fibres fail. (From A. Kelly and G. J. Davies (1965) *Metallurgical Reviews*, **10**, 37.)

of failure. It will also be noted that there is a wide minimum for values of ϕ from 40° to 90° where the strength is roughly one tenth of its optimum value.

Where strength is necessary in more than one axis, two-dimensional isotropy may be achieved by the random orientation of fibres in a plane, to give a 'mat' of fibres, but the strength in the plane is only about one third of the strength of a uniaxially aligned composite for equivalent V_f values. For three-dimensional isotropy the strength drops to one sixth of the uniaxial composite. It should also be noted that the maximum value for V_f drops as we depart from uniaxial alignment; for a planar mat, the theoretical maximum for V_f is 0.78 as compared with 0.91, and the practical limit appears to be of the order of 0.3. Thus isotropy can only be achieved at the expense of high strengths, and careful fibre alignment is seen to be a very important part of manufacture. A better method, however, of achieving planar isotropy, is to adopt the 'plywood' principle, in which thin sheets of uniaxially aligned composite are bonded together at varying orientations (Fig. 16.37).

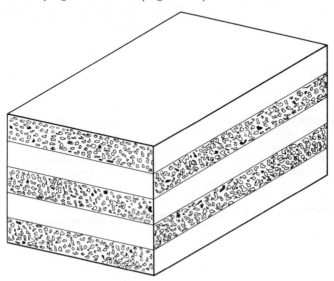

Figure 16.37   Plywood design to give two-dimensional strength to fibre-reinforced composites.

The most important engineering feature of fibre-reinforced composites is not so much their high strength but their toughness, particularly in the presence of notches, compared with other high-strength materials. This is achieved either by the crack-blunting ability of a soft matrix (Fig. 16.38(a)) or a relatively weak interface between matrix and fibre in the case of a brittle (e.g. resinous) matrix. Delamination and extreme crack blunting then occur as a result of the tensile stress developed in front of a crack (Fig. 16.38(b)). Obviously, to work satisfactorily this latter technique calls for careful control of the fibre-matrix interface: if the interface is too weak, insufficient stress transference will occur; if it is too strong, cracks may not be arrested between fibres and brittle failure becomes a possibility.

The above mechanisms will work satisfactorily until the applied stress reaches the failure stress of the fibres. When this occurs, there are again various possibilities. If the fibres are continuous and all of (the same) uniform strength along their length, the first one to fail

543

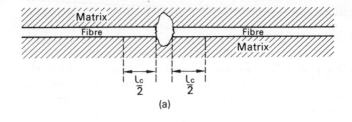

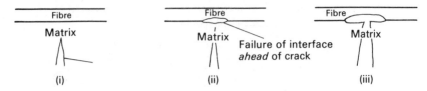

The applied stress is parallel to the fibre

(b)

Figure 16.38 Crack-blunting mechanisms in composites. (a) Rounding of crack-tip by plastic deformation after fibre failure has occurred. The matrix takes the additional load over a distance equal to the transfer length on either side of the fibre failure. (b) If the interface between fibre and matrix is moderately weak, the stress component in front of an advancing crack tip may lead to interface failure in front of the crack, with a very pronounced blunting effect. The effect is known as *delamination*.

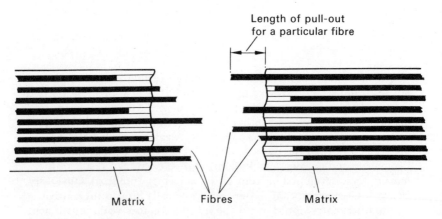

Figure 16.39 Schematic representation of the pull-out effect when a fibre-reinforced composite is fractured. The average degree of pull-out approximates to $l_c/4$ (see text), where l_c is the transfer length of the fibres. If N is the number of fibres of mean radius r per unit cross-section in a composite of cross-sectional area A, then the frictional drag acts on an area given by $NA \cdot \pi r l_c/2$, which can result in an appreciable increment to the toughness of the composite if it contains a high volume fraction of fibres with long transfer lengths.

will throw an unacceptable extra load on its neighbours at that point, and total fracture will follow immediately. If the fibres have varying strengths along their lengths (as a result of surface damage, variations in diameter or structural imperfections) fractures will occur at random points along the fibres as the stress approaches the ultimate value, and the result will be a change from the continuous to the discontinuous fibre mode of deformation. The fibres will break up into shorter and shorter portions until eventually the broken parts are individually shorter than the critical length. Once this point has been

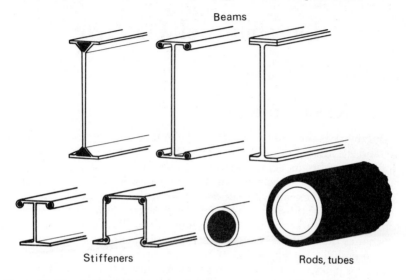

Beams

Stiffeners   Rods, tubes

Figure 16.40   AVCO boron-epoxy reinforcement of conventional structural sections. (From N. J. Mayer (1974) *Engineering Applications of Composites, Vol. 3* (Ed. B. R. Noton), by permission of Academic Press Ltd.)

reached, full load transference to the fibres is not possible and matrix failure will occur.

The important feature of this failure mode is that fibres which lie across the fracture surface but whose ends are within the distance $l_c/2$ from that surface *will not break*, since they are not supporting the full load in that region. Instead, they are pulled out from the other half of the composite (Fig. 16.39) and the frictional resistance to 'pull out' affords a useful increment to fracture toughness. In adjusting the internal structure of the composite to this end, however, a compromise is again necessary, since high friction between the fibres and the matric will result in a short transfer length.

Many materials have been investigated as possible fibre materials, but as far as the civil engineer is concerned high-strength steel wires are likely to be the most useful, if this technique becomes feasible on economic grounds for structures. Alternatively, the use of conventional steel girders has been suggested, using lighter sections to save weight and increasing strength and stiffness with fibre-epoxy resin fillers (Fig. 16.40).

CHAPTER 17

Durability of Concrete

17.1 INTRODUCTION

Durability is the ability to endure. A durable concrete is able to withstand the attacks of destructive processes of chemical, physical or
mechanical origin, acting externally or arising internally.

Deterioration is the result of disruptive processes. Internally, they
can consist of expansive reactions between the chemicals in the
aggregates and those in the cement. In their common form these are
referred to as alkali-aggregate reactions. Internal disruption may also
occur as the result of changes of temperature, either downwards so
as to mobilise the expansion of free water as it freezes, or upwards,
perhaps in a fire, when the thermal incompatibility between h.c.p.
and aggregate may become too great. The external threat arises from
the penetration of aggressive agents that come into contact with the
surface of the concrete. They may be natural chemicals such as
sulphates in clay soils, or industrial chemicals such as acids, and they
may attack either the h.c.p. or the aggregate, the deterioration
becoming more serious the farther they penetrate. Concrete is often
reinforced with steel, and the prevention of steel corrosion is an additional concern. The steel is protected by the surrounding concrete,
and it is important that agents that threaten the protection of the
steel, such as carbonic acid, should be kept out by the concrete
cover. Special kinds of mechanical load can also be classified as
external threats to durability. These include such actions as surface
friction causing abrasion, or negative water pressures causing
cavitation.

Testing for durability is difficult because the only sure proof
comes from observing the concrete over its full life in the potentially
disruptive environment. Fortunately, this normally takes a very long
time, and a satisfactory conclusion is hardly relevant to the structure
observed and may not easily be related to other circumstances. However, experience has led to the identification of many (perhaps most)
of the likely aggressive agents and disruptive processes, and from this
some desirable properties for durable concrete have been deduced.

546

This is useful in that it suggests possible short-term measurements, and hence some attempt can be made to rank different concretes in order of their durabilities; but it does not give any quantitative assessment of durability itself — it is not possible to predict that a given material will crumble after x years.

Clearly, the effect of external agents is minimised if the concrete is impermeable, so that permeability is an important property in relation to the limitation of attack from outside. The topic of permeability is presented in Chapter 8, but particular aspects relevant to durability are discussed in the next section.

17.2 PERMEABILITY AND DURABILITY

The line of reasoning pursued above is often developed to the more general conclusion that a concrete of low permeability will be durable. It follows that a durable concrete should have a low porosity and that it should be capable of full compaction without segregation. The low porosity implies that the water/cement ratio should be low and that curing should be effective so as to achieve the maximum possible degree of hydration. The need for full compaction implies that workability should be appropriate to the means of compaction available and that the mix should not lack cohesion. It is a safeguard towards satisfying these conditions if the aggregate/cement ratio is not too high, and, with this in view, a minimum cement content may be set in specifications for durable concrete.

If the requirements for low permeability are followed the concrete produced will also have high strength. This is a desirable property anyway, in that a stronger concrete is better able to resist the disruptive forces causing deterioration. However it should be noted that permeability is the more basic of the two properties, since it is possible to make strong concretes which are permeable, and hence of potentially poor durability.

17.2.1 Initial Surface Absorption Test (ISAT)

Normal permeability tests (see Section 8.2.2.1), from which the values of the coefficient of permeability are found, are performed by subjecting a saturated concrete specimen to a considerable external water pressure, and observing the steady-state flow through the specimen. This is sensible enough when the concrete is to be used in structures such as dams or offshore platforms, in which the water is driven by an external head. It is less clearly related to the more usual circumstances in which the aggressive agents are not pushed in from outside but are drawn in by capillary attraction or transfer within a concrete that is less than saturated. That is, penetration tests (see Section 8.2.4) are more appropriate to durability than the straight steady-state permeability tests.

The ISAT was developed to represent better the more usual conditions of penetration, and it is a measure of the ingress of water in a

547

specimen dried in a standard manner, the external water pressure being very small. It has the further advantage that it can be applied nondestructively to concrete in situ. The capacity of the concrete to take in water is characterised by the rate of flow through the outer layers immediately after the water is brought into contact with the surface; that is, it is given by rate of the initial surface absorption.

The apparatus is shown in Fig. 17.1 and the test is described in

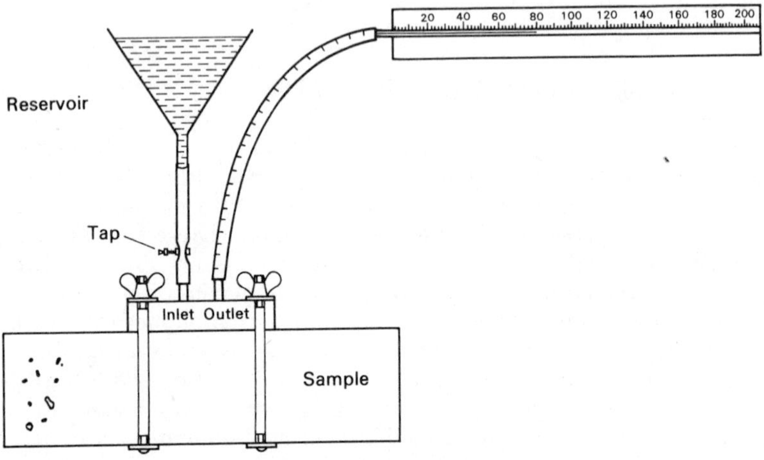

Figure 17.1   Diagram of the ISAT apparatus. (From BS 1881 (1970) Part 5, reproduced by permission of BSI, 2 Park Street, London W1A 2BS, from whom complete copies can be obtained.)

detail in BS 1881 'Methods of Testing Concrete', Part 5. Methods of testing hardened concrete for other than strength (see BSI, 1970 and Bibliography). In brief, a cap is clamped to the surface of the concrete and a reservoir set up with a constant head of 200 mm of water. The reservoir is connected through the cap to a capillary tube set level with the water surface in the reservoir. At the start of the test, water is allowed to run through the cap (thus coming in contact with the concrete surface) and to fill it and the capillary tube. The rate of absorption is then found by closing off the reservoir and observing the rate of movement of the water meniscus along the capillary tube. Readings are taken at standard times after the start of the test (10 min, 30 min, 1 h and 2 h). As shown in Fig. 17.2, the rate drops off with time in a regular manner. It naturally increases with the permeability of the concrete, and can thus be taken as a measure of durability. It is possible to specify maximum rates which should not be exceeded in concretes intended for particular conditions of exposure. For example, for concrete kerbs a maximum rate of 0.4 ml/m^2/s at 10 min (from the start of the test) has been suggested (Lebitt, 1971), and this can be backed up by an additional requirement that the rate should not exceed 0.15 ml/m^2/s at one hour.

548

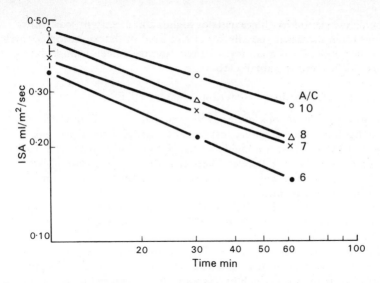

Figure 17.2 Relation between the initial surface absorption rate and time after the start of the test, for various aggregate/cement ratios (A/C). (From M. Levitt (1971) *British Journal for Non-Destructive Testing*, **13**, by permission of the British Institute of Non-Destructive Testing.)

17.3 ALKALI-AGGREGATE REACTIONS

It is probably true that all aggregates react with the cement paste to some extent. This may be beneficial in improving the bond between the aggregate and the h.c.p., but it may also be deleterious in forming expansive products that disrupt the concrete structure. It is with this latter aspect that this section is concerned. The failure usually takes months or years to develop and it is made evident in the form of pop-outs (the spalling of small circular areas from the concrete surface) or by the development of pattern or map cracking (similar to crazing). The phenomenon of alkali-aggregate reaction is associated with high levels of alkalis in the h.c.p., usually deriving from the cement, though the presence of alkalis in the aggregate can also contribute. The chemical reactions are complicated and the expansive mechanisms far from easy to understand; the explanations given below are interpretations or speculations rather than proven facts. It is not possible to determine beforehand with certainty the rocks that are vulnerable, but experience has indicated certain classes of material that should be treated with caution. As little as 0.5% of a defective aggregate may be enough to cause damage to the concrete, and sensible precautions are to use low-alkali cements and to avoid the use of any suspect aggregate.

There is evidence of reactive aggregates in the UK, as well as in other parts of the world where the problem has been faced for some time. The trend towards the employment of aggregates of less good quality, together with cements of higher alkali content, can only lead to an increasing possibility of this kind of deterioration occurring in the future. Two main types of reaction have been found, the alkali-silica and the alkali-carbonate.

17.3.1 Alkali-Silicate Reaction

Top of the list of rock minerals for reactivity is opal, and others on the list include chalcedony, crystobalite, tridymite and volcanic glasses. All of these are highly siliceous, but many quartz-rich rocks are not reactive and it seems likely that the physical state of the rock (high permeability for instance) is an important factor in promoting the reaction.

The large expansion that can occur is demonstrated in Fig. 17.3 for mortars containing 5% and 10% of opal. The reaction takes place at the surface of the aggregate and reaction rims of new products are often observed around the aggregate surface. The effect of the reaction is to form an alkali silicate which gradually takes up water to form a gel which exerts a swelling pressure on the surrounding material. In support of this view it has been shown that over a few millimetres away from the aggregate surface the surrounding gel has a decreasing alkali (sodium and potassium) content.

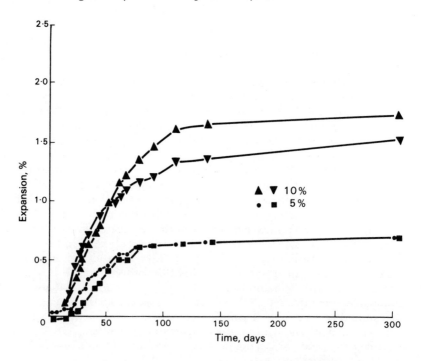

Figure 17.3   Expansion caused by alkali-silicate reaction in mortars mixed with 5 and 10% of opal of size >125 μm in the aggregate. (From S. Diamond and N. Thaulow (1974) *Cement and Concrete Research*, **4**, 591, by permission of Pergamon Press.)

550

17.3.2   Alkali-Carbonate Reaction

Many carbonate rocks react with alkalis but only a few of the reactions are expansive. The practical evidence indicates that it is
mostly the dolomites that cause trouble, particularly those with a
fine-grained matrix and containing clay. It seems that it is the dolomites and limestones which contain excess magnesium and calcium
ions in their crystal structure that are more likely to be reactive.

The known chemical reaction is that between carbonate and alkali
hydroxide, for example:

$$CaMg(CO_3)_2 + 2NaOH \rightarrow Mg(OH)_2 + CaCO_3 + Na_2CO_3$$
$$\text{dolomite} \qquad \text{alkali}$$

This clearly shows the breakdown of the dolomite, but in reality
it is a simplification and other compounds, sulphates for instance,
may intervene. The overall effect is to develop reaction rings around
the aggregate which may be as much as 2 mm thick, and from which
cracks grow. There is doubt about whether the above reaction is, of
itself, sufficient to cause expansive disruption, and the significance
of fine grain and the presence of clay have lead to the suggestion that
the reaction produces a porous structure which enables the included
clay to take up water and swell.

17.4   INGRESS OF AGGRESSIVE AGENTS

Concrete is attacked, at least to some extent, by a variety of chemicals
coming into contact with it. A list of potentially destructive agents
has been published by the Portland Cement Association of America
and it is reproduced in the ACI Monograph No. 4 on Durability of
Concrete Construction (see Bibliography). We are told that concrete
disintegrates in cider, but not in beer or wine, peanuts disintegrate
the surface slowly, but sauerkraut has little if any effect. In general
it is acids and sulphates that do the most damage. Acids in significant
concentrations are encountered only in special applications (such as
cider or sewage), whereas sulphates occur in dangerous concentrations
in many soils in most parts of the world (including the UK). The
engineer should be aware of the threat from sulphates, and also from
the ingress of other agents that may promote the corrosion of any
reinforcement. These two cases are dealt with below.

17.4.1   Sulphates

Sulphates in solution react with the constituents of the h.c.p.
Sodium sulphate attacks the C_3A hydrate to produce calcium sulphoaluminate (ettringite) (see Section 3.2.2) and the reaction is accompanied by a considerable increase in volume. It also attacks the
calcium hydroxide to produce gypsum, again with an accompanying
increase in volume. Magnesium sulphate is even more aggressive as it
attacks not only the C_3A but also the C_3S to give gypsum, magnesium

hydroxide and silica gel. The chemical equation is:

$$3Ca0.2SiO_2 \cdot aq + MgSO_4 \cdot 7H_2O \rightarrow CaSO_4 \cdot 2H_2O + Mg(OH)_2 + SiO_2 \cdot aq$$

C_3S	magnesium sulphate	gypsum	magnesium hydroxide	silica gel

The severity of the sulphate attack thus depends on the particular sulphate, and it also increases with the strength of the sulphate solution. The rate of attack increases most rapidly up to about 1% concentration, and at a slower rate thereafter. As shown in Fig. 17.4, loss

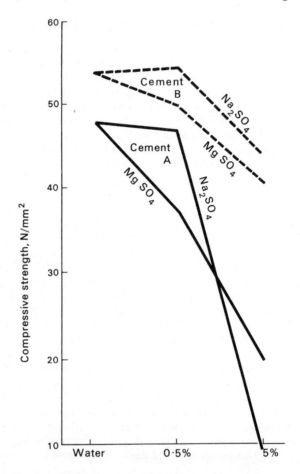

Figure 17.4 The effect on strength of concrete of storage for a year in water and in 0.5% and 5% sulphate solutions. (From F. M. Lea (1970) *The Chemistry of Cement*, by permission of Edward Arnold.)

of strength can be very rapid, but in the field it is more normal for the breakdown of the concrete to take place over a number of years.

Measures can be taken to minimise the effects of sulphate attack. Firstly, sulphate-resisting cement with a low proportion of C_3A can be used (see Section 3.3) and, secondly, the extent of penetration can be reduced by ensuring that the concrete has a low permeability. Both of these precautions are covered by the recommendations of the Code of Practice for the Structural Use of Concrete (CP 110) (see Bibliography). A third possibility is the introduction of high-pressure steam curing, which removes the calcium hydroxide, and encourages the formation of more stable silicate and aluminate compounds.

17.4.2   Steel Protection

The electrochemical corrosion of iron and steel is dealt with in detail elsewhere (see Chapter 19), and in this section attention is confined to the problems of reinforcing bars in concrete. The steel is embedded both to ensure composite action between the two materials and to provide the steel with a protective surround. If the concrete protection is inefficient the steel rusts with the unfortunate loss of some of the cross-sectional area, and a rise in the stress that the remaining steel has to carry. In addition the increase in volume that takes place in the steel when it changes to rust is another source of internal expansion leading to disruption and spalling of the concrete, and hence to an acceleration in the rate of corrosion.

For corrosion to occur both water and oxygen must be present. Hence there should be little corrosion in a really dry concrete, nor should there by any great threat to concrete permanently submerged well below the surface of the sea. The high alkalinity of concrete (a pH of about 13) provides a chemical protection in addition to the physical protection of the concrete surround itself. It causes the formation of a thin oxide film on the surface of the steel, which acts as a barrier to the electrolyte ions engaged in galvanic action. The steel is thus rendered passive. Its passivity may be impaired by a reduction of the pH to about 11.5, so that the ingress of fluids that reduce alkalinity also reduces the ability of the steel to resist corrosion.

There are two main agents that attack the oxide film. The first is carbon dioxide, which even in the normal atmosphere is in sufficient concentration to react with the calcium hydroxide of the h.c.p. to form calcium carbonate. Steel undoubtedly corrodes more readily in carbonated concrete, but in a normal good quality concrete the CO_2 penetrates very slowly, and only the outer 10—20 mm is affected after years of exposure.

Secondly, soluble chlorides reduce the passivity of the steel. Further, variations in the concentrations of chlorides reaching the steel cause variations in the degree of passivity along the bar, and thus accelerate the galvanic action. The concentration gradient can come from differences in permeability of the concrete, or from concentration gradients imposed on the concrete surface, such as might occur in the splash zone just above the surface of the sea. In fact, the splash zone is one of the worse natural conditions for corrosion in that the concrete is neither saturated nor completely dry, both water

553

and oxygen are present, and the chloride concentration is high on the concrete surface just above the water, and diminishes as the height above the water increases. Chlorides can also be introduced by human volition. De-icing salts are an incidental surface application to concrete, while the use of calcium chloride as an accelerating additive (see Section 3.4.1) is a deliberate introduction of a chloride into the body of the concrete.

The simple way to minimise the risk of corrosion is to ensure that the concrete cover is of a thickness to suit the expected exposure, that the concrete is of low permeability and that any cracks, whether load-induced or not, are of the smallest possible width. In addition, extra precautions can be taken by introducing admixtures to inhibit corrosion, such as sodium nitrite, or by coating the steel, by galvanizing for example. Not too much should be expected of such precautions as their effects may be short-lived, or they may introduce other undesirable behaviour, such as loss of bond strength between the concrete and a coating.

17.5 FREEZING AND FIRE

The internal disruption of concrete associated with low temperature is a direct result of the change of state of water at and below the freezing point. At high temperatures too the change of state from liquid to vapour contributes to failure under conditions of fire, though the temperatures rise well above the boiling point, and other mechanisms, such as incompatibility of thermal expansion between aggregate and paste, are important. These physical processes of deterioration are discussed in this section.

17.5.1 Freezing

When the temperature of h.c.p. is lowered the water in the larger capillary pores begins to freeze at the normal freezing point of bulk water of 0 °C, but the temperature has to drop below this level to freeze the water in the smaller pores, and it is estimated that a temperature of −78 °C must be reached before freezing can occur in the small gel pores. The formation of ice crystals causes an increase in volume of some 9% and the effect is to put the capillary water under pressure. If there are no air voids, an expansion of the kind shown in Fig. 17.5 (dashed line) results as the water pressure reacts against the structure of the paste. If the pressure becomes too great disruption of the structure follows. Subsequent thawing allows additional water to enter the material, with a consequent further deterioration when freezing occurs again. Thus, in due course, a complete disintegration of the concrete is possible.

The alternative to internal damage is to ensure that the pressure in the capillary water is dispersed by flow into air voids which may be the boundary of the member or internal cavities. This relief of pressure is promoted if the concrete is not too impermeable to the

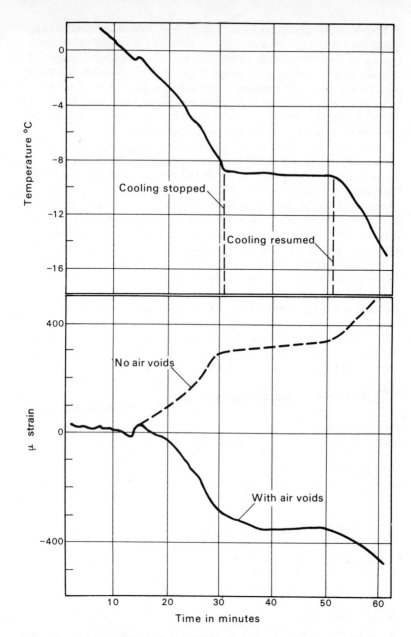

Figure 17.5 Dimensional changes of cement paste with and without air voids. (From T. C. Powers and R. A. Helmuth (1953) *Proceedings of the Highway Research Board*, 32, 285, by permission of the Transportation Research Board.)

flow, if the air boundary is not too far from the freezing site and if the rate of freezing is not too great. A successful relief operation produces a dramatic change in behaviour. As shown by the lower line of Fig. 17.5, instead of expansion there is a large shrinkage.

The decrease in volume is not simply a matter of the thermal contraction of the solid. In addition, water moves from the gel to the capillaries thus resulting in drying shrinkage of the kind discussed in Chapter 6. The driving force arises from the loss of thermodynamic equilibrium when the capillary water freezes, giving rise to excess free energy in the gel water. In addition, the gradient may be steepened by the osmotic pressure resulting from the increasing concentration of any salts in the freezing capillaries.

Much the same mechanism operates in the aggregate; disruption can occur if there are insufficient air voids to accommodate the expansion of the absorbed water, or if there is insufficient time for the water to reach the boundary of the aggregate particle. Some aggregates have higher tensile strength than h.c.p. and can sustain large internal pressures without breaking down, but others, such as shales, are of limited strength, and absorb large amounts of water into a fine pore structure, and are thus very susceptible to damage.

The evidence of disruption caused by freezing includes pop-outs where individual aggregate particles expand and cause local failure, producing a pockmarked surface to the concrete, pattern cracking, showing a developed system of general disintegration, and scaling in which the whole surface breaks away from relatively sound concrete underneath. This is associated with a weak porous h.c.p. just below the surface resulting from the upward flow of water in the fresh concrete during and just after casting (bleeding). The weak zone is often overlaid by a denser, stronger layer formed by finishing operations such as floating off with a trowel.

There are some obvious precautions that should be taken when concreting in severe climates such as that of Canada. The right materials should be used, and, in particular, it is wise to avoid highly absorptive aggregates which are found, in general, to be the most vulnerable to frost damage. The h.c.p. should be strong, so that the water/cement ratio should be low and bleeding should be avoided. Poor curing can make things worse and so can water and salts coming into contact with the concrete after it is in service. Last, but certainly not least, it is valuable to introduce air-entraining agents (see Section 3.4.4) which produce well distributed air bubbles up to 0.1 mm in diameter and thus encourage the rapid diminution of the capillary water pressure. It is this kind of treatment that provides the air voids, and hence the shrinkage, shown in Fig. 17.5.

17.5.2 Fire

In conditions of fire the temperature of the concrete can rise to over 1 000 °C which is enough virtually to destroy the original material. In the early stages of heating the evaporable water is lost over the range 20—110 °C, and if escape is impeded the water vapour thus formed can exert a very considerable pressure within the material. Above 110 °C the cement hydrates decompose, the calcium hydroxide is broken down and the calcium carbonate suffers decarbonation so that by about 700 °C there is little of the initial chemical

556

structure left. The aggregate also suffers changes which contribute to the general loss of structure.

The volume changes can be great and are caused by three main mechanisms, the effects of which are shown schematically in Fig. 17.6. Firstly, the loss of evaporable water leads to internal

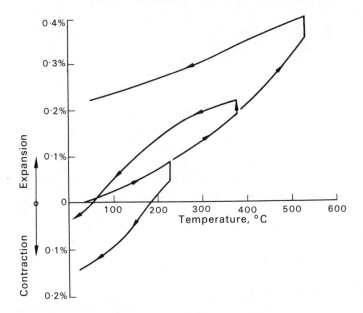

Figure 17.6 Trends in thermal expansion of concrete. (From J. A. Purkiss, 'A Study of the Behaviour of Concrete to High Temperature under Restraint and Compressive Loading', Ph. D. thesis, by permission of the University of London.)

drying of the h.c.p. of the kind discussed in Chapter 6, and there is a consequent shrinkage of the concrete, which is observed only as a residual strain after cooling and only then if the temperature has not exceeded about 300 °C. It is not seen during heating because it is masked by the second mechanism, that of normal thermal expansion. Thirdly, the coefficients of thermal expansion of aggregates are usually lower than those of the h.c.p.s and this thermal incompatibility, together with the shrinkage incompatibiliy, puts a great strain on the interfaces between the two components. Internal cracking is inevitable at higher temperatures and this results in an irreversible expansion. Thus, cooling after heating to, say, 500 °C leaves a residual expansion, in contrast to the contraction remaining after heating to 200 °C.

The decomposition, internal vapour pressure and interface cracking cause a steady deterioration of properties as the temperature is raised. In particular, the strength diminishes and so does the rigidity. The decrease in the elastic modulus is shown in Fig. 17.7, which reports tests in which the concrete was heated to its reference temperature before any load was applied. In view of the drastic changes

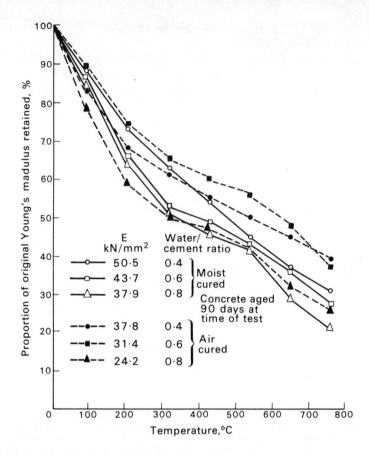

Figure 17.7 Effect of temperature on Young's modulus of concrete. (From R. E. Philleo (1958) *Proceedings of the American Concrete Institute*, **54**, by permission of the American Concrete Institute.)

that take place it is remarkable that concrete can, at 800 °C, retain as much as 35% of its initial stiffness. The stress—strain curve at high temperatures naturally has a shallower loading slope and a lower peak stress. It is also in keeping with its behaviour at lower temperatures that the creep rate is much greater at high temperature, with a consequent advantage in relief of induced stresses. More surprising is that the material is much more ductile and this comes from the descending branch of the stress—strain curve, as well as the ascending one being much more shallow, than at the lower temperatures.

Complete failure can occur in a fire because the weakened concrete can no longer sustain the existing load, which may indeed be enhanced by external restraint of the thermal expansion. The breakdown usually consists of spalling, that is the separation of surface material from the main body of the concrete. This may be in the form of local pitting or blistering, and aggregate particles may split as part of the failure plane. In a more general manner the gradual separation of layers of surface material can occur, and this is referred to as sloughing off.

The performance of concrete in fires is greatly affected by the type of aggregate. The traditional preference for calcareous aggregates such as limestone is justified in that the concretes made with them are less likely to spall than other kinds. On the other hand, siliceous aggregates like the widely used flint gravels suffer an expansive quartz transformation of 575 °C, and this is a potent contribution to aggregate splitting and spalling. Lightweight aggregates with their low rigidity are also attractive in that they can be expected to more readily accommodate the differential thermal strains. In addition, they have low thermal conductivities, which means that the time taken for the fire to reach the vulnerable steel is increased, and structural collapse can be delayed by precious minutes.

The requirements for good fire resistance are that the concrete should be strong, but should have a high creep and an aggregate of low rigidity. The aggregate and h.c.p. should have compatible thermal expansions, and the thermal conductivity should be as low as possible. Understanding of the phenomenon is far from complete and this list of desirable properties is likely to be extended and amended as knowledge increases.

17.6 REFERENCES

BSI (1970) *Methods of Testing Concrete, Part 5, Methods of Testing Concrete for Other Than Strength*, BS 1881, British Standards Institute, London.

Diamond, S. and Thaulow, N. (1974) A study of expansion due to alkali-silicate reaction as conditioned by the grain size of the reactive aggregate, *Cement and Concrete Research*, 4, 591.

ISE/CS (1975) *Fire Resistance of Concrete Structures*, Report of a Joint Committee of the Institution of Structural Engineers and the Concrete Society.

Lea, F. M. (1970) *The Chemistry of Cement and Concrete*, Arnold, London.

Levitt, M. (1971) The ISAT — a nondestructive test for the durability of concrete, *The British Journal of Non-Destructive Testing*, 13.

Philleo, R. E. (1958) Some physical properties of concrete at high temperatures, *Proceedings of the American Concrete Institute*, 54, 857.

Powers, T. C. and Helmuth, R. A. (1953) Theory of volume changes in hardened Portland cement paste during freezing, *Proceedings of the Highway Research Board*, 32, 285.

CHAPTER 18

Durability of Timber

18.1 INTRODUCTION: THE PHYSICAL, CHEMICAL AND BIOLOGICAL AGENCIES

Durability is a term which has different concepts for many people: it is defined here in the broadest possible sense to embrace the resistance of timber to attack from a whole series of agencies whether physical, chemical or biological in origin.

By far the most important are the biological agencies, the fungi and the insects, both of which can cause tremendous havoc given the right conditions. In the absence of fire, fungal or insect attack, timber is really remarkably resistant and timber structures will survive, indeed have survived, incredibly long periods of time, especially when it is appreciated that it is a natural organic material with which we are dealing. Examples of well preserved timber items now over 2 000 years old are to be seen in the Egyptian tombs.

The effect of hydrolytic, oxidative and photochemical reactions are usually of secondary importance in determining durability. On exposure to sunlight the colouration of the heartwood of most timbers will lighten, e.g. mahogany, afrormosia, oak, though a few timbers will actually darken, e.g. Rhodesian teak. Indoors the action of sunlight will be slow and the process will take several years: however, outdoors the change in colour is very rapid taking place in a matter of months and is generally regarded as an initial and very transient stage in the whole process of *weathering*.

In weathering the action not only of light, but also of rain and wind render the timber silvery-grey in appearance: part of the process embraces degradation of the cellulose by ultraviolet light, which erodes the cell wall and in particular the pit aperture and torus. However, the same cell walls that are attacked act as an efficient filter for those of the cells below and the rate of erosion from the combined effects of UV, light and rain is very slow indeed; in the absence of fungi and insects the rate of removal of the surface by weathering is of the order of only 1 mm in every 20 years. Nevertheless, because of the continual threat of biological attack, it is unwise to leave most timbers completely unprotected from the weather; it should be appreciated that during weathering the integrity of the surface layers is markedly reduced thereby adversely affecting the performance of an applied surface coating. In order to effect good adhesion the weathered layers must first be removed (see Section 20.4).

As a general rule, timber is highly resistant to a large number of chemicals and its continued use for various types of tanks and containers, even in the face of competition from stainless steel, indicates that its resistance, certainly in terms of cost, is most attractive. Timber is far superior to cast iron and ordinary steel in its resistance to mild acids and for very many years timber was used as separators in lead-acid batteries. However, in its resistance to alkalies timber is inferior to iron and steel: dissolution of both the lignin and the hemicelluloses occurs under the action of even mild alkalis.

Iron salts are frequently very acidic and in the presence of moisture result in hydrolytic degradation of the timber; the softening and darkish-blue discolouration of timber in the vicinity of iron nails and bolts is due to this effect.

Timber used in boats is often subjected to the effects of chemical decay associated with the corrosion of metallic fastenings, a condition frequently referred to as *nail sickness*. This is basically an electrochemical effect, the rate of activity being controlled by oxygen availability. Areas of different polarity are set up, the salt water which has permeated the timber acting as the electrolytic bridge, while the wet timber assumes the role of the conductor; permeability of the timber is therefore a significant factor and care must be exercised in the selection of timbers to use only impermeable species. Alkali will be produced at the cathodic surfaces which, as noted above, will cause the timber to become soft and spongy, impairing its ability to hold the fastenings. In the anodic areas ions pass into solution and form a soluble metallic salt with the negative ions of the electrolyte. As noted above, in the case of iron fastenings, the iron salts so formed cause degrade of the timber resulting in considerable loss in its strength and marked discolouration.

Generally when durability of timber is discussed reference is being made explicitly to the resistance of the timber to both fungal and insect attack. This resistance is termed *natural durability*, but as the former agent of attack has a much higher significance the term natural durability is frequently and inaccurately applied to the natural resistance of timber to only fungal attack at ground contact.

18.2   NATURAL DURABILITY

Recalling that timber is an organic product it is surprising at first to find that it can withstand attack from fungi and insects for very long periods of time, certainly much greater than its herbaceous counterparts. This resistance can be explained in part on the basic constituents of the cell wall, and in part on the deposition of extractives (Sections 4.2.1 and 4.2.3.3; Table 4.2).

561

The presence of lignin which surrounds and protects the crystal-line cellulose appears to offer a slight degree of resistance to fungal attack: certainly the resistance of sapwood is higher than that of herbaceous plants. Fungal attack can commence only in the presence of moisture and the threshold value of 20% for timber is about twice as high as the corresponding value for nonlignified plants.

Timber has a low nitrogen content being of the order of 0.03—0.1% by weight and, since this element is a prerequisite for fungal growth, its presence only in such a small quantity contributes to the natural resistance of timber.

The principal factor conferring resistance to biological attack is undoubtedly the presence of extractives in the heartwood. The far higher durability of the heartwood of certain species compared with the sapwood is attributable primarily to the presence in the former of toxic substances, many of which are phenolic in origin. Other factors such as a decreased moisture content, reduced rate of dif-fusion, density and deposition of gums and resins also play a role in determining the higher durability of the heartwood.

Considerable variation in durability can occur within the heartwood zone: in a number of timbers the outer band of the heartwood has a higher resistance than the inner region, owing, it is thought, to the progressive degradation of toxic substances by enzymatic or micro-bial action.

Durability of the heartwood varies considerably among the dif-ferent species, being related to the type and quantity of the extrac-tive present: the heartwood of timbers devoid of extractives has a very low durability. Sapwood of all timbers is susceptible to attack owing not only to the absence of extractives, but also the presence in the ray cells of stored starch which constitutes a ready source of food for the invading fungus.

In the UK, timbers have been classified into five durability groups which are defined in terms of the performance of the heartwood when buried in the ground. Examples of the more common timbers are presented in Table 18.1. Such an arbitrary classification is in-formative only in relative terms though these results on 2 inch x 2 inch ground stakes can be projected linearly for increased thick-nesses. Timber used externally, but not in contact with the ground, will generally have a much longer life, though quantification of this is impossible.

18.2.1 Nature of Fungal Decay

In timber some fungi. e.g. the moulds, are present only on the surface and although they may cause staining they have no effect on the strength properties. A second group of fungi, the sap-stain fungi, live on the sugars present in the ray cells and the presence of their hyphae in the sapwood imparts a distinctive colouration to that region of the timber. One of the best examples of sap stain is that found in recently felled Scots pine logs. In temperate countries the presence of this type of fungus results in only inappreciable losses in bending

Table 18.1 Durability Classification (Resistance of heartwood to fungi in ground contact) For the More Common Timbers*

Durability class	Perishable	Nondurable	Moderately durable	Durable	Very durable
Approximate life in contact with ground (years)	<5	5–10	10–15	15–25	>25
Hardwoods	Alder Ash, European Balsa Beech, European Birch, European Horse chestnut Poplar, black Sycamore Willow	Afara Elm, English Oak, American red Obeche Poplar, grey Seraya, white	Avodire Keruing Mahogany, African Oak, Turkey Sapele Seraya, dark red Walnut, European Walnut, African	Agba Chestnut, sweet Idigbo Mahogany, American Oak, European Utile	Afrormosia Afzelia Ekki Greenheart Iroko Jarrah Kapur Makore Opepe Purpleheart Teak
Softwoods		Hemlock, western 'Parana pine' Pine, Scots (redwood) Pine, yellow Podo Spruce, European (whitewood) Spruce, Sitka	Douglas fir Larch Pine, maritime	Pine, pitch Western red cedar Yew	

*Note that the sapwood of all timber is perishable.

strength, though several staining fungi in the tropical countries cause considerable reductions in strength.

By far the most important group of fungi are those that cause decay of the timber by chemical decomposition; this is achieved by the digesting action of enzymes secreted by the fungal hyphae. Two main groups of timber-decaying fungi can be distinguished:

(a) *The brown rots*, which consume the cellulose and hemi-celluloses but attack the lignin only slightly. During attack the wood usually darkens and in an advanced stage of attack tends to break up into cubes and crumbles under pressure. One of the best known fungi of this group is *Merulius lacrymans* which causes *dry rot*. Contrary to what its name suggests, the fungus requires an adequate supply of moisture for development.

(b) *The white rots*, which attack all the chemical constituents of the cell wall. Although the timber may darken initially, it becomes very much lighter than normal at advanced stages of attack. Unlike attack from the brown rots, timber with white rot does not crumble under pressure, but separates into a fibrous mass.

In very general terms, the brown rots are usually to be found in constructional timbers, whereas the white rots are frequently res-ponsible for the decay of exterior joinery.

Decay, of course, results in a loss of strength, but it is important to note that considerable strength reductions may arise in the very early stages of attack; toughness is particularly sensitive to the presence of fungal attack. Loss in weight of the timber is also characteristic of attack and decayed timber can lose up to 80% of its air-dry weight.

18.2.2 Nature of Insect Attack

Although all timbers are susceptible to attack by at least one species of insect, in practice only a small proportion of the timber in service actually becomes infested. Some timbers are more susceptible to attack than others and generally the heartwood is much more resist-ant than the sapwood: nevertheless the heartwood can be attacked by certain species.

Insect attack can take one of two forms. In certain insects the timber is consumed by the adult form and the best known example of this mode of attack are the *termites*. Few timbers are immune to attack by these voracious eaters and it is indeed fortunate that these insects cannot survive the cooler weather of this country. They are to be found principally in the tropics but certain species are present in the Mediterranean region including southern France.

In this country insect attack is always by the second form of attack, namely by the grub or larval stage of certain beetles. The adult beetle lays its eggs on the surface of the timber, frequently in surface cracks, or in the cut ends of cells; these eggs hatch to produce grubs which tunnel their way into the timber, remaining there for periods of up to three years. The size and shape of the tunnel and the type of detritus left behind in the tunnel (frass) are characteristic of the species of bettle. Well known examples of beetle larvae attacking timber in this country are the *furniture* and *death watch* beetles, but considerable damage also occurs from the *powder-post* and *longhorn* beetles.

18.2.3   Marine Borers

Timber used in salt water is subjected to attack by marine-boring animals such as the shipworm (*Teredo* sp.) and the gribble (*Limnoria* sp.). Marine borers are particularly active in tropical waters; nevertheless around the coast of Great Britain *Limnoria* is fairly active and *Teredo*, though spasmodic, has still to be considered a potential hazard. The degree of hazard will vary considerably with local conditions and there are relatively few timbers which are recognised as having heartwood resistant under all conditions: the list includes ekki, greenheart, okan, opepe and pyinkado.

The sapwood and heartwood of many species of timber can have its natural durability increased by impregnation with toxic chemicals; the preservative treatment of timber is considered in Chapter 20, which is concerned with the mechanical and chemical processing of timber.

18.3   PERFORMANCE OF TIMBER IN FIRE

The performance of materials in fire is an aspect of durability which has attracted much attention in recent years, not so much from the research scientist, but rather from the material user who has to conform with recent legislation on safety and who is influenced by the weight of public opinion on the use of only 'safe' materials. While various tests have been devised to assess the performance of materials in fire there is a fair degree of agreement in the unsatisfactory nature of many of these tests, and an awareness that certain materials can perform better in practice than is indicated by these tests.

Thus, while no one would doubt that timber is a combustible material showing up rather poorly in both the 'spread of flame' and 'heat release' tests, nevertheless in at least one aspect of performance, namely the maintenance of strength with increasing temperature and time, wood performs better than steel.

There is a critical surface temperature below which timber will not ignite. As the surface temperature increases above 100 °C volatile gases begin to be emitted as thermal degradation slowly commences; however, it is not until the temperature is in excess of 250 °C that there is a sufficient build-up in these gases to cause ignition of the timber in the presence of a pilot flame. Where this is absent the surface temperature can rise to about 500 °C before the gases become self-igniting. Ignition, however, is related not only to the absolute temperature but also to the time of exposure at that temperature, since ignition is primarily a function of heat flux.

Thermal degrade certainly occurs at temperatures down to 120 °C and it has been suggested that degrade can occur at temperatures as low as 66 °C when timber is exposed for long periods of time.

The performance of timber at temperatures above ignition level is very similar to that of certain reinforced thermosetting resins which have been used as sacrificial protective coatings on space-return capsules. Both timber and these ablative polymers undergo thermal decomposition with subsequent removal of mass, leaving behind enough material to preserve structural integrity.

The onset of pyrolysis in timber is marked by a darkening of the timber and the commencement of emission of volatile gases; the reaction becomes exothermic and the timber reverts to a carbonised char popularly known as charcoal (Fig. 18.1). The volatiles, in moving

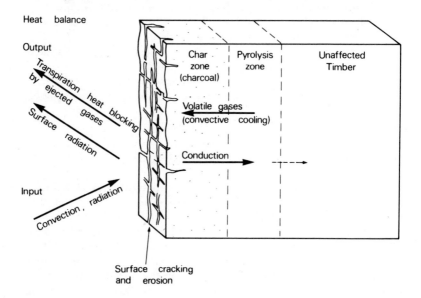

Figure 18.1 Diagrammatic representation of the thermal decomposition of timber. (From the Princes Risborough Laboratory, Building Research Establishment. © Crown copyright.)

to the surface, cool the char and are subsequently ejected into the boundary layer where they block the incoming convective heat: this most important phenomenon is known as *transpirational cooling*. High surface temperatures are reached and some heat is rejected by thermal radiation: the heat balance is indicated in Fig. 18.1. The surface layers crack badly both along and across the grain and surface material is continually but slowly being lost.

A quasi-steady state is reached, therefore, with a balance between the rate of loss of surface and the rate of recession of the undamaged wood. For most softwoods and medium-density hardwoods the rate at which the front recedes is about 0.64 mm/min: for high density hardwoods the value is about 0.5 mm/min (Hall and Jackman, 1975).

The formation of the char, therefore, protects the unburnt timber which may be only a few millimeters from the surface. Failure of the beam or strut will occur only when the cross-sectional area of the unburnt core becomes too small to support the load. By increasing the dimensions of the timber above those required for structural consideration, it is possible to guarantee structural integrity in a fire for a given period of time. This is a much more desirable situation than that presented by steel, where total collapse of the beam or strut occurs at some critical temperature.

18.3.1 Current Methods of Assessing Performance

Four standard tests are applicable to the evaluation of the performance of timber in fire (TRADA, 1976). The first is the *Noncombustible Test for Materials* (BS 476, Part 4, 1970) where a small sample of material is subjected to a temperature of 750 °C; timber and timber products, even when treated with fire retardants, are classified as *combustible*. The second test is a crude measure of ignitability where a small pilot flame is used to determine whether the sample will ignite easily (BS 476, Part 5, 1968); using this test, timber and most timber-based products are rated as *not easily ignitable*.

Following ignition, the development of a fire is dependent on a number of factors, one of the more important being the rate of spread of flame. Many test methods have been developed throughout the word to quantify *flamespread* and there is much disagreement not only on the relative values obtained by the different tests, but also on the significance of the test in practical terms. Using BS 476, Part 7, 1971, timber and board materials over 400 kg/m^3 are rated as Class 3 while timber products with a lower density are rated as Class 4.

For many applications, current regulations call for wall and ceiling linings to conform to Class 1; timber and timber products can be upgraded either by the application of intumescent paints to the surface, or by the incorporation of, or impregnation by, flame-retardant chemicals. These products influence the mechanism of decomposition, lower the temperature of onset of decomposition and increase the thickness of the char layer. The process of impregnation of timber by flame-retardant chemicals will be discussed in Chapter 20, which deals with timber processing.

Although a very great deal of attention has recently been paid to flamespread as a means of assessing the fire performance of materials, it must never be forgotten that it represents only one of several factors involved in the assessment of fire safety.

The rate at which a combustible material contributes heat to a developing fire is a most important aspect and one in which timber and board materials do not show up very well, especially so when compared with alternative materials such as plasterboard. The *fire propagation test* (BS 476, Part 6, 1972) provides some measure of the rate of heat release though a more specific test is required.

Thus, in three out of the four standard 'fire' tests currently carried out, timber and board products do not fair at all well. None of the four tests demonstrates the predictability of the performance of timber in fire, nor do they indicate the guaranteed structural integrity of the material for a calculable period of time. The performance of timber in the widest sense is certainly superior to that indicated by the present set of standard tests.

18.4 REFERENCES

Hall, G. S. and Jackman, P. E. (1975) Performance of timber in fire, *Timber Trades Journal*, 15 Nov 1975, 38—40.

TRADA (1976) *Timber and Wood-Base Sheet Materials in Fire*, Timber Research and Development Association, High Wycombe, Wood Information sheet, Section 4 no. 11.

CHAPTER 19

Corrosion

19.1 INTRODUCTION

The working life limit of an engineering component — excluding
obsolescence — is normally dependent on three factors acting either
separately or in combination: corrosion, mechanical breakdown due
to fatigue or creep, and wear. Corrosion is the most important and
universal of these factors, representing a threat to all metal
structures irrespective of environment, whereas the other two are of
relatively minor significance in civil, chemical and electronic engineering. In a sense, corrosion may be likened to disease, both in its
omniscience and in the need for constant precautions to keep it
under control. All metals corrode, but the severity of corrosion
depends considerably on details of their immediate environment,
and it cannot be too strongly emphasised that *corrosion problems
begin on the drawing board.* The adage that 'prevention is better
than cure' is nowhere more true than when coping with corrosion
problems, and in addition it is cheaper; good design can represent a
substantial increase in working life or reduction in maintenance
costs for a structure. In this connection it is worth noting the
economic significance of corrosion: the Hoar Report (1971) estimated the national cost of corrosion at about £1 365 000 000 per
annum, of which approximately £310 000 000 might be saved by
better design and better thought-out protective measures. There is
thus a powerful economic incentive to combat corrosion effectively,
which has been reinforced in recent years by an increasing awareness
of the need to conserve supplies of raw materials. Corrosion may be
regarded as the way in which metals seek to return to their naturally
occurring state, but whereas geological ore deposits are conveniently
concentrated (relatively speaking) for exploitation, the rust which
forms on existing structures and the scrap from obsolete and
defunct structures is all too frequently far too wide spread to be
economically worth recovering. It is also worth pointing out that,
even if scrap recovery techniques achieve a sophisticated undreamt
of today, their processing to re-create engineering alloys must require

a large energy input — and our nonrenewable energy resources are being depleted all too rapidly as it is.

The figure quoted above includes costs of preventative measures, but cannot cost many of the penalties resulting from bad practice. For example, steel allowed to rust on site before erection may lead to delays while (usually inadequate) cleansing techniques have to be employed prior to painting. Once the surface of structural steel has rusted it is virtually impossible to clean it sufficiently for paints to adhere thoroughly (any car owner will confirm this statement); the result may be inability to meet the requisite standards or a deadline, and possibly considerably raised maintenance costs. In more extreme instances corrosion may be the direct cause of breakdown or failure, which at best can lead to loss of trade and goodwill, and may even involve injury or loss of life.

Good design could represent an appreciable decrease in corrosion costs, even at the expense of a slightly higher initial outlay. At the same time, a sense of proportion must be kept when considering corrosion problems in design; metals must be selected for their engineering qualities as well as their corrosion resistance, and it is pointless to spend large amounts of money on a structure designed to resist corrosion for 20 years if it is likely to fail mechanically or become obsolete in ten. It is also pointless to incur a higher outlay by the use of corrosion-resistant materials than would be spent in maintenance costs or replacement of cheaper materials. These points sound very obvious, but waste can and has occurred from over-protection as well as under-protection, and the above points all serve to emphasise the importance of a careful survey of corrosion possibilities at the design stage. More specific points in connection with design will be considered in Section 19.10.2.

Corrosion may be very roughly divided in four categories, all capable of many subdivisions:

(a) dry corrosion, including high temperature effects;
(b) moist or atmospheric corrosion;
(c) corrosion of immersed metals and alloys; and
(d) miscellaneous varieties of corrosion, including effects of corrosion combined with stress or irradiation.

All categories with the exception of the first are largely dependent on electrochemical mechanisms; for that reason, a very elementary introduction to some of the principles of electrochemistry is given in Section 19.3. It need hardly be said that the above divisions overlap very considerably; they are introduced arbitrarily as an aid to exposition rather than from any fundamental differences.

19.2 DRY CORROSION OF METALS AND ALLOYS

Since the principal result of dry corrosion is the oxidation of metals, we shall confine ourselves in this section to reactions between metals and oxygen, only referring to other reactions insofar as they affect oxidation resistance. Oxidation problems have become increasingly important in recent years with advances in the design and use of gas turbines, but for the civil engineer the major import-ance is that oxide films may profoundly affect the response of an alloy to other varieties of corrosion attack.

It is probably not generally appreciated how universal is oxidation; all metals (with the exception of gold) oxidise readily at room tem-perature and the only reason that metals do not all slowly 'burn' away is that in many cases the metal oxide itself acts as a protective film. The widely differing oxidation behaviour of metals is thus due more to the properties of their oxides than to the chemical behaviour of the metals themselves. Let us examine more closely the stages involved in the oxidation process of pure metals at room temperature.

On exposure to oxygen, nuclei of oxide form on the surface of the metal; normally they form preferentially on surface imperfec-tions such as grain boundaries or impurity atoms. For high-purity systems, the density of nuclei also depends on the crystallographic orientation of the surface, but since this only affects the initial rate of oxidation it has little practical engineering significance. After the nuclei have formed, lateral growth of the oxide film proceeds rapidly until the surface is completely covered. With completion of the initial stage, direct access of oxygen to the metal surface is no longer possible and further oxidation can only occur if the oxide film is permeable either to oxygen ions or to metal ions.

Permeability of the film can occur in two ways. If the relative volume of the oxide is less than that of the metal (i.e. if the volume of oxide produced is less than the volume of metal consumed in producing it), the oxide film will be porous, and access of oxygen to the metal surface is not prevented. Alternatively, the oxide film may be nonporous on a microscopic scale, but contain vacancies on the atomic scale which are sufficient for diffusion of metallic or oxygen ions to take place. The first of these criteria is known as the *Pilling—Bedworth Principle*, and according to it we should expect all engineering metals to have oxides whose relative volume is equal to or greater than that of the metal. With the exception of magnesium, this is in fact the case; iron, copper, aluminium, nickel, beryllium and titanium all have favourable metal-oxide ratios, whereas the reactive light metals such as sodium and calcium do not. The picture is now considered to be rather more complex than that presented by Pilling and Bedworth in 1923, although it is certainly adequate to explain the general behaviour of metals. Even the metals in the 'por-ous oxide' group initially produce a thin continuous film of oxide over their surfaces; the film, however, is in a considerable state of tension, and breaks down to give the observed porous oxide as soon as it reaches a certain width. Oxidation then proceeds by diffusion through a very thin film of constant thickness on top of which grows the gross porous layer; the reaction rate is therefore constant, depending on the permeability of the film (Fig. 19.1(a)). Magnesium has a relatively impermeable film, which accounts for its reasonable corrosion resistance at ambient temperatures. In conditions of high purity, calcium and sodium show surprisingly slow oxidation rates, but under normal conditions the film is disrupted by secondary reactions with atmospheric moisture and ceases to have any protective value at all. At high temperatures, diffusion through the film becomes more rapid and if the heat evolved in the reaction is not dissipated both temperature and reaction rate rise quickly until the metal catches fire. This of course, has always been a hazard with magnesium, but the protective nature of the thin film can be strikingly enhanced by the addition of 0.2% beryllium and 0.1% calcium,

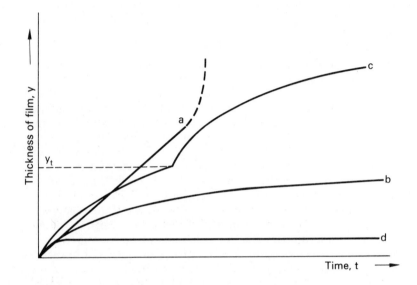

Figure 19.1 The principal oxidation curves shown by metals. (a) Linear growth (porous or nonadherent oxide). The upward swing at the end of the line takes place if the reaction rate increases because the heat of reaction cannot escape fast enough. Ultimately the metal may ignite under these conditions. (b) Parabolic growth. (c) Parabolic growth with cracking when the film reaches a critical thickness y_t. (d) Logarithmic growth. After a short time oxidation virtually ceases.

giving rise to the Magnox alloys which show good oxidation resistance up to 600 °C.

The mere fact of a favourable oxide/metal volume ratio according to the Pilling—Bedworth Principle does not of itself mean that a metal will exhibit good oxidation resistance. If the oxide film has a larger volume than the metal producing it, the film is usually laid down in a state of compression. As the film thickens the stresses increase and eventually the film buckles and cracks, allowing direct ingress of oxygen to the metal surface and starting the process afresh. At room temperature the growth of oxide becomes very slow because diffusion rates through the film are slow, and the film seldom reaches the 'cracking thickness'. Under these conditions the oxidation

572

rate is inversely proportional to the thickness of the oxide film, i.e.

$$\frac{dy}{dt} = \frac{k}{y} \qquad (19.1)$$

where y is the thickness of the film at time t, and k is a constant depending on temperature, which gives a parabolic rate of growth (Fig. 19.1(b)). Providing that the film does not crack, the oxidation eventually becomes very slow, which explains the perfectly adequate oxidation resistance of iron and copper at ordinary temperatures. If the temperature is raised, however, the growth rate increases, the cracking thickness is exceeded, and the oxidation process appears as for the curvec in Fig. 19.1. Obviously, if cracking occurs very frequently, the growth rate approaches a straight line relationship as shown in Fig. 19.1(a). It follows that the metals with an oxide/ metal volume ratio near to unity should show the best oxidation resistance.

We must now consider the second criterion for oxidation resistance, the atomic structure of the oxide film. Only relatively few oxides have the exact chemical formula assigned to them, i.e. are *stoichiometric*; these oxides have all sites in their crystallographic lattice filled with the appropriate atoms and there are no vacancies available for diffusion through the film. Provided that such a film has a favourable oxide/metal volume ratio, it will give excellent oxidation resistance; elements in this category are aluminium, chromium and silicon, all widely used for their powers of protection against oxidation in engineering alloys. It is interesting to note in passing that it is their very affinity for oxygen (which contributes to their stoichiometry) that makes them so resistant to attack. Their growth rate curve obeys a logarithmic rather than a parabolic relationship, and is shown in Fig. 19.1(d); after initial rapid growth, the rate rapidly slows down and ceases completely well below any 'cracking thickness'.

The majority of oxide films are nonstroichiometric, having a deficiency of either metal ions or oxygen ions. In the first category are the oxides Cu_2O, FeO and NiO, the metals of which all have alternative higher valencies. Vacancies in this case arise from the substitution of (say) an Fe^{3+} ion for an Fe^{2+} ion; to preserve the electronic balance of the oxide one of the metal sites in the lattice must remain unfilled (Fig. 19.2). Oxidation proceeds in these cases by the diffusion out of metal ions through the film to react with oxygen on the outer surface (Fig. 19.3(a)). Oxides deficient in oxygen ions are CdO, ZrO_2, TiO_2 and Fe_2O_3; here growth occurs by the inward diffusion of oxygen, the reaction taking place at the metal-oxide interface (Fig. 19.3(b)). A few oxides do not fall into either category; zinc oxide contains excess metal in the form of interstitial zinc ions, and oxidation occurs by movement of the metal ions, although stoichiometrically the oxide belongs to the second group. Vanadium dioxide VO_2 allows the interstitial diffusion of oxygen, diffusion of oxygen in the atomic rather than the ionic form probably occurs in MgO, and with cobalt both metal and

573

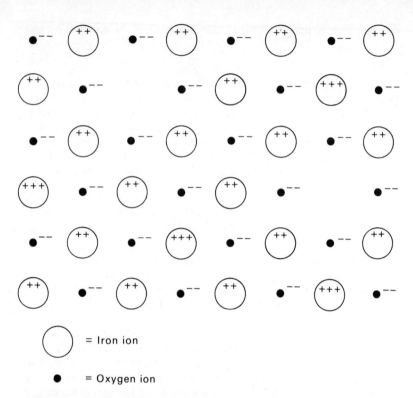

○ = Iron ion

● = Oxygen ion

Figure 19.2 Defective lattice of ferrous oxide (schematic). Each vacant site for a metal ion is balanced by the presence of two ferric ions. The vacancies permit movement of the metal ions through the film when sufficient thermal energy is available.

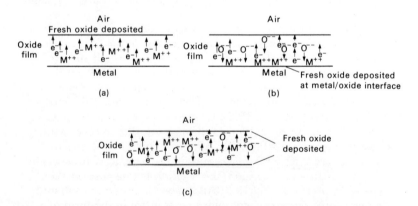

Figure 19.3 Migration of metal ions, oxygen ions, and electrons through the oxide film during oxidation. (a) Metal-deficient film; metal ions and electrons move out (Fe, Cu, Ni). (b) Oxygen-deficient film; oxygen ions move in and electrons move out (Cd, Nb, Ti). (c) Transport of anions and cations. metal ions and electrons move out and oxygen ions move in. Oxide is deposited simultaneously on both the upper and the lower surfaces of the oxide film (Co).

oxygen ions pass through the film in opposite direction, giving rise to two layers of oxide, though both have the composition CoO (Fig. 19.3 (c)).

The immediate results of oxidation differ somewhat according to whether it is the metal or the oxygen ions which are mobile. If it is the metal ions, there is a steady transport of material outwards through the oxide film; the fresh oxide forms without constraint, but lattice vacancies and eventually cavities occur beneath the oxide. The usual result is the collapse of the oxide from lack of support and a rise in oxidation rate; the growth curve produced is as in Fig. 19.1(c). Where oxygen diffuses inwards and the reaction occurs at the oxide — metal interface, the oxide is formed under constraint and considerable compressive stresses may result. If film growth slows down while the film is sufficiently thin, films of this nature can give excellent corrosion resistance, since the compressive stresses serve to close up any cracks in the film. Zirconium and titanium are examples of this behaviour; in both cases, however, the sudden removal of the protective film (by abrasion or impact, for example) from these highly reactive metals in the presence of moisture can result in explosions. Great care should be taken when handling either scrap or powder of these metals. If the film exceeds the critical thickness, the compressive stresses cause cracking and a similar incremental parabolic growth curve to oxides in which the metallic ions are mobile.

We may thus conclude that a metal has good oxidation resistance only if the oxide/metal volume ratio is favourable and then only if the oxide film produced is stoichiometric in nature. Nonstoichiometric films may be adequate at ambient temperatures, but they are unsatisfactory at elevated temperatures when diffusion rates through the film become fast. It is important to note that these arguments are only applicable to oxidation at a steady temperature; where temperature changes occur, the difference in thermal expansion coefficients between oxide and metal usually leads to spalling and loss of the protective film. This danger is accentuated if either the metal or the oxide undergoes any phase changes during heating or cooling.

19.2.1 Oxidation of Alloys

Even the addition of small amounts of alloying elements can alter the oxidation behaviour of metals considerably (c.f. magnesium, Section 19.2); it is therefore important to examine the effects of alloying elements on oxidation behaviour, especially as the results can frequently be very harmful. Let us consider the possibilities of a minor alloying element Y in solid solution in a metal X of moderate oxidation resistance. If Y oxidises preferentially to X, one of two results may occur, depicted schematically in Fig. 19.4. If oxygen can diffuse through X faster than Y can diffuse to the surface, internal oxidation of Y occurs together with normal surface oxidation of X; this is less a problem in oxidation resistance than a technique for producing dispersion-hardened alloys (see Section 16.2). On the

575

other hand, if Y diffuses through X faster than oxygen, oxidation of X and Y takes place simultaneously on the surface. From the point of oxidation resistance, the best results are obtained where Y gives rise to a stoichiometric oxide immiscible with the oxide of X. A layer of oxide of Y forms preferentially on the surface of X; since Y is a minor alloying element this film may well be incomplete (Fig. 19.4(b)), but, providing it is impermeable itself, the oxidation rate of X should be considerably reduced and may even be completely halted, e.g. beryllium in copper.

Where the two oxides are miscible the result depends on the effect of Y on the stoichiometry of the oxide film. If X develops an

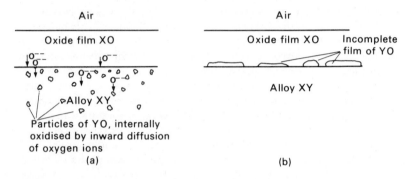

Figure 19.4   Oxidation patterns in the presence of a minor alloying element Y with a greater affinity for oxygen than metal X. (a) Diffusion of oxygen ions through X is faster than diffusion of Y ions. (b) Diffusion of Y ions through X is faster than diffisuon of oxygen ions. In the example shown the oxides of X and Y remain as separate phases.

oxygen-deficient oxide film, the vacancies will be filled if Y is a metal of higher valency than X, whereas, if Y is of lower valency, additional vacancies are introduced into the oxide film and resistance is impaired. For metal deficient oxides the reverse holds true; this leads to the somewhat surprising conclusion that small additions of chromium (tervalent) to nickel (divalent) actually lower the oxidation resistance. It is only when the chromium is present in sufficient quantity to form a coherent film of chromium oxide that oxidation resistance is enhanced. On the other hand, traces of lithium oxide in the scale definitely improve the resistance of nickel (lithium is monovalent).

Of the two ways of improving oxidation resistance, the first mentioned (preferential film formation of a protective oxide on the surface of the alloy) is probably the more efficacious and widely used — e.g. the use of chromium in stainless steels and aluminium in aluminium bronzes — but the second (improvement of lattice structure of the film) can be achieved with lower additions of alloying elements.

Oxygen can also penetrate down the grain boundaries of metals, forming intergranular films of oxide. Providing that penetration is

576

slight and the grains of metal are not encircled by oxide film, this can provide a valuable anchor for an oxide film required to withstand thermal cycling. It is known as the 'pegging in' process, and has been used with considerable success in the Nichrome alloys based on 80% Ni 20% Cr, which are used for electric heating elements and components in furnaces. If the pegging in process proceeds too far, however, the surface grains are encircled and spalling will occur, the oxide film taking the immediate surface of the alloy with it as well (Fig. 19.5).

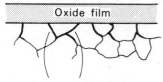

Penetration of oxide film a short way down grain boundaries. This anchors the oxide film to the metal surface and discourages spalling

(a)

Excessive grain boundary penetration leads to spalling of the oxide film together with the surface grains that have been surrounded by- oxide

(b)

Figure 19.5   The 'pegging-in' principle for improving the adhesion of oxide films: (a) correct (b) incorrect.

19.3   ELEMENTARY PRINCIPLES OF ELECTROCHEMISTRY

Some knowledge of electrochemical principles is essential for an appreciation of corrosion at ambient temperatures since, as will be seen, almost all varieties of corrosion in the presence of moisture are dependent, at least in part, on electrochemical mechanisms. The following account makes no attempt to be exhaustive or rigorous, but merely aims to set out the basic features of electrochemical reactions. These are best approached by consideration of the features of a simple electrochemical cell.

If we take crystalline copper sulphate, $CuSO_4$, and dissolve it in water, the salt goes into solution in the ionised form. If unlimited solid $CuSO_4$ is available, solution will occur until saturation is reached, at which point equal amounts of $CuSO_4$ are going into solution and being deposited from solution, according to the equation

$$CuSO_4 \rightleftharpoons Cu^{2+} + SO_4^{2-}$$

At the same time, ionisation of the solvent also occurs:

$$H_2O \rightleftharpoons H^+ + OH^-$$

Solutions of ionised salts are known as electrolytes, and can carry an electric current by means of the mobility of the ions, especially the solvent ions.

If a strip of pure copper is dipped into a solution of copper sulphate, there is a tendency for the copper to go into solution in ionic form, leaving two electrons per atom in the body of the metal:

$$Cu \rightleftharpoons Cu^{2+}(\text{in solution}) + 2e^-(\text{in the metal})$$

The result is that a potential gradient is set up between the metal and the solution (see Fig. 19.6), which opposes the further passage of copper ions into the solvent and halts solution whilst it is still negligible. The potential set up depends partly on the concentration of the electrolyte, but for equivalent concentrations all metals develop different *electrode potentials*, as they are known. Measurement of the potential in the system shown in Fig. 19.6 (known as a 'half

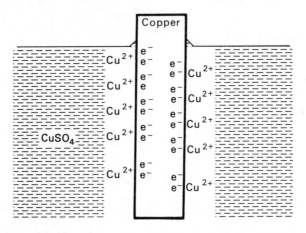

Figure 19.6   Surface 'double layer' developed on a copper rod dipped into copper sulphate solution.

cell') is not possible directly since it would involve the introduction of a second electrode and a measuring device, which in themselves would change the conditions at the electrode being investigated. However, by setting up two half cells using different metals, the *difference* in potential between the two half cells can be measured very accurately using a potentiometer. The e.m.f. of such a combination is, of course, a standard source of electricity; the Daniell Cell, illustrated in Fig. 19.7, is one of the best-known examples. Here the zinc electrode goes into solution and becomes the *anode*, whilst the copper is deposited on the copper electrode which becomes the cathode. The reactions involved are respectively:

$$Zn \longrightarrow Zn^{2+}(\text{in solution}) + 2e^-(\text{in the metal})$$

$$Cu^{2+}(\text{in solution}) + 2e^{2-}(\text{in metal}) \longrightarrow Cu \text{ (deposited on the electrode)}$$

578

The amount of copper deposited (or zinc dissolved) is related to the current passed by *Faraday's Law*, which states that the amount of electricity required to deposit 1 gram-equivalent of metal (atomic weight in g divided by its valency) is about 96 500 amp s or coulombs. This quantity is known as 1 Faraday. Since the electrons in the zinc anode are free to pass along the wire to cathode and take part in the deposition of copper, no electrical double layer is set up and the reaction will proceed, with the steady generation of an e.m.f. until all the zinc has dissolved or all the copper has been deposited. Conventionally the metal deposited (i.e. the cathodic metal) is said

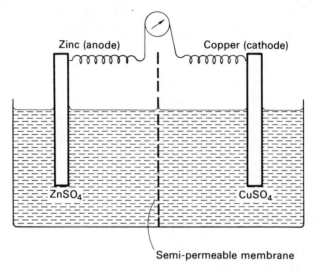

Figure 19.7 A Daniell cell. (From A. A. Smith (1965) Electrochemical processes, *Chemical Engineering Practice, Vol. 8* (Ed. H. W. Cremer and S. B. Watkins) by permission of Newnes—Butterworths.)

to be *positive* with respect to the metal passing into solution (the anode).

Since only relative measurements of electrode potentials can be made, it is convenient to measure them against a standard electrode, arbitrarily taken to have zero e.m.f., and the results can then be arranged to give the elements in order of an *electropotential series*. The standard taken is the hydrogen electrode and potentials are measured for metals against a concentration of their ions in solution; some values are given in Table 19.1. The significance of this series is that if any two metals are in contact in the presence of an electrolyte (which means virtually all naturally occurring moisture), there is the probability of a galvanic cell being set up, with the more electropositive element becoming the cathode, and the more electronegative element becoming anodic and going into solution, i.e. *being corroded away*. The greater the difference in electrode potentials between the two metals, the more rapid will be the corrosion of the anode.

Table 19.1 Some Normal Electrode Potentials at 25 °C

No.	Equilibrium	Volts	No.	Equilibrium	Volts
1	$F_2(gas)/F^-$	+2.87	28	Sb^{3+}/Sb	+0.1
2	Co^{3+}/Co^{2+}	+1.84	29	H^+/H	0.0000 (Standard)
3	Au^+/Au	+1.68	30	Fe^{3+}/Fe	−0.04
4	Ce^{4+}/Ce^{3+}	+1.61	31	Pb^{2+}/Pb	−0.126
5	Mn^{3+}/Mn^{2+}	+1.51	32	Sn^{2+}/Sn	−0.136
6	Au^{3+}/Au	+1.4	33	Mo^{3+}/Mo	−0.2
7	$Cl_2(gas)/Cl^-$	+1.358	34	Ni^{2+}/Ni	−0.25
8	Pt^{2+}/Pt	+1.2	35	Co^{2+}/Co	−0.28
9	Ir^{3+}/Ir	+1.0	36	Tl^+/Tl	−0.336
10	$2Hg^{2+}/Hg_2^{2+}$	+0.92	37	Ti^{3+}/Ti^{2+}	−0.37
11	Hg^{2+}/Hg	+0.85	38	Cd^{2+}/Cd	−0.402
12	Pd^{2+}/Pd	+0.83	39	Cr^{3+}/Cr^{2+}	−0.41
13	Rh^{3+}/Rh	+0.8	40	Fe^{2+}/Fe	−0.44
14	Ag^+/Ag	+0.799	41	Ga^{3+}/Ga	−0.52
15	Hg_2^{2+}/Hg	+0.798	42	Cr^{3+}/Cr	−0.71
16	Fe^{3+}/Fe^{2+}	+0.771	43	Zn^{2+}/Zn	−0.763
17	Tl^{3+}/Tl	+0.72	44	Mn^{2+}/Mn	−1.18
18	I_2/I^-	+0.536	45	Ti^{2+}/Ti	−1.63
19	Cu^+/Cu	+0.52	46	Al^{3+}/Al	−1.66
20	O_2/OH^-	+0.401	47	Be^{2+}/Be	−1.85
21	$Fe(CN)_6^{3-}/Fe(CN)_6^{4-}$	+0.36	48	Mg^{2+}/Mg	−2.38
22	Cu^{2+}/Cu	+0.34	49	Na^+/Na	−2.713
23	$Hg_2Cl_2/Hg, 1.0. NKCl$	+0.2810	50	Ca^{2+}/Ca	−2.87
24	$AgCl/Ag, 1.0. NHCl$	+0.2224	51	Ba^{2+}/Ba	−2.90
25	Bi^{3+}/Bi	+0.2	52	K^+/K	−2.92
26	Sn^{4+}/Sn^{2+}	+0.154	53	Li^+/Li	−3.01
27	Cu^{2+}/Cu^+	+0.153			

From A. A. Smith (1965) Electrochemical Processes, *Chemical Engineering Practice, Vol. 8* (Ed. H. W. Cremer and S. B. Watkins) by permission of Newnes—Butterworths.

19.3.1 Limitations of the Electropotential Series

The fundamental importance of the electropotential series is difficult to overestimate, but as a pointer to practical conditions it must be used with discretion. The general indications from consideration of the series are usually valid, but its predictions may be masked by secondary effects at the site of corrosion. There are two reasons for this. Firstly, the potentials given in the series are necessarily measured under very carefully controlled standard conditions of temperature, pressure and solution concentration, and they are measured for a reversible reaction *under equilibrium conditions*, i.e. when the electrode reaction takes place infinitely slowly and there is therefore no passage of current. Under practical conditions, the solution strength and temperature may vary considerably from the standard, thus altering the relative positions of metals in the series. This effect is relatively small, and is only likely to be significant where the two metals lie very close together in the series; however, the effect of departure from equilibrium conditions can be considerable. Under practical conditions, the reaction proceeds at an appreciable rate and becomes irreversible; as a result the electrode potentials are displaced from their reversible values and the cell is said to be polarised. The faster the reaction proceeds, the greater is the degree of polarisation and consequent shift in potentials; the potential of hydrogen is particularly prone to polarisation effects, and can develop a high 'over potential', i.e. excess potential needed to liberate hydrogen at the electrode. This is the reason for the immunity of lead to attack by hydrochloric acid: the hydrogen overpotential developed is sufficient to reverse their relative positions in the series.

The second reason for departure from the electrochemical series is the possible effect of secondary chemical reactions, which may conceal the true electrode potential of the metal concerned. One simple instance is the oxide film normally present on aluminium. As a result aluminium behaves in a much more electropositive manner than its position in the series would suggest; it behaves cathodically to zinc, for example, despite their true positions in the series. This sort of behaviour may be due to a naturally occurring film on the surface of one metal (as for aluminium) or it may occur because of film formation during the corrosion reaction. An interesting domestic example is the fact that zinc becomes cathodic to iron at temperatures approaching 100 °C, which can disastrously accelerate the rusting of galvanized hot water pipes and cisterns. The reversal of relative positions is due to the formation of an oxide film on the zinc at high temperatures, which behaves as a cathode with respect to the iron. Hence, a system must be very carefully examined for the possibility of secondary reactions before its behaviour can be confidently predicted from the electropotential series.

19.4 GALVANIC CORROSION

We must now apply the principles set out in Section 19.3 to systems occurring in practical engineering. Some degree of simplification is still necessary in order to emphasise the more important features, since corrosion even in apparently simple aspects is an extremely complex process; as many as six different types of corrosion may occur simultaneously in an iron pipe with water flowing through it. The mechanism of rusting is worth considering first, being common and illustrating several important points. Figure 19.8 shows the system involved when rusting of mild steel occurs under a drop of electrolyte, in this case taken as weak sodium chloride solution. The process is as follows. Attack commences by direct chemical action at random points on the steel surface, in which very minute quantities of iron go into solution as ferrous chloride. At the same time, the liquid at the edges of the drop shows a tendency to become alkaline because of diffusion in of atmospheric oxygen, which is in more plentiful

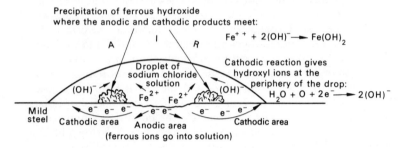

Precipitation of ferrous hydroxide where the anodic and cathodic products meet:

$Fe^{++} + 2(OH)^- \longrightarrow Fe(OH)_2$

Cathodic reaction gives hydroxyl ions at the periphery of the drop: $H_2O + O + 2e^- \longrightarrow 2(OH)^-$

Figure 19.8 The mechanism of rusting of mild steel.

supply at the edge than the centre. The two reactions, which are interdependent, are:

$$Fe \longrightarrow Fe^{2+} + 2e^-$$

$$H_2O + O\,(\text{atmospheric}) + 2e^-(\text{in steel}) \longrightarrow 2OH^-$$

A flow of electrons (i.e. an electric current) is thus generated in the steel and the process, after the initial chemical attack, becomes entirely electrochemical with the steel at the outside of the drop being the cathode and the centre becoming the anode. At this stage any initial foci of attack in the cathodic regions will disappear, the attack being concentrated entirely at the anode. The principle that regions of metal with relatively freer access of oxygen become cathodic is general: it is known as the *differential aeration* effect and we shall meet it frequently when discussing examples. In the present instance it demonstrates that both water and air are necessary for rusting; in the absence of either no attack will take place. The result is a build-up of ferrous chloride at the centre of the drop and of sodium hydroxide at the periphery. Where the two liquids meet, ferrous hydroxide is precipitated, which changes to the complex hydrated oxide commonly known as rust.

$$FeCl_2 + 2NaOH \longrightarrow Fe(OH)_2\downarrow + 2NaCl$$

There are two important points in connection with this reaction: the regeneration of sodium chloride and the fact that precipitation of rust occurs *between* the anodic and cathodic regions. From the first point it can be seen that the reaction is cyclic as far as the chloride is concerned and therefore a very little chloride can, in the absence of evaporation, cause a great deal of rusting. From the second point, the end product of the reaction cannot stifle the re-action as it might well do if precipitation occurred *at* the anode or

582

the cathode. (This principle is used in the selection of inhibitive pigments in paints.) With hard water and, to a lesser extent, sea water rusting is partly counteracted by the increased alkalinity in the cathodic regions which causes precipitation of calcium carbonate on the cathode, thereby reducing contact between cathode and electrolyte and slowing the whole process down.

Where different metals or alloys of different compositions are in contact in the presence of moisture, the attack is rather simpler in mechanism — and frequently faster. Typical instances are at the joints between pipes and tanks, or where an alloy is protected by a surface layer of another metal (as in galvanized iron), or between different phases in an alloy system, or where particles of a metal are deposited on the surface of another metal (a common cause of distress in domestic water systems). In all these examples one of the most important variables is the relative area of the anode and cathode. The current passed during corrosion is usually limited by the conditions at the cathode, and thus the larger the cathode compared to the anode, the greater the current density at the anode. Where the anode is large compared to the cathode, current density is low and the attack, although unsightly, is spread over a large area, with slight reduction in thickness of the anode. If the anode is small, however, the same weight of material is removed from a much smaller area and the attack results in pitting and possibly perforation — usually a much more serious situation. The larger the ratio of cathode to anode, the more rapidly does pitting progress. Diagrams of the various types of attack instanced above are shown in Fig. 19.9.

Joints between dissimilar metals are best protected by the insertion of a insulating gasket or by the complete exclusion of moisture using some protective covering such as paint. The second case, that of a metallic surface layer, introduces the important principle of *sacrificial protection*, which is widely used to combat corrosion. If an alloy is to be protected by a metallic surface layer, the layer may be either anodic or cathodic with respect to the alloy. At first sight a cathodic layer might appear preferable, since the more electropositive metals are naturally more corrosion resistant, but in practice it must be remembered that very few coatings are initially completely free of pin-holes and all coatings are liable to develop defects from impacts or abrasions during service. It is therefore not the corrosion resistance of the coating per se which is important but the state of affairs which is likely to be set up at any pinholes in the presence of an electrolyte, and Fig. 19.9(b) shows the danger of cathodic coatings under these conditions. The cathodic area is very large compared to the area of the anode exposed at the pin-hole, and hence rapid pitting and undermining of the coating sets in. By contrast, with an anodic coating the metal exposed at the pin-hole becomes cathodic and remains unattacked although bare to the electrolyte; admittedly the attack on the coating is accelerated, but the current density will be low and hence the attack is slow and evenly distributed. Hence the use of zinc rather than copper as a

583

protective coating for steels, and indirectly the danger of copper pipes in a system containing galvanized tanks. Even where the two are well separated, if any copper passes into solution it will be displaced by the zinc or iron in a tank with deposition of copper on the tank surface, thus setting up galvanic cells with anodes around the small regions of deposited copper. Copper powder or filings left behind by plumbers and washed into the tank by the flow of water are obviously even more dangerous because of the greater mass of cathode metal involved (Fig. 19.9(c)).

Figure 19.9(d) illustrates the attack round particles of precipitate in an alloy; whether the particles be anodic or cathodic to the body

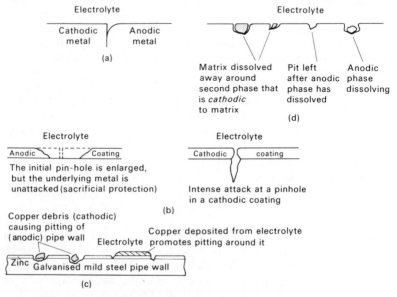

Figure 19.9 Examples of galvanic attack. (a) Simple galvanic attack at the junction of two dissimilar alloys. (b) Metallic coatings on metals: the pattern of corrosion at breaks in anodic and cathodic coatings. (c) Attack promoted by copper debris and precipitated copper film on a galvanized iron pipe. (d) Corrosion intensified by the presence of anodic and cathodic second phases in an alloy.

of the metal, attack is accelerated and the ultimate result is complete loss of the particles. If anodic, they are lost by direct solution; if cathodic, a crevice is eaten round them in the anodic matrix until they fall out. This form of attack is particularly liable on age-hardened alloys and accounts for the great drop in corrosion resistance of many aluminium alloys (e.g. duralumin) compared with the pure metal. Since precipitation occurs preferentially at grain boundaries, attack tends to be intercrystalline, especially in cases where extra dense precipitation leaves the matrix more anodic in these regions than in the body of the grains.

Pitting due to the combination of small anode and large cathode can also be dangerous in painted structures. On ships' hulls, for

584

example, severe pitting can occur from the cell: steel hull/sea/bronze propeller, the hull being the anode. Although the hull is very large compared to the propeller, the operative area of the anode is confined to scratches and imperfections in the paintwork and is in fact very small. Protection is usually effected by a sacrificial bar of zinc attached near the propeller which draws the current to itself and can be renewed when it has been eaten away. Similarly a copper pipe in a large steel tank may have a negligibly corrosive effect (providing that deposition of copper on the tank does not occur), but the same pipe in a painted steel tank might cause disastrous pitting. In this sense, painting can indirectly prove more dangerous than beneficial.

19.4.1   Crevice Corrosion

Crevices are an inevitable consequence of design in all engineering components where two sections are bolted or riveted together; they are also potent initiators of corrosion, and as such their location and protection must be given special attention in any assembly. Corrosion may commence from a variety of causes. To take the simplest first: crevices are not always easy to reach with paint unless the components are painted before assembly — even then scratches during assembling may break down protective measures at a crevice. Paintwork at the edges of a crevice (especially where these are sharp) is likely to be thinner and more easily damaged or worn away. There is thus a likelihood that protective measures will fail preferentially at crevices. Secondly, crevices always retain moisture for longer than flat surfaces, thereby allowing the corrosive reactions longer to attack the metal. Where the corrosion product is porous, moisture may be retained for even longer and, as build up of (say) rust occurs, high pressures may be set up in the crevice. This means the crevice may be opened up, allowing further and deeper penetration of moisture, and in extreme cases the rivets may even be fractured. Thirdly, conditions in a crevice are invariably favourable to differential aeration: the oxygen supply at the bottom of the crevice, whether completely immersed or merely moist, is scanty compared with that on the bulk surface of the metal, and thus the bottom of the crevice becomes anodic. The result is a typically intense large cathode-small anode attack which is all the more dangerous because (a) it may proceed unseen for a considerable time if the crevice is very narrow, and (b) normally inert metals and alloys which depend on an oxide film for their corrosion resistance, such as *aluminium, titanium and stainless steel*, are prone since the oxide film at the base of the crevice cannot be replenished easily if it is damaged. Crevices may also occur accidentally, as well as being a feature of design; overlying gaskets or washers, stones resting against immersed metal work, deposited grit in a circulating water system can all give rise to rapid attack, even against stainless steel. Note that the other half of the crevice need not be a conductor; it is the *geometry* rather than the second material of the crevice which is the cause of the trouble, although of course galvanic action between dissimilar metals at a crevice still further intensifies the attack. Some examples of crevice corrosion are shown in Fig. 19.10.

A slightly different form of corrosion may occur when chemically abnormal conditions are set up in a crevice. The classic example is the caustic cracking to which faulty boilers are susceptible, but it can occur in a variety of ways in chemical engineering plant. The origin of caustic cracking lies in the small amounts of sodium hydroxide which are present in most boiler waters. In very dilute solution this actually helps to prevent corrosion, but should there be a crevice with access to the outer atmosphere — as, for example, around a faulty rivet — water will penetrate and evaporate at the

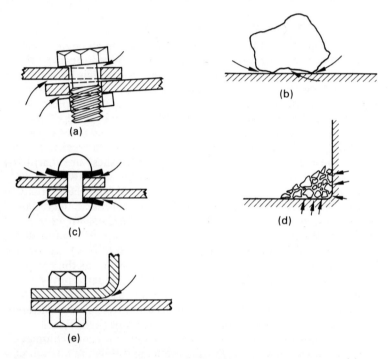

(a)

(b)

(c)

(d)

(e)

Figure 19.10   Some conditions that may give rise to crevice corrosion. (a) Badly aligned and fitted bolted or riveted joints. (b) Stone resting against immersed metal work. (c) Badly fitting washers or gaskets. (d) Grit or sand deposited in cisterns or pipes. (e) Flange joint. The arrows indicate possible crevices.

outer surface, leaving the sodium hydroxide behind. In this way, the concentration of sodium hydroxide rises until direct chemical attack on the boiler plates occurs, and complete failure may result. Localised concentrations of this nature may also occur under deposits in boilers and chemical engineering plant, but for serious results the presence of residual stresses in the metal is necessary and the effect is partly an example of stress corrosion (Section 19.7.1); badly aligned rivets are obviously ideal for this form of attack, combining a crevice with possibly severe localised stresses.

586

19.4.2 Passivity

Metals such as aluminium, chromium, titanium and zirconium naturally develop surface oxide films which confer remarkable resistance to further corrosion. All metals, however, can under certain conditions of wet corrosion become immune by the formation of oxide films; in this state the metal is said to be *passive*, and will remain unattacked indefinitely as long as the conditions remain unchanged. A well known example is the case of iron in concentrated nitric acid. After a slight initial reaction, during which the oxide film is formed, all attack ceases and will only recommence if the film is damaged in some way. Weakly alkaline conditions promote passivity in most metals, whereas acid conditions stimulate corrosion; the presence of chloride and sulphate ions is particularly damaging to passive films. In neutral solutions both results are possible simultaneously, and the result is usually a scattered galvanic attack. The balance between attack and passivity depends on the acidity of the corrodant and magnitude of any galvanic potential acting on the metal; charts plotting the relation between these variables have now been prepared for all engineering metals and are known as *Pourbaix diagrams*. An example is shown in Fig. 19.11. Metallic behaviour is mapped under three headings: immunity (attack is thermodynamically impossible); passivity (attack is prevented physically by a protective film); and corrosion (attack proceeds). These charts provide a valuable pointer to metallic behaviour but, as with the electropotential series, they are based on very carefully controlled experiments and must be used with great caution in making predictions about practical systems, in which many factors (e.g. crevices, the nature and location of any corrosion product, temperature fluctuations) may complicate the overall picture.

19.5 CORROSION OF BURIED METALS

It is not difficult to see that much of the corrosion which afflicts buried metal work is of the galvanic types already discussed, such as rusting, or at joints, or where stones or lumps of clay or chalk rest against the metal causing crevice corrosion. There are, however, certain particular varieties of corrosion to which buried metals are particularly prone, and the conditions under which corrosion commences also need comment. Galvanic cells causing severe rusting and pitting may be set up where a pipe passes from one soil to another, especially if one of the soils is acid or contains sulphides, or if one soil has a much higher oxygen content. In fact, from the principle of differential aeration, a varying oxygen content can cause galvanic corrosion even where only one soil is present. The electrical conductivity of the soil is also important; the higher the conductivity the further apart anodic and cathodic regions may be spaced and still give rise to corrosion. In highly conducting soils distances over a mile may occur. Buried pipe lines are also liable to attack from 'stray'

587

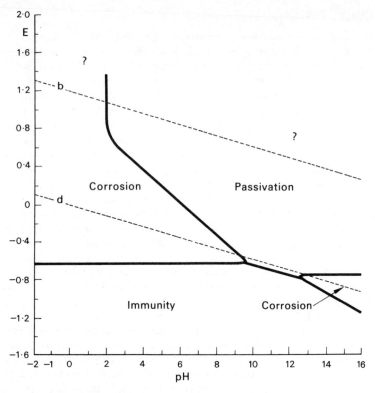

Figure 19.11   Simplified Pourbaix diagram for iron. (From U. R. Evans, *The Corrosion and Oxidation of Metals*, by permission of Edward Arnold (Publishers) Ltd.)

currents where they pass near to badly insulated DC electric rails on the surface. (AC currents may abet naturally occurring processes, but the effect has not been found to be serious.) The system of attack is shown in Fig. 19.12; anodic regions develop where a small amount of current leaves the rails, and where it leaves the pipes to return to the rails. Corrosion of the rails is relatively simple to detect, but the real danger is to the pipeline since the attack is concealed and extremely difficult to locate. Protection is normally carried out by the use of accessible sacrificial bars of zinc connected to the pipe with which good earth contact is ensured; these bars then become the anodes and can be replaced as they get exhausted. Alternatively, cathodic protection using an external e.m.f. may be employed (see Section 19.9.4). Problems of stray currents have diminished considerably with the passing of trams from the city scene.

Bacterial attack is also a possible source of trouble where pipes run through waterlogged clay; in addition, bacteria have caused damage to petrol tanks in which there is a trace of water. (More serious than the attack on the container, however, is the contamina-

588

tion of the contents with hydrogen sulphide which can lead to breakdown of engines.) The attack commences chemically, the iron
forming ferrous ions and liberating small quantities of hydrogen. In
the absence of bacteria, equilibrium is quickly attained and the
attack ceases while still negligible. The bacteria, however, live by
'catalysing' the reaction

$$SO_4^{2-} + 8H \longrightarrow S^{2-} + 4H_2O$$

thereby removing the hydrogen and preventing equilibrium from
being reached. This permits dissolution of the iron to continue,

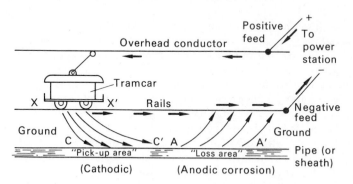

Figure 19.12 Corrosion of buried pipes by stray electric currents (the arrows
show the direction of movement of 'positive' electricity). (From U. R. Evans,
The Corrosion and Oxidation of Metals, by permission of Edward Arnold
(Publishers) Ltd.)

whilst the liberated sulphide ions react with the pipe to form ferrous
sulphide. Eventually the entire wall thickness may be transformed to
weak and porous ferrous sulphide, with resultant bursting of the
pipe. Protection is usually achieved by coating the pipe with tar or
bitumen containing about 30% of inorganic filler (asbestos, talc,
pumice) or by wrapping the pipe with some fabric impregnated with
tar or bitumen. Inorganic fabrics should be used, as organic fabrics
such as hessian may rot and the corrosion products serve to stimulate
the sulphate reducing bacteria.

19.6   CORROSION OF STEELS IN CONCRETE

This is a specialised field of particular interest to the civil engineer,
since steels are increasingly used in the stressed and unstressed state
in large-scale concrete structures. Fortunately concrete normally
offers an excellent protective environment; the alkali liberated
during the setting process serves to produce a protective film of
ferric oxide on the surface of the reinforcing rods. As always, however, it is perilous to assume therefrom that corrosion *cannot* occur
within concrete. Dangers arise if the concrete is porous; if the depth
of coverage of the reinforcing rods is insufficient; or if chloride ions

589

are present within the concrete. Each of these circumstances may result in the alkaline environment gradually changing to an acid one, so that immunity ceases; if the effect occurs at a few isolated points (perhaps in the vicinity of surface cracks) the attack may be localised and enhanced by the development of the dangerous large cathode/small anode combination.

Most concretes are porous to some degree and it is the porosity of the concrete that principally determines its susceptibility. A porous concrete allows slow penetration of atmospheric oxygen and carbon dioxide and (in industrial areas) sulphur dioxide; it is the latter two gases that bring about the change from an alkaline to an acid environment. This has little or no effect on the concrete itself, but the protective film on the reinforcing rods is broken down and rusting will ensue. Since oxygen tends to set up cathodic conditions whilst carbon dioxide and sulphur dioxide stimulate anodic behaviour, the likelihood is the setting up of both anodes and cathodes with consequent acceleration of attack. The formation of rust with its consequent volume expansion then generates powerful compressive stresses at the cement-reinforcement interface. If the concrete covering the reinforcing rod is sufficiently thick and sufficiently strong, it is just possible that the highly compressed rust may block further ingress of water and stifle corrosion. This is fortunate but rare. More usually the stresses are sufficient to cause cracks to develop in the concrete, permitting further ingress of water and acidic gases, and acceleration of attack. Ultimately the concrete may spall away from the reinforcing rod, leaving it completely exposed to atmospheric attack. Structurally, the result usually looks worse than it is; the rusting involved is not sufficient to impair the strength of the bars, and it has been remarked that the greatest immediate risk is of discomfort to passers-by from dropping fragments of cement. However, it is exceedingly unsightly, and if allowed to proceed will eventually lead to widespread loss of bond, serious structural weakening, and failure.

Much more serious consequences may accrue from the addition of calcium chloride to concrete to accelerate hardening. The effect is accelerated by the presence of oxygen and moisture and is therefore particularly pronounced in the vicinity of cracks, whilst the expansion resulting from the formation of the corrosion products will serve to open up the cracks. Further dangers arise in marine atmospheres, when cracked or porous cements can expose reinforcing bars to chloride attack even in the absence of calcium chloride. In harbour installations also, the water content of the cement varies considerably depending on whether it is above the splash level, between the tide marks, or permanently immersed. Regions with higher water content become relatively anodic, and galvanic corrosion currents are generated.

Prestressed concrete has a further element of risk in that if corrosion starts it will be accelerated by the high tensile stress in the steel rods (see below). Because of the potential severity of stress-corrosive attack, the use of calcium chloride in prestressed concretes is not permitted.

Where there is the risk of attack by sodium chloride (e.g. in marine structures or bridge decks where salt may be laid down in icy conditions) the use of galvanized prestressing bars has been recommended; tests indicate that the steel-concrete bond is not impaired by zinc coating. Any imperfections in the coating arising from cutting or bending the bars on site should be remedied with zinc paint, or there is a risk of severe local attack.

Undoubtedly the best and simplest way of avoiding the trouble is to ensure that the concrete is dense, well cured and free from chloride ions; half an inch of high-quality concrete is preferable to six inches of porous mix.

19.7 COMBINED ACTION OF CORROSION AND STRESS

The combination of corrosion and stress, either as a steady stress or as a fatigue stress, is of great importance since the two factors acting together may cause very much more damage than the sum of their effects acting separately. Not all alloys are susceptible to stress corrosion cracking, but all are vulnerable to corrosion fatigue under which conditions no alloy shows a fatigue limit.

19.7.1 Stress Corrosion

The visible result of stress corrosion (if the metal is sectioned and examined under a microscope) is the spreading of cracks across the specimen roughly at right angles to the axis of stress (Figs. 19.13 and 19.14). The cracks may follow transgranular or intergranular paths depending on the alloy and corrodant concerned, and the exact mechanism of crack propagation is still doubtful in many cases. Crack propagation may be principally mechanical with the corrodant merely serving to aid failure at the tip of the crack (Al–Zn alloys) or it may be a matter of electrochemical corrosion with the stress acting to open the crack so formed and allow free passage of the corrodant to the crack tip (stainless steels); in many instances corrosive and mechanical mechanisms have a roughly equal effect on the spread of the cracks. The presence of high-density regions of dislocations has also been shown to have a considerable effect on crack propagation. They may act by causing sudden brittle failure over short distances as the crack grows or lead to an intensification of electrochemical action as the crack reaches them. The presence of impurities or a precipitated phase in the grain boundaries can render an alloy susceptible to stress corrosion cracking; the effect is particularly marked if, as a result of precipitation, the grain boundaries become anodic to the main body of the grains or if regions of residual stress are set up around the precipitate. Examples of this include the age-hardening Al–Cu alloys, and the Al–Mg alloys which are not susceptible as normally fabricated but which can become sensitised by grain-boundary precipitation on heating. The

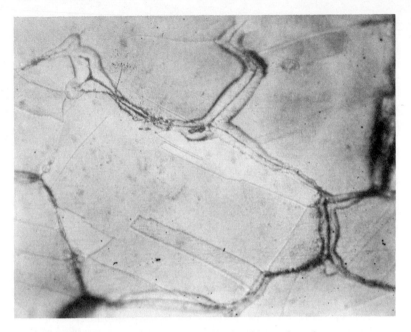

Figure 19.13 Intergranular stress-corrosion cracking in brass by mercurous nitrate (× 450).

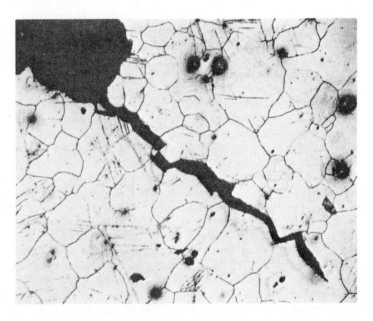

Figure 19.14 Transgranular stress-corrosion cracking in a magnesium alloy (× 200). (From A. R. Bailey (1967) *The Role of Microstructure in Metals*, by permission of Metallurgical Services Ltd.)

592

effect is worse as the magnesium content increases, and with amounts in excess of 9% can occur even on exposure to the sun; the 5% Mg–Al alloy (which is used for aircraft rivets) is normally safe but may become dangerously susceptible after prolonged exposure in the tropics. Other well known examples of stress corrosion are caustic cracking in boilers (Section 19.4), weld decay of stainless steels (Section 16.4.3) and season cracking in brasses. The latter occurs as a result of intergranular corrosion in atmospheres containing ammonia and ammonium salts in the presence of residual tensile stresses remaining in the brass after fabrication. Careful low-temperature stress-relief annealing at about 180–200 °C is usually sufficient to remove the dangerous residual stresses without affecting work hardness. Steels may suffer from stress corrosion cracking in concrete if there is appreciable (>1%) chloride present, and also in hot aqueous nitrate solutions; the degree of susceptibility, as in the age-hardening light alloys, depends strongly on the state of heat treatment. Any form of grain-boundary precipitation (e.g. of carbides after long anneals) or residual stress (as in tempered steels) greatly increases the likelihood of attack.

19.7.2 Corrosion Fatigue

Under conditions of corrosion fatigue, where no fatigue limit exists, the resistance of alloys is specified as the *endurance limit*, which gives the highest loading that can be supported for a given number of cycles (Fig. 19.15). The mechanism of corrosion fatigue would appear to be fairly simple in its essentials. The alternating stresses cause rupture at a few places in the protective film on the surface of the metal; pitting by ordinary galvanic corrosion occurs at these points giving rise to stress raisers, and thereafter the growth of the cracks occurs principally by ordinary fatigue mechanisms (see Section 15.3), although aided to some extent by corrosion at the tip of the crack. In contrast to ordinary fatigue a great many cracks grow in the initial stages of the attack, which usually gives the fracture surface a more facetted appearance. Methods of prevention are based on either exclusion of the corrosive agent or cathodic protection (combined with any of the standard measures to prevent the onset of fatigue). One interesting point arises: apparently the multiplicity of cracks in the early stages of attack to some extent help to spread the load and slow penetration down. It follows that, if protection is applied imperfectly or applied after one or two pits have had time to form, breakdown may occur more rapidly than in the complete absence of protective measures. Thus protection must be applied *before* components go into service and be rigorously maintained. In the civil engineering field, any coastal, harbour or marine installations suffering impressed vibrations are particularly at risk.

19.8 CORROSION UNDER MOVING LIQUIDS

Although much of the corrosion resulting from immersion in

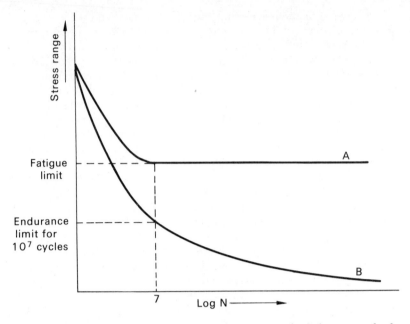

Figure 19.15 The effect of a corrosive environment on the fatigue strength of steel. Curve A indicates fatigue response in the absence of corrosion; Curve B shows the result of corrosion attack and the elimination of a safe stress range for the steel.

dynamic conditions is due to effects previously discussed (e.g. galvanic corrosion and crevice corrosion), there are some special features which must be borne in mind under these conditions. On the credit side, the tendency to crevice corrosion is lessened (although not stifled) since the flow of water helps to keep crevices supplied with oxygen and in general reduces the difference between conditions inside and outside the crevice. (This does not necessarily apply if fouling of the metal surface takes place; conditions may then approach stagnancy in regions close to the fouling and crevice corrosion can proceed unchecked.) The commonest sources of damage are due to the impingement of bubbles or solid matter in suspension, and the result is twofold. The constant impingement of solid matter, which usually occurs at exactly the same spot, may abrade the surface so that there is a slow physical removal of matter; more damaging, however, is the fact that impingement of sand or bubbles prevents the retention of any protective film on the surface of metal. Thus there is a strong probability that points of impingement become anodic with respect to the rest of the tube walls, giving rise to large cathode—small anode conditions and resulting in severe pitting attack. Further, the flow of liquid usually prevents the build up of any corrosion products at the site of attack which might otherwise stifle the reaction.

It has been noted that large bubbles, breaking up into smaller bubbles as they strike the metal surface, are very much more

594

damaging than small bubbles; this is probably due to the turbulence and swirl accompanying the break-up, which has a greater effect in prising loose any protective film. The design of pumps, inlets and sluices in civil engineering works should therefore aim to minimise entrapped air, and the degree of which this is successful may play a major part in determining the life of the associated piping. Erosive damage in its most extreme form may occur when slurries or dry powders (e.g. Portland cement) consisting of abrasive particles are pumped through pipelines. In these circumstances physical attack may be extremely rapid; sharp corners in the pipeline must be avoided and, if necessary, the pipe walls should be reinforced at bends.

Cavitation erosion is a more severe variety of bubble attack. It occurs under conditions of turbulence when low-pressure bubbles are formed in the liquid and collapse against the metal surface; the suction side of propellers and turbine blades are prone to attack. The collapse of these low-pressure bubbles is quite violent, exerting impact loads up to 135 MN/m^2 and sending out shock waves which may damage metal near to as well as at the point of collapse, and of course completely prevent the formation of any protective film. Once pitting has commenced, the attack may continue at least in part by fatigue corrosion, with cracks spreading into the material from the bottom of the erosion pits. Thus, while the damage is initially the result of bubble 'hammering', it becomes increasingly dependent on galvanic action, as outlined in Section 19.7.2. Under these conditions penetration, which takes the form of a honeycomb effect, may be very rapid indeed; cast iron half an inch thick has been perforated in 300 h in a diesel engine, and there have been numerous reports of damage to propeller blades.

Two further forms of attack associated with water pipes must be mentioned, although they do not in fact result from the motion of the water. Copper pipe, immediately after forming, will contain a thin film of lubricant on its surface. If this is not completely removed prior to annealing, the heat will cause it to char and a carbonaceous film is left which acts as a cathode with respect to the underlying metal. Intense attack of the large cathode—small anode variety may then occur at any breaks in the film (which will crack when the pipe is bent and may suffer erosive attack), and pitting and perforation will ensure. Once under way, the attack becomes more a form of crevice corrosion, which will be aggravated by the deposition of voluminous copper carbonate corrosion products that cause conditions to become virtually stagnant in the vicinity of the crevice. Multiple pinhole perforation has occurred from this cause within three years of pipework installation. Adequate cleansing by the manufacturers is the only real answer to this problem; alternatively, hard-drawn pipe may be used which, being unannealed, cannot have a carbonaceous film. This last solution, however, requires the use of separate fittings for all but the gentlest of bends, and may prove somewhat cumbersome and expensive.

Further problems in circulating systems have arisen in power stations, which involve the circulation through condensers of large quantities of cooling water. For a modern station the requirement may be of the order of 50×10^6 gallons/h (the equivalent of an hourly discharge of a 50 mile long queue of 4 000 gallon tanker lorries); because of this, many modern installations are situated to draw their cooling water from estuaries of the sea. Condenser tubing must thus withstand circulation of sea or estuarine waters, and the highest reliability is called for: a 0.1 mm diameter perforation in the tubing will lead a shut-down within 48 h and entail a heavy financial loss while the failure is located and replaced. Where the sea water is clean, the use of aluminium brass (approx. 77Cu/21Zn/2Al) has proved remarkably effective. The aluminium appears to repair a protective film sufficiently rapidly to prevent the onset of galvanic attack. The alloy is, however, vulnerable to sand erosion, and for this reason cooling water should be rigorously filtered before entering the system.

If the water is contaminated (as is frequently the case with estuarine or harbour sources) by the products of bacterial action and organic decay, hydrogen sulphide and cystine may both be present. Both these substances may promote rapid attack on condenser tubing: hydrogen sulphide by direct formation of a cathodic sulphide film which concentrates attack at points where the film cannot form, and the cystine by catalysing galvanic action so that it can continue in the absence of oxygen. Copper nickel alloy (70Cu/30Ni) has proved the best material in this aggressive environment; it is expensive but justified in view of the overriding economic need for reliability. Additional resistance has been conferred by injecting ferrous sulphate into the cooling water intake, which results in the deposition of a protective film of $FeO \cdot OH$ on the surface of the tubing.

19.9   TECHNIQUES OF PROTECTION

19.9.1   Introduction

We must now consider how the underlying principles of corrosion may be applied to protect metals against its insidious attack. At the outset the *economic* importance of suitable protective measures must be emphasised, since this is an inseparable part of the national bill for corrosion, and a part which might well be reduced if more attention were paid to protective measures at the earliest stages of project design. Frequently the extra expense in producing a well protected structure is more than offset by the reduction in maintenance costs and increase in working life. Even where there is not direct monetary advantage, the additional cost of good protection is necessary if corrosion is liable to reduce safety factors (e.g. in the aircraft industry or nuclear reactors), or become a hazard to health (as in the food industry). It may also be justified if it is necessary to preserve a pleasing surface finish on an article; many car owners must have felt at some time that more intelligent design against corrosion would be well worth a modest increase in the initial price of the vehicle.

In parallel to that argument, it is often true that a more costly maintenance specification may save money if it increases the interval between maintenance operations. Put in simple terms, if the annual upkeep cost of a structure is £c, then any additional sum up to £c spent on a superior maintenance specification will result in an overall saving if it extends the maintenance period by a year. To give some notional figures, let us say that upkeep costs comprise labour (l) and materials and ancillary costs (m), and maintenance is carried out every five years. Then $l + m = 5c$ and Δm, the maximum permissible increase in materials costs, is given by $\Delta m = (1 + m)/5$. If the cost of labour is ten times the cost of the materials and ancillaries, then $\Delta m = 11m/5$ and a specification costing *three* times as much as the initial one would save money if it increases the maintenance period to six years. Such a calculation is valuable in demonstrating that 'penny-wise' may indeed be 'pound-foolish' in the economics of protection; however, it takes no account of a number of variables, and in practice each case must be decided on its own merits. For example, the maintenance period of any structure is very dependent on location: whether it is in the country, or in an industrial setting, or exposed to marine atmospheres. Also it is unlikely that a more exacting specification whilst raising m would leave l unchanged; it might require the application, say, of three extra coats of paint and a more lengthy preparative treatment, all of which would increase labour costs. However, the underlying principle is valid and points to the necessity of considering suitable protective schemes at the design stage, since the practicality of various schemes may well be influenced by the design itself.

19.9.2 Corrosion, Design and Commonsense

Before considering how scientific principles can be applied to combat corrosion, it is pertinent to examine the ways in which design and procedure can make a contribution. This involves the application less of science than of commonsense, which is a cheaper and frequently more effective technique if available. The following points refer in the main to structural steelwork and may seem so simple as to be virtually platitudes. They can, however, contribute as much to the satisfactory life of a structure as the most scientific techniques of protection (which is not to say that scientific techniques are superfluous).

It should always be remembered that moisture is the essential pre-requisite for corrosion at ambient temperatures; dry surfaces do not corrode. Therefore the surfaces of structures should be exposed as little as possible to moisture and arranged to dry out quickly after wetting. Protective measures must be applied most rigorously in those areas which cannot dry rapidly. In practice, all surfaces are at risk from moisture; vertical surfaces suffer 'run-off'; flat surfaces retain moisture on their upper side and can attract dew or condensa-

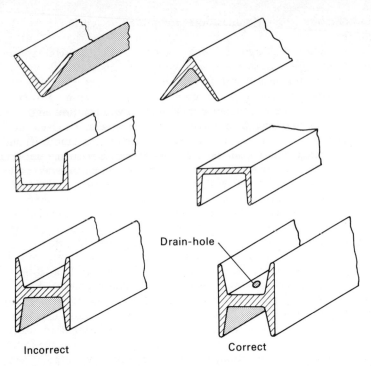

Drain-hole

Incorrect   Correct

Figure 19.16   Arrangement of structural sections to avoid water retention.

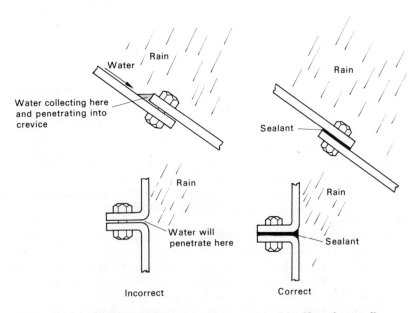

Rain

Water

Water collecting here
and penetrating into
crevice

Rain

Sealant

Rain

Water will
penetrate here

Rain

Sealant

Incorrect   Correct

Figure 19.17   Design of joints to minimise corrosion risks. Note that small
relative movements at joints can quickly abrade away any protective coatings of
primer or paint.

tion droplets on their underside; north-facing surfaces and surfaces screened from sunlight will dry off more slowly. All these are inevitable consequences of normal exposure; however, the situation can be improved if V, H and channel sections can be positioned to avoid becoming water-traps (Fig. 19.16). Where a channel cannot be so positioned, a drainage hole is a simple and effective expedient provided that it is mechanically acceptable. Overlaps and joints should be arranged to avoid the formation of water channels; the danger here is that retention of water (Fig. 19.17) can all too easily initiate crevice corrosion. Ideally such points should be blocked off with a sealant; at the least they should be meticulously painted. If nonmetallic materials are used as gaskets in joints (e.g. to prevent fretting) they should not overlap the joint or crevice corrosion may ensue (Fig. 19.18); aluminium alloys are particularly prone to this

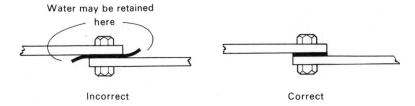

Figure 19.18   The use of gaskets or washers at joints: the gasket must be cut to the correct size or there is a risk of crevice corrosion.

form of attack. Porous material used for soundproofing or mechanical damping should not be in contact with metal in positions where it can take up water, since the water will be retained in contact with the metal long after the rest of the surface has dried out, and will provide a near-permanent active corrosion site. Dirt, accumulating on flat surfaces, in channels, crevices and corners, will act in an exactly similar fashion, retaining moisture and preserving an aggressive environment. Regular cleaning of all such danger spots (particularly in industrial areas where dirt may contain an appreciable content of acid-retaining soot and sulphate ions) will therefore contribute to the life of the structure; all dirt of course must be rigorously cleaned off prior to repainting.

Condensation poses a problem which is best dealt with at the design stage, since it can collect on surfaces not directly exposed to rainfall, corroding them and also trickling down sloping surfaces and collecting on ledges which might be thought immune from the danger of retaining water. It is also a danger where waste heat extraction from boiler exhausts may lower the temperature of the exhaust gases to the dew-point, when a highly corrosive condensate can form and trickle down the inside surface of the flue. The interiors of totally enclosed structural members, such as box girders, can suffer from condensation and should be sealed when the humidity is low; the danger here is that such attack is invisible and inaccessible. Where small holes connect with the outer air such members will 'breath' at night, sucking in cooler air which leads to internal condensation and corrosion. This danger is, if anything, greater in recent years, as stress-analytical techniques have become more sophisticated and engineers use thinner sections in consequence.

One of the commonest causes of trouble 'built in' to structures arises from the unprotected storage of materials on site. Structural steelwork may well be left unprotected for a period if the intention is to apply protective measures after erection, an approach which has some logic in view of the difficulties of erection without damaging any existing protective coating. Unfortunately, if rusting once commences, it is virtually impossible to cleanse the steel sufficiently for a subsequent coating to adhere satisfactorily. With the techniques available on site, some rust is inevitably left at the bottom of the pits that the rusting has dug, and this will contain sufficient active corrosive agents (analysis shows that a typical rust may contain 4% sulphate) to continue working under the paint until blistering and failure of paintwork occurs and rusting is free to continue unhindered. Thus will careless procedure on site lead to either structural weakness or greatly increased maintenance costs.

The only real remedy for this state of affairs is never to let rusting start. Ideally, steelwork should be well protected with priming costs as an integral part of the manufacturing process. Primers may then be applied under suitably controlled conditions (e.g. warmth and low humidity, which cannot be guaranteed when painting is done on site). It should then be handled *carefully* at all times until erection is completed, and any damage made good as quickly as possible, for example on cut edges and at welds or rivet holes. If steelwork must be stored for any length of time on site, it should be kept clear of the ground and well protected from dirt and moisture, but not stored under impermeable plastic covers which are liable to generate considerable condensation.

Finally, structures require maintenance; therefore good design should permit this to be carried out as easily as possible. Protective coatings can be readily applied in paint shops or on site at ground level, but this may not be the case by any means when the steelwork is incorporated into a structure, and it is very important that adequate access for maintenance should be allowed for at the design stage. Awkward and concealed corners and surfaces are very apt to receive scant protection. At ground level steelwork is best set in concrete or at least given a thick bituminous coating for at least a foot above the earth; this will protect the steel against attack by moisture from the ground — or stray dogs — and enables repainting to be done to the limit of the exposed metal.

19.9.3   Inhibitors

For galvanic corrosion to proceed freely it is necessary for both the anodic and cathodic regions to have good electrical contact with the electrolyte. If the corrosion products at either region are insoluble or sparingly soluble, contact is quickly lost and the process tends to be self-stifling. Inhibitors are chemicals introduced into the system which will provide an insoluble precipitate at anodic or cathodic regions in the event of a galvanic corrosion cell being set up. They may be added to fluids in cooling systems or applied in the form of a pretreatment to metal surfaces prior to painting, or they may be present as an inhibitive pigment in paints. When added to cooling systems, the inhibitors used are naturally soluble (zinc sulphate, potassium chromate, sodium phosphate, sodium benzoate or sodium nitrite), but when used in the solid form as in priming paints or pre-treatments, near-insoluble inhibitors are used. Red lead is the best known of these and in many ways is still the most effective, with a history of use dating back to Roman times, and calcium plumbate (equally effective but rather less heavy) has been increasingly used in recent years. Now, however, lead primers are being superseded by other pigments; not because they are less efficient but because of the health hazards involved in spraying or in welding metal that has already a coat of lead primer. Zinc chromate, zinc phosphate and metallic zinc powder are perhaps the best of the nontoxic inhibitive coatings, although the first-named is rather soluble and vulnerable to leaching by rain water unless protected with a good top-coat of paint.

Inhibitors are normally classified as *anodic* or *cathodic* according to the region at which they react; a few, such as red lead and zinc chromate, react at both the anode and the cathode. In general cathodic inhibitors are more foolproof in use than anodic inhibitors since the amount added to the system is less critical. If too little cathodic inhibitor is added, the effective area of the cathode is diminished, less current flows between the anode and the cathode, and the result is a slowing-up of corrosion in the anodic regions; thus, although the situation is not ideal, no great harm results from inadequate inhibition. If, however, too little anodic inhibitor is added, the effective anodic area is decreased whilst that of the cathode remains the same or increases if the inhibited surface can also act as the cathode. The result is the large cathode—small anode combination discussed earlier with localised intensification of the attack and pitting, and (if the cathodic area is enlarged) an increase in the total amount of metal corroded. The effect gets more and more intense until the inhibitor concentration is sufficient to block off all the anodic area, when the attack suddenly ceases. Figure 19.19 illustrates this point. It should be noted that, even after a mistake of this nature has been rectified and the correct amount of anodic inhibitor added, the pits remaining may continue to cause trouble by setting up conditions for crevice corrosion.

The chemical reactions resulting in cathodic inhibition are normally straightforward precipitations which block off the cathodic area from the electrolyte. Zinc salts precipitate zinc hydroxide (on steel surfaces this contains an appreciable amount of iron from the anodic reaction product); magnesium salts precipitate magnesium hydroxide, and calcium bicarbonate (present in many hard waters) precipitates calcium carbonate. Anodic inhibitors are not so simple

601

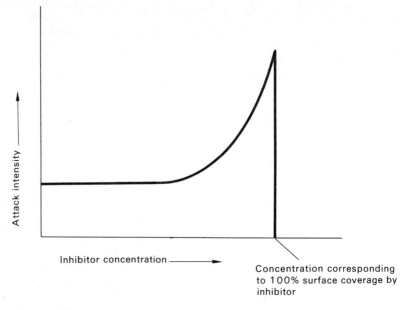

Figure 19.19   Intensification of corrosion attack in the presence of insufficient inhibitor. Note that the *area* of attack diminishes with increasing additions of inhibitor.

in their behaviour. The formation of an insoluble iron salt might be expected (e.g. iron phosphate or iron chromate) but in fact the protective film (which is usually invisibly thin) has been shown to consist primarily of ferric oxide containing only small pockets of phosphate or chromate. It is probable that the inhibitor stimulates the formation of the protective oxide film, i.e. it encourages passivation of anodic regions, and it then plugs any gaps or pores in the film by precipitation as soon as the anodic reaction commences. Dissolved oxygen alone in liquids has a pronounced inhibitive effect as might be expected and the action of anodic inhibitors may be considered akin to a 'catalytic' effect promoting the formation of an oxide film under conditions where the oxygen concentration is insufficient to be effective by itself.

Not all anions act as anodic inhibitors; chloride and sulphate ions accelerate corrosion strongly in aqueous solutions. Oxidising anions, such as the chromate and nitrate anions, promote the formation of the protective oxide film and are used as inhibitors to prevent aqueous corrosion in closed or semiclosed systems. Phosphates act partly by the formation of sparingly soluble salts as indicated above, but the effectiveness — or otherwise — of the inhibiting ion varies considerably with the pH of the liquid. In general, acid solutions tend to favour corrosion and alkaline solutions to favour inhibition, and the balance may be extremely delicate. One of the uses of Pourbaix diagrams is to give information on the correct conditions for adequate inhibition.

Finally, there are the organic inhibitors; some, such as sodium benzoate and cinnamate, probably act by a dynamic adsorption on the metal surface. That is to say, the organic ions are adsorbed onto sites on the metal surface, but are constantly changing sites. Thus only a small portion of the surface needs to have an oxide film at any one moment, which may be supplied by a low concentration of oxygen. The advantage of these inhibitors is that because of their dynamic action no part of the surface is available long enough at a time for serious pitting to commence. Hence, even when there is not quite enough inhibitor present the result is merely a mild form of overall attack rather than the intense pitting which can occur with anodic inhibitors. Oily fluids which form emulsions with water are also used; the negatively charged oil particles are naturally attracted to anodic areas and are deposited or absorbed there as an oily film. In complex cooling systems, such as those used in cars where a variety of metals and solders are encountered, it is usually necessary to employ a combination of inhibitors to prevent corrosion by anti-freeze agents. Sodium benzoate-sodium nitrite and triethanolamine phosphate-sodium mercaptobenzthiazole (TEP + NaMBY) have been used with success.

Inhibitors are also used in pickling baths to prevent acid attack on areas of metal where the oxide has already been removed. They are organic compounds based on thiourea or toluidine, which adsorb strongly onto metal but not onto oxide, giving a monomolecular protective film. This cuts down hydrogen evolution and lessens the risk of hydrogen embrittlement by absorption of nascent hydrogen into the metal.

19.9.4 Cathodic Protection

The principle of cathodic protection is widely used to protect buried or immersed metalwork; one method of applying it (sacrificial protection) has already been considered (Section 19.4), but other methods may be used. Where the metalwork has an inert coating applied to it (e.g. paint, bitumen or tar), some form of cathodic protection is often used to protect the metal at any flaws in the coating. There are two ways of achieving this.

(a) Sacrificial anodes may be placed some distance from the metal and connected to it by wire (Fig. 19.20). Such anodes are normally either a 91Mg/6Al/3Zn alloy or high-purity zinc. The absence of any cathodic impurities (iron, copper, nickel, tin) is important if excessive wastage of the anodes is to be avoided.

(b) Instead of using a system which generates its own e.m.f. as in (a), an external power source can be applied to the system such that the metal becomes cathodic to its surroundings (Fig. 19.21). The result of the *impressed current* is to shift the position of the metal in the appropriate Pourbaix diagram from the vulnerable to the immune region. In this case inert anodes are frequently used; platinum, silicon, iron, titanium. lead, stainless steel and graphite have been employed, although graphite anodes may undergo slow wastage

603

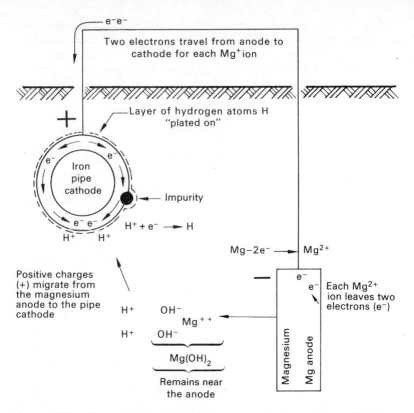

Figure 19.20   The principle of cathodic protection by means of a sacrificial anode. (From L. M. Applegate, *Cathodic Protection*. Copyright 1960 McGraw-Hill. Used with permission of McGraw-Hill Book Company.)

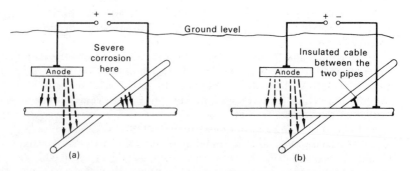

Figure 19.21   Impressed current cathodic protection. (a) Showing the danger of secondary corrosion circuits when several buried metal components are within range of the anode. The danger area depends in part on the conductivity of the soil and may therefore vary seasonally. (b) Secondary corrosion circuit eliminated by making good electrical contact between the pipes.

604

and eventually need replacement. In many cases the cost of the impressed power supply is less than the cost of replacement of sacrificial anodes, but not all corrosion sites have an available source of power. There is also the possibility of hydrogen evolution at the cathode:

$$2H^+ + 2e^- \longrightarrow 2H \longrightarrow H_2$$

in which case some of the atomic hydrogen may diffuse into the metal and embrittle it. Sometimes both sacrificial and impressed current techniques are used in combination, whereby cheap scrap iron sacrificial anodes have an external e.m.f. applied. The power requirements are less than for graphite anodes, whilst the cost of replacement is less than that for magnesium or zinc anodes. Metals to be protected by either technique are normally given some form of protective coating as well; this greatly reduces the necessary power since protection is then only active at imperfections in the coating.

When protecting immersed metalwork in hard-water areas or sea water the application of a high initial current density will assist by laying down a cathodic film of calcium carbonate at any gaps in the protective coating, thus affording an additional barrier to corrosion. Careful shaping and distribution of the anodes to maintain an adequate protective potential over the entire structure is important for economic operation. If some areas are remote from the anodes, the necessary potential at these places can only be maintained by applying a very much greater potential than is necessary in the areas closest to the anodes. This is wasteful of both power and anode material (if the anodes are of scrap iron) and may contribute to a further danger if other buried metalwork is nearby. The higher the protective potential, the greater is the risk that some of the power may be diverted to neighbouring metals and initiate stray current corrosion there.

Cathodic protection has been widely used for marine environments, both on ships' hulls and harbour installations; the technique has received a further impetus from the need for the highest quality of protection, on offshore oil rigs. For many years jetties were protected anodically by an external e.m.f. at a current density of about 10 mA/ft^2 by means of an anode or anodes on the sea bed. Now, however, with the advent of super-tankers displacing 300 000 tons and over, jetties have had to be built further out, exposing them to more severe buffeting from wind and waves; they must also withstand accidental impact from tankers if mooring is to be done in awkward winds or tides. Jetties are therefore much larger and more complex in structure, so that the relatively simple former system is no longer adequate. Multiple anodes are now used, shielded and clamped to the structure at suitable points; with improvements in the application of protective coatings the average level of current density to give protection is now about 2 mA/ft^2, and the use of multiple anodes permits different levels of protection to be maintained in different areas, according to the particular needs of the environment.

Similar problems are encountered on off-shore oil rigs, where both sacrificial anodes and external e.m.f. techniques have been used, sometimes in combination. Zinc is an approved sacrificial metal, but the sheer amount needed to confer adequate protection constitutes a problem. Some 300—400 tons may be necessary; not only does this involve a very considerable investment (high purity zinc at £1 000 per ton — 1974 price — is necessary), but the increase in total weight of the structure and impedance to tidal currents may mean a loss of up to 700 tons top weight. Because of these disadvantages, impressed-current protective systems are coming into favour, although these require continuous monitoring, and there is the possibility, referred to above, of hydrogen embrittlement.

Sacrificial anodes must, of course, be inspected at intervals and renewed where necessary; impressed-current anodes should in theory be permanent, but in practice should also be inspected periodically for damage. Over-enthusiastic maintenance may result in a thick overlay of paint; it is possible to clean paint off a new anode, but old anodes inevitably become pitted and then paint removal becomes a virtual impossibility. Anodes on jetties and harbour structures may suffer mechanical damage from ships or detritus being flung against them by the sea. Welders, attracted by the metallic surface of an anode, may use it for the purpose of striking an arc. There is also the possibility of damage to the cables carrying the power supply.

Inspection of Fig. 19.11 will show that protection of a metal may be achieved by raising its potential into the passivity zone instead of lowering it into the immunity zone; this technique is known as *anodic protection*. It has proved valuable in the protection of metals against corrosive acids, but can only be applied to those metals which *passivate* (i.e. develop an inert film, usually oxide) when made the anode of a cell at a sufficiently high current density. For this reason it is limited to iron, aluminium, nickel, chromium and titanium of the commoner metals and is primarily of use to chemical rather than civil engineers.

19.9.5   Pretreatment of Metals for Coatings

Before considering the various protective coatings which may be applied to metals it is necessary to stress the importance of adequate cleansing of the metal surface. Slovenly preparation may show quickly (as in galvanising or electroplating) or slowly (as frequently in painting) but in the long run the quality of preparation makes or mars the quality of protection as much as the nature of the coating itself.

The major sources of trouble on metal surfaces are grease and oil films, rust and mill scale on steel surfaces, and moisture (where paint is to be applied). Moisture is perhaps the commonest of the three, together with rust; it is also fairly simple to remove, but on large structures its avoidance is largely a matter of common sense. Ideally, paint should be applied when the temperature is above

40 °F and the relative humidity below 86%; therefore wherever possible, especially in the winter months, painting should be done indoors, whilst outdoor painting should be done in dry weather and preferably the latter part of the day to allow the metal to warm up and any dew to evaporate. These points are most important when applying undercoat or primer, but should be given consideration for all coats of paint.

Millscale, the relatively thick adherent layer of oxide on hot-rolled steel, and rust have always been a problem. In theory a perfect overall layer of scale might provide a reasonable base for painting; in practice, however, the brittle nature of millscale leads to cracks and imperfections so that the only safe course is its entire removal. In the past, weathering the steel was employed to remove the scale; rust forming at cracks worked under the scale and lifted it off. However, the method, if cheap, was slow and presented a rather doubtful surface when finished. Removal of both scale and rust is now effected mechanically, chemically or by heat. Mechanical methods include chipping, wire brushing (either with hand brushes or power driven rotary brushes) and sand, grit or shot blasting. Except for small areas, hand methods are impracticable and wire brushes, whatever the motive power, tend to polish rather than remove tightly adherent millscale. They are more suitable for removing rust. All the blasting methods are effective in removing both scale and rust, but sand blasting can give rise to silicosis in workers, which makes shot and grit blasting preferable. Small sizes of grit are preferable to prevent excessive roughening of the surface which might make complete coverage by paint difficult to achieve. 'Grit' in this context may mean alumina particles, but is more often powdered white cast iron. Recently, wet grit blasting and recirculating blast systems (in which the grit is recovered by suction after hitting the surface and re-circulated) have been coming into favour since they reduce health hazards and dust. A further advantage of wet blasting is that chromates or phosphates can be added to the water, giving an immediate protective coating on the steel surface and reducing the risk of rusting prior to painting.

Flame cleaning is also employed quite extensively. Oxyacetylene flames are passed over the surface of the metal, whereupon differential expansion of scale and steel causes the former to flake off. Rust patches are also dried out, making for easier removal by brushing, and the procedure leaves a warm dry surface which is advantageous if the undercoat can be applied straight away. Care must be taken, however, not to let the intense heat of the flames cause sintering of the scale, or even melting; for this reason the flames must be kept moving. Flame cleaning does not give as clean a finish as grit blasting at its best, but the equipment is more mobile and can be used on structures during erection, and for small areas which might be inaccessible to grit blasting.

Chemical methods of scale removal are based on acid *pickling*. 5–10% sulphuric acid at 70 °C, or 5–15% hydrochloric acid at 35 °C are the commonest solutions, with an adequate amount of restrainer present to minimise the risk of hydrogen embrittlement and also to cut down acid spray (caused by hydrogen evolution). Pickling cleans the metal surface well, and is very suitable as a pretreatment for electroplating or hot dipping. Any trace of salts left on the metal surface after removal from the pickling bath must be vigorously removed or rapid rusting will result, with the speedy removal of any paint or plating. Rinsing with weak alkali after pickling removes the worst danger of rusting, but traces of alkali left after rinsing can soften and loosen a coat of paint applied over them. The Footner Process gives the best results, in which the steel is pickled in sulphuric acid, rinsed, and given an immersion in hot (85 °C) 2% phosphoric acid for about 5 min. A film of iron phosphate is thus left on the surface, enhancing the protection given by the paint, and once the warm metal has dried it is in the optimum condition for applying the undercoat. For the removal of rust in the absence of heavy scale, e.g. on car bodies, 10—25% phosphoric acid alone at 65 °C is used, together with some restrainer to prevent excessive attack on any patches of bare metal.

19.9.6   Metallic Coatings

Protective metallic coatings can be applied in two ways: by dry, or diffusion, coatings in which article is dipped, sprayed, clad or heated in a powder of the protective metal, and by wet, electrodeposited coatings. The protection they afford may be either sacrificial (if the coating is anodic to the protected metal) or purely physical (if the coating is cathodic to the protected metal). The dangers arising from imperfections in a cathodic coating have already been discussed in Section 19.4, but even so cathodic coatings are used in some instances — chromium plating on steel being one example. In general terms, the advantages of metallic coatings are that they are strong, and insensitive to light and moderate heat; for the most part they are more wear resistant than paints, and amenable to soldering. On the other hand, if corrosion once starts it will proceed faster because of the galvanic conditions which are set up and because it is not possible to impregnate metallic coatings with the inhibitive compounds that are present in paints. The equipment for applying metallic coatings is more elaborate than that for painting, which puts up costs and may make it difficult to replace coatings which have failed on structures — as compared with the relatively straightforward task of repainting.

19.9.6.1   *Hot Dipping*

In this technique, the article to be coated is lowered into a bath of the molten protective metal and withdrawn carrying a film of the protective metal on its surface. The commonest instances are coating steel with zinc (galvanising) and tin (tin plate). Adhesion results from alloying at the interface and as usual is dependent on complete cleanliness of the metal surface. To this end, after the standard cleansing procedures have been carried out, the article is passed through a layer of flux which is maintained on the surface of the molten dip. However, since some of the alloy phases developed are brittle, the time of immersion must be kept to the minimum necessary for an adequate thickness of coating, in order to prevent excessive growth of brittle intermetallic phases. The addition of a very small amount of aluminium to the molten zinc has been found beneficial in suppressing diffusion of zinc into the steel and thereby maintaining a reasonably ductile coating. Galvanised steel is widely used for sheeting in buildings, for wire-mesh fencing and for water cisterns and storage tanks. The technique for tinning is similar in principle to that for galvanising, but the product has relatively little importance for the civil engineer, being used principally for food containers.

Aluminium may also be applied to steel surfaces by hot dipping, although the commonest method of application is by spraying. It is not as popular as zinc for coating steels, but weight for weight it has better general corrosion resistance, except in marine conditions. The high temperature of the dipping bath also means that the process is unsuitable for work-hardened steels. Painting (as with zinc coatings) helps to prolong the life of the coating.

19.9.6.2   *Coatings from Metal Powders*

Zinc and aluminium may also be applied to steels by heating the object to be coated in zinc or aluminium powder. The processes are known respectively as Sherardising and Calorising. To prevent sintering of the powder during the process, it is normally mixed with oxide (this is effective even above the melting point of the metal powders). In the case of aluminium, 2% ammonium chloride is also added, which breaks down at the working temperature (about 900 °C) giving a reducing atmosphere, whilst the chloride acts as a 'carrier' for the aluminium. Volatile aluminium chloride is formed, which decomposes on the steel surface with the deposition of aluminium which diffuses into the steel. The principal uses of these processes are for the coating of small objects (nuts, bolts, small springs, etc.) that can be packed in quantity in drums containing the powder, and they give high-quality coatings.

19.9.6.3   *Sprayed Metal Coatings*

The particular advantage of spraying is that it is performed by means of a reasonably portable 'pistol' using metal wire or powder, which means that spraying can be done on site after a structure has been erected and can also be used to patch coatings which have become faulty with the passage of time. Spray coatings are porous; this is due to their structure of overlapping flakes of metal which have solidified instantly on contact with the surface. As a result, there is no alloying with the protected surface and adhesion depends on the keying of the sprayed metal into irregularities on the base metal. To this end, grit blasting with angular grit as a surface preparation serves the double purpose of removing rust and scale and keying the surface.

609

Alloying, however, can be obtained by annealing after spraying, as in the Aluminising process in which the sprayed coating is protected from oxidation by bitumen, and then heated. Zinc and aluminium are the two metals which are most commonly applied by spraying; lead and cadmium with their low melting points are physically suitable but their toxic character necessitates special protection for workers, which increases costs and makes the process less flexible. Sprayed ceramic coatings have increased very considerably in importance in recent years, and have a civil engineering potential in their high level of erosion resistance. They are usually sprayed onto metal surfaces as solid particles and then sintered on by the passage of a high-temperature flame.

19.9.6.4   *Clad Coatings*

Whereas all the previously considered metallic coatings are used almost entirely on steels, cladding is used extensively on light alloys. The process is carried out as a variant of the ordinary rolling of sheet metal, in which thin sheets of corrosion-resistant alloy are rolled onto a centre sheet of the basic material. Alloying between the layers is minimal, the process being a form of pressure welding; cleanliness of the surfaces is vitally important for good adhesion. Aluminium-coated duralumin (Alclad), which gives the corrosion resistance of pure aluminium coupled with the strength of duralumin, is the best known instance of cladding; however, care must be taken to avoid heating the alloy, or there is a risk that the copper will diffuse into the aluminium surface layers and reduce their corrosion resistance to that of ordinary duralumin. For containing corrosive chemicals, mild steel may be clad with nickel or stainless steel to give a highly corrosion-resistant sandwich that is consideraly cheaper than the pure cladding materials would be. Such composites are obviously unsuitable for welding.

19.9.6.5   *Electroplating*

This affords excellent (and frequently decorative) protection if carried out with due care, and can be used for laying down all the common metals (except aluminium) and a wide variety of alloys. Unfortunately the need for high purity, cleanliness and careful monitoring of the process makes it unsuitable as a large-scale protective technique for civil engineering structures. It is, however, valuable as a finishing process for smaller components, particularly since a number of platings are abrasion as well as corrosion resistant.

The article to be plated is made the cathode in an electrochemical cell; the electrolyte contains the correct concentration of plating ions, M^+, at the correct pH (both these conditions are usually critical) and the cathode reaction for a monovalent metal is $M^+ + e^- \longrightarrow M$, the amount deposited corresponding (ideally) to Faraday's Laws. The anode may consist of pure M, in which case it is termed a *soluble anode* and plating is essentially a matter of transferring metal from the anode to the cathode. Alternatively, if for example the electrolyte would corrode pure M too rapidly or if M would become passive in the plating bath, an *inert anode* may be used such as lead, stainless steel, or graphite. Oxygen is then liberated at the anode, $2H_2O \rightarrow 4H^+ + O_2 + 4e^-$, and plating proceeds, but the electrolyte becomes increasingly acidic (from the anode reaction) and impoverished in M^+ ions (from the cathode reaction) making chemical control of the electrolyte difficult.

Metals to be plated must always be very carefully cleaned of all oxide and grease, otherwise the plating will be nonadherent or non-existent in patches. Since no alloying with the base metal occurs during electroplating, some form of keying of the surface is ad-vantageous. In theory, grit blasting would be beneficial, but since electroplating follows the contours of the base metal so exactly it is necessary to have a high degree of polish on the base metal if a smooth plating is desired. Mechanical or electrolytic polishing is therefore carried out, followed by etching to improve the keying of the surface.

Various nonmetallic additions are frequently made to plating baths to improve the quality or the appearance of the plating. According to their action, they are known as grain refiners, levellers or brighteners, but the way in which they act is rather similar. Grain refiners act by temporary preferential adsorption on the most suit-able sites for metal deposition; the action is dynamic, in which all sites on the cathode are blocked and unblocked in rapid succession and only a small proportion of them are blocked at any one time. No correlation between the plated metal and the most efficient grain refiner has yet been deduced; glue is used for lead, gelatin and β-naphthol for tin, dextrin for zinc, and gelatin, molasses or liquorice for cadmium. Small amounts of refiners are deposited in the plating (about 0.04%) and, providing the correct concentration in the bath is not exceeded, there is no danger of embrittlement of the plating from this cause.

Levellers and brighteners are used where the plating has a decora-tive purpose; levellers result in a smoother plating surface than that of the underlying metal, whilst brighteners give a highly reflective surface. As with refiners, the mechanism is one of adsorption on sites on the surface of the growing plating. Levellers act by preferential absorption on any ridges or humps on the basis metal so that build up of the plating occurs relatively more rapidly in hollows and valleys. Choice of leveller may depend on the final appearance required; the best levelling agents do not necessarily confer a bright finish. Brighteners probably act by a dynamic adsorption which encourages completely random deposition of plating atoms on the cathode. In the absence of a brightener, atoms are deposited pre-ferentially at sites provided by incomplete rows of atoms, which leads to a microscopically uneven surface and a matt finish. Brighten-ing can also be induced by the codeposition of a second metal. Cobalt-nickel and tin-nickel alloys plate more brightly than pure nickel, probably because the second metal distorts the lattice of the principal metal and thereby favours random deposition. Phosphorus,

611

although not a metal, produces bright cobalt-phosphorus and nickel-phosphorus platings for the same reason.

19.9.6.6 *Problems in Electroplating*

(a) *Thickness of plating.* It is important both for economy and for adequate protection that the thickness of the plating should be uniform, irrespective of varying distance from the anode or the presence of crevices and corners. The ability of a plating bath to achieve this is known as the *throwing power* of the bath. No bath ever has perfect throwing power since the plating ions always tend to travel by the path of least resistance between the anode and the cathode. Thus the parts of the cathode nearest the anode tend to be plated preferentially and the preferential deposition gets more pronounced as the plating thickness increases. It is therefore important for any crevices or 'awkward' spots to be covered early in the process; a bath which is satisfactory in this respect is said to have good *covering power*. Baths in which the plating metal is contained as a complex ion, e.g. a silver plating bath in which the silver occurs as $[Ag(CN)_2]^-$, have a better throwing power, and incidentally give finer-grain deposits, than those where the metal is in the simple M^+ form. Where the throwing power is poor, or the shape of cathode is complex, multiple anodes are used, carefully arranged so that all parts of the cathode are roughly equidistant from an anode.

(b) *Porosity.* All electroplatings contain pores, and the only certain way of eliminating them is to plate to an uneconomic thickness when they will eventually all be bridged over. Their occurrence is minimised by careful filtering of the electrolyte and careful surface preparation of the cathode. Any solid impurities settling on the cathode from the electrolyte and any scratches or nonmetallic inclusions on the cathode surface act as nucleating points for pores. For practical purposes the thickness of a plating is a compromise between quality of protection and cost of production; it is obviously more important to minimse the pores in plating which is cathodic to the basis metal than those in a plating which affords sacrificial protection. Thicknesses normally lie in the range 0.25×10^{-3} to 50×10^{-3}.

(c) *Residual stresses.* Many platings are deposited in a state of residual stress, which can be considerable (up to 225 MN/m^2) if conditions in the plating bath are not properly adjusted. Such stresses may cause the plating to break away, or may even result in distortion of the plated component. Of the common plating metals, lead and zinc are deposited in compression and cadmium, chromium, cobalt, copper, iron and nickel in tension, which may have dangerous consequences if the plated article is then subject to tension, fatigue or corrosion during service. The occurrence of these stresses is probably due in part the codeposition of hydrogen with the plating, some of which remains in the plating, and in part to misfit between the atom spacing of the plating and that of the basic metal.

19.9.7 Paints

Perhaps the most widely employed — and in many ways the cheapest and best — technique of protection against atmospheric corrosion is by painting, particularly for large-scale structures. Paints are normally made up of four components which vary in proportion depending on whether the paint is to be used as a primer, a barrier coat, an undercoat, or a top coat. These components are: the binder, i.e. the oil or resin which polymerises to form the solid body of the paint coat; an inhibitor; a colouring pigment; and, because the first three components would in many instances be far too stiff by themselves for convenient application, a thinner. Both the inhibitor and the binder must be carefully selected according to the working environment of the structure; a paint admirably suited for the guttering of a suburban house would not necessarily be appropriate for the underwater legs of an offshore oil rig.

Inhibitive pigments have already been briefly considered in Section 19.9.3; it is the purpose of the priming paint to carry them in sufficient quantity that protective reactions may occur at the site of any exposed metalwork. Barrier coatings contain inhibitors but, as their name implies, their prime function is to exclude moisture and aggressive ions from the metal surface. They therefore contain flake pigments which by their overlaps will lengthen the diffusion path that ions must traverse to reach the underlying metal. Micaceous iron ore (MIO) is the cheapest and most popular of these pigments, but graphite or aluminium powders are also used; if a particularly hard or abrasion-resistant barrier is required the paint may contain powdered silica, silicon carbide or tungsten carbide.

For centuries the only binder in use was based on linseed oil; now, with the great expansion of polymer science, the number of available binders has multiplied greatly so that there is a wide choice available for differing conditions of use. Within this diversity there are seven main groups. The 'natural' organic binders are based on linseed oil and tung oil; these harden by polymerisation from the take-up of atmospheric oxygen, usually aided by a catalyst known as a 'dryer'. In addition there are now many 'synthetic' binders: the *alkyd* resins which harden by a condensation mechanism between phthalic acid and various complex alcohols; the *phenolic* resins which harden by a similar process based on phenols; *polystyrene* and *polyvinyl* binders which harden by the evaporation of a solvent-thinner; the *epoxy*

esters which harden by cross-linkage when the group $\overset{O}{\overset{\displaystyle\diagup\diagdown}{CH_2\text{——}CH\text{——}}}$ is broken open to form $\overset{\mid}{CH_2}\text{——}\overset{\mid}{C(OH)}\text{——}$; *chlorinated rubber* binders; and, for heat-resistant paints, *silicone* binders.

It is difficult to summarise the general virtues and limitations of each group concisely. 'Synthetic' paints tend to give thinner coatings than the 'natural' linseed oil products; This means either the need for extra coatings or the acceptance of a possibly shorter lifetime. On the other hand synthetic paints harden more rapidly, thus permitting speedier completion of the process. Alkyd resins give good general-purpose paints, but for polluted industrial environments the phenolic, polyvinyl or chlorinated rubber varieties show greater durability. Where really high corrosion resistance is required (e.g. for structures in chemical works and containers of chemicals) the epoxy paints are good, but the best of these require mixing with a curing agent immediately prior to application, which may pose problems when large surfaces are to be covered — particularly as the paint, once mixed, will not 'keep'. For abrasion resistance, polyurethane and two-pack epoxy paints show the highest performance. Silicone paints have extremely good all-round qualities, but their cost makes their use economically unjustifiable except where other binders are definitely inadequate. Phenolic binders are reliable in marine conditions, but linseed oil should be avoided as it is insufficiently resistant to the alkali that is formed at cathodic areas under sea water. Anti-fouling preparations should be included in marine paints intended for total immersion, or there is a serious risk of crevice corrosion between and under attached growths, particularly if the crevices become contaminated with decay products.

Once again it must be emphasised that all paint coatings, even of the highest quality and most painstaking application, are only as good as the quality of surface preparation that precedes them. Rust and millscale must be scrupulously removed — preferably the former should never be allowed to commence — and any scale around welds done on the site should be chipped off prior to priming. Sandblasting (preferably wet blasting for health reasons) for flame cleaning are probably the best cleansing techniques for site work; if protection can be applied before delivery to site, flame cleaning or pickling followed by a phosphate treatment to ensure good paint adhesion followed by two priming coats gives a high-quality result. Extra primer should be applied at vulnerable points such as edges, joints, bolts and welds. In terms of quality it is better, although slower, to apply two thin coatings rather than one thick one; some thin patches are a likelihood in any paint coat but it is highly improbable that the thin patches in successive coats will coincide. A simple expedient which helps to ensure adequate coverage at each successive stage of painting is to specify successive coatings of slightly different colour; in this way any areas of scant coverage are easily seen and remedied.

Application of paint should always be made on dry surfaces and in conditions of low humidity. Next to lack of adequate preparation, failure to observe these two essentials is the most frequent cause of early failure. Such conditions are easy enough to obtain in a paint shop but the outdoor painter is very much at the mercy of the weather: hence the general superiority of factory-applied coatings. Brushing is a more expensive technique than spraying, but is the only practicable method for small pipes and thin struts, or if the paint contains toxic lead pigments or anti-fouling compounds. Spraying is quick and reliable for large flat surfaces but care must be taken to

614

ensure that awkward angles and 'shaded' spots are well covered (e.g. behind rivets or bolt heads).

19.9.8 Bituminous Coatings

Coatings based on pitch and bitumen — the residues left from the distillation of coal tar and petroleum respectively — show excellent resistance to industrial pollution. Both are complex mixtures of organic compounds containing a high percentage of ring-structured hydrocarbons that are not readily attacked by chemicals, although some are vulnerable to breakdown by ultraviolet light. By adjusting the proportions of the various constituents these materials may be prepared with a sufficient degree of plasticity to prevent cracking in cold weather and a sufficiently low melting point that they may be applied to metalwork by hot dipping at about 150—200 °C. As they are not mechanically tough enough in themselves to withstand much wear or stress they are usually applied by 'hot wrapping' in which a fibreglass swathe is passed through the cauldron of molten bitumen and then wound round the object to be protected. Inorganic or polymer weaves should be used for wrapping; ordinary sacking or hessian is subject to decay and in so doing creates conditions suitable for both direct acidic attack and bacterial corrosion. This mode of protection is particularly suited to the exterior of pipelines both above and below ground level, and by it we can achieve much thicker coatings than is possible with unsupported bitumen. A second technique for producing thick coatings is by the addition of about 30% of inorganic filler such as powdered limestone or asbestos to the bitumen; the resulting composition is known as a *mastic* coating.

The principal limitation of these materials is their vulnerability to heat; many will soften and creep on inclined surfaces if exposed to strong sunshine and are therefore also unsuitable for applying to pipes carrying hot liquids. They should also be used with caution in conjunction with oil paints, as for example when a bituminous coating might be applied on top of a red lead primer. The reason for this is that chemicals in the bitumen can have the effect of preventing hardening in oil-bound paints and the result may be a failure in adhesion of the combined coating. It should also be noted that they are unlovely in appearance, which restricts their use to buried metalwork and industrial protection systems in which aesthetic values are not a high priority.

CHAPTER 20

Processing of Timber

20.1 INTRODUCTION

After felling, the tree has to be processed in order to render the timber suitable for man's use. Such processing may be basically mechanical or chemical in nature or even a combination of both. On the one hand timber may be sawn or chipped, while on the other it can be treated with chemicals which markedly affect its structure and its properties. In some of these processing operations the timber has to be dried and this technique has already been discussed in Chapter 7 on water relationships and will not be referred to again in this chapter.

The many diverse mechanical and chemical processes for timber have been described in great detail in previous publications and it is certainly not the intention to repeat such description here: readers desirous of such information are referred to the excellent and authoritative texts listed in the Bibliography. In looking at processing in this chapter the emphasis is placed on the properties of the timber as they influence or restrict the type of processing. For convenience the processes are subdivided below into mechanical and chemical but frequently their boundaries overlap.

20.2 MECHANICAL PROCESSING

20.2.1 Solid Timber

20.2.1.1 *Sawing and Planing*

The basic requirement of these processes is quite simply to produce as efficiently as possible timber of the required dimensions having a quality of surface commensurate with the intended use. Such a requirement depends not only on the basic properties of the timber, but also on the design and condition of the cutting tool; many of the variables are inter-related and it is frequently necessary to compromise in the selection of processing variables.

In Chapter 4 the density of timber was shown to vary by a factor of ten from about 120 to 1 200 kg/m³. As density increases so the

616

time taken for the cutting edge to become blunt decreases: whereas it is possible to cut over 10 000 feet of Scots pine before it is necessary to resharpen, only one or two thousand feet of a dense hardwood such as jarrah can be cut. Density will also have a marked effect on the amount of power consumed in processing. When all the other factors affecting power consumption are held constant, this variable is highly correlated with timber density as illustrated in Fig. 20.1.

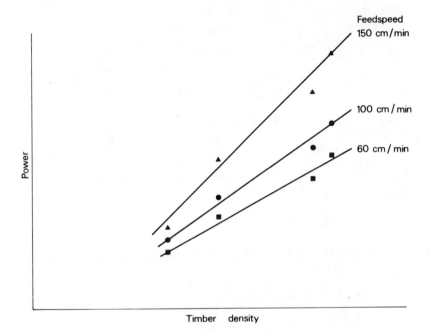

Figure 20.1   Effect of timber density and feedspeed on the consumption of power using a circular saw to cut along the grain (*rip-sawing*). (From the Princes Risborough Laboratory, Building Research Establishment. © Crown copyright.)

Timber of high moisture content never machines as well as that at lower moisture levels. There is a tendency for the thin-walled cells to be deformed rather than cut because of their increased elasticity in the wet condition. After the cutters have passed over, these deformed areas slowly assume their previous shape resulting in an irregular appearance to the surface which is very noticeable when the timber is dried and painted; this induced defect is known as *raised grain*.

The cost of timber processing is determined primarily by the cost of tool maintenance, which in turn is related not only to properties of the timber, but also to the type and design of the saw or planer blade. In addition to the effect of timber density on tool life, the presence in certain timbers of gums and resins has an adverse effect because of the tendency for the gum to adhere to the tool thereby causing overheating; in saw blades this in turn leads to loss in tension resulting in saw instability and a reduction in sawing accuracy.

A certain number of tropical hardwood timbers contain mineral inclusions which develop during the growth of the tree. The most common is silica which is present usually in the form of small grains within the ray cells (Fig. 20.2). The abrasive action of these inclusions

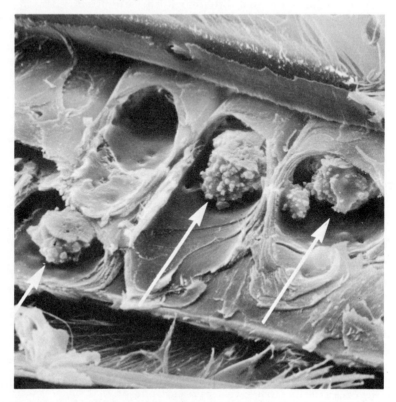

Figure 20.2   The presence of silica grains (arrowed) in the ray cells of *Parinari* species; scanning electron micrograph (x 15 000). (From the Princes Risborough Laboratory, Building Research Establishment. © Crown copyright.)

is considerable and the life of the edge of the cutting tool is frequently reduced to almost $\frac{1}{100}$ of that obtained when cutting timber of the same density but free of silica. Timbers containing silica are frequently avoided unless they possess special features which more than offset the difficulties which result from its presence.

One or two timbers are recognised as containing occasionally large deposits of calcium carbonate, a feature usually referred to as *stone*: the timber iroko is a well known example. Not only stone, but also the presence in some timbers of nails, wire and even bullets does little to extend the life of expensive cutting tools.

Moisture content of the timber also plays a significant role in determining the life of cutting tools. As moisture content decreases,

618

so there is a marked reduction in the time interval between re-sharpening both saw and planer blades. The fibrous nature of tension wood (Section 4.2.4) will also increase tool wear.

Service life will also depend on the type and design of the tool. Although considerably more expensive than steel, the use of tungsten-carbide-tipped saws and planer blades extends the life of the cutting edge especially where timbers are either dense or abrasive. Increasing the number of teeth on the saw or the number of planer blades on the rotating stock will increase the quality of the surface provided that the feedspeed is sufficient to provide a minimum bite per revolution; this ensures a cutting rather than a rubbing action, which would accelerate blunting of the tool edge.

One of the most important tool design variables is the angle between the edge and the timber surface. As discussed in Section 4.3.2.1, timber is seldom straight-grained, tending in most cases to be in the form of a spiral or low pitch; occasionally, the grain is interlocked or wavy as discussed previously. Under these circumstances there is a strong tendency for those cells which are inclined towards the direction of the rotating cutter to be pulled out rather than cut cleanly, a phenomenon known as *pick-up* or *tearing*. The occurrence of this defect can be removed almost completely by reducing the cutting angle (*rake angle*) of the rotating blades, though this will result in increased power consumption.

The cost of processing though determined primarily by tool life will be influenced also by the amount of power consumed. In addition to the effect of density of timber previously discussed, the amount of energy required will depend on the feedspeed (Fig. 20.1), tool design and, above all, on tool sharpness.

20.2.1.2 *Bending*

Steam bending of certain timbers is a long-established process which was used extensively when it was fashionable to have furniture with rounded lines. The backs of chairs and wooden hat stands are two common examples from the past, but the process is still employed at the present time, albeit on a much reduced volume. The handles for certain garden implements, walking sticks and various sports goods are all produced by steam bending.

The mechanics of bending involves a presteaming operation to soften the lignin, swell the timber, and render the timber less stiff. With the ends restrained, the timber is usually bent round a former, and after bending the timber must be held in the restrained mode until it dries out and the bend is *set*. In broad terms the deformation is irreversible, but over a long period of time, especially with marked alternations in humidity of the atmosphere, a certain degree of recovery will arise especially where the curve is unrestrained by some fixing. Although most timbers can be bent slightly, only certain species, principally the hardwood timbers of the temperate region, can be bent to sharp radii without cracking. When the timber is bent over a supporting but removable strap, the limiting radius of curvature is reduced appreciably. Thus it is possible to bend 25 mm thick ash to a radius of 64 mm and walnut to a radius of only 25 mm. The significance of the anatomy of the timber in determining the limiting radius of curvature is poorly understood.

20.2.2   Board Materials

The total value of board materials used in the UK in 1973 was approximately £200 million, a value equivalent to about a third that of solid timber (Fig. 20.3(a)). In terms of the volume and value of board materials used, it would be easy to justify compiling a book solely on these: whilst this is possible in terms of testing procedures and properties of these boards, it would be so very difficult to perform in terms of structure-property relationships as has been attempted for solid timber in this book. Compared with the long history of timber utilisation, the use and understanding of behaviour of board materials is still in its infancy.

As a material timber has a number of deficiencies.

(a)   It possesses a high degree of variability.
(b)   It is strongly anisotropic in both strength and moisture movement.
(c)   It is dimensionally unstable in the presence of changing humidity.
(d)   It is available in only limited widths.

Such material deficiencies can be lowered appreciably by reducing the timber to small units and subsequently reconstituting it, usually in the form of large flat sheets, though moulded items are also produced, e.g. trays, bowls, coffins, chair backs. The degree to which these boards assume a higher dimensional stability and a lower level of anisotropy than is the case with solid timber is dependent on the size and orientation of the component pieces of timber and the method by which they are bonded together. There are an infinite variety of board types though there are only three principal ones — plywood, chipboard and fibre-building board: the production of these and their major properties are discussed in detail in subsequent sections.

In comparison with timber, board materials possess a lower degree of variability, lower anisotropy, and higher dimensional stability: they are also available in very large sizes. The reduction in variability is due quite simply to the random repositioning of variable components, the degree of reduction increasing as the size of the components decrease.

In Table 20.1 a comparison is made between the bending strength of timber and that of the three major types of board. When material of the same density is compared, the bending strength and stiffness of fibre-building board and chipboard are considerably lower than those of plywood, which in turn are slightly lower than those for solid timber along the grain, though greater than for solid timber across the grain. Thus boards do not possess the high levels of aniso-

Table 20.1 Strength Properties of Timber and Boards

	Thickness (mm)	Density (kg/m³)	Bending strength (N/mm²)		E (N/mm²)	
			∥	⊥	∥	⊥
Solid timber						
Douglas fir	20	500	80	2.2	12 700	800
Plywood						
Douglas fir	4.8	520	73	16	12 090	890
Douglas fir	19	600	60	33	10 750	3 310
Chipboard						
UF	18.6	720		11.5		1 930
PF	19.2	680		18.0		2 830
MF/UF	18.1	660		27.1		3 460
Fibre-building board						
Tempered	3.2	1 030	69	65	4 600	
Standard	3.2	1 000	54	52	–	
Medium density	9–10	680	18.7	19.2	–	

tropy characteristic of timber: in both chipboard and fibre-building board, anisotropy is almost absent and even in plywood with its higher strength and stiffness the degree of anisotropy is very much lower than for solid timber.

The dimensional stability in the plane of the board appears to be remarkably constant within a single board and even between boards of different types (Table 20.2); however, there is considerable dif-

Table 20.2   Dimensional Stability of Timber and Boards

Percentage change in dimensions from 30% to 90% relative humidity.

	Direction to grain or board length		Thickness
	Parallel	Perpendicular	
Solid timber			
Douglas fir	negligible	2.0—2.4	2.0—2.4
Beech	negligible	2.6—5.2	2.6—5.2
Plywood			
Douglas fir	0.24	0.24	2.0
Chipboard			
UF	0.33	0.33	4.7
PF	0.25	0.25	3.9
MF/UF	0.21	0.21	3.3
Fibre-building board			
Tempered	0.21	0.27	7—11
Standard	0.28	0.31	4—9
Medium density	0.24	0.25	4—8

ference in stability across the thickness of the board among the three types. The stability of all board types is poorer than that of timber along the grain, but vastly superior to that of timber across the grain; it is primarily for this latter reason that the various board materials are used so extensively where large widths are required, e.g. furniture, flooring and wall panelling.

Over the last decade there has been a marked annual increase in the amount of boards used in the UK, much of this being due to a rapid development in chipboard usage. By 1974 the value of chipboard used had increased exponentially over the previous decade to a value of £80 million at which point it equalled that for plywood (Fig. 20.3(b)). However, since the cost of plywood is considerably higher than that for chipboard, the mass of chipboard used in 1974 represented 60% of the total mass of board materials, with plywood accounting for a further 30%.

In subsequent sections these three board materials are further discussed: a fourth product is also included, namely the wood—wool cement board. Although produced in only very small quantities its inclusion in this text is justified technically on the grounds that they represent an interesting composite of timber strands in a cement matrix.

20.2.2.1 Plywood

Logs, the denser of which are softened by boiling in water, may be sliced into thin veneer for surface decoration by repeated horizontal or vertical cuts, for plywood, or peeled by rotation against a slowly advancing knife to give a continuous strip. After drying, sheets of veneer for plywood manufacture are coated with adhesive and are laid up and pressed with the grain direction at right angles in alternate layers: plywood should contain an unequal number of plies so that the system is balanced around the central veneer.

As the number of plies increase, so the degree of anisotropy in both strength and movement drops quickly from the value of $40 : 1$ for timber in the solid state. With three-ply construction and using veneers of equal thickness the degree of anisotropy is reduced to $5 : 1$, while for nine-ply this drops to $1.5 : 1$. However, cost increases markedly with number of plies and for most applications a three-ply construction is regarded as a good compromise between isotropy and cost.

The mechanical and physical properties of the plywood will depend not only on the species of timber selected, but also on the type of adhesive used. Up to recent times more emphasis has been placed on characterising the quality of plywood in terms of the property of the adhesive, but this situation is changing with the recognition of the significance of the timber as playing an equally important role in determining ultimate performance.

Both softwoods and hardwoods within a density range of $400-700$ kg/m^3 are normally utilised. Plywood for internal use is produced from the nondurable species and urea-formaldehyde adhesive, while plywood for external use should be manufactured using phenol-formaldehyde resins and durable timbers, or permeable nondurable timbers which have been preservative treated, as specified in BS 1455, 1088 and 4079.

Plywood is the oldest of the timber sheet materials and for many years has enjoyed a high reputation as a structural sheet material. Its use in the Mosquito aircraft in the forties, and its subsequent performance for small boat construction, for sheathing in timber-frame housing, and in the construction of web and hollow-box beams all bear testament to its suitability as a structural material.

It is not possible to talk about strength properties of plywood in general terms since not only are there different strength properties in different grain directions, but that these are also affected by configuration of the plywood in terms of number, thickness, orientation and quality of the veneers and by the type of adhesive used. The factors which affect the strength of plywood are the same as those set out in Chapter 14 for the strength of timber, though the effects are not necessarily the same. Thus the intrinsic factors, such as knots and density, play a less significant part than they do in the case of timber, but the effect of the extrinsic variables such as moisture content, temperature and time is very similar to that for timber.

In theory, any strength property of any plywood should be calculable from a detailed knowledge of the properties of the constituents, but in practice, because of the great range of variables affecting plywood strength, this is impossible to achieve and recourse is made to standard tests as set out in BS 4512.

In principle, the *working stresses* for plywood are derived in a manner similar to that for solid timber prior to 1973 (Section 14.9.2.1). The basic stress for a particular property is first determined from the mean, standard deviation and a safety factor, and is subsequently reduced by a factor which allows for the occurrence of defects to give, at least in theory, the working stress. However, considerable difficulty occurs in ensuring that all the variables are in fact covered and it is fairly common practice, therefore, to carry out tests on structural sizes of plywood in order to obtain or to confirm the working stress.

One approach to the calculation of working stresses is to consider the plywood cross-section as homogeneous and to work out an equivalent stress for the complete cross-section — an approach which fails to provide the stress in each lamina, but one which allows the comparison of the thickness required for a particular application with that of other materials. Such an approach, however, requires the testing of all available types of plywood and can only be successful when a small number of types are standardised. Such an approach has been adopted in the current edition of the British Standard Code of Practice CP 112, to which reference has previously been made in Chapter 14. With the increasing variety of plywood types including mixtures of species such a simplified system may have to be replaced by a more comprehensive system when the Code is revised in the near future.

Elasticity in plywood was examined in the early forties when it was shown that the values were property dependent, unlike the case with solid timber. Working with veneers of the same thickness and same species fairly simply generalised equations were derived, but as variation is built into the analysis the equations become complex (see Hearmon, 1948, to which reference was made in Section 11.2.4). As noted earlier, analysis of deformation in plywood items can be calculated using the plane-stress system of orthotropic elasticity.

Plywood possesses high strength and stiffness, especially so when expressed in terms of specific gravity. It is unfortunate that it is now becoming a rather expensive material and is being replaced by other board materials in a number of applications.

624

20.2.2.2 *Chipboard*

The chipboard industry dates from the mid-forties and originated with the purpose of utilising waste timber. After a long, slow start, when the quality of the board left much to be desired, the industry has grown tremendously over the last decade far exceeding the supplies of waste timber available and now relying to a very large measure on the use of small trees for its raw material. Such a marked expansion is due in no small part to the tighter control in processing and the ability to tailor-make boards with a known and reproducible performance.

Although the value of chipboard used in the UK is similar to that of plywood (Fig. 20.3(b)) the volume of chipboard used is almost double: the annual percentage increase in consumption of chipboard is now running considerably higher than all the other board materials. About a quarter of our consumption is produced in the UK.

In the manufacture of chipboard the timber, which is principally softwood, is cut by a series of rotating knives to produce thin flakes which are dried and then sprayed with adhesive. Usually the flakes or chips are blown onto flat plattens in such a way that the smaller chips end up on the surfaces of the board and the coarse chips in the centre. The fibre matt is first cut to length before passing into the press where it is held for 0.10–0.20 min per mm of board thickness at temperatures up to 200 °C. The density of boards produced range from 450 to 750 kg/m^3, depending on end-use classification, while the resin content varies from about 11% on the outer layers to 5% in the centre, averaging out for the board at about 8% on a dry weight basis.

Instead of using the batch platten process, chipboard can be made continuously using either the Mendé or an extrusion process. The former is applicable only in the manufacture of thin chipboard, i.e. 6 mm or less, and the process is analogous to that of paper manufacture in that the board is sufficiently flexible to pass between and over large heated rollers. In the extrusion process the chipboard matt is forced out through a heated die, but this results in the orientation of the chips at right angles to the plane of the board which reduces both the strength and stiffness of the material.

Direct comparison of the mechanical properties of chipboard and plywood are not available since their properties are assessed by different techniques set out in different international and national standards; in the UK the relevant chipboard standard is BS 5669 (1979). However, it is fairly safe to say that in most strength properties chipboard will be weaker than plywood of the same thickness. The strength of chipboard will increase with increasing length of chip for a given adhesive content, and the recent introduction of waferboard where the chips are about 30 mm in length and frequently as wide is an attempt to compete with plywood as a structural material.

The performance of chipboard, like that of plywood, is very dependent on the type of adhesive used. Most of the chipboard used in the UK contains urea-formaldehyde which, because of its sensitivity to moisture, renders this type of chipboard unsuitable for use where there is a risk of the material becoming wet, or even being subjected to marked alternations in relative humidity over a long period of time. More expensive boards possessing some resistance to the presence of moisture are manufactured using either melamine-formaldehyde or phenol-formaldehyde adhesives: however a true external-grade board has not yet been produced commercially.

Chipboard, like timber, is a viscoelastic material and an example of the deformation over an extended period of time has already been presented (Fig. 11.14). However, the rate of creep in chipboard is considerably higher than that in timber though it is possible to reduce it by increasing the amount of adhesive or by modifying the chemical composition of the adhesive. At the present time it appears unlikely that the existing brands of chipboard could be used as a full structural material comparable to the structural grades of plywood. However, chipboard continues to be used extensively in a semi-structural mode in floors and roofing where it derives support from the timber framework.

Similar boards to chipboard are produced from a wide variety of plant material and synthetic resin of which flaxboard and bagasse board are the best-known examples. These, along with chipboard, are collectively called particleboards in the UK, the term chipboard being restricted to the product produced from timber.

20.2.2.3 Fibre-Building Board

Although much smaller quantities of fibre-building board are used in the UK than either chipboard or plywood (Fig. 20.3(b)) it is nevertheless a most important panel product, used extensively in the UK for insulation and the linings of doors and backs of furniture, and in Scandinavia as a cladding and roofing material.

The process of manufacture is quite different from that of the other board materials in that the timber is first reduced to chips which are then steamed under very high pressure in order to soften the lignin which is thermoplastic in behaviour. The softened chips then pass to a defibrator which separates them into individual fibres, or fibre bundles without inducing too much damage.

The fibrous mass is frequently mixed with hot water and formed into a matt which is cut into lengths and, like chipboard, pressed in a multi-platten hot press at a temperature of from 180 °C to 210 °C. In the more modern dry-forming process the fibrous mass is conveyed in an air stream to the matt-forming station and, in order to obtain boards of adequate strength, small quantities of phenolic resin are added to supplement the bonding by the softened lignin.

By modifying the pressure applied in the final pressing, boards of a wide range of density are produced ranging from the insulation boards with a density of about 250 kg/m^3 to hardboard with a density of 950 kg/m^3. Fibre-building board, like the other board products, is moisture sensitive, but a certain degree of resistance can

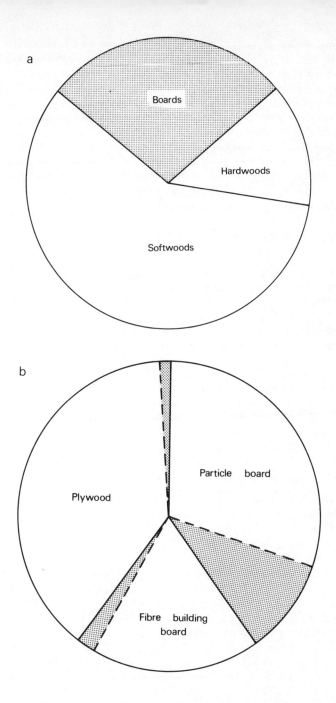

Figure 20.3   (a) Relative values of timber and board materials used in the UK in 1973. (b) Relative values of the three major types of board materials used in the UK in 1973. Shaded areas represent home production. (From the Princes Risborough Laboratory, Building Research Establishment. © Crown copyright.)

be obtained by the passage of the material through a hot oil bath thereby imparting a high degree of water repellency: this material is referred to as *tempered* hardboard.

The assessment of the properties of the fibre-building boards is carried out according to BS 1142: once again, because of the lack of uniform testing methods, a direct comparison of properties with those of the other board materials is not possible. Up to the present time hardboard has not been used as a load-bearing material, but recent tests on structural units have indicated a performance for tempered hardboard similar to the structural grades of plywood and there is now the probability of the inclusion in the revised CP 112 of working stresses for this material.

20.2.2.4 *Wood—Wool Cement Boards*

A composite board with low thermal conductivity and enhanced durability but with lower strength than the other board types can be produced by cementing together a random distribution of very long thin strands of timber, popularly known as wood—wool and at one time an important packaging material. Care must be exercised in the selection of timbers since the extractives of some can cause cement poisoning resulting in retarded setting of the cement and loss in board strength.

The high durability of these boards is due to the high level of alkalinity present in the matrix. Wood—wool cement slabs are used extensively in flat-roof construction and in terms of their resistance to the effects of moisture ingress are considerably superior in performance to the other board materials; nevertheless the volume used is very small compared with the other boards.

20.2.3 Laminated Timber

Where timber beams of a particular shape or excessive length are required these can be fabricated to order using a laminating process. Either urea-formaldehyde or resorcinol-formaldehyde adhesive is used depending whether the material is required for internal or external conditions.

In the process thin strips of timber are glued and laid parallel to one another in a jig, the whole assembly being clamped until the adhesive has set. Where the individual laminae are thin they are end-jointed using a scarf-joint, but where the laminae are thicker than 10 mm finger jointing techniques are usually employed (Fig. 20.4).

Laminated construction is to be found in such diverse items as tennis racquets, skis, hulls of wooden ships and arched beams in halls and sports stadiums. Information on the derivation and application of working stresses for laminated timber is to be found in BS 4978 and CP 112 to which reference was made in Section 14.9.2.2.

20.2.4 Mechanical Pulping

The pulping industry is the single largest consumer of wood. In the

628

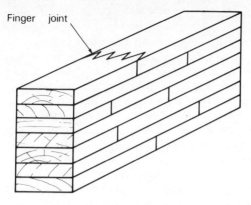

Finger joint

Figure 20.4 The construction of a laminated timber beam, the laminae of which are end-jointed using finger joints. (From the Princes Risborough Laboratory, Building Research Establishment. © Crown copyright.)

UK the value of pulp and paper imports for 1974 was £1 030 million, compared with £847 million for timber and timber products; pulp and paper account for about 4.5% of the total UK imports.

Pulp may be produced by either mechanical or chemical processes and it is the intention to postpone discussion on the latter until later in this chapter. In the original process for producing mechanical pulp logs with a high moisture content are fed against a grinding wheel which is continuously sprayed with water in order to keep it cool and free it of the fibrous mass produced. The pulp so formed, known as stone groundwood, is coarse in texture, comprising bundles of cells rather than individual cells, and is mainly used as newsprint.

To avoid the necessity to adopt a costly bleaching process only light-coloured timbers are accepted. Furthermore, because the power consumed on grinding is a linear function of the timber density, only the low-density timbers with no or only small quantities of resin are used.

More recent developments in mechanical pulping have centered on disc-refining. Wood chips, softened in hot water, by steaming, or by chemical pretreatment, are fed into the centre of two high-speed counter-rotating, ridged, metal plates; on passing from the centre of the plates to the periphery the chips are reduced to fine bundles of fibres or even individual fibres. This process is capable of accepting a wider range of timbers than the traditional stone groundwood method.

20.3 CHEMICAL PROCESSING

20.3.1 Impregnation

Although impregnation embraces both mechanical and chemical

629

aspects of processing, it is included within this section on chemical processing purely for convenience.

The degree of impregnation of timber by chemical solutions is related directly to the permeability of timber which was discussed in some detail in Chapter 9 on flow. In that chapter the pathways of flow were described and it will be recalled that permeability was shown to be a function not only of moisture content and temperature, but also of grain direction, sapwood/heartwood, earlywood/ latewood and species. Longitudinal permeability is usually about 10^4 x transverse permeability owing principally to the orientation of the cells in the longitudinal direction. Heartwood, owing to the deposition of both gums and encrusting materials, is much less permeable than the sapwood, and earlywood in the dry condition also has a lower permeability than the latewood owing to aspiration of the bordered pits.

Perhaps the greatest variability in impregnation occurs between species. Within the softwoods this can be related to the number and distribution of the bordered pits and to the efficiency of the *residual* flow paths which utilise both the latewood bordered pits and the semi-bordered ray pits. Within the hardwoods variability in impregnation is related to the size and distribution of the vessels and to the degree of dissolution of the end walls of the vessel members.

Four arbitrary classes of impregnation are recognised, the timbers being apportioned to these according to the depth of penetration of the solution after a fixed period of time at a particular pressure. These classes are *permeable, moderately resistant, resistant* and *extremely resistant.* Whilst this classification is used more frequently with reference to the impregnation of artificial preservatives it is equally applicable to impregnation by flame retardants, or dimensional stabilisers for, although differences in viscosity will influence degree of penetration, the species will remain in the same relative order.

20.3.1.1 *Artificial Preservatives*

Except where the heartwood of a naturally durable timber is being used, timber, provided it is permeable, should always be preservative treated if there is any significant risk that its moisture content will rise in excess of 20%; nondurable timber resistant to impregnation should not be used in these conditions. In theory, it should not be necessary to protect internal woodwork which should remain dry, but, because of water spillage, and leakage from pipes or from the roof, the moisture content of internal woodwork can also rise above this critical level.

Short-term dipping and surface treatments by brush or spray are the least effective ways of applying a preservative because of the small loading and poor penetration achieved. In these treatments only the surface layers are penetrated and there is a risk of deep splits occurring in service, thereby exposing untreated timber. Such treatments are not suitable for timber that is to be in ground con-

630

tact, but they are used for window and door joinery.

The most effective treatment of timber is by pressure impregnation, which ensures deeper penetration and much higher loadings. Two variants of the process are available and the choice is usually determined by the preservative selected. In the first, the timber is placed in a sealed tank and a vacuum drawn: the preservative is then allowed to enter the tank and is taken up by the timber. The same procedure occurs in the second process except that pressure is applied at the final stage to assist penetration. Pressure treatment of timber is specified in BS 913, 1954 and BS 4076, 1966.

Although it is not an impregnation process as defined above, it is convenient to examine here the diffusion process of preservation. The timber must be in the green state and the preservative must be water soluble. Usually the preservative, commonly borax, is applied to the surface in the form of a paste and over a period of some weeks diffuses into the green timber: only very small quantities of timber are treated in this way.

There are three main types of preservative in general use which are specified in BS 1282. The first group are the tar oils of which coal-tar creosote is the most important: its quality is covered by BS 144 and BS 3051. Its efficacy as a preservative lies not in any natural toxicity, but rather in its supreme water repellency. It has a very distinctive and heavy odour and treated timber cannot be painted unless primed with a metallic paint.

The second group are the water-borne preservatives which are suitable for both indoor and outdoor uses. The most effective formulations are those containing copper, chromium and arsenic salts and these are usually applied by a vacuum/pressure treatment; the chemicals are *fixed* in the timber, i.e. they are not leached out in service.

The third group are the solvent-type preservatives, which tend to be more expensive than those of the first two groups, but they have the advantage that machined timber can be brush or dip treated without the grain being raised, as would be the case with aqueous solutions. The formulations of the solvent type are usually based on pentachlorophenol, tributyltin oxide or chlorinated naphthalenes, and are applied by vacuum impregnation or by immersion.

For many years there was some difference of opinion as to whether these preservatives merely lined the walls of the cell cavity or actually entered the cell wall. More recently it has been shown by electron microanalysis that whereas creosote only coats the cell walls the water-borne and solvent type of preservatives impregnate the cell wall. However, because of the capillary nature of the cell wall, it was thought for some time that selective filtration might occur owing to the disparity in sizes of the constituent chemical groups present. Such doubts, however, were removed by the results of recent work using the electron microscope microanalyser, EMMA 4, fitted with a probe only $0.2 \mu m$ in diameter which allowed the distribution of copper, chromium and arsenic across the cell wall following impregnation to be obtained (Chou, Chandler and Preston,

1973). All three elements were present in all the regions examined and the distribution of this fine deposit is dictated by the distribution of the microfibrils; these appear to be coated individually with a layer some 1.5–2.0 nm in thickness. In some parts of the cell wall a coarser deposit is also present. The preservative was also found to be present on the surface of the cell cavity with concentrations of the elements some 2–5 times those within the cell wall; thus previous disagreements on distribution of preservatives appear to be resolved.

In those timbers which can be impregnated, it is likely that the durability of the sapwood after pressure impregnation will be greater than the natural durability of the heartwood, and it is not unknown to find telegraph and transmission poles the heartwood of which is decayed while the treated sapwood is perfectly sound.

Mention has been made already of the difficulty of painting timber which has been treated with creosote. This disadvantage is not common to the other preservatives and not only is one able to paint the treated timber, but it is also possible to glue together treated components.

20.3.1.2 Flame Retardants

Flame-retardant chemicals may be applied as surface coatings or by vacuum-pressure impregnation, thereby rendering the timber less easily ignitable and reducing the rate of flame spread. Intumescent coatings will be discussed later and this section is devoted to the application of fire retardants by impregnation.

The salts most commonly employed in the UK for the vacuum-pressure impregnation process are monoammonium phosphate, di-ammonium phosphate, ammonium sulphate, boric acid and borax. These chemicals vary considerably in solubility, hygroscopicity and effectiveness against fire. Most proprietary flame retardants are mixtures of such chemicals formulated to give the best performance at reasonable cost. Since these chemicals are applied in an aqueous solution it means that a combined water-borne preservative and fire-retardant solution can be used which has distinct economic considerations. Quite frequently, corrosion inhibitors are incorporated where the timber is to be joined by metal connectors.

Considerable caution has to be exercised in determining the level of heating to be used in drying the timber following impregnation. The ammonium phosphates and sulphate tend to break down on heating giving off ammonia and leaving an acidic residue which can result in degradation of the wood substance as described in Chapter 18. Thus it has been found that drying at 65 °C following impregnation by solutions of these salts results in a loss of bending strength of from 10% to 30%. Drying at 90 °C, which is adopted in certain kiln schedules, results in a loss of 50% of the strength and even higher losses are recorded for the impact resistance or toughness of the timber. It is essential, therefore, to dry the timber at as low a temperature as possible and also to ensure that the timber in service is never subjected to elevated temperatures which would initiate or

632

continue the process of acidic degradation. Most certainly, timber
which has to withstand suddenly applied loads should not be treated
with fire retardants, and care must also be exercised in the selection
of glues for construction. The best overall performance from timber
treated with a flame retardant is obtained when the component is
installed and maintained under cool and dry conditions.

20.3.1.3   *Dimensional Stabilisers*

In Chapter 7 on movement, timber, because of its hygroscopic
nature, was shown to change in dimensions as its moisture content
varied in order to come into equilibrium with the vapour pressure of
the atmosphere. Because of the composite nature of timber such
movement will differ in extent in the three principal axes.

Movement is the result of water adsorption or desorption by the
hydroxyl groups present in all the matrix constituents. Thus it
should be possible to reduce movement (i.e. increase the dimensional
stability) by eliminating or at least reducing the accessibility of these
groups to water. This can be achieved by either chemical changes or
by the introduction of physical bulking agents.

Various attempts have been made to substitute the hydroxyl
groups chemically by less polar groups and the most successful has
been by acetylation. In this process acetic anhydride is used as a
source of acetyl groups while pyridine is added as a catalyst. A very
marked improvement in dimensional stability is achieved with only a
marginal loss in strength.

Good stabilisation can also be achieved by reacting the wood
with formaldehyde which then forms methylene bridges between
adjacent hydroxyl groups. However, the acid catalyst necessary for
the process causes acidic degradation of the timber.

Most of the successful stabilising processes involve the impregnation of the cell wall by chemicals which hold the timber in a swollen
condition even after water is removed, thus minimising dimensional
movement. In the mid-forties some solid timber, but more usually
wood veneers, were impregnated with solutions of phenol-
formaldehyde. The veneers were stacked, heated and compressed to
form a high-density material with good dimensional stability which
found wide usage as insulation in the electrical industry prior to the
era of plastics.

Considerable success has also been achieved using polyethylene
glycol (PEG), a wax-like solid which is soluble in water. Under controlled conditions, it is possible to replace all the water in timber by
PEG by a diffusion process, thereby maintaining it in a swollen
condition. The technique has found application among other things
in the preservation of waterlogged objects of archaeological interest,
the best example of which is the Swedish wooden warship *Wasa*,
which was raised from the depths of Stockholm harbour in 1961
having foundered in 1628. From 1961 the timber was sprayed continuously for over a decade with an aqueous solution of PEG which
diffused into the wet timber gradually replacing the bound water in

the cell wall without causing any dimensional changes.

PEG may also be applied to dry timber by standard vacuum impregnation using solution strengths of from 5 to 30%. Frequently, preservative and/or fire-retardant chemicals are incorporated in the impregnating solution. It will be noted from Fig. 20.5 that the amount of swelling has been reduced to one third following impregnation.

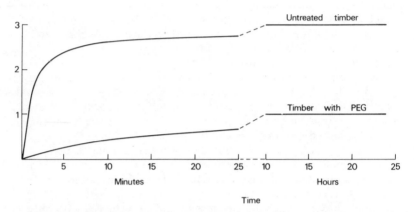

Figure 20.5 The comparative rates of swelling in water of untreated pine timber and timber impregnated with a 50% (by weight) solution of PEG (polyethylene glycol): this is equivalent to 22% loading on a dry wood basis. (Adapted from R. E. Morén (1964).)

Within the last two decades attention has been centred on the possibility of impregnating timber with low-viscosity liquid polymers which can then be polymerised within the timber to convert them to solid polymers. Timber so treated is referred to as *polymer-impregnated wood* or *wood-plastic composite* (WPC). A whole range of monomers have been tried, but most success has been achieved with methyl methacrylate, or a mixture of styrene and acrylonitrile (60 : 40).

After the timber is impregnated with the liquid monomer using standard vacuum-pressure impregnation techniques, the monomer is polymerised by either gamma irradiation or by the use of free-radical catalysts and heat. The former method is usually selected since a greater bulk of timber can be treated than is the case with heating; however, special equipment is required and stringent safety precautions must be enforced. The radiation dose must be controlled carefully to avoid degrade of the timber (see Section 14.7.8).

Swelling of the timber occurs during impregnation, which is indicative of the penetration of the cell wall by the monomer; after polymerisation of the monomer the timber possesses enhanced dimensional stability (Fig. 20.6). The degree of penetration, and consequently the amount of dimensional stability, can be increased

634

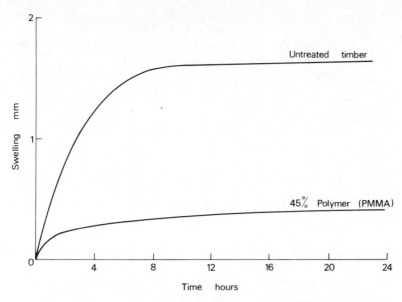

Figure 20.6 The comparative rates of swelling in water of untreated birch and a composite of the same wood with PMMA (polymethyl methacrylate). (From the Princes Risborough Laboratory, Building Research Establishment. © Crown copyright.)

by adding swelling solvents, such as dioxan, to the impregnant.

It is possible that the polymer is acting in more than a bulking role in that it could form copolymers with the various chemical constituents of the cell wall. A limited amount of evidence is available to support this hypothesis of *grafting* though some workers have ruled out such a possibility. Not only is the dimensional stability improved by this process, but a number of the strength properties are also increased. Modulus of rupture and compression parallel to the grain are usually increased slightly while the hardness is raised almost three-fold. On the other hand, the shock resistance and elasticity tend to fall slightly (Table 20.3).

Table 20.3 Values of Strength and Stiffness of Wood—
Plastic Composite as a Percentage of Untreated Timber

MMA = Methyl methacrylate on v/v basis; γ radiation dose = 4 Mrad.

	MOR %	Impact %	Compression along grain %	Radial hardness %	MOE (Bending) %
Birch + 45% MMA	116	68	119	228	94
Podo + 70% MMA	111	100	116	371	89

WPC has a most attractive appearance, but unfortunately tends to cost some three to four times as much as untreated timber. Consequently it has been used for only small and specialised items such as cutlery handles and brush backs. The material has much potential as a flooring material for heavy-duty areas such as dance halls, but so far, because of its cost, it has found little acceptance. The exception is to be found in Helsinki where the floor of the terminal building at the international airport comprises wood plastic composite.

Recent developments in the production and use of water-repellent preservatives based on resins dissolved in low-viscosity organic solvents have resulted in the ability to confer on timber a low but nonetheless important level of dimensional stability. Their application is of considerable proven practical significance in the protection of joinery out-of-doors and is discussed further in Section 20.4.

20.3.2 Chemical Pulping

The magnitude of the pulping industry has already been discussed as has also the production of mechanical pulp. Where paper of a higher quality than newsprint or corrugated paper is required a pulp must be produced consisting of individual cells rather than fibre bundles. To obtain this type of pulp the middle lamella has to be removed and this can be achieved only by chemical means.

There are a number of chemical processes which are described in detail in the literature. All are concerned with the removal of lignin, which is the principal constituent of the middle lamella. However, during the pulping process lignin will be removed from within the cell wall as well as from between the cells: this is both acceptable and desirous since lignin imparts a greyish colouration to the pulp which is unacceptable for the production of white paper.

It is not possible to remove all the lignin without also dissolving most of the hemicelluloses which not only add to the weight of pulp produced but also impart a measure of adhesion between the fibres. Thus a compromise has to be reached in determining how far to progress with the chemical reaction and the decision is dependent on the requirements of the end product. Frequently, though not always, the initial pulping process is terminated when a quarter to a half of the lignin still remains and this is then removed in a subsequent operation known as bleaching, which, though expensive, has relatively little effect on the hemicelluloses. The yield of chemical pulp will vary considerably depending on the conditions employed, but it will usually be within the range of 40—50% of the dry weight of the original timber.

The yield of pulp can be increased to 55—80% by semi-chemical pulping. Only part of the lignin is removed in an initial chemical treatment designed to soften the wood chips: subsequent mechanical treatment separates the fibres without undue damage. These high-yield pulps usually find their way into card and board-liner which are extensively used for packaging where ultimate whiteness is not a prerequisite.

20.3.3 Other Chemical Processes

Brief mention must be made of the *destructive distillation* of timber, a process which is carried out either for the production of charcoal alone or for the additional recovery of the volatile by-products such as methanol, acetic acid, acetone and wood-tar. The timber is heated initially to 250 °C, after which the process is exothermic: distillation must be carried out either in the complete absence of air, or with controlled small amounts.

Timber can be softened in the presence of ammonia vapour as a result of plasticisation of the lignin. Timber can therefore be bent or moulded using this process, but, because of the harmful effects of the vapour, the process has never been adopted commercially.

20.4 FINISHES

Finishes have a combined decorative and protective function. Indoors they are employed primarily for aesthetic reasons though their role in resisting soiling and abrasion is also important; outdoors, however, their protective function is vital. In Chapter 18, the natural weathering process of timber was described in terms of the attack of the cell wall constituents by ultraviolet light and the subsequent removal of breakdown products by rain; the application of finishes is to slow down this weathering process to an acceptable level, the degree of success varying considerably among the wide range of finishes commercially available.

In Chapter 4 the complex chemical and morphological structure of timber was described while in Chapter 7 the hygroscopic nature of this fibre composite and its significance in determining the movement of timber was discussed. The combined effects of structure and moisture movement have a most profound effect on the performance of coatings. For example in the softwoods the presence of distinct bands of early and latewood with their differential degree of permeability results not only in a difference in sheen or reflectance of the coating between these zones, but also in marked differences in adhesion; in Douglas fir, where the latewood is most conspicuous, flaking of paint from the latewood is a most common occurrence. In addition the radial movement of the latewood has been shown to be as high as six times that of the earlywood and consequently the ingress of water to the surface layers results in differential movement and considerable stressing of the coatings. In those hardwoods characterised by the presence of large vessels the coating tends to sag across the vessel and it is therefore essential to apply a paste filler to the surface prior to paint; even with this, the life of a paint film on a timber such as oak (see Fig. 4.6) is very short. The presence of extractives in certain timbers (see Section 4.2.3.3 and Table 4.2)

637

results in the inhibition in drying of most finishes; with iroko and Rhodesian teak clear varnish never dries.

Contrary to general belief, deep penetration of the timber is not necessary for good adhesion, but it is absolutely essential that the weathered cells on the surface are removed prior to repainting. Good adhesion appears to be achieved by molecular attraction rather than by mechanical keying into the cell structure.

Although aesthetically most pleasing, fully exposed varnish, irrespective of chemical composition, has a life of only a very few years, principally because of the tendency of most types to become brittle on exposure, thereby cracking and disintegrating because of the stresses imposed by the movement of the timber under changes in moisture content. Ultraviolet light can readily pass through the majority of varnish films, degrading the timber at the interface and causing adhesion failure of the coating.

A second type of natural finish which overcomes certain of the drawbacks of clear varnish is the water-repellent preservative strain or exterior wood stain. There are many types available, but all consist of resin solutions of low viscosity and low solids content: these solutions are readily absorbed into the surface layers of the timber. Their protective action is due in part to the effectiveness of water-repellent resins in preventing water ingress, and in part from the presence of finely dispersed pigments which protect against photo-chemical attack. The higher the concentration of pigments the greater the protection, but this is achieved at the expense of loss in transparency of the finish. Easy to apply and maintain these thin films, however, offer little resistance to the transmission of water vapour in to and out of the timber. Compared with a paint or varnish the water-repellent finish will allow timber to wet up and dry out at a much faster rate, thereby eliminating problems of water accumulation which can occur behind impermeable paint systems; the presence of a preservative constituent reduces the possibility of fungal development during periods of high moisture uptake. The films do require, however, more frequent maintenance, but nevertheless have become well established for the treatment of cladding and hardwood joinery.

By far the most widely used finish, especially so for external softwood joinery, is the traditional opaque alkyd gloss or flat paint system embracing appropriate undercoats and primers; a four-coat system is usually recommended. Multiple coats of oil-based paint are effective barriers to the movement of liquid and vapour water; however, breaks in the continuity of the film after relatively short exposure constitute a ready means of entry after which the film will act as a barrier to moisture escape, thereby increasing the likelihood of fungal attack. The effectiveness of the paint system is determined to a considerable extent by the quality of the primer. Quite frequently window and door joinery with only a priming coat is left exposed on building sites for long periods of time. Most primers are permeable to water, are low in elasticity and rapidly disintegrate owing to stresses set up in the wet timber; it is therefore essential that only a high quality of primer is used.

Recent test work has indicated that the pretreatment of surfaces to be painted with a water-repellent preservative solution has a most beneficial effect in extending the life of the complete paint system, first by increasing the stability of the wood surface thereby reducing the stresses set up in primers on exposure, and second by increasing adhesion between the timber surface and the primer. This concept of an integrated system of protection employing preservation and painting, though new for timber, has long been established in certain other materials; thus it is common practice prior to the coating of metal to degrease the surface to improve adhesion.

One specialised group of finishes for timber and timber products is that of the flame-retardant coatings. These coatings, designed only to reduce the spread of flame, must be applied fairly thickly and must neither be damaged in subsequent installation and usage of the material nor their effect negated by the application of unsuitable coverings. Nearly all the flame retardants on the UK market intumesce on heating and the resulting foam forms a protective layer of resistant char.

20.5 REFERENCES

Chou, C. K., Chandler, J. A. and Preston, R. D. (1973) Microdistribution of metal elements in wood impregnated with a copper-chrome-arsenic preservative as determined by analytical electron microscopy, *Wood Science and Technology*, 7, 151—160.

Morén, R. E. (1964) Some practical applications of polyethylene glycol for the stabilisation and preservation of wood, paper presented to the British Wood Preserving Association annual convention.

Bibliography

MATERIALS IN GENERAL

Freudenthal, A. M. (1950) *The Inelastic Behaviour of Engineering Materials and Structures*, John Wiley. It's quite a time since this book was written, but it is still outstanding in treating material properties in relation to their structure. The later results may not be there, but most of the principles are, and, although it is not easy reading, the effort to do so is very rewarding.

Gordon, J. E. (1976) *The New Science of Strong Materials, 2nd edition,* Pelican Books. The subtitle is 'Why you don't fall through the floor, and the book is both entertaining and informative. Many students will learn more from reading this on the bus than they will from hours of study in a library.

Holliday, E. (Ed.) (1966) *Composite Materials*, Elsevier. This collection of articles on the structure and properties of composite materials is a prototype for the comparison of different materials by juxtaposition. The writers are authoritative and this makes for an excellent reference text. It includes, among others, concrete, metals, and paper and board.

Evans, R. C. (1964) *Crystal Chemistry, 2nd edition*, Cambridge University Press. Part I of this textbook gives a good account of the general principles of bonding between atoms; some prior knowledge of chemistry is advisable though it need not be to any great depth. Part II deals with individual elements and compounds and is very much more specialised; of doubtful relevance to engineers.

Hume-Rothery, W. *Atomic Theory for Students of Metallurgy*, Institute of Metals Monograph No. 3. A more advanced treatise written primarily for the metallurgist, but no less valuable on that account to engineers. It is particularly good on the more modern theories of matter and the way in which these are related to electrical properties.

Hume-Rothery, W. and Raynor, G. V. (1964) *The Structure of Metals and Alloys, 3rd edition*, Institute of Metals Monograph No. 1. Another advanced but very readable text; perhaps rather theoretically biased, but part VI on the structure of iron alloys is excellent for anyone wishing to understand the basis of the structures encountered in steels.

Cottrell, A. H. *Mechanical Properties of Matter*, Edward Arnold. A first class account of the scientific basis of materials; it covers a far wider field than the purely 'mechanical properties' implied in the title. It is particularly useful for its treatment of surfaces and of viscosity which are not normally included in materials science texts. I feel it should be required reading for all civil (and mechanical) engineers; strongly recommended.

Eley, D. D. (Ed.) (1961) *Adhesion*, Oxford University Press. Covers a good deal of the physical chemistry of surfaces and surface reactions as well as useful chapters on polymer properties and adhesive joints.

Fast, J. D. (1962) *Entropy*, Phillips Technical Library. Chapter 2 (the first half, p. 47–62) gives a very good introduction to statistical thermodynamics, and shows that this is a vitally important and universal subject which need not be anything like as forbidding as most engineers believe, or as its name suggests. The rest of the chapter is well worth reading too.

Angrist, S. W. and Hepler, L. (1973) *Order and Chaos*, Pelican Books. A discursive, anecdotal and entertaining account of the development and principles of thermodynamics. The treatment is nonmathematical, but the concepts are clearly set forth, and chapters seven and eight are particularly relevant to the later parts of this book. Strongly recommended bedtime reading.

CONCRETE

General

Neville, A. M. (1972) *Properties of Concrete*, Pitman. This book of nearly 700 pages gives a very comprehensive review of all aspects of concrete properties. It mentions, but does not dwell on, h.c.p. structure, and concentrates on engineering tests and methods. If you want to know anything about concrete this is the first book to look at; if the information is not there, you'll most likely find a suitable reference.

Neville, A. M. *Hardened Concrete; Physical and Mechanical Aspects*, American Concrete Institute, Monograph No. 6. If the previous reference is for reference, this one, being much shorter and crisper, is excellent for first reading.

Brooks, A. E. and Newman, K. (Ed.) *The Structure of Concrete*, Proceedings of an International Conference, London, 1965, Cement and Concrete Association. The conference was an important landmark in the course of convincing engineers that concrete as well as metal has a structure, and that it is helpful to associate fundamental theoretical ideas with engineering properties. The excellence of some of the papers is enhanced by the first class presentation of the proceedings. It is a joy to re-read.

Structure, Solid Mechanics and Engineering Design, (1971) Proceedings of the Southampton International Conference, 1969, Wiley-Interscience. As in the previous reference there are a number of papers on strength, but this conference was much wider in scope, covering materials other than concrete, and allowing much greater freedom in the type of material presented. There are many more papers, of more uneven quality, with less significant discussion.
British Standards Institute (1970) BS 1881, *Methods of Testing Concrete*, in six parts. Most of the important tests for concrete are included, with the wealth of detailed description necessary for establishing a standard for comparison.
(a) British Standards Institution (1972) *The Structural Use of Concrete*, CP 110, and
(b) Comité European du béton — Federation Internationale de la Precontrainte (1970) *International recommendations for the design and construction of concrete structures.*

These two documents are the considered conclusions of concrete experts (a) in Great Britain, and (b) from Europe and North America. They provide a wide-ranging guide to the design and construction of concrete structures.
American Concrete Institute. The ACI has an intricate system of committees and meetings which provide a variety of documents to help the practitioner by sieving and interpreting research and experience. The output includes:
    monographs, presenting state-of-the-art reports on selected topics;
    special reports containing a number of papers presented at
        meetings called to satisfy current needs in particular areas;
    committee reports, often in the form of notes on good practice
        for designers, published in the Proceedings.

Structure of Concrete

Powers, T. C. and Brownyard, T. L. (1946—47) Studies of the physical properties of hardened Portland cement paste (in nine parts), *Proceedings of the American Concrete Institute*. These original classical studies of the structure of cement paste were neglected until the end of the 1950's. The approach is that of the physicist, and it's hard going for the unsuspecting civil engineer. Nevertheless, perseverance is rewarded and it's worth looking out for the many later Powers publications.
Lea, F. M. (1970) *The Chemistry of Cement and Concrete*, Arnold. The author draws on his own wealth of experience and on that of his colleagues at the Building Research Station in Great Britain to present a comprehensive account of the scientific approach to cement and concrete. The cement chemist is provided with a mass of detail, but the treatment also includes invaluable information on practical behaviour. It is a readable reference book.
Czernin, W. (1962) *Cement Chemistry and Physics for Civil Engineers*, Crosby Lockwood, London. This is a short book to help the civil engineer over his initial difficulties with the science of cements.

642

In spite of recent advances it still contains much useful information.

Taylor, H. F. W. (1964) *The Chemistry of Cements*, Academic Press, New York. The stated editorial objective was 'to set out the chemistry and to show how this may be applied'. To this end expert coverage is provided by a variety of authors writing on their own speciality. The two volumes give an excellent overview of the topic at the time of publication.

Hydraulic Cement Pastes: their Structure and Properties, Proceedings of a Conference held at the University of Sheffield, 1976, Cement and Concrete Association. This is an interesting collection of papers presented at a Conference that set out to link the scientific investigations into structure with properties and engineering applications.

Shirley, D. E. (1975) *Introduction to Concrete,* Cement and Concrete Association, Wexham Springs. This short (24 pages) booklet is intended as an introduction to concrete technology for students. It serves its intended purpose admirably.

Environmental Volume Change in Concrete

The Shrinkage of Hydraulic Concretes (1968) Proceedings of an International Colloquium, RILEM/Cembureau, Madrid. Conferences on just shrinkage are rare, and this one contains a wide range of short snappy papers giving a very good overall view of the attitudes to and the findings on shrinkage current at the time.

Flow of Moisture and Heat in Concrete

Crank, J. (1957) *Mathematics of Diffusion*, Oxford University Press, London. Solutions to the problems of diffusion are presented clearly, for a variety of boundary conditions, and using both closed form and numerical methods.

Pihlajavaara, S. E. (1963) *Notes on the Drying of Concrete,* The State Institute for Technical Research, Finland. An admirable account of the application of scientific methods, both experimental and analytical, to the investigation of the structure and state of cement paste and concrete. It contains the precision of science and extends to the selected simplicity of engineering.

Response of Concrete to Stress

Neville, A. M. (1970) (with chapters 17 to 20 written in collaboration with W. Dilger), *Creep of Concrete: Plain, Reinforced and Prestressed*, Amsterdam, North-Holland. This book treats creep of concrete in all its detail, from the basic mechanisms causing creep to design methods for structural concrete members. The coverage is impressive, even awe-inspiring; it is the unchallenged authoritative book on creep.

The Concrete Society (1973) *The Creep of Structural Concrete*, Concrete Society Technical Paper No. 101. This report has the patchiness to be expected of a committee production. Its intention is to provide practitioners with a brief rounded view of creep, and to show them how to assess creep affects in their structures.

Jones, R. (1962) *Nondestructive Testing of Concrete*, Cambridge University Press. Written by a pioneer in the field of NDT, the detailed descriptions of the early techniques are somewhat dated. Nevertheless it still provides a valuable statement on the dynamic modulus of elasticity of concrete and its measurement.

Strength and Failure of Concrete

Newman, K. and Newman, J. B. (1971) *Failure Theories and Design Criteria for Plain Concrete*, Proceedings of the Conference on Structure, Solid Mechanics and Engineering Design, Southampton, Wiley — Interscience. A useful survey of the engineering approach to concrete strength, including suggestions on how to present and apply information in design. The article is written by the leaders of one of the most experienced groups of workers in this field.

Special issue on fracture, *Cement and Concrete Research*, Vol. 3, No. 4, 1973. In addition to a review paper on fracture of h.c.p. and concrete, this issue includes glimpses of the various lines of research currently being followed. The papers are all taken from the Third International Congress on Fracture held at Munich in April 1973.

Fagerlund, G. (1973) Strength and porosity of concrete, IUPAC/ RILEM International Symposium on pore structure and properties of materials, Vol. 2, Prague. The paper contain a thoughtful and developed discussion of the relations between porosity and strength, the writer suggesting, for instance, that the effects due to changes of porosity resulting from continuing hydration are different from those caused by different water/cement ratios. Attention to other papers in the Symposium could be rewarding.

Durability of Concrete

Idorn, G. M. (1967) *Durability of Concrete Structures in Denmark; a study of field behaviour and microscopic features*, Danish Technical Press, Copenhagen. Deterioration of concrete is difficult to measure, and this book is particularly valuable in that it gives descriptions of techniques of examination including many excellent illustrations. The treatment is all the more effective because it is related to first-hand experience in the field.

Woods, H. (1968) *Durability of Concrete Construction*, American Concrete Institute, Monograph No. 4. The American Concrete Institute, with its well developed committee structure, provides the most comprehensive series of collated views on concrete topics. The monograph is one of the forms that they have adopted, and this one gives a wide coverage of the known agencies that attack concrete. It is written from the American standpoint and is suitable for the con- crete technologist. It includes no less than 195 further references.

Cordon, W. A. (1966) *Freezing and Thawing of Concrete — Mechanisms and Control*, American Concrete Institute Monograph No. 3. This second monograph expands on one of the topics covered by Woods, above.

The Institution of Structural Engineers and the Concrete Society, *Report of a Joint Committee on the Fire Resistance of Concrete*, 1975. The fire resistance of concrete has been a neglected area, and this Report sets out to present the current knowledge in an orderly manner, with the intention of encouraging the future rational design of concrete structures to resist fire.

TIMBER

General

Farmer, R. H. (1974) *Materials and Technology*, Vol. VI, Chapter 1: Wood. Although containing little information on the mechanical properties of timber, especially deformation and strength, this book presents a comprehensive review of the production and processing of both timber and board materials: as such, it compliments the subject matter of the present text.

Kollmann, F. F. P. and Coté, W. A. (1968) *Principles of Wood Science and Technology, I, Solid Wood*, Springer Verlag, 592 pp. A most comprehensive text — initial chapters on the anatomical structure, chemical composition and physical properties of timber are followed by a set of chapters dealing in considerable depth with stiffness and the various strength parameters. Final chapters are concerned with the drying and mechnical processing of timber.

Findlay, W. P. K. (1975) *Timber: Properties and Uses*, Crosby Lockwood Staples, 224 pp. Although limited in the treatment of the performance of timber in terms of its structure and in the presence of graphical evidence throughout the text, this book nevertheless constitutes a very valuable introduction to the study of timber and timber products.

Gordon, J. E. (1976) *The New Science of Strong Materials, 2nd edition,* Penguin. Surely the most readable book on material science, containing not only a delightful range of anecdotes but also a wealth of information on a wide range of materials including timber and plywood.

Latham, B. (1957) *Timber: Its Development and Distribution — A Historical Survey*, Harrap, London. A fascinating and most readable account of the origin and development of the timber trade in different parts of the world; the final chapter traces the development over the centuries of various types of wooden structures.

Panshin, A. J. and de Zeeuw, C. (1970) *Textbook of Wood Technology, 3rd edition, Vol. 1, Structure, Identification, Uses and Properties of the Commercial Woods of the United States and Canada* 705 pp. Somewhat limited in value since the last third of the book is devoted to identification of North American timbers and the physical and mechanical properties of timber receive scant attention. However, its strength lies in the first part of the book dealing in considerable depth with the growth of the tree and wood formation, and with the anatomical structure of both softwoods and hardwoods.

645

Wood Handbook: Wood as an Engineering Material (1974) US Dept of Agriculture, Forest Products Laboratory, Madison. An essential reference text on timber technology particularly from the engineering rather than the material science viewpoint; part of it is a manual of data for design work.

Dinwoodie, J. M. (1975) 'Timber — A Review of the Structure — Mechanical Property Relationship, *Journal of Microscopy*, **104**, 3—32. A condensed version of the chapters on timber in the present text and containing an extensive bibliography of research papers.

Timber Structure

Coté, W. A. (Ed.) (1965) *Cellular Ultrastructure of Woody Plants*, Syracuse University Press, 603 pp. A series of papers primarily but not exclusively on the ultrastructure of timber; a wealth of knowledge, little of which has become outdated over the last decade. Essential reading for any student concerned with detail at the ultrastructural level.

Jane, F. W. (1970) *The Structure of Wood, 2nd edition*, Adam & Charles Black, London, 478 pp. A most comprehensive treatment of wood primarily at the cellular level with useful and informative chapters on variability and the relationship of structure to properties.

Preston, R. D. (1974) *The Physical Biology of Plant Cell Walls*, Chapman & Hall, London, 491 pp. A classic in reference texts with exhaustive treatment of the chemical and ultrastructural aspects of both cellulosic and noncellulosic plants. The relationship of structure, stiffness and permeability to the basic structure of timber is discussed in some detail.

Tsoumis, G. (1968) *Wood as Raw Material*, Pergamon Press, Oxford, 276 pp. Possessing a somewhat misleading title, this text is concerned primarily with the structure and variability of timber; these are described in a very readable form.

Philipson, W. R., Ward, J. M. and Butterfield, B. G. (1971) *The Vascular Cambium: its development and activity*, Chapman & Hall, London, 182 pp. A comprehensive text on a specialised area: much of the variation in timber structure is described in terms of cambial activity.

Rendle, B. J. (revised by Brazier, J. D.) (1971) *The Growth and Structure of Wood*, Bulletin No 56, Forest Products Research, Princes Risborough, HMSO, 15 pp. A short but informative introductory text on the subject.

Zimmerman, M. H. (Ed.) (1964) *The Formation of Wood in Forest Trees*, Academic Press, New York, 562 pp. A collection of articles concerned primarily with timber structure and the relation of structure to growth variables.

Reaction Wood (1972) Technical Note No 57, Princes Risborough Laboratory, Building Research Establishment, 20 pp. Describes the anatomical, physical and mechanical properties of both tension wood and compression wood.

Handbook of Hardwoods, 2nd edition (1972) HMSO, 243 pp. Prepared by the Princes Risborough Laboratory (formerly the Forest Products Research Laboratory) this volume contains detailed descriptions of the appearance, physical and mechanical properties of 117 hardwood timbers while a further 103 hardwoods are described in lesser detail. A very necessary reference text for anyone concerned with the purchasing and utilisation of hardwood timbers. *Handbook of Softwoods, 2nd edition* (1977) HMSO, 63 pp. A companion volume to that described above but containing less mechanical data on the timbers. 49 softwood timbers are described fully and a further six are described briefly: another most useful reference book.

Moisture Content, Drying, Shrinkage, Movement

Skaar, C. (1972) *Water in Wood*, Syracuse University Press, 212 pp. A fundamental text on wood—water relationships embracing both practical and theoretical aspects. Hygroscopic shrinkage and swelling is explained in terms of anatomical structure, and a wide range of techniques for measuring moisture is presented: thermodynamics of moisture sorption and the various theories explaining equilibrium moisture content are fully discussed.
The following Technical Notes (obtainable free from the Princes Risborough Laboratory, Princes Risborough, Aylesbury, Buckinghamshire) will be found useful by those readers new to the subject.

No. 32 — A small electrically heated timber dryer
No. 36 — Air seasoning of sawn timber
No. 37 — Kiln drying schedules
No. 38 — The movement of timbers
No. 46 — The moisture content of timber in use

Flow, Permeability, Diffusion, Conductivity

Siau, J. F. (1971) *Flow in Wood*, Syracuse University Press, 131 pp. Permeability to liquids and gases, thermal conductivity, and the diffusion of water vapour through timber are treated as analogous phenomena frequently with the use of geometric models; both steady-state and unsteady-state conditions are considered.
Thomas, R. J. and Kringstad, K. P. (1971) The role of hydrogen bonding in pit aspiration, *Holzforschung* 25(5), 143—149. An important research contribution in this field.

Deformation: Elasticity, Viscoelasticity

Hearmon, R. F. S. (1961) *An Introduction to Applied Anisotropic Elasticity*, Oxford University Press, 136 pp. The text, now regarded as one of the classics in this field, is concerned primarily with the solution of problems encountered in anisotropic elasticity; the first two chapters are devoted to general principles including the effect of symmetry on elastic constants.
Jayne, B. A. (1972) *Theory and Design of Wood and Fibre Composite*

647

Materials, Syracuse University Press, 418 pp. A series of papers of which the following are particularly relevant to readers concerned with understanding deformation in timber; orthotropic elasticity, elastic behaviour of the wood fibre, mechanical behaviour of the molecular component of fibres, mechanical response of cellular materials.

Schniewind, A. P. (1968) Recent progress in the study of the rheology of wood, *Wood Science and Technology*, 2, 188—206. An excellent and comprehensive review of the literature on all aspects of the rheological behaviour of timber that were published during the decade 1957—67. Although the review is now over ten years old, the information it contains has not been nullified by the results of more recent research.

Ward, I. M. (1971) *Mechanical Properties of Solid Polymers*, Wiley. A comprehensive text covering both linear and nonlinear visco-elastic behaviour, but containing only passing reference to aniso-tropic materials. Nonetheless, it is a most useful background text in which the mechanics of behaviour are described first and are followed by discussion on the molecular interpretations of behaviour.

Strength, Structural Applications, Design

Booth, L. G. and Reece, P. O. (1967) *The Structural Use of Timber — A Commentary on BS, CP 112*, Spon. Although Amendment No. 1 to CP 112 published in 1973 has rendered out of date large parts of this text, the information on joints and fixings in timber structures, and the design of beams and struts is still most relevant to modern practice.

Reece, P. O. (1949) *An Introduction to the Design of Timber Structures*, Spon. As an introductory book to the subject it is an exceedingly clear and concise exposition of the properties of timber, elementary mechanics, application of elasticity theory to timber, strength and grading, design and stresses, and the analyses of beams, columns and struts.

Ozelton, E. C. and Baird, J. A. (1976) *Timber Designers Manual*, Crosby Lockwood Staples, 518 pp. This is the first British book to deal comprehensively with the practical design of common engineering components in timber and is highly recommended. Although it relates specifically to the application of the British Code of Practice CP 112 to structural design the book is nevertheless of considerable merit internationally where other design codes are used. The design of beams, columns and trusses, using different types of connectors, is treated in great detail: attention is paid to such difficult areas as shear deflection, partial restraint and stiffener design.

American Institute of Timber Construction (1974) *Timber Construction Manual, 2nd edition*, Wiley. An authoritative text for the designer and specifier with a comprehensive selection of data for designing in timber. Strength data for timber are restricted to North American timbers and the strength values for plywood are not compatible with UK practice: these aspects apart, this is a most

648

informative and useful manual.

Design of Timber Members (1967) Timber Research and Development Association, High Wycombe, Buckinghamshire. Includes stresses and typical calculations for beams and columns in accordance with the revised Code of Practice CP 112, 1967. Beam-span tables for a range of constructional timbers are also incorporated in the text.

Dinwoodie, J. M. (1971) Brashness in timber and its significance, *Journal of the Institute of Wood Science* 28(5), 3–11. A review of the numerous anatomical, physical and chemical reasons for the abnormal, brittle behaviour in timber.

Stress Grading Timber, Technical Information Sheet No. 25 of the Princes Risborough Laboratory.

Home-Grown Softwoods for Building, Digest No. 72, Building Research Establishment, Garston, Watford, and obtainable from HMSO.

Durability, Decay, Fire

Cartwright, K. St. G. and Findlay, W. P. K. (1958) *Decay of Timber and its Preservation*, HMSO, 332 pp. A comprehensive text describing the principal wood-rotting fungi and the conditions under which they flourish; recommended procedures in seasoning, storage and preservative treatment to prevent such damage are presented.

Hall, G. S., Saunders, R. G., Allcorn, R. T., Jackmann, P. E., Hickey, M. W. and Fitt, R. (1972) *Fire Performance of Timber*, Timber Research and Development Association, High Wycombe, Buckinghamshire. A most comprehensive review of literature on methods of test, theories of ignition, effect of temperature on both the chemical and mechanical properties of timber, fire performance, and flame-retardant treatments.

The Natural Durability Classification of Timber, Technical Note No. 40, Princes Risborough Laboratory.

Timber Decay and its Control, Technical Note No. 57, Princes Risborough Laboratory.

Decay in Buildings – Recognition, Prevention and Cure, Technical Note No. 44, Princes Risborough Laboratory.

The Implications of Using Inorganic Salt Flame-Retardant Treatments with Timber, Information sheet No 13/74, Princes Risborough Laboratory.

Timber Fire Doors, Digest No. 155, Building Research Station, Garston, Watford and obtainable from HMSO.

Processing

Kollmann, F. F. P., Kuenzi, E. W. and Stamm, A. J. (1975) *Principles of Wood Science and Technology II, Wood-Based Materials*, Springer–Verlag, 703 pp. An exhaustive treatment of board materials and the adhesives used in their production; the text is weighted towards the processing of these materials though a considerable amount of information on their properties is nevertheless included.

Stevens, W. C. and Turner, N. (1970) *Wood Bending Handbook*, HMSO. Principally a handbook of practice, the text sets out to describe the principal procedures used in wood bending: theoretical background information is kept to a minimum.

Panshin, A. J., Harrar, E. S., Bethel, J. S. and Baker, W. J. (1962) *Forest Products, their Sources, Production and Utilization*, McGraw-Hill, 538 pp. Contains considerable detail on the processing and properties of the very wide range of timber products including veneers, wood flour, mine timber, pulp and paper, and turpentine.

Curry, W. T., *The Strength Properties of Plywood*. HMSO:

> Part 1: Comparison of three-ply woods of a standard thickness, *Forest Products Research Bulletin*, 29 (2nd edition) 1964.
> Part 2: Effect of the geometry of construction, *Forest Products Research Bulletin*, 33, 1954.
> Part 3: The influence of the adhesive, *Forest Products Research Bulletin*, 39, 1967.
> Part 4: Working stresses, *Forest Products Research Bulletin*, 42 (2nd edition) 1969.

Gibson, E. J., Laidlaw, R. A. and Smith, G. A. (1966) Dimensional stabilisation of wood, I: Impregnation with methyl methacrylate and subsequent polymerisation by means of gamma radiation, *Journal of Applied Chemistry*, 16, 58–64.

Preservative Treatment for External Joinery Timber, Technical Note No. 24, Princes Risborough Laboratory.

Painting Wood Work, Digest 106, Building Research Station, Garston, Watford — obtainable from HMSO.

Natural Finishes for Exterior Timber, Digest 182, Building Research Station, Garston, Watford — obtainable from HMSO.

METALS

Crystal Structure of Metals

Cottrell, A. H. (1965) *Theoretical Structural Metallurgy, 2nd edition*, Edward Arnold. The whole book is a valuable adjunct to a study of materials. Chapters VII–XI give a very clear and comprehensive treatment of equilibrium diagrams and their thermodynamic basis.

Chadwick, G. A. (1972) *Metallography of Phase Transformations*, Butterworth. A more leisurely and detailed treatment of the subject matter covered by Cottrell (above). It is useful in that it gives a profusion of diagrams and photographs of actual structures, which are rather sparse in Cottrell's book. The sections on steels in Chapter 7 are good for the understanding of heat treatment processes.

Hansen, M. (1958) *Constitution of Binary Alloys, 2nd edition*, McGraw-Hill. A comprehensive and meticulously annotated reference work of all the equilibrium diagrams any user of metals is ever likely to need.

Reed-Hill, R. E. (1973) *Physical Metallurgy Principles, 2nd edition*, Van Nostrand. An excellently written textbook, the greater part of which is devoted to the principles underlying the mechanical properties of metals. The two chapters on dislocation behaviour are very well illustrated, and there are good sections on the heat treatment of alloys. Strongly recommended for those who wish to probe deeper into this aspect of materials.

Dieter, G. E. (1976) *Mechanical Metallurgy, 2nd edition*, McGraw-Hill Kogakusha. Parts II and III, dealing with metallurgical fundamentals and materials testing, are valuable for the engineer; Part I on stress analysis and Part IV on plastic forming are interesting and well written but of doubtful relevance for the present text.

Honeycombe, R. W. K. (1968) *The Plastic Deformation of Metals*, Edward Arnold. A more specialised and advanced treatise, dealing with fundamental physical principles, but there are plenty of examples taken from real alloy systems. The illustrations and diagrams are plentiful and excellent.

Brick, R. M., Pense, A. W. and Gordon, R. B. (1977) *Structure and Properties of Engineering Materials, 4th edition.* Concerned primarily with strengthening mechanisms, this book is valuable in that it treats alloy systems individually, after a short initial section on general principles, and also includes chapters on polymers and glasses.

Frost, N. E., Marsh, K. J. and Pook, L. P. (1974) *Metal Fatigue*, Oxford University Press. A comprehensive work which covers all aspects of fatigue failure, both theoretical and practical. Of particular interest are the sections discussing the effect of pin joints, shrink fits, screws and welds on the fatigue life of structures. There are copious references at the end of each chapter for any reader who wishes to delve deeper into any particular branch of the subject.

Teed, P. L. (1960) 'Fretting', *Metallurgical Reviews* No. 19 (Vol. 5). A good general introduction to the significance of fretting, particularly as an instigator of fatigue failure. There have been more comprehensive treatments (by Waterhouse, for example), but this affords an excellent first reading.

Kelly, A. (1973) *Strong Solids*, Oxford University Press. Recommended in particular for the chapters on fibre reinforcement and fibrous solids, but there is also a very readable treatment of the behaviour of cracks and notches, and the discrepancies between the actual and theoretical strengths of materials.

A. S. M. Symposium, *Fracture of Engineering Materials* (1964) American Society of Metals. Although it is now 14 years old, this symposium presents a valuable survey of the various aspects of fracture. All 12 papers are interesting and well presented, and at least eight of them are of direct importance to the civil engineer.

Wyatt, O. H. and Dew-Hughes, D. (1974) *Metals, Ceramics and Polymers*, Cambridge University Press. A first class general textbook on materials, which goes to some depth into the modern theories of deformation and strengthening in metals. Particularly recommended are Chapters 5, 6 and 9–12, but the whole book is excellent for the student who wishes to grasp the finer details of the subject.

Kelly, A. and Davies, G. J. (1965) The principles of the fibre reinforcement of metals, *Metallurgical Reviews* No. 37 (Vol. 10). A good general account of the fibre reinforcement; written some years ago but it still presents a concise picture of the basic principles of this important technique.

Tweeddale, J. G. (1969) *Welding Fabrication, Vols. 1 and 3*, Iliffe. Volume 1 deals with the metallurgy of welds and metallurgical problems arising therefrom; Volume 3 is concerned with design and fabrication of welded structure and associated problems. Valuable for structural engineers; the books throughout approach the topics from the engineering rather than purely metallurgical viewpoint. (Volume 2, dealing with different welding techniques, is interesting but less relevant as further reading.)

Corrosion

Shreir, L. L. (Ed.) (1976) *Corrosion* (2 Vols.), Newnes-Butterworth. A mammoth tome of reference, totalling 3 370 pages and covering all aspects of corrosion. Volume 1 is concerned primarily with the theory and scientific principles of corrosion, volume 2 with control and testing. Individual sections are valuable to the Civil Engineer faced with a particular problem, but the sheer size of the book makes it unsuitable for general background reading. All engineers should be aware of its existence, however,

Evans, U. R. (1963) *An Introduction to Metallic Corrosion, 2nd edition*, Edward Arnold. An excellent, readable and reasonably short text on the scientific principles of corrosion, less than one tenth the length of Shreir (above) and written by the greatest of all the pioneers in corrosion science. The approach is more metallurgical than engineering, but it is strongly recommended for general background reading.

Evans, U. R. (1960) *The Corrosion and Oxidation of Metals*, Edward Arnold. First Supplementary Volume, 1976. Size and cost, alas, confer on this also the status of reference work. It covers the full range of corrosion topics, principally from the academic standpoint, but with no lack of practical examples. There is a particularly valuable and interesting chapter in the 1960 book on the application of probability and statistical theory to corrosion systems. Instead of introducing further editions of the original text, Dr. Evans has brought out two supplementary volumes that set forth the latest ideas on the subject, but the original work is by no means outdated and remains a classic textbook on the subject.

Clark, W. D. (1958) 'Design from the Viewpoint of Corrosion', *Metallurgical Reviews* Vol. 3, No. 11. Despite its age, this remains a very valuable paper which should be made compulsory reading for all engineers.

Rogers, T. H. (1968) *Marine Corrosion*, Newnes. A good textbook covering both the principles and protective techniques associated with marine corrosion; this aspect of the subject has rapidly increased in importance with the siting of power stations around our coasts and the development of the North Sea oilfield. The section on control of corrosion by structural design is well illustrated, and has a relevence that goes well beyond the scope of the title of the book.

Index